Seismic Stratigraphy — applications to hydrocarbon exploration

Memoir 26

Seismic Stratigraphy — applications to hydrocarbon exploration

Edited by
CHARLES E. PAYTON

Published by The American Association of Petroleum Geologists
Tulsa, Oklahoma, U.S.A., 1977

Published December 1977
Second printing, June 1978
Third printing, February 1979
Fourth printing, December 1980
Library of Congress Catalog Card No. 77-91023
ISBN: 0-89181-302-0

The AAPG staff responsible:
Gary D. Howell, Managing Editor
Ronald L. Hart, Project Editor
Sally B. Hunt, Production Coordinator
Regina Landon, Production
Carol Short, Production
Nancy G. Wise, Production

Printed by Edwards Brothers, Inc., Ann Arbor, Michigan

Contents

Foreword

This *Memoir* is the result of plans made after the first Research Symposium on Seismic Stratigraphy presented at the 1975 national convention of the American Association of Petroleum Geologists. Selected reports from technical meetings since that time are also included.

Seismic stratigraphy is one of the fastest growing geoscience disciplines. The basic concepts of seismic response to thin transitional beds and the synthesis of seismograms from stratigraphic sequences were explained more than 20 years ago, but routine use of these concepts had to await modern electronic technology. Likewise, only in the last few years has the quality of seismic data been adequate to interpret reservoir conditions and depositional facies with some accuracy.

The discipline has evolved along two different paths. One approach seeks to recover stratigraphic information from qualitative analysis of reflections. Variations in reflection amplitude, continuity, and concordance are used to group regions of distinctive appearance. These are assigned stratigraphic meaning by comparison to subsurface information. The other approach attempts to duplicate a seismogram by numerical modeling. A reflection coefficient model of the strata thickness, velocity, density and absorption is constructed either from subsurface data or from the explorer's imagination. The model is convolved with a seismic pulse to produce a synthetic seismogram which, in turn, is compared for similarity to the field record.

The articles in this *Memoir* are grouped into three sections. The first describes principles that both permit and also limit interpretations. The second section presents sixteen articles that describe the qualitative approach to stratigraphic interpretations of reflection records, and the final section presents techniques and examples of modeling.

Better knowledge of petrophysics and continued experimentation with processing techniques will sustain the development of seismic stratigraphy. Future advances in data acquisition technology, principally improved recovery of high frequency signals, and routine recording of shear-wave reflections promise to provide even more accurate stratigraphic interpretations.

Charles E. Payton
Houston, Texas
Special Editor

June 6, 1977

section 1:
fundamentals of stratigraphic
interpretation of seismic data

Limitations on Resolution of Seismic Reflections and Geologic Detail Derivable from Them[1]

ROBERT E. SHERIFF[2]

Abstract Stratigraphic conclusions from seismic data depend on the data being sufficiently free of noise so that the seismic response is predominantly that of the sediments. Thus good recording and processing are essential. Given a reasonably noise-free response, seismic wavelength limits the detail which can be seen in two dimensions: vertical, or the thickness of stratigraphic units; and horizontal, or the areal size of features.

Most reflection events seen on a seismic section are composites of reflections from individual interfaces. Calculating the waveform from a sequence of interfaces helps in understanding and interpreting waveform shape. This process is called *synthetic seismogram construction* where the input information is derived from well logs, and *modeling* where lateral variation is the principal concern. A comparison between synthetic seismograms and well logs illustrates the resolving power of seismic data and the limitations in trying to invert the process and derive logs from seismic data—thus, the *seismic log process*. The ease with which stratigraphic significance can be derived from seismic data also depends on the type of display; those which enhance different aspects of data are helpful in appreciating geologic significance.

INTRODUCTION

Modern seismic sections often bear such striking resemblance to stratigraphic cross sections that they invite direct interpretation by people who do not appreciate geophysical limitations. Seismic sections show the response of the earth to seismic waves, and the position of geologic bedding is only one of several factors which affect the response. Analysis and processing can attenuate the many events present in addition to simple reflections. Reflections have to be repositioned before geologic interpretation because their travel paths commonly have a horizontal component. Because most reflections are interference composites, there is no one-to-one correspondence between seismic events and interfaces in the earth.

Notwithstanding these limitations, geophysicists often can interpret subtle changes in waveshape into stratigraphic terms. One should be careful not to assume that, in noise-free areas with high-quality data, every waveshape variation has a geologic meaning which only needs definition.

Stratigraphic significance commonly is apparent only where seismic data are tied with well data. Once seismic waveshape can be related to known stratigraphic features, seismic data can be used to show how far the particular stratigraphy extends. Accurate recognition of distinctive features may also help in recognizing similar stratigraphy elsewhere.

Even where seismic data have been recorded and processed so that the interpreter can be confident that he is interpreting primary reflection energy, interpretational ambiguities remain, many of which stem from the magnitude of seismic wavelength. Physical principles involved in reflection will later be discussed so the interpreter may better appreciate the limitations and therefore better assess the degree of reliability of his stratigraphic conclusions.

ATTENUATING NOISE BEFORE INTERPRETATION

Data must be essentially noise-free before stratigraphic interpretation can begin; variations in waveshape must represent variations in the subsurface rather than changes in the noise. (A glossary of geophysical terms is given as an appendix.) Careful recording and processing are essential.

The key in developing methods to remove noise effects is understanding the noise sources (Fig. 1). Differences in the vicinity of the seismic source or geophones may cause waveshape changes. Surface-consistent static correction programs often remedy differences in arrival time because of near-surface variations; signature correction processing remedies variations in the source waveshape; and surface-consistent amplitude processing and divergence corrections remove amplitude variations not relevant to the subsurface geology. Velocity filtering and the redundancy provided by recording with high degrees of multiplicity permit the attenuation of many types of coherent wavetrains, as well as random noise. Predictive deconvolution and common-depth point stacking attenuate multiples. Deconvolution helps remove near-surface reverberation and broadens the frequency spectrum to sharpen the seismic wavelet.

Migration helps position data elements in proper spatial relations rather than underneath the locations at which the data are observed. Wave-equation migration clarifies stratigraphic evi-

[1]Manuscript received, October 1, 1976; accepted, January 1, 1977.

[2]Seiscom Delta Inc., Houston, Texas 77036.

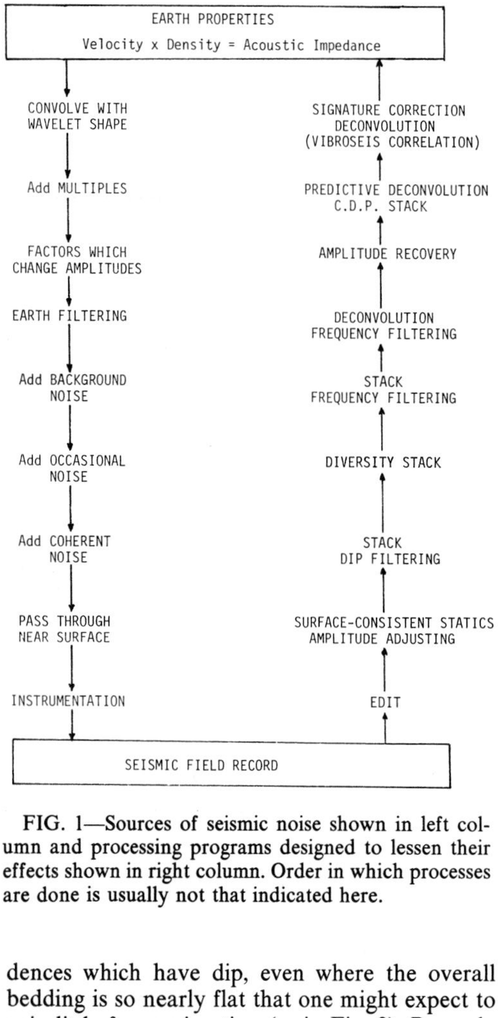

```
                    EARTH PROPERTIES
         Velocity x Density = Acoustic Impedance

   CONVOLVE WITH              SIGNATURE CORRECTION
   WAVELET SHAPE                  DECONVOLUTION
                             (VIBROSEIS CORRELATION)

   Add MULTIPLES             PREDICTIVE DECONVOLUTION
                                 C.D.P. STACK

   FACTORS WHICH             AMPLITUDE RECOVERY
   CHANGE AMPLITUDES

   EARTH FILTERING              DECONVOLUTION
                             FREQUENCY FILTERING

   Add BACKGROUND                   STACK
   NOISE                     FREQUENCY FILTERING

   Add OCCASIONAL            DIVERSITY STACK
   NOISE

   Add COHERENT                     STACK
   NOISE                       DIP FILTERING

   PASS THROUGH             SURFACE-CONSISTENT STATICS
   NEAR SURFACE                AMPLITUDE ADJUSTING

   INSTRUMENTATION                  EDIT

              SEISMIC FIELD RECORD
```

FIG. 1—Sources of seismic noise shown in left column and processing programs designed to lessen their effects shown in right column. Order in which processes are done is usually not that indicated here.

dences which have dip, even where the overall bedding is so nearly flat that one might expect to gain little from migration (as in Fig. 2). Properly done, wave-equation migration does not distort waveshape or amplitude, both of which are vital to stratigraphic interpretation.

WAVELENGTH AND RESOLUTION

The seismic "measuring stick" is wavelength. Whether or not features can be seen depends on their magnitude compared with wavelength. Most seismic energy is contained in a band of frequencies centered near the dominant frequency (the reciprocal of the dominant period, or the time between successive points where the phase is the same). We can measure the dominant period on our records. The basic relation for wavelength is simply

$$\lambda = \text{wavelength} = \text{velocity} \times \text{period} = \text{velocity/frequency}$$

Velocities in the shallow part of the earth (below the water table) are usually in the 1,500 to 2,000 m/sec range and the dominant frequency of reflections from this zone might be approximately 50 Hz, giving wavelengths in the 30 to 40 m range. In the deep part of the earth, velocities are three to four times greater, in the 5,000 to 6,000 m/sec range, and deep reflections commonly have low dominant frequencies, perhaps 20 Hz, and give wavelengths of 250 to 300 m. Thus we see that the wavelengths involved in conventional seismic exploration are in the general range of 30 to 300 m. Wavelength generally increases with depth because: (1) velocity increases, and (2) frequencies become lower. Because the wavelength limits resolving power, deep features have to be much larger than shallow features to produce the same expression.

Resolution is the ability to distinguish separate features, and is commonly expressed as the minimum distance between two features, such that two can be defined rather than one. Seismic interpretation is concerned with resolution in two directions, vertical (in time or in depth) and horizontal (from trace to trace).

The only variable affecting wavelength which we can control is the frequency. We can improve resolution if we can record higher frequencies and a broader band of frequencies. We might be able to do this by using sources rich in high frequencies, and by not discriminating against high frequencies with our recording and processing practices. To maximize resolution, we must sample more than every 4 msec, which involves throwing away frequencies above 70 Hz in anti-alias filtering. Minor time differences between the elements of source or detector arrays (such as poor geophone planting might produce) cannot be tolerated either; random time differences of only 2-msec magnitude effectively remove frequencies above 70 Hz.

VERTICAL RESOLUTION

Vertical resolution may be rephrased as the minimum separation between reflectors needed to define separate interfaces rather than a single interface (note: this is not asking, "How accurately can we time an event?"). Resolution is somewhat subjective, depending on background noise and the interpreter's sensitivity to minor waveshape changes, etc. Generally, resolution is about 1/8 to 1/4 wavelength. Under ideal conditions, such as simple structure and noise-free situations where a good reference is available, interfaces closer than this may be resolved.

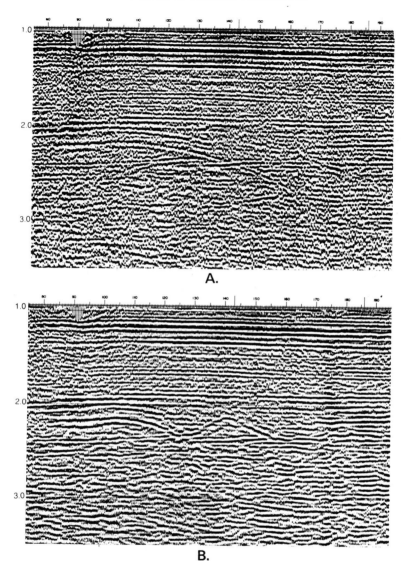

A.

B.

FIG. 2—A generally flat section before (**A**) and after (**B**) wave equation migration.
Note how migration has clarified salt solution edge and residual salt pod. Courtesy
Seiscom Delta Inc.

Consider Figure 3A, a pinchout of a wedge of
sand. The wavelet assumed for this example is 1.5
cycles long with a dominant frequency of 50 Hz.
Where the wedge is about ¼ wavelength thick
(12.4 m in this example), reinforcement occurs
and more than one interface is involved. The decrease in amplitude as the wedge becomes thin
can be used to determine the thickness for thinner
wedges, given a reference amplitude.

Resolving power is improved by increasing the
frequency as illustrated in Figure 4. These are
land data and there are variations in surface elevation, weathering thickness, and irregularities in
coverage because of access and permit problems.
The conventional section (at top) has been processed to remove most of these variations. The
major reflectors correspond to well-defined contacts, and extensive drilling in the area has veri-

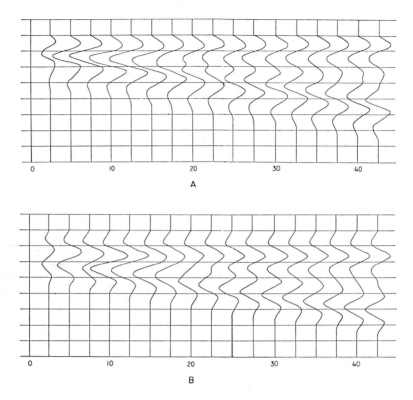

FIG. 3—Reflection from thin bed where the overlying and underlying lithology is same. Velocity is assumed to be 2,500 m/sec; for the 50-Hz wavelet, wavelength is thus 50 m. Values indicate thickness of the bed in meters; timing lines are 10 msec apart. (**A**) minimum-phase wavelet shape. (**B**) zero-phase wavelet with same frequency content.

fied that this seismic work is reliable. The major objective has been gas trapped in pinchouts of the prograding Woodbine sandstones. The prograding pattern on the seaward side of Edwards reef can be seen clearly, as can prograding in the lower Midway section. These data were reprocessed with a wavelet equalization and contraction program to reduce trace to trace variations, bring up higher frequencies, and shorten the wavelet. The result has been to improve the resolution.

INTERFERENCE OF REFLECTIONS FROM CLOSELY SPACED REFLECTORS

The amplitude of a reflected wave compared to the amplitude of an incident wave is called the *reflection coefficient* or *reflectivity*. In the general case, a single incident wave on an interface results in four waves: reflected and transmitted shear waves, and reflected and transmitted compressional waves. The situation is simplest at perpendicular incidence (when the wavefront is tangent to the interface); then only reflected and

transmitted waves of the same kind as the incident wave may occur.

This perpendicular incidence situation can be expressed in terms of the density (ρ) and velocity (V) of the media on opposite sides of the interface:

$$
\begin{aligned}
\text{reflection} \atop \text{coefficient} &= \frac{\text{amplitude of reflected wave}}{\text{amplitude of incident wave}} \\
&= \frac{\rho_2 V_2 - \rho_1 V_1}{\rho_2 V_2 + \rho_1 V_1} \qquad (1) \\
&= \frac{\text{change in acoustic impedance}}{\text{twice the average acoustic impedance}}
\end{aligned}
$$

The product of density and velocity is called *acoustic impedance*. If the incident angle is small (up to, perhaps, 20°), the departure from this simple relation is small. Most reflection seismology involves small incident angles.

The seismic trace from a sequence of reflecting interfaces can be obtained by summing the effects

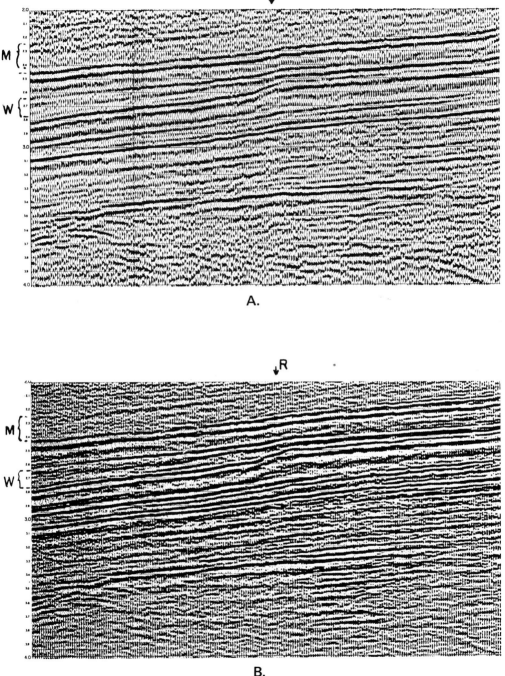

FIG. 4—A dip line in East Texas. Major reflections generally show how section breaks into its major units. The Midway (M) and Woodbine (W) are prograding sediments. Note Edwards Reef edge (R). (A) Conventional processing; (B) Processing (PULSE) designed to shorten the equivalent source wavelet and make it constant in shape. Resolution of Woodbine pinchouts has been increased by this processing. Courtesy Seiscom Delta Inc.

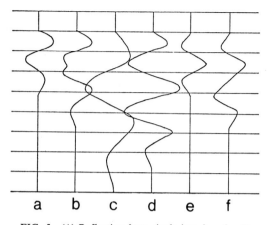

a   b   c   d   e   f

FIG. 5—(**A**) Reflection from single interface for 50-Hz minimum-phase wavelet. (**B**) Reflection from layer 20 m thick. (**C**) Reflection from same layer for 25-Hz wavelet. (**D**) Reflection from two layers each 10 m thick, separated by layer through which travel time is same. (**E**) Same as (**A**) except wavelet is zero-phase. (**F**) Same as (**B**) except wavelet is zero-phase; this is one of the wavelets shown in Fig. 3B.

of each interface in appropriate time relations; such a summing process is called *convolution*.

Let us calculate the composite wavelet shape from a sandstone layer 20 m thick with a velocity of 2,500 m/sec, assuming reflectivity at the top and base of the sandstone of 0.1 and −0.1, respectively. Assume a basic wavelet (Fig. 5A) which has the following amplitudes at successive 4-msec sample points: 10, 9, −8, −9, 0, 5, and 3. The wavelet reflected from the top of the sandstone will also have this shape, although diminished in amplitude by 0.1. The wavelet from the base of the sandstone will traverse the unit twice and hence will arrive 40 m/2,500 m/sec = 16 msec, or four sample intervals later than the reflection from the top, and will be multiplied by −0.1 factor[3], the negative sign indicating phase reversal. Summing these two gives the composite wavelet shape:

$$
\begin{array}{llllllll}
1.0, & .9, & -.8, & -.9, & 0, & .5, & .3 \\
 & & & -1.0, & -.9, & .8, .9, 0, & -.5, & -.3 \\
\hline
1.0, & .9, & -.8, & -.9, & -1.0, & -.4, 1.1, .9, 0, & -.5, & -.3
\end{array}
$$

This wavelet is plotted in Figure 5B. The waveshape shows that two component reflections are present so the top and base of the sandstone are

"resolved." This is one of the wavelets shown in Figure 3A, where the waveshapes for other layer thicknesses also can be seen. If the wavelet had half the frequency (i.e.: 5, 10, 12, 9, 0, −8, −11, −9, −4, 0, 4, 5, 4, 3, 1); then the composite waveshape would have been that of Figure 5C and the top and base of the sandstone would not have been resolved. Resolution depends on frequency.

Sometimes component reflections from several reflectors interfere and give a tuned response. If a shale bed is in the middle of the sandstone body in the previous problem, with a thickness such that the travel time through it is the same as through the sandstones, then the four component reflections can be summed:

$$
\begin{array}{l}
1.0, .9, - .9, -.8, 0, .5, \quad .3 \\
\quad -1.0, -.9, .9, .8, \quad 0, -.5, -.3 \\
\quad\quad 1.0, .9, - .9, -.8, \quad 0, .5, .3 \\
\quad\quad\quad -1.0, -.9, \quad .9, .8, 0, -.5, -.3 \\
\hline
1.0, .9, -1.9, -1.7, 1.9, 2.2, -1.6, -2.2, \quad .6, 1.3, .3, -.5, -.3
\end{array}
$$

This wavelet is shown on Figure 5D; the ringing character results from the natural resonance.

The shape of the elemental wavelet is critical to the detailed appearance of the composite, although not particularly relevant to resolution if it has the same frequency content. Actual seismic wavelets are commonly minimum phase (or nearly so). Because the reflection event does not even begin until the time corresponding to the interface, the amplitude buildup occurs after the travel time to the reflector. Zero-phase wavelets are sometimes used in making models. Such wavelets are symmetrical and have the advantage that their peak amplitude occurs at the round-trip travel time to the interface. However, they are anticipatory, and half of the wavelet precedes the travel time to the interface. They also will not match real impulsive-source seismic records unless the records have been subjected to processing which converts the seismic record to appear as it would if the source had been zero-phase. The model of a sandstone layer shown on Figure 5B has been recalculated using a zero-phase wavelet of the same frequency content (Fig. 5E) and is displayed as Figure 5F. Although the waveshape differs, the resolution is the same. Figure 3B shows the effect of the same wedge as Figure 3A except for the use of a zero-phase wavelet.

Many (perhaps most) reflections are the interference composites resulting from several interfaces. The constancy of reflection character which is generally observed can be attributed to the fact that bedding does not change rapidly in the horizontal direction, so the interference remains nearly the same. In many places reflections change their character slowly and eventually disappear, and only "phantoms" (drawn parallel to nearby reflections but not necessarily oriented to any single reflection) can be mapped over extensive areas. The constancy of reflection character is sometimes taken as a measure of whether an interface is regionally constant as opposed to being the composite of closely spaced reflectors which change over the area.

Reflections generally follow depositional time surfaces rather than facies boundaries. This is mainly a matter of scale. Consider a series of sandstones which are shaling out one by one (Fig. 6). Seismic traces are commonly spaced every 25 to 50 m so that adjacent traces usually see the same reflecting sequence. Since lateral continuity over this distance is common, reflections will mainly parallel the bedding. On the other hand, analysis based on wells is likely to show the top of a different sandstone at each well because the wells are spaced much farther apart than seismic traces. Therefore a "top of sands" map based on well logs will cross depositional time lines and generally disagree with a seismic phantom map.

RELATING WELL LOGS TO SEISMIC TRACES

Reflectivity depends on the velocity and density of rocks, two quantities measured by sonic and density logs, so the relations between well logs should relate to seismic traces. A seismic trace calculated from well-log data is called a *synthetic seismogram* and a well log calculated from a seismic trace is called a *seismic log*. These calculations can be done only approximately; differences should be expected for the following reasons:

1. Logs are plotted in depth; seismic traces in two-way travel time. This difference can be remedied easily—the well log may be replotted in terms of equivalent two-way travel time or the seismic traces may be replotted in equivalent depth. A seismic sample interval of 2 msec corresponds to a depth sample interval of about 3 m (at 3,000 m/sec velocity).

2. Logs respond to the magnitudes of velocity and/or density where reflectivity depends on differences in the product of velocity and density. The relation is somewhat like that between a function and its derivative. A closer comparison would be achieved if the seismic trace were integrated or if the log were differentiated.

3. The frequencies of well logs are very high whereas those of seismic energy are very low. Wavelengths seen on well logs are fractions of meters; on seismic traces tens or hundreds of meters. If we integrate a seismic trace as indicated in (2) above, we make a too-low signal even lower because integration emphasizes low frequencies.

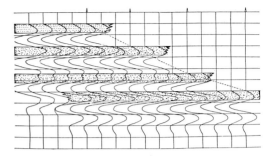

FIG. 6—Reflections from a series of sandstone pinchouts. Wavelet shape is that of Fig. 5A. Seismic reflections tend to follow depositional (time) lines where wells would be apt to see top-of-sands (facies) line which is dashed. The difference is essentially one of spacing of data points.

Likewise, if we differentiate a well log, we make a too-high signal even higher because differentiation emphasizes high frequencies. There is nothing we can do to restore high frequencies if they are not present above the noise level. The best match is likely to be a seismic trace with a heavily filtered sonic or density log.

4. Logs "see" only a short distance away from the borehole whereas seismic response is of *Fresnel-zone* dimensions (see subsequent section).

5. different sorts of noise affect each. Well logs are subject to hole size variations, rock alteration by drilling fluid invasion, cycle skipping, etc. Seismic traces are subject to interference with non-reflection energy, multiples, etc. Noise effects are removed from both well logs and seismic traces by processing.

6. The seismic trace at the well location may relate to a different portion of earth than that seen by the well unless *migration* has been performed.

SYNTHETIC SEISMOGRAMS

Synthetic seismograms provide a means for linking borehole logs with actual seismic records. Their principal use is in identifying reflections—in determining which event relates to a particular interface or sequence of interfaces. Comparison between actual field records and synthetic seismograms made for primaries-only and primaries-plus-multiples allows determination of which events are primaries. The input data can be changed to illustrate the effect on the seismic record if the geologic section changes; a variation in the thickness of units, removal of units, or assuming changes in the lithology also affects the seismic record. Such a study can be a valuable guide in knowing for what changes in seismic waveshape to look. Synthetic seismograms are a simple form of seismic modeling and can be combined with seismic modeling concepts to produce more realistic model seismic records.

The input to make a synthetic seismogram should be complete velocity and density logs, and the end product is usually the zero-offset seismic trace which is expected: (1) velocity and density values are multiplied to get an acoustic impedance log, (2) a reflection coefficient log is generated from the acoustic impedance values, and then (3) the reflection coefficient log is convolved with a seismic wavelet. Synthetic seismograms are sometimes generated for primary reflection events only, sometimes for primaries plus certain classes of multiple reflections, and sometimes effort is made to include all multiples. Seismic events other than reflections and multiples are seldom included. Source-to-geophone offset distance is sometimes allowed for, and a multitrace seismic record is constructed rather than the single zero-offset trace.

The principal deficit in making synthetic seismograms is that the input data commonly are incomplete. Reliable density data are often missing over most of the borehole, but this is not an especially severe handicap because the density usually varies in the same way as the velocity so that the use of velocity data alone gives a reasonable approximation. Sometimes an empirical relation between velocity (V) and density (ρ) is used, such as $\rho = KV^{1/4}$, but such a relation does not apply universally and its use will produce some errors. The upper part of the velocity log is generally not available because logs are not often measured over the portion of a hole covered by the first casing string. Major velocity variations which are important in generating multiples often occur within this region, so that multiples may not be calculated correctly. The base of the weathering is often the dominant generator of multiples. Gas in the shallow section, even in minor amounts, also may give rise to large reflection coefficients and hence be important in multiple generation.

Where velocity information is not available, a synthetic seismogram can be made from an electric log using an empirical relation between resistivity and velocity. Faust's Law ($V = K[ZR]^{1/6}$, where Z = depth and R = resistivity) is one such relation, especially used in clastic sections. Local empirical relations derived from wells where both electrical and velocity logs are available can be used for calculations at nearby wells where velocity logs are lacking. Such a relation might be used in studying how seismic character would be expected to vary as the stratigraphy changes across a basin.

Synthetic seismogram manufacture often involves other problems. Check shots may not be available and the sonic log may involve small systematic errors which accummulate in the integration. Hence there may be error in calculating the arrival time of an event even though the prediction of the waveshape is good. Discrepancies between the arrival time of an event on a synthetic seismogram and one observed on a seismic record also can be produced by filter delays in recording and/or processing, by the use of different reference datums, or for other reasons. The filtering on the seismic record and on the synthetic seismogram may be different so that one may be inverted with respect to the other or may have some other phase shift.

SEISMIC LOGS

Whereas synthetic seismograms provide a means for calculating a seismic trace from well data, a *seismic log* is an attempt to calculate the equivalent of a well log from the seismic trace. Equation (1) can be solved for the acoustic impedance of the lower medium, $\rho_2 V_2$:

$$\rho_2 V_2 = \rho_1 V_1 \frac{(1+R)}{(1-R)}, \qquad (2)$$

where R is the reflection coefficient. This expression can be used to calculate the acoustic impedance below a reflector in terms of the acoustic impedance above the reflector. If the complete sequence of reflection coefficients can be determined, then the acoustic impedance of the nth interface can be determined in terms of the shallowest impedance:

$$\rho_n V_n = \rho_0 V_0 \prod_n^{i=1} \frac{(1+R_i)}{(1-R_i)} \qquad (3)$$

The goal of the seismic log process generally is a synthetic-sonic log rather than an acoustic-impedance log. An empirical relation between velocity and density, such as referred to above, permits this. More often it is assumed that the density does not vary.

The inversion process assumes that the seismic trace is a good approximation of a reflection coefficient log; only primary reflection energy must be present and it must be present in proper proportions. Nonreflection energy, including multiples, must have been removed; amplitude must have been strictly preserved; and the equivalent wavelet must have been reduced to an impulse. Since these cannot be done completely, the process is at best an approximation. However, in some instances seismic logs can be produced

FOR SPHERICAL WAVES:

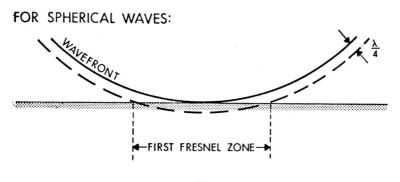

A.

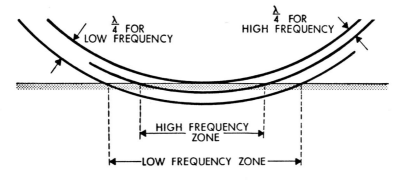

B.

FIG. 7—First Fresnel zone for (**A**) spherical waves reflected from plane interface. (**B**) shows how size of the Fresnel zone depends on frequency.

which are reasonable approximations to filtered velocity logs.

SPATIAL RESOLUTION CONSIDERATIONS

Seismic waves are often viewed as analogous to light—thin pencils of energy which travel from source to reflector along a "ray path," and which behave irrespective of adjacent pencils of energy. Following this point of view, the reflection involves only a point on the reflector.

Useful as this point of view is, it is more realistic to think in terms of a *wavefront,* the location of the disturbance resulting from a source. Wavefronts have physical reality and move with time; detectors buried throughout the earth would activate as the wavefront passes. As the wavefront reaches a reflecting interface, part of it will be reflected. Where a seismic wave commonly consists of approximately 1.5 cycles, the disturbance will continue for a region behind the wavefront. Consider Figure 7A which shows the region ¼-wavelength behind the wavefront, tangent to the reflector. The portion of the reflector between points of contact with the wavefront is the area which effectively produces the reflection; it is called the *first Fresnel zone.* Energy from the periphery of the first Fresnel zone will reach a detector at the source location ½-wavelength later than the first reflected energy, allowing for two-way travel time. All the energy reflected from the first Fresnel zone will arrive within ½-wavelength and therefore will interfere constructively. If the reflecting point is removed, for example by cutting a small hole in the reflector, a reflection will nevertheless be observed (Fig. 8). The concept of an area rather than a point on the reflector being involved in reflection is the essence of understanding spatial resolution.

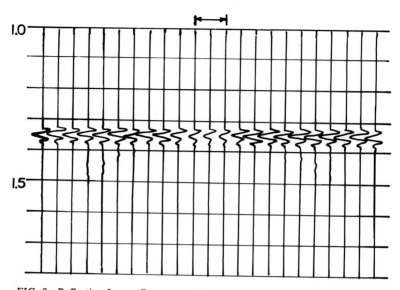

FIG. 8—Reflection from reflector containing a hole. Reflection is observed at hole because hole is smaller than Fresnel zone. Location and hole dimensions are indicated by arrow.

The Fresnel-zone concept can be extended to higher orders but the net contributions of successive Fresnel-zones are small and commonly only the first Fresnel zone is considered. (The Fresnel zone for spherical wavefronts as described above differs from that generally given in books on physical optics, where the description is usually framed in terms of plane waves. Fresnel zones are then defined in terms of a ½-wavelength criterion rather than a ¼-wavelength one. The effective reflector area for plane-wave seismic recording, such as is simulated in the SIMPLAN™ method, is larger than that for point-source seismic recording.)

Since wavelength depends on frequency, the dimensions of the Fresnel zone depend on frequency (Fig. 7B). Hence different portions of the reflector are effective for the different frequency components which together make up the seismic waveform. For example, for a plane reflecting interface at a depth of 1,000 m and an average velocity of 2,000 m/sec, the first Fresnel zone has a radius of 130 m for a 60-Hz component and 183 m for a 30-Hz component. The size of the Fresnel zone also obviously depends on the distance from the observing point and the curvature of the wavefront. For a deeper reflection, say one at 4,000 m with an average velocity of 3,500 m/sec, the first Fresnel zone has a radius of 375 m for a 50-Hz component and 594 m for a 20-Hz component. Thus spatial resolving power deteriorates with depth; a deep feature has to be larger in

areal extent to produce the same effect as a smaller shallow feature.

If the reflector is not regular, different frequency components may be reflected in different proportions. Consider the edge of a reflector (a "fault" model or perhaps the edge of a reef) and how the Fresnel zone is affected as the edge is approached. Where the observing point is remote from the edge (Fig. 9A), the Fresnel zone areas for low- and high-frequency components have a certain "normal" ratio. As the edge is approached (Fig. 9B), the low-frequency zone "sees" the edge before the high-frequency zone and hence a smaller portion of low-frequency energy is reflected, changing the frequency spectrum and the reflection waveshape. From a point directly over the edge (Fig. 9C), the Fresnel zones for high- and low-frequency components each have half the area they had when remote from the edge, and thus the reflection wavelet contains half the energy but the same waveshape. As the reflection (now called a *diffraction*) is observed from beyond the edge (Fig. 9C), the area of the high-frequency zone decreases relative to that of the low-frequency zone and hence the waveshape changes. The net effect is that the edge is seen over a region rather than only directly overhead.

Another way to consider spatial resolution is from the viewpoint of diffractions. A diffraction represents the effect of a point in the subsurface. Where two points are separated by the Fresnel zone diameter and where the noise is low, two

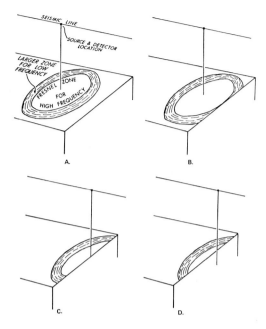

FIG. 9—Fresnel zone explanation for changes in wave-shape produced by edge of feature.

points are seen rather than only one and hence they are resolved. Certainly there are indications even where the points are closer, that sensitivity to interference effects can resolve these closer points, so that there are some subjective aspects to the definition of horizontal resolution just as there are to vertical resolution.

Migration helps with spatial resolution. Wave equation migration involves calculating what geophones would see if they were lowered through the earth. As the geophones approach features, the Fresnel zone dimensions become smaller and consequently resolution improves. Thus migration helps in defining features more clearly. Migration is sensitive to noise, and on migrated sections migration noise can become the equivalent limitation.

DISPLAY

An important aspect relating to the ability to identify stratigraphic features in seismic data is the scale and scale ratio of displays. A compressed horizontal scale commonly is helpful in seeing stratigraphic changes, although it may make structural interpretation more difficult because of the distortion which vertical exaggeration produces. No single display is ideal for all needs, so displays should be optimized for specific objectives.

Auxiliary seismic-derived quantities such as amplitude, polarity, frequency, etc., likewise required several different displays. Commonly a section is plotted at several amplitudes because the full range of significant changes can not otherwise be comprehended. Likewise a section may be replotted with reverse polarity to emphasize different aspects. The use of color provides a method of conveying variations in auxiliary measurements to an interpreter in a way which permits greater comprehension of interrelations.

CONCLUSIONS

Is stratigraphic interpretation of seismic data art or science? Today it is mostly art—recognizing patterns and exercising imagination. However, it is a constrained art, limited by fundamental considerations. Successful stratigraphic interpretation of seismic data has to be a combination of three elements: principles, experience, and imagination.

APPENDIX: GLOSSARY OF GEOPHYSICAL TERMS

Acoustic Impedance—the product of velocity and density.

Anti-alias Filter—a filter used before sampling to remove high frequencies which otherwise would cause ambiguities in information content.

Common-Depth Point Stacking—the combination of data which have in common the same midpoint between source and detector.

Compressional Wave—a seismic wave in which the particles move in the direction of the wave.

Convolution—filtering, which changes waveshape.

Deconvolution—removing the effect of an earlier filtering action.

Diffraction—the energy returned from a point.

Divergence Correction—correction to amplitude values to compensate for the decrease in energy density with distance from the source.

Frequency Spectrum—the amplitude of sine waves of different frequencies which would add up to a particular waveform.

Fresnel Zone—the portion of a reflector responsible for a reflection event seen at a point. Those reflection ray paths from source to detector which differ by less than $\frac{1}{2}$ wavelength can interfere constructively; the portion of the reflector from which they can be reflected constitutes a "Fresnel zone."

Hz—Hertz or cycles per second.

Impulsive Source—a source of very short duration.

Interference—the superposition of waveforms.

Migration—repositioning of reflected energy so that it indicates location of the reflecting point.

Minimum Phase—a characteristic of waveforms which have their energy concentrated early in the waveform.

Multiples—energy which is reflected more than once.

Noise—any unwanted energy.

Offset—the distance between source and detector.

Phase—the amount of rotation in circular motion.

Predictive Deconvolution—using early arrivals to predict and remove multiples.

Reverberation—energy which bounces back and forth within a layer.

Ringing Character—a wavelet which has more cycles than usual.

Shear Wave—a seismic body wave in which particles move at right angles to the direction of the wave.

Signature Correction—a process to change recordings where the wavelet shape was different and known, into what would have been obtained with a desired wavelet shape.

Stacking—combining the data from different records.

Static Correction—a correction for variations in arrival time because of near-surface elevation, or datum variations.

Surface Consistent—assignment of time delays or amplitude attenuation to source or detector locations.

Velocity Filtering—attenuating events on the basis of dip move-out.

Wave Equation Migration—downward continuation of the seismic wavefield by numeric solution of the wave equation.

Zero Phase—a characteristic of waveforms which are symmetrical.

Aspects of Rock Physics From Laboratory and Log Data that are Important to Seismic Interpretation[1]

A. R. GREGORY[2]

Abstract This paper summarizes some relations between rock physical properties and the influence of subsurface environmental conditions that commonly are encountered in seismic stratigraphy problems. Many of these relations are empirical correlations based on laboratory and field data, but theory also provides useful guidelines for explaining observed relations. Correlations between velocity, porosity, density, mineral composition, and geologic age, and their dependence on pressure and temperature are documented. Permeability remains an elusive parameter that is not directly measurable by geophysical methods.

Techniques for measuring velocity and attenuation of rock samples in a laboratory environment are discussed to show the capabilities and limitations of these methods. Predictions based on the theory of Biot indicate that viscous losses caused by fluid motion in rocks are of minor importance at low frequencies compared with losses caused by solid friction. Evaluation of the elastic constants of grossly anisotropic rocks requires specialized laboratory techniques. Differences between elastic moduli derived from static and dynamic measurements appear to be related to the presence of microcracks in rocks at low pressures.

The effect of temperature on elastic properties is too large to be ignored in many reservoirs and especially in those located in geothermal zones. The theory of Gassmann is used to show that velocities and reflection coefficients are relatively independent of the type of pore fluid at depths greater than about 6,000 ft (1,830 m) in Miocene sediments in a Gulf Coast area. Generally, when both overburden pressure and formation fluid pressure are varied, only the difference between the two (the effective overburden pressure) has a significant influence on velocity.

Results of laboratory studies show that fluid saturation effects on compressional wave velocity are much larger in low porosity than in high porosity rocks. Shear-wave velocities of sedimentary rocks fully saturated with gas or water do not always agree with the Biot theory; agreement is dependent on pressure, porosity, fluid-mineral chemical interactions, and presence of microcracks in the cementing material. The presence of gas in sedimentary rocks reduces the elastic moduli, and the effect is greatest at low pressures. Elastic moduli and ratios of compressional and shear-wave velocities have significant diagnostic value for differentiating between gas and liquids in sedimentary rocks.

VELOCITY-POROSITY RELATIONS IN SEDIMENTARY ROCKS

Time-Average Relationship

The influence of pressure on velocity becomes small at pressures corresponding to the deeper sediments in situ (Wyllie et al, 1956). Under this condition, the velocity is determined primarily by porosity and mineral composition. The time-average equation (1) empirically relates velocity and rock parameters for a fairly wide range of porosities:

$$\frac{1}{V} = \frac{\phi}{V_F} + \frac{1-\phi}{V_M}, \qquad (1)$$

where ϕ = fractional porosity, V = liquid-saturated rock velocity, V_F = pore-fluid velocity, and V_M = rock matrix velocity. Porosity can be determined from sonic log parameters:

$$\phi = \frac{\Delta t_{log} - \Delta t_M}{\Delta t_F - \Delta t_M}, \qquad (2)$$

where Δt_{log} = transit time from log (μ sec/ft), Δt_M = transit time of matrix material, and Δt_F = transit time of pore fluid. Wide experience with both in situ determinations and laboratory experimentation supports the general applicability of this relation for most consolidated sedimentary rocks, particularly where the pore fluid is water or brine (Figs. 1, 2).

The parameter V_M characterizes the mineral content of the rock and is equal to the value of V as ϕ approaches zero. The theories of Voigt (1928) and Reuss (1929) can be used to estimate the velocity for the extreme case of zero porosity. Numerical values of V_M that have been found useful are listed below.

Rock Type	V_M (ft/sec)	V_M (m/sec)
Sandstones	18,000–19,500	5,486–5,944
Limestones	21,000–23,000	6,401–7,010
Dolomites	23,000–24,000	7,010–7,315

As a special illustration of the use of the time-average equation, laboratory data are plotted in Figure 3 for cores from a depth of about 5,000 ft (1,524 m). The cores were confined at a pressure of 3,000 psi (211 kg/sq cm) with brine in the pores to simulate the original environmental conditions. The two principal minerals in the rock (calcite and quartz in the form of tripolite) are mixed in relative proportions ranging from approximately 50% calcite–50% quartz to 80% calcite–20% quartz. A petrographic analysis of the cores indicated that the samples with lower porosity (also from the upper

[1] Manuscript received, October 26, 1976; accepted, May 13, 1977. Publication authorized by the Director, Bureau of Economic Geology, University of Texas at Austin.

[2] Bureau of Economic Geology, University of Texas at Austin.

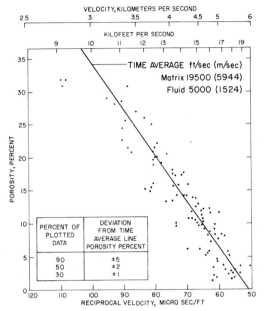

FIG. 1—Velocity-Porosity data determined in laboratory for water-saturated sandstones compared with time-average relation for quartz-water system (Wyllie et al, 1962).

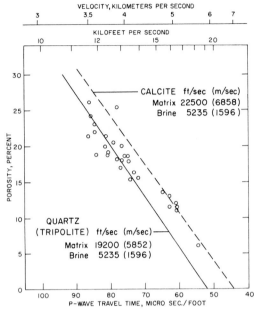

FIG. 3—Comparison of P-wave velocities as function of porosity for brine-saturated tripolite samples under confining pressure of 3,000 psi or 211 kg/sq cm (Gardner et al, 1974).

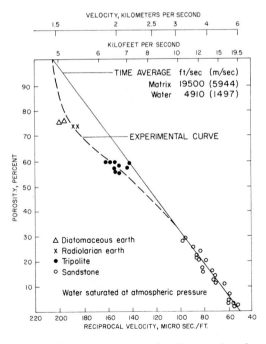

FIG. 2—Velocity versus porosity for silicious rocks under axial pressure (Wyllie et al, 1958).

part of the formation) have a continuous calcite matrix, whereas the samples with higher porosity have a continuous quartz matrix. The lower porosity data points can be approximated by a time-average line with V_M = 22,500 ft/sec (6,858 m/sec), which is an appropriate velocity for a calcite matrix, and the higher porosity data points by a time-average line with V_M = 19,200 ft/sec (5,852 m/sec), which is an appropriate velocity for a quartz matrix. Interestingly, the data appear to separate on these two lines, according to the mineral that is predominantly the continuous phase. No correlation was found between velocity and the concentration by volume of the minerals.

It is also of interest that when the travel times at high pressure (10,000 psi; 703 kg/sq cm) were plotted against porosity, no separation of the data along two lines could be detected. These results may indicate that velocity measurements on cores, at pressures appropriate to the depth of the formation, contain more useful information than measurements at an arbitrarily high pressure.

The pore-fluid velocity V_F varies with the compressibility of the fluid, which is influenced by temperature and pressure. The velocity of brine depends on the salinity also and ranges from 4,910 ft/sec (1,497 m/sec) at 25°C (distilled water) to about 5,575 ft/sec (1,699 m/sec) for an NaCl concentration of 200,000 mg/l at the same temperature (Fig. 4). The velocity of seawater at 2°C with an NaCl concentration of 37,000 mg/l increases from about 4,790 to about 5,130 ft/sec (1,460 to 1,564 m/sec) as pressure increases from 0 to 9,000 psi (633 kg/sq cm).

For formations at shallow depths, the influence of pressure on microcracks can cause significant errors in the time-average prediction. The equation can still be useful if V_M is regarded as an empirical constant with a value less than the Voigt-Reuss values. It can be assumed that the transit time Δt is a linear function of porosity at any depth, with the coefficients of the linear form chosen from suitable data.

The more severe case is that of sands which are uncompacted or geopressured. Transit times in such sands may be much greater than the time-average predicts. One way to approach the problem is to use an empirical correction factor C_p to obtain a corrected porosity ϕ_c.

$$\phi_c = \frac{\Delta t_{log} - \Delta t_M}{\Delta t_F - \Delta t_M} \cdot \frac{1}{C_p}. \tag{3}$$

The value of C_p can be obtained by measuring the deviation from the relation between true porosity and ϕ computed from the uncorrected time-average (Equation 2). The true porosity can be obtained from core analysis data or from log-derived relations (Schlumberger, 1972).

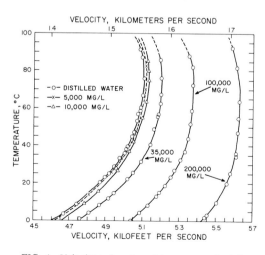

FIG. 4—Velocity as function of temperature for brines at various concentrations of NaCl, at atmospheric pressure (Wyllie et al, 1956).

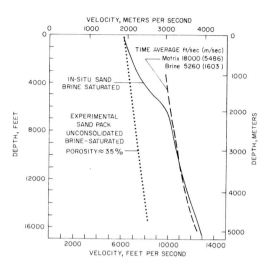

FIG. 5—Velocity as a function of depth showing consolidation effect for in-situ Tertiary sandstones. For comparison, velocities of experimental sand packs at pressures corresponding to these depths are also shown. (Gardner et al, 1974).

An illustration of the deviation of the time-average relation from well log velocity data in a shallow young sedimentary basin is given in Figure 5. In this basin the uppermost sedimentary layers are unconsolidated and the porosity varies mainly with the grain-size distribution and clay content. The velocity is only slightly greater than that of seawater. With increasing depth, the velocity increases partly because the pressure increases and partly because cementation occurs at the grain to grain contacts. Cementation is the more important factor.

The rapid increase of velocity with depth normally continues until the time-average velocity is approached. Beyond that depth, the layers behave like well-consolidated rock and the velocity depends mainly on porosity.

The solid curve in Figure 5 shows representative velocity versus depth for brine-saturated in-situ sandstones based on some sonic and electric log data. The dotted curve is from laboratory data for fresh unconsolidated packings of quartz sand grains saturated with water at pressures corresponding to the depth. Thus, the dotted curve indicates what would happen to sand if it were buried without consolidation or cementation, and the divergence of these curves is attributable to consolidation. The dashed curve shows the time-average velocity calculated using the average porosity read from well logs. At the shallower depths, the actual velocity is less than the time-average, but below about 8,000 ft (2,438 m) the agreement is close.

PERMEABILITY

Permeability can be determined indirectly from analysis of well log data but cannot be measured directly by present geophysical methods. Empirical relations of core data show that permeability normally increases as porosity increases. The type of porosity found in sedimentary rocks influences permeability. Isolated pore spaces (vugs) that are not interconnected with flow channels, microcracks in cement, pores within kaolinite clay, and material in pore fillings, do not contribute to the effective permeability. Permeability data from unconfined specimens in the laboratory can be expected to overestimate the permeability of deep reservoirs. Alterations of permeability, porosity, and elastic properties caused by pressure and heat can have a substantial influence on the bulk volume, pore-fluid volume, and deliverability of a reservoir. The reduction in permeability associated with increases in the effective overburden pressure caused by pressure decline in a producing reservoir is of particular importance to the permeability and long-range deliverability of geopressured reservoirs. McLatchie et al (1958) showed that low-permeability rocks are more sensitive to changes in the effective overburden pressure than high-permeability rocks. Reductions in permeability approach 90% when low-permeability rocks are subjected to effective overburden pressures of 5,000 psi (351 kg/sq cm) or more. In a flowing water well, clay particles can be dislodged from the rock, obstruct or plug flow channels, and reduce both permeability and porosity. Similarly, gas that has been released from solution in a pressure-reduced reservoir will decrease the effective permeability to liquid.

The aqueous permeability of some sandstones is very sensitive to temperature because of the combined influence of thermal expansion of grains into pores and pore throats, mechanical stresses caused by differential expansion of different minerals along different crystallographic axes, and fluid-rock surface interactions. Casse and Ramey (1976) observed a 28% reduction in absolute permeability to water as the temperature increased from 24°C to 143°C for Berea Sandstone under a constant confining pressure of 4,000 psi (281 kg/sq cm).

The most realistic permeability data for subsurface reservoirs are obtained from well production flow tests. Methods have been developed for evaluating permeability from pressure buildup and drawdown data, absolute open-flow potential test data, and from the behavior of a producing well during flow tests (Matthews and Russell, 1967).

VELOCITY-DENSITY-LITHOLOGY RELATIONS

The types of sedimentary rocks that are encountered most often in a wide range of basins, geologic ages, and depths to 25,000 ft (7,620 m) are shown to have a wide range of velocities and a lesser range of bulk densities (Fig. 6). Two empirical relations also are plotted in the illustration and serve as reference curves: the time-average curve for sandstone, and the 0.25 power relation between velocity and bulk density. Hence it is possible to estimate the density or velocity of rocks of different lithology if only one parameter (density or velocity) is known.

Figure 7 illustrates velocity differences for rocks of different lithology. Velocity increases as pressure on the rock frame increases. The increase in velocity with pressure is attributed to the presence of microcracks at low pressure which are diminished at higher pressures. In the case of the packing of sand, the microcracks are presumably the contacts between grains.

AGE-VELOCITY-DEPTH RELATION

The velocity of sedimentary rocks has a one-sixth power dependence on age and depth as

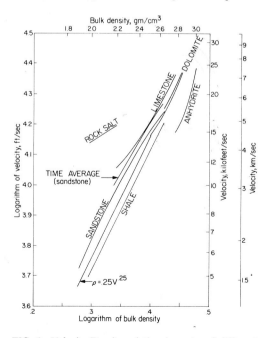

FIG. 6—Velocity-Density relation in rocks of different lithology (Gardner et al, 1974).

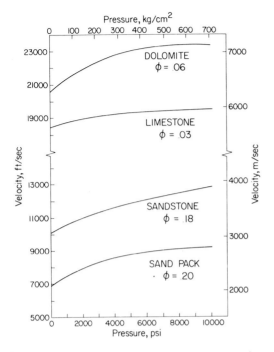

FIG. 7—P-wave velocity versus confining pressure for brine-saturated carbonates, sandstone and sand pack (Gardner et al, 1974).

shown by the empirical expression of Faust (1951).

$$V = K(ZT)^{1/6}, \qquad (4)$$

where K = 125.3, Z = depth, ft, T = age, years, and V = velocity, ft/sec. Figure 8 shows how velocity varies with depth for sandstones and

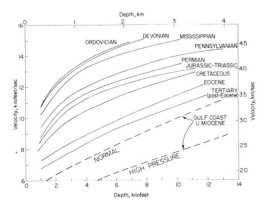

FIG. 8—Interval velocity versus depth for clastic sediments of different age groups (Faust, 1951).

shales of different age as determined by Faust (1951).

LABORATORY VELOCITY MEASUREMENT SYSTEMS

Laboratory studies of rocks have a definite advantage over field studies because of the amount of control that can be exercised on physical and environmental parameters that are pertinent to the investigation. Parameters such as pressure, temperature, mineral composition, porosity, permeability, density, type of pore fluid, and degree of saturation can be precisely determined. Control over such parameters is virtually non-existent in field work. There are, of course, limitations on the size of rock samples that can be accommodated by laboratory facilities. The possibility that velocity dispersion exists in the frequency gap between laboratory and field methods has not been fully explored in the geophysical literature.

Most laboratory methods use some type of electromechanical transducers for generating and detecting elastic waves. The actions of the transducer and the particle motions of elastic waves are depicted schematically in Figure 9. The piezoelectric elements in a transducer assembly may consist of quartz or ceramic compositions such as barium titanate, lead zirconate titanate, and lead metaniobate. Resonant

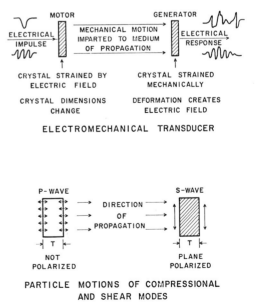

FIG. 9—Electromechanical interactions and particle motions of transducer elements.

frequencies are related to the shape and size dimensions of the element.

Relation of System Components

A typical experimental setup for generating and detecting an elastic disturbance in a solid sample consists of five units (Fig. 10). The arrows on the connecting lines indicate that the units are mutually interacting. For example, the shape of the electric pulse is influenced by both Units 1 and 2. If the type of transducer in Unit 2 is changed, the shape and amplitude of the electric pulse will be affected. Similarly, the frequency response of Unit 2 is influenced by the sample, Unit 3, as well as by the source impedance of Unit 1. The response of Unit 3 will be determined by the acoustic impedance and dissipation of the sample material, its geometry, and the loading produced by the driver and detector. The response of Unit 4, in turn, will be influenced by the loading caused by the sample and the input impedance of the amplifier. Unit 5, the amplifier and oscilloscope, has its own response which to some extent is influenced by the source impedance of the detector transducer. It is evident that a change in the overall response can be affected by a change in any one of the five units or their interactions.

Many of the modern laboratory methods devised for ultrasonic velocity measurements in rock samples use broadband high-voltage electrical excitation pulses and quartz or ceramic transducers with resonant frequencies ranging from about 0.2 to 1 MHz. Transit times are measured by locating the first break of the received signal from its baseline as observed on an oscilloscope, or in some cases, by locating the first positive or negative peak voltage of the signal. It is found that transit times of compressional waves (P waves) determined from the first break of the first arrival can be measured in homogeneous rocks with a resolution of about 0.1 microsecond. The arrivals through different rocks may have fast or slow rise times depending on the shape of the input pulse and the frequency filtering characteristics of the material. When the rise time is slow, the location of the first break becomes ambiguous and errors in velocity

may range from one to several percent. These difficulties are magnified for shear waves (S waves) because many S-wave transducers also generate P waves that interfere with the location of the first break of the signal.

The velocity measurement method described by Gregory (1967) uses narrowband tone-burst excitation pulses and a mode conversion technique for generating S waves. The method is intended to eliminate some of the difficulties with the broadband measurement system. The advantages of the narrowband method are: (1) transit times are measured to a resolution of 0.01 microsecond by using an electronic counter-timer and locating the first voltage peak of the received signal, (2) human errors associated with the broadband system in locating the first break of the received signal are eliminated, and (3) there are no P-wave components in the S-wave signal response to cause interference with locating the shear arrival. The disadvantages are: (1) the possibility exists for errors caused by frequency distortion of the narrowband pulse in highly attenuating rocks, and (2) less penetration of the low-voltage pulse in rocks restricts the method to relatively short samples.

The dual-mode ultrasonic velocity apparatus described by Gregory and Podio (1970) further develops the capability of the tone-burst method described above to permit consecutive measurements of P-wave and S-wave velocities of rock samples in the same stress cycle. Hysteresis effects caused by making separate measurements of P-wave and S-wave velocities in two different stress cycles are therefore avoided. The time required for testing a given sample is reduced by approximately one half. The apparatus also is readily adapted to automated transit time measurements in rock samples at various stress levels. On-line computer processing of the input data allows rapid step by step evaluation of the dynamic elastic properties of the rock as the experimental test progresses.

Narrowband Dual-Mode Tone-Burst System

The transducer cells are assembled for measuring elastic-wave velocities through rock samples under triaxial loading as shown in Figure

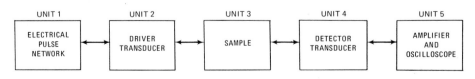

FIG. 10—Components of velocity measurement system.

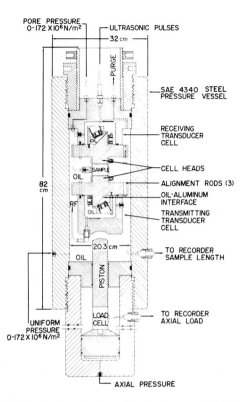

FIG. 11—Dual-mode ultrasonic apparatus assembled for triaxial loading of sample (Gregory and Podio, 1970).

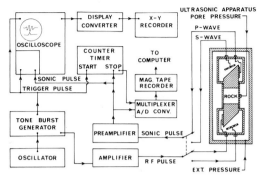

FIG. 12—Schematic diagram of electronic-acoustic system (Gregory and Podio, 1970).

11. A neoprene sleeve with a wall thickness of 0.125 in. (0.317 cm) jackets the sample and serves as a flexible impermeable barrier to the hydraulic fluid. The fluid exerts pressure uniformly in the lateral and axial directions on the external surface of the sample. Additional pressure can be placed on the sample in the axial direction by the piston located at the bottom of the pressure vessel. Pressure on the pore fluid is transmitted through pore pressure lines, which are connected directly through the neoprene sleeve. The apparatus is designed for independent control of both external and pore pressures between 0 and 25,000 psi (1,758 kg/sq cm). Total axial loads up to 50,000 psi (3,515 kg/sq cm) can be achieved. The designed pressure limits of the apparatus exceed those required to restore subsurface stress conditions on cores recovered from the deepest existing well bores, i.e., about 32,000 ft (9,754 m).

Figure 12 shows the electronic system associated with the ultrasonic apparatus. The output of the oscillator is switched on and off by a tone-burst generator to form phase-coherent repetitive sinusoidal pulses. The tone of the pulses matches the resonant frequency of the piezoelectric transducers. Phase coherence is necessary to obtain consistent results from Fourier spectrum analysis. If the tone-burst is incoherent, the signal will drift between sine and cosine tone-bursts, and the phase correction will cause a corresponding drift in the spectrum. The tone-burst spectrum has the band-width properties of a pulse and can be adjusted from very wide to very narrow by using bursts containing one cycle or many cycles. Since a fairly narrow band width is desired, the pulse from the generator is adjusted to contain four cycles and the interval between pulses is 10 to 20 msec. These pulses are amplified to 50 to 120 volts (peak/peak) and applied alternately through a coaxial relay switch to the P-wave and S-wave emitter transducers, which in turn generate ultrasonic pulses. After passing through the sample and receiving transducer, these pulses are amplified and displayed by the oscilloscope.

The apparatus is calibrated in the pressure vessel with the heads of the transducer cells placed face to face with no sample between them (Fig. 11).

Figures 13 and 14 show typical S-wave pulses obtained with and without a sample of limestone between the heads of the transducer cells. Calibrated marker outputs from the counter-timer were manually adjusted to indicate visually the exact parts of the waveforms that are being used to start and stop the counter (Fig. 14). The leading edges of the markers have rise times of 2 nannoseconds. The travel time of the S wave through the rock is the difference between ΔT_{S2} and ΔT_{S1} and is determined to a resolution of 10^{-2} microseconds.

The frequency spectra of a P-wave tone-burst

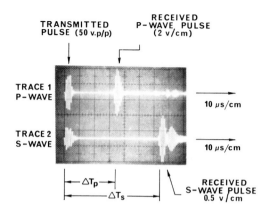

FIG. 13—P- and S-wave pulses obtained through dual-mode apparatus with cell heads placed face to face; no sample (Gregory and Podio, 1970).

pulse is shown in Figure 15. An example of velocity and compressibility data obtained for dry Boise sandstone at various confining pressures is shown in Figure 16.

Broadband Velocity Measurement System

If the rock samples are too long or too highly attenuating, a broadband system may be necessary to make velocity measurements. This system makes use of a high-voltage pulser which develops a 1,200-volt V-shaped electrical pulse with a rise time of about 0.5 microsecond (left part of oscilloscope traces shown in Fig. 17). Some systems use an electrical pulse with a much lower voltage. Frequently the transducers are highly damped. Most of the energy of the frequency spectra of the electrical pulse is between 0 and 0.5 MHz. The frequency spectra of the acoustic signal detected after traveling

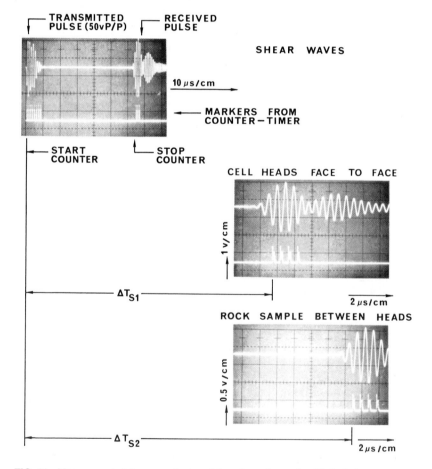

FIG. 14—Measurement of S-wave pulse travel time through sample of Solenhofen Limestone (Gregory and Podio, 1970).

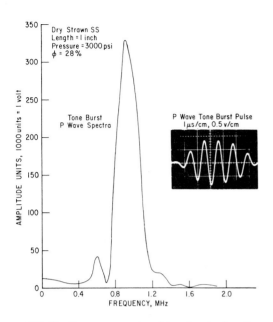

FIG. 15—Frequency spectra of tone burst pulse after passing through dry sample of Strawn sandstone under a confining pressure of 3,000 psi (211 kg/sq cm).

though a rock sample may have an amplitude which peaks somewhere between 0.2 and 0.3 MHz, depending on the characteristics of the rock. The spectra for a waveform that was obtained through a sample of aluminum is shown in Figure 18.

ATTENUATION MEASUREMENTS IN FLUID-SATURATED ROCKS

Attenuation is an important material property that can be related to the dynamic elastic moduli (Volarovich et al, 1969). Some attempts have been made in seismic exploration to relate attenuation to the type of reservoir fluids contained in subsurface geologic structures. Although many types of loss mechanisms (White, 1965) are possible for elastic waves propagating in solids, viscous and solid friction losses have received the most attention in studies related to rocks and pore saturants. It appears that solid friction is the primary loss mechanism in low-permeability rocks. Viscous losses caused by relative motion between pore fluid and solid material in porous and permeable rocks can be determined from theory (Biot, 1956). If viscous losses are large enough to be measured, then some possibility exists for estimating in-situ permeability from well log or seismic data. Existing theory appears to be inadequate for calculating solid friction losses, hence they must be measured.

The results of field measurements of attenua-

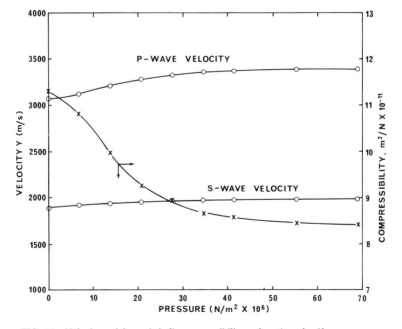

FIG. 16—Velocity and dynamic bulk compressibility as function of uniform pressure for dry Boise sandstone (Gregory and Podio, 1970).

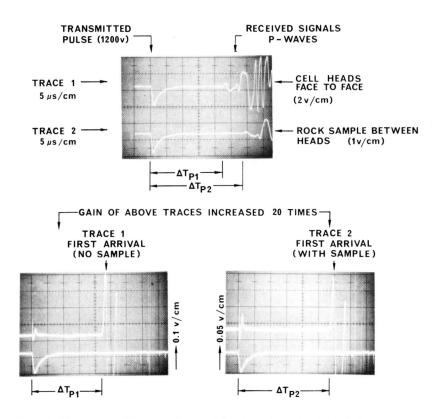

FIG. 17—Measurement of P-wave pulse travel time through sample of Solenhofen Limestone
using broadband pulse method (Gregory and Podio, 1970).

tion are seldom published, but the data reported for a thick section of Pierre Shale are notable (McDonal et al, 1958). Many investigators have undertaken laboratory attenuation measure-

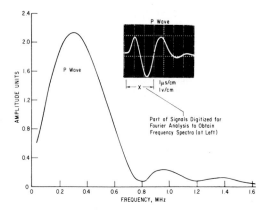

FIG. 18—Frequency spectra of broadband P-wave signal (inset) after passing through sample of aluminum.

ments on rock samples using highly selective frequency band widths and a variety of techniques.

Meaningful attenuation data for rocks are not plentiful, but a few generalizations can be made regarding the influence of diagenesis and evironmental factors. The attenuation in highly consolidated rocks is lower than in poorly cemented or unconsolidated aggregates. The application of uniaxial or biaxial stress on rocks reduces the attenuation. The presence of pore fluids tends to increase attenuation. The logarithmic decrement of subsurface formations probably ranges from 0.001 to 0.10 for frequencies below 20 KHz (Wyllie et al, 1962).

Attenuation Due to Fluid Motion in Rocks

Calculation of logarithmic decrements are made using Biot's (1956) theory. S-wave decrement, Δ_s:

$$\Delta_s = \pi \frac{M_2}{M_1 + M_2} \frac{f}{f_c}; \quad f/f_c < 0.15; \quad (5)$$

$$f_c = \frac{\mu_f \phi^2}{2\pi K(\rho_{12} - \rho_{22})}; \qquad (6)$$

where M_1 = mass of dry rock (g), M_2 = mass of fluid in saturated rock (g), f = frequency (Hz), f_c = critical frequency (Hz), μ_f = fluid viscosity (poise), ρ_{12} and ρ_{22} = density coefficients (g/cc), ϕ = fractional porosity, and K = permeability (sq cm; 1 darcy = 0.9869 $\times$ 10^{-8} sq cm). P-wave decrement, Δ_p:

$$\Delta_p = \pi \left.\frac{M_1 + M_2}{M_2}\right| (Z_1 - 1)(Z_2 - 1) \Big|$$

$$\cdot (\sigma_{11}\sigma_{22} - \sigma_{12}{}^2) \, f/f_c,$$

$$\text{for } f/f_c < 0.15, \qquad (7)$$

where Z_1 and Z_2 are the roots of the quadratic equation,

$$(\sigma_{11}\sigma_{22} - \sigma_{12}{}^2)Z^2 - (\sigma_{22}\gamma_{11} + \sigma_{11}\gamma_{22}$$

$$- 2\sigma_{12}\gamma_{12})Z + (\gamma_{11}\gamma_{22} - \gamma_{12}{}^2) = 0. \qquad (8)$$

The components σ_{ij} and γ_{ij} are elastic and density coefficients.

Some definitions and unit conversions are: Logarithmic Decrement (Δ) = attenuation per cycle, Δ/λ (cm) = nepers/cm, Δ/λ (ft) = nepers/ft, 8.868 Δ/λ (ft) = db/ft, Q = π/Δ, and λ = V/f; where V = elastic wave velocity (cm/sec or ft/sec), f = frequency (Hz), and λ = wavelength (cm or ft).

P-wave Attenuation due to Fluid Motion in Rocks

The P-wave logarithmic decrement Δ_p was computed as a function of absolute viscosity μ_f of the pore fluid for Boise sandstone which has a permeability K of 1,500 md (1.48 μsq m). The operating frequency ranged from 0.1 to 2000 KHz. The pore fluids consisted of water with a viscosity of 0.01 poise and eight types of crude oil ranging in viscosity from 0.03 to 120 poise.

The data in Figure 19 are for Boise sandstone under a uniform pressure of 5,000 psi (352 kg/sq cm) and at several frequencies from 0.1 to 2,000 KHz. Boise sandstone has a higher attenuation because of fluid motion than do most consolidated rocks. The high attenuation is caused primarily by its large permeability and its pore-size distribution. The pore throat diameter of Boise sandstone determined by mercury injection ranges from 0.12 to 60 μ and a diameter of 34 μ (34 $\times$ 10^{-4} cm) is found for pores that

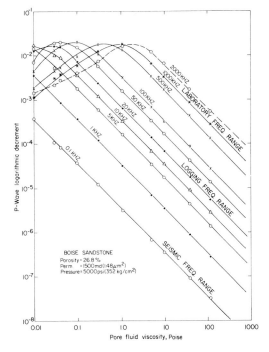

FIG. 19—P-wave attenuation versus pore fluid viscosity for Boise sandstone at several frequencies.

are filled when 50% of the pore volume is occupied by mercury. Peak values of the decrement of about 0.02 are obtained at several of the higher frequencies and also are dependent on the viscosity of the pore fluid. The P-wave decrement tends to be a linear function of fluid viscosity on a log-log plot at lower frequencies. The attenuation increases as the fluid viscosity decreases and as the operating frequency F increases. At higher frequencies, however, the assumption of Poiseuille flow breaks down, and a reversal of the data trend occurs when the frequency ratio f/f_c exceeds a critical value of about 0.15. The frequency ratio is related to rock permeability K, as indicated in Equation 6.

Hence, the P-wave decrement of a rock with high permeability (1,500 md; 1.48 μsq m) ranges from 3 $\times$ 10^{-7} to 0.02 (equivalent Q values are 10^7 and 157) depending on the operating frequency and the pore fluid viscosity. The P-wave decrement of a rock with low permeability (0.1 md; 9.86 $\times$ 10^{-5} μsq m) ranges from less than 10^{-8} to 1.9 $\times$ 10^{-3}. Equivalent Q values are 3 $\times$ 10^8 and 1.6 $\times$ 10^3.

The S-wave decrement exceeds the P-wave decrement by a factor which ranges from 2 to

9 depending on the characteristics of the rock and the operating frequency.

High rock permeability, low pore-fluid viscosity, and high operating frequencies lead to large decrements. Low rock permeability, high pore-fluid viscosity, and low operating frequencies lead to low decrements.

Increasing the static pressure on the frame of a rock reduces Q for the P-wave but Q for the S-wave is not affected (Fig. 20).

Conclusions Regarding Attenuation

1. The small magnitude of viscous losses predicted by theory at relatively low frequencies casts doubt on the practical significance of this type of attenuation in subsurface investigations using well logging and seismic methods.

2. The results challenge the widely held belief that oil-saturated rocks are more highly attenuating than water-saturated rocks. This possibility in a subsurface environment is not completely ruled out, however, because of inherent limitations in the theory. For example, the possible effect on attenuation of chemical reactions that alter subsurface diagenetic processes, and hence the elastic properties of rocks in a water-saturated environment as opposed to the absence or reduction of this effect in an oil-saturated environment, is outside the scope of Biot's mathematical model.

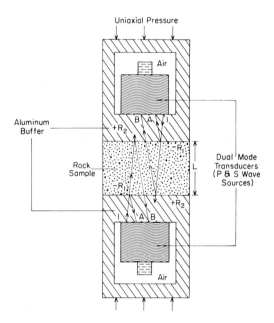

FIG. 21—Schematic configuration for pulse-echo attenuation measurements under uniaxial stress.

3. The magnitude of viscous losses predicted at high frequencies for rocks with relatively high permeability may be large enough to measure with suitable experimental techniques. Definitive answers regarding the high-frequency range can only be obtained by further experiments designed to measure and separate the viscous and solid friction components of attenuation under simulated subsurface conditions.

LABORATORY MEASUREMENTS OF SOLID FRICTION LOSSES

Pulse-Echo Method

The apparatus (Fig. 21) contains two highly damped piezoelectric elements with resonant frequencies near 1 MHz. One element vibrates in the compressional mode, the other in the shear mode. Because attenuation is to be determined over the widest possible range of frequencies, the transducers are alternately excited by a broadband pulse of 50 to 100 volts and 2 to 5 microseconds duration. These tranducers also can be used for velocity measurement from transmission delay. However, each transducer acts as both transmitter and detector for pulse-echo attenuation measurements. Thus, two measurements in opposite directions are obtained reducing errors due to sample inhomogeneity by averaging.

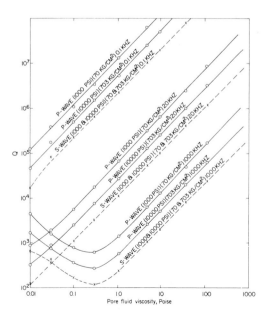

FIG. 20—Q for P waves and S waves as a function of pore fluid viscosity for Berea Sandstone under low and high pressures and at low and high frequencies.

For the configuration in Figure 21 the logarithmic decrement Δ and the attenuation parameter Q are given by the relation:

$$\Delta = \frac{\pi}{Q} = \frac{\lambda}{2L} \ln \frac{A R_2 (1 - R_1^2)}{B R_1}, \qquad (9)$$

where L = sample thickness, A = amplitude of reflection A, B = amplitude of reflection B, λ = wavelength of signal, and R_1, R_2 = normal incidence reflection coefficients.

The ratio of reflection coefficients R_2/R_1 in Equation 9 is equal to unity when the same buffer material is present on both sides of the specimen as in this case. Therefore,

$$\Delta = \frac{\lambda}{2L} \ln \frac{A}{B} T^2, \qquad (10)$$

where $T^2 = 1 - R^2$ is the transmission coefficient and

$$R = \frac{\rho_2 V_2 - \rho_1 V_1}{\rho_2 V_2 + \rho_1 V_1}, \qquad (11)$$

is the normal incidence reflection coefficient with ρ_2 = rock density, V_2 = wave velocity in rock, ρ_1 = density of buffer material, and V_1 = wave velocity of buffer material. Examples of waveforms for echoes A and B and their frequency spectra are given in Figure 22 for a dry sample of Berea sandstone at atmospheric pressure.

As seen in Equations 9 and 10, the attenuation is defined for a particular frequency (wavelength λ). Therefore the values of λ, A, and B have to be determined for particular frequencies propagating in the specimen. The time domain signals, such as those in Figure 22, are digitized with a transient recorder which is capable of sampling at 0.01-microsecond intervals to avoid aliasing effects. The digitized waveforms are transformed to the frequency domain through a discrete Fourier Transform program. This operation provides amplitude and phase components of the signal at multiples of Δf over the effective band width of the pulse. Δf is defined as $1/NT$ where N is the number of samples and T the sampling time interval.

Amplitudes from the frequency spectra are corrected for diffraction effects (Papadakis, 1966) and used in the attenuation equation. Typical values of Q for compressional waves in sandstone are shown in Figures 23 and 24.

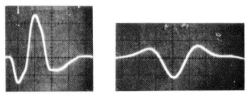

SIGNAL A   SIGNAL B

0.5 μ SEC/CM, 0.5 V/CM

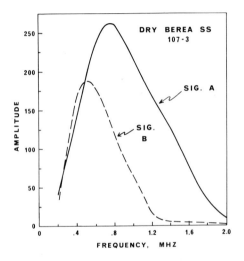

FIG. 22—Photographs (above) and frequency spectra (below) of signals A and B for dry Berea Sandstone at atmospheric pressure.

Other methods for obtaining Q from pulse echoes have been suggested. Methods based on pulse broadening (Ramana and Rao, 1974) and

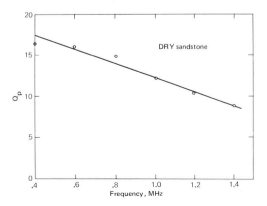

FIG. 23—Q_p versus frequency for dry sandstone sample.

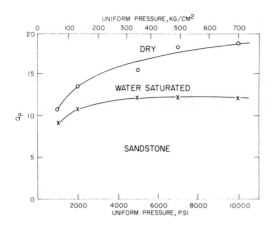

FIG. 24—Q_p versus uniform pressure for sample of dry and water-saturated sandstone.

rise time variations with travel distance (Gladwin and Stacey, 1974) should be investigated further.

ANISOTROPY OF ROCKS

The most unsymmetrical material has a maximum of 21 elastic constants. As the degree of symmetry increases, the number of independent elastic stiffnesses decreases until there are *nine* for an orthotropic material (Fig. 25), *five* for

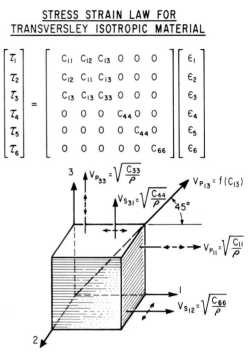

FIG. 26—Relations between elastic coefficients and wave velocities for transversely isotropic materials (Podio, 1968).

a transversely isotropic material (Fig. 26), and *two* for an isotropic material.

To determine the elastic properties of these materials by wave propagation methods, it is first necessary to identify the orientation of the axes of symmetry. In sedimentary rocks, it is assumed that one of the principal axes is perpendicular to the bedding plane. The other principal axes are then located as shown in Figure 27. Except for specialized studies, the evaluation of elastic constants of orthotropic rocks on a routine basis is too complex to be practical. The dynamic elastic properties of transversely isotropic rocks can be determined by measuring P-wave and S-wave velocities in the directions required for evaluating the coefficients C_{11}, C_{12}, C_{13}, C_{33}, and C_{44} (Podio, 1968). Calculation of the coefficients requires the measurement of five propagation velocities, i.e., two P-wave and two S-wave velocities along the axes of symmetry and one P-wave velocity along a general orientation.

For the generalized stress-strain relation,

$$\tau_i = C_{ij}\epsilon_j$$
$$i, j = 1, 6. \tag{12}$$

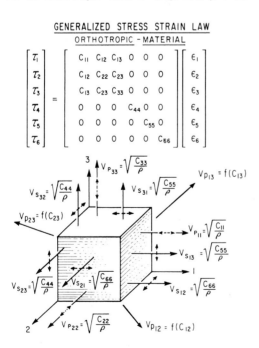

FIG. 25—Relation between elastic coefficients and wave velocities for orthotropic materials (Podio, 1968).

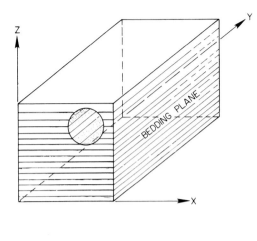

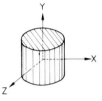

FIG. 27.—Location of principal axes in sedimentary rock.

These coefficients are associated with the propagation velocities of shear and longitudinal waves as shown in Figures 25 and 26. For example, V_{Sij} indicates the velocity of an S wave propagating in the i direction with particle vibration in the j direction.

Although numerous rocks can be classified as transversely isotropic, the routine measurement of velocity in five different directions for large numbers of cores in formation evaluation work presents formidable problems. The multiaxial dynamic testing apparatus developed and fabricated by A. L. Podio (personal commun., 1976) shows promise as a tool for studying the complex elastic behavior of anisotropic rocks. A cross section of the apparatus is shown in Figure 28. The apparatus consists of a system of six hydraulic rams housed in a cylindrical vessel. Each ram contains a set of piezoelectric transducers for generating and detecting P and S waves while stresses up to 25,000 psi (1,758 kg/sq cm) can be applied to each side of the cubic rock sample.

An approximation to the behavior of real rocks is often made by assuming that a rock is a homogeneous, isotropic, elastic medium. The velocity of propagation can be expressed in

terms of Hooke's constants. Hence for isotropic bodies:

$$K = \rho (V_p^2 - 4/3 \, V_s^2), \qquad (13)$$

$$E = (9\rho \, V_s^2 \, R_2^2)/(3R_2^2 + 1), \qquad (14)$$

$$\mu = \rho V_s^2, \qquad (15)$$

$$\nu = 1/2 \left[\frac{(R_1 - 2)}{(R_1 - 1)} \right] \qquad (16)$$

where K = bulk modulus, E = Young's modulus, μ = rigidity modulus, ν = Poisson's ratio, $R_1 = (V_p/V_s)^2$, $R_2^2 = K/(\rho V_s^2)$, and β = bulk compressibility = $1/K$. The dynamic elastic moduli above can be evaluated by measuring the velocities of P waves and S waves propagating in one direction only through the rock. Absolute values of the moduli for sedimentary rocks are highly dependent on porosity, pressure, and fluid saturation. The moduli decrease as porosity increases and increase as pressure increases. The effect of pore fluids (water, gas, oil, and mixtures of these fluids) on elastic moduli is discussed later.

Poisson's ratio tends to increase as porosity increases but may increase, remain constant, or decrease as a function of pressure. The relation between Poisson's ratio and V_p/V_s is shown in Figure 29.

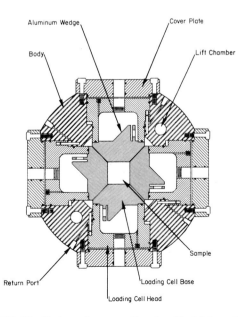

FIG. 28—Horizontal cross section of multiaxial dynamic testing apparatus.

Vp/Vs
POISSON'S RATIO
- | +
0
1
2
3
4
5
6
-0.1
0
0.1
0.2
0.3
0.4
0.5

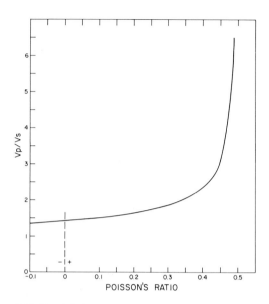

FIG. 29—Relation between Poisson's ratio and V_p/V_s.

CORRELATION OF DYNAMIC AND STATIC MODULI

Several investigators (Onodera, 1962; Sutherland, 1962; Morgans and Terry, 1957; Simmons and Brace, 1965; and Gregory, 1962) have confirmed that elastic parameters of rocks derived from dynamic measurements are greater than those from static measurements. Figure 30 shows some static and dynamic values of Young's modulus for various rocks (Panek, 1965). The dynamic/static ratios for Young's modulus range from 0.86 to 2.90. Unfortunately, the stress levels of the measurements are unspecified. There seems to be general agreement that discrepancies are associated with the presence of cracks in the rock at low pressure.

At elevated pressure the difference between static and dynamic moduli decreases, as shown for Young's modulus and bulk modulus of Berea Sandstone (King, 1969) in Figure 31.

Despite research in this area, there remains considerable doubt about whether properties measured dynamically are the same as those measured statically. For example, compressibility computed from wave velocity data is adiabatic whereas static values measured directly are essentially isothermal (Simmons and Brace, 1965). Also dynamic measurements usually are carried out at frequencies in the low megahertz range whereas static measurements are at near-zero frequency. Wave velocities derived from acoustic propagation are less likely

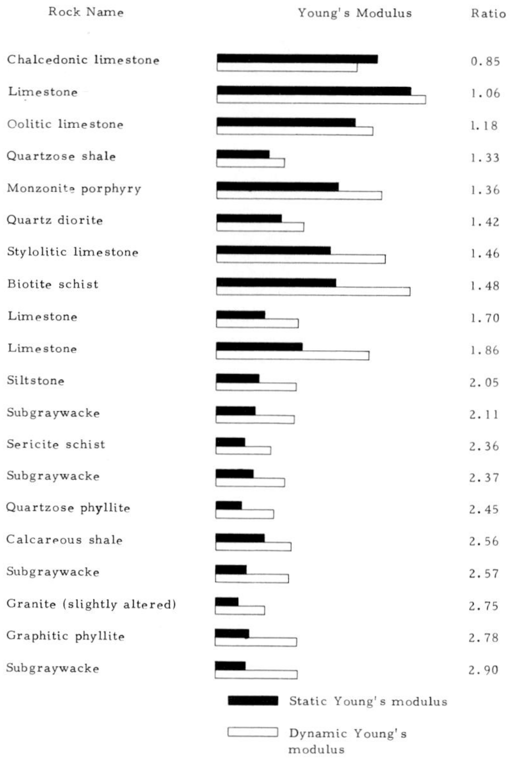

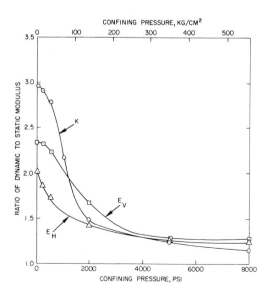

Rock Name
Young's Modulus
Ratio
Chalcedonic limestone 0.85
Limestone 1.06
Oolitic limestone 1.18
Quartzose shale 1.33
Monzonite porphyry 1.36
Quartz diorite 1.42
Stylolitic limestone 1.46
Biotite schist 1.48
Limestone 1.70
Limestone 1.86
Siltstone 2.05
Subgraywacke 2.11
Sericite schist 2.36
Subgraywacke 2.37
Quartzose phyllite 2.45
Calcareous shale 2.56
Subgraywacke 2.57
Granite (slightly altered) 2.75
Graphitic phyllite 2.78
Subgraywacke 2.90
Static Young's modulus
Dynamic Young's modulus

FIG. 30—Static and dynamic values of Young's modulus (Panek, 1965).

CONFINING PRESSURE, KG/CM²
0
100
200
300
400
500
RATIO OF DYNAMIC TO STATIC MODULUS
1.0
1.5
2.0
2.5
3.0
3.5
K
E_V
E_H
0
2000
4000
6000
8000
CONFINING PRESSURE, PSI

FIG. 31—Ratios of dynamic to static elastic moduli for Berea Sandstone (King, 1969).

to be affected by microcracks than strains developed by static stress (Ide, 1936) because a considerable amount of acoustic energy may bypass many cracks, particularly those oriented parallel to the direction of propagation.

Regardless of the differences between static and dynamic measurements, in the majority of the cases where other parameters are taken into account (such as stress level, sample type, etc.) a definite correlation can be established between the moduli. This aspect offers the possibility of in-situ determination of elastic moduli by wave propagation measurements as successfully implemented in the study of dam foundations and tunneling (Onodera, 1962).

An apparatus has been constructed and techniques developed[3] for measuring both static and dynamic elastic moduli of rocks as a function of confining pressure. Both measurements are made simultaneously in the same pressure cycle, thus avoiding hysteresis effects that are encountered when the two measurements are made in separate pressure cycles.

HEAT ALTERATION OF ROCK PHYSICAL PROPERTIES

One of the important technical problems associated with hot reservoirs is understanding how heat alters the physical characteristics of the rock. Alterations of permeability, porosity, and elastic properties caused by heat can have a substantial influence on the bulk volume, pore-fluid volume, and flow potential of the reservoir. Heat effects on elastic properties in particular may have an important impact on the interpretation of seismic and well log data that are used for finding, producing, and evaluating such reservoirs. Some errors that are possible in well logging analysis and formation evaluation because of the common practice of neglecting temperature effects have been pointed out by Sanyal et al (1974) for certain rocks. A review of published work indicates serious gaps in our knowledge about how temperature influences the rock properties that are most important in evaluating reservoirs. The available data establish some trends showing the effect of temperature on sedimentary rock properties (Table 1).

Both compressional- and shear-wave velocities were observed to decrease with elevation of temperature to 200°C and pressures up to 10,000 psi (703 kg/sq cm). Figure 32 shows the result for dry Berea Sandstone. The bulk volume of Berea Sandstone was found to decrease by about 1.5% at 200°C and 18,006 psi (1,266 kg/sq

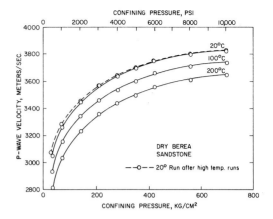

FIG. 32—P-wave velocity as function of confining pressure and temperature (Mobarek, 1971).

cm), the amount of decrease caused by heat alone was about 0.2% (Lobree, 1968; Fig. 33).

The elastic moduli of most sedimentary rocks appear to be reduced by heat. However, the formation resistivity factor increased with temperature. Heat effects on porosity are difficult to separate from pressure and pore fluid effects, hence little definitive information is available. Heat is believed to increase the number of microcracks, however, and these may contribute to the porosity at low pressures.

Thermal effects caused by differential expansion of different minerals along different crystallographic axes may result in structural damage which can substantially reduce the acoustic velocity and thermal conductivity of rocks. Some minerals in sedimentary rocks undergo phase or related changes when heated to sufficiently high temperatures. Loss of water of

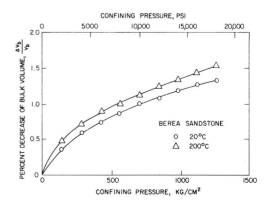

FIG. 33—Variation of bulk volume with pressure at 20°C and 200°C (Lobree, 1968).

[3]Center for Earth Sciences and Engineering, Rock Mechanics Laboratory, Balcones Research Center, The University of Texas at Austin.

Table 1.   Effect of Temperature on Properties of Sedimentary Rocks

Rock Property	Effect of Temperature	Reference
Permeability	Decreases up to 60%	a,b,c,d,e
Porosity	No definitive results have been reported	f
Formation Resistivity Factor	Increases	g,h,i
Breaking Strength	Decreases up to 60% at 800°C	d
Fracture Pressure	Reduced	j
Fracture Index	Number of fractures increased	d
Compressibility of Rock Solid	Increases	k
Pore Volume Compressibility	Increases by 21% in range 24-204°C	l
Bulk Compressibility	Increases in range 24-204°C	k
Elastic Moduli	Decreases	m
Compressional Wave Velocity	Decreases up to 40% at 800°C; reduction greater in water-saturated rocks at low pressure than in dry rocks. Velocity increases in water-saturated sediments.	d,n,o,p
Shear Wave Velocity	Decreases	n
Internal Friction	High loss peak near room temperature in sandstones but peak is reduced in limestones. There is some evidence that Q is insensitive to heat above room temperature.	q,r,s,t

a. Casse & Ramey, 1976; b. Weinbrant & Ramey, 1972; c. Afinogenov, 1969;
d. Somerton et al.,1965; e. McLatchie et al., 1958; f. Sanyal et al., 1974;
g. Helender & Campbell, 1966; h. Sanyal et al., 1972; i. Brannan & Von Gonten,
1973; j. Clark, 1964; k. Lobree, 1968; l. Von Gonten & Choudhary, 1969; m.
Griggs et al., 1960; n. Hughes & Cross, 1951; o. Shumway 1958; p. Gregory,
1976a; q. Kissell 1972; r. Attewell & Brentnall, 1964; s. Volarovich 1957;
t. Savage 1966.

crystallization and dissociation also may contribute to change in physical properties. Some of these reactions are reversible, but many of the major reactions are irreversible. Volume expansions of three common sandstones and quartz were computed from linear expansion data by Somerton and Selim (1961). The expansion in sandstones seems to be controlled by the quartz content and amounts to 4 or 5% at 600°C (Fig. 34). The sharp change in the curves is caused by inversion of quartz from the alpha to beta phase at 573°C. For geothermal temperatures of about 260°C, the volume expansion of the sandstones in Figure 34 would be about 1%.

The effect of water saturation on heat-treated Brazilian gabbro was investigated by Gregory (1976a). Heating the gabbro to 750°C increased

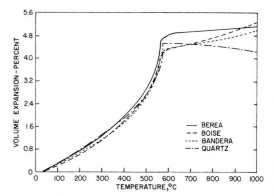

FIG. 34—Volume expansion of sandstones as function of temperature (Somerton and Selim, 1961).

the porosity and caused reductions in velocity of as much as 40% (Fig. 35). The intense heat caused unequal expansion of the minerals in the rock, increased the bulk volume, and initiated a new system of cracks. The velocities increased substantially when heat-treated samples were saturated with water, but failed to regain the original values measured on unheated water-saturated samples.

SPECIAL STUDY OF RECENT SEDIMENTS

Rock Composition and Gassmann's Theory

The three components which characterize the composition of rocks are: (1) the solid matter of which the skeleton or frame is built (index ^), i.e., grains or crystals; (2) the frame or skeleton (index -), i.e., empty porous rock; and (3) fluid occupying the pore space of skeleton (index ~). Properties of the whole rock are indicated by symbols without superscripts.

Gassmann (1951) showed that when a rock is a closed system, grossly isotropic and homogeneous, the use of elementary elastic theory yields the following interrelation between the rock parameters:

$$k = \hat{k}(\bar{k} + Q)/(\hat{k} + Q), \qquad (17)$$

where;

$$Q = \tilde{k}(\hat{k} - \bar{k})/\phi(\hat{k} - \tilde{k}).$$

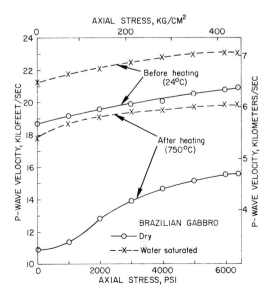

FIG. 35—Effect of heat and axial stress on P-wave velocities of dry and water-saturated Brazilian gabbro (Gregory, 1976a).

He also noted that $\mu = \bar{\mu}$ and that $M = k + 4/3\,\mu$. White (1965) gave these relations in the form;

$$M = \bar{M} + (1 - \bar{k}/\hat{k})^2/(\phi/\tilde{k}$$
$$+ (1 - \phi)/\hat{k} - \bar{k}/\hat{k}^2) \qquad (18)$$

It is also known that:

$$\rho = (1 - \phi)\,\hat{\rho} + \phi\,\tilde{\rho}, \qquad (19)$$

and

$$\text{P-Wave Velocity} = \sqrt{M/\rho}, \qquad (20)$$

where M = space modulus (or P-wave modulus), μ = rigidity modulus (or S-wave modulus), k = bulk modulus (or compressibility reciprocal), ν = Poisson's ratio, ρ = bulk denisty, ϕ = fractional porosity, and $\bar{F}$ = pressure on skeleton = total external confining pressure less the internal fluid pore pressure = effective overburden pressure.

The parameters of the solid matter of the frame that enter Equations 17, 18, and 19 are $\hat{k}$ and $\hat{\rho}$. Some typical values are listed in Table 2. The parameters of typical fluids also are known and some are listed in Table 3.

Calculation of P-Wave Velocity Through Rock, Filled with any Fluid, when the Velocity Through Rock Filled with Brine, and Density, Are Given

To make the calculation, Poisson's ratio of the rock is needed when there is no fluid in the pores, $\bar{\sigma}$. For most dry rocks and unconsolidated sands, $\bar{\sigma}$ is about 0.1 and is independent of pressure. Fortunately, the calculated P-wave velocity is not very sensitive to the value of $\bar{\sigma}$, and no great error is made by guessing its value. If the value of the shear-wave velocity is available, $\bar{\sigma}$ can be calculated.

Since the frequencies in seismic records are low, Biot's (1956) theory of wave propagation in the form and notation given by White (1965) can be used.

The plane wave modulus of the empty skeleton of the rock is related to the bulk modulus and Poisson's ratio by

$$\bar{M} = \frac{3(1 - \bar{\sigma})}{1 + \bar{\sigma}}\,\bar{k},$$

$$= S\bar{k} \;\text{(Definition of S)}. \qquad (21)$$

White's (1965, p. 132) equation (3–53) is used to solve for $\bar{k}$. Specifically, let

A. R. Gregory

Table 2. Bulk Modulus and Density of Some Minerals

Solid	Bulk Modulus, $\hat{k}$ dynes/cm^2 x 10^{10}	Density, ρ g/cm^3
α- Quartz	38	2.65
Calcite	67	2.71
Anhydrite	54	2.96
Dolomite	82	2.87
Corundum	294	3.99
Halite	23	2.16
Gypsum	40	2.32

Table 3. Bulk Modulus and Density of Some Fluids

Fluid	Bulk Modulus, $\tilde{k}$ dynes/cm^2 x 10^{10}	Density, $\tilde{\rho}$ g/cm^3
Water, 25°C (Distilled)	2.239	0.998
Sea Water, 25°C	2.402	1.025
Brine, 25°C (100,000 mg/L)	2.752	1.0686
Crude Oil:		
(1)	0.862	0.85
(2)	1.740	0.80
Air, 0°C (Dry, 76 cm Hg)	0.000142	0.001293
Methane, 0°C (76 cm Hg)	0.0001325	0.0007168

$$y = 1 - \bar{k}/k_s, \tag{22}$$

then

$$y^2(s-1) + y\{\phi s(k_s/k_f - 1) - S + M/k_s\}$$
$$- \phi(S - M/k_s)(k_s/k_f - 1) = 0. \tag{23}$$

The value of M is given by:

$$\text{P-Wave Velocity} = \sqrt{M/\rho}. \tag{24}$$

Having solved Equation 23 for y, substitute in Equation 25 to obtain M for any other fluid, then find the P-wave velocity from Equation 24.

$$M = s\bar{k} + \frac{(1 - \bar{k}/k_s)^2}{\phi/k_f + (1 - \phi)/k_s - \bar{k}/k_s^2}. \tag{25}$$

Equation 25 is, again, White's (1965) equation (3–53).

Table 4 illustrates the use of these equations. The shear-wave velocity for sandstones filled with oil and gas also has been computed and listed. Table 5 lists the average velocity and density for brine-filled shales. From these tables it is observed that the reflection coefficient for P waves may depend significantly on the nature of the fluid saturant, whereas the reflection coefficient for S waves does not. Additional discussion of the computation method is given by Gardner et al (1974). Curves in Figure 36 show the influence of fluid saturation on the bulk density of shale and poorly consolidated sandstone. Well log data are shown for shale and sandstones containing brine. Curves showing the bulk density of sandstones containing oil and gas are computed from Equation 19. Curves in Figure 37 for oil- and gas-saturated sandstone were computed from mathematical

Table 4. Properties of Average Gulf Coast Sands (or Sandstones).
Porosity, Density (ρ), Longitudinal Velocity (V_L), and Shear Velocity (V_S)
listed as a function of depth for sandstones containing different fluids
$(k_s=38\times10^{10}$ dynes/cm^2; $k_{brine}=2.81\times10^{10}$ dynes/cm^2

Depth	Porosity*	Brine-Saturated Sand			Oil-Saturated Sand**			Gas-Saturated Sand		
		ρ*	V_L*	V_S	ρ	V_L	V_S	ρ	V_L	V_S
m	%	gm/cm^3	m/sec	m/sec	gm/cm^3	m/sec	m/sec	gm/cm^3	m/sec	m/sec
61	39	2.010	1935	539	1.93	1686	549	1.59	905	604
305	38.5	2.022	2012	658	1.94	1786	671	1.61	1103	735
610	37	2.042	2103	768	1.97	1902	783	1.65	1283	856
914	36	2.065	2210	896	1.99	2036	914	1.68	1490	994
1219	34.6	2.083	2393	1097	2.01	2249	1116	1.71	1811	1207
1524	33.2	2.104	2652	1356	2.04	2545	1378	1.75	2228	1484
1829	32	2.125	2957	1640	2.06	2883	1664	1.78	2682	1786
2134	31	2.144	3109	1771	2.08	3051	1798	1.81	2886	1923
2438	29.3	2.167	3170	1814	2.10	3115	1841	1.85	2941	1960
2743	28	2.189	3231	1859	2.13	3182	1884	1.89	3002	2002
3048	27	2.210	3292	1908	2.15	3249	1932	1.92	3069	2045
3353	25.5	2.230	3399	1993	2.18	3362	2015	1.96	3188	2124
3658	24.2	2.248	3475	2051	2.20	3444	2073	1.99	3267	2179
3962	23	2.270	3551	2109	2.22	3527	2131	2.02	3350	2231
4267	21.8	2.290	3642	2179	2.25	3621	2201	2.06	3450	2298
4572	20.2	2.312	3749	2262	2.27	3734	2280	2.10	3560	2374
4877	19.0	2.332	3871	2359	2.29	3862	2377	2.13	3700	2466
5182	17.8	2.350	3962	2426	2.31	3956	2448	2.16	3798	2533

* Data derived from well logs; the remainder of the data was computed.

** Oil was Gulf Pride SAE-20, 25°C, density = .865 gm/cm^3

Table 5. Properties of Average Gulf Coast Shales

Depth		Shale Porosity	Brine-Saturated Shale		
			ρ	V_L	
Feet	Meters	%	gm/cm^3	ft/sec	m/sec
200	61	50	1.87	5000	1524
1000	305	38	2.05	6200	1890
2000	610	32.2	2.15	7100	2164
3000	914	29	2.20	7800	2377
4000	1219	27	,2.24	8200	2499
5000	1524	25	2.27	8700	2652
6000	1829	24	2.29	9000	2743
7000	2134	22.5	2.31	9400	2865
8000	2438	21.4	2.33	9700	2957
9000	2743	20.5	2.34	10000	3048
10000	3048	19.8	2.36	10400	3170
11000	3353	19	2.37	10600	3231
12000	3658	18.4	2.38	10900	3322
13000	3962	17.8	2.39	11200	3414
14000	4267	17	2.40	11450	3490
15000	4572	16.4	2.41	11700	3566
16000	4877	16	2.42	11900	3627
17000	5182	15.3	2.43	12150	3703

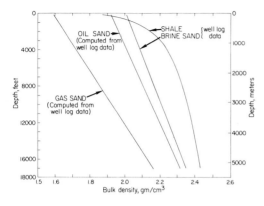

FIG. 36—Density as function of depth for Miocene in-situ shales and sandstones containing different pore fluids.

expressions discussed above. Note that pore fluid has relatively little influence on velocity at depths greater than about 6,000 ft (1,830 m).

Reflection coefficients (Fig. 38) were derived from well log and computed data discussed above. Note that R becomes relatively independent of the type of pore fluid at considerable depths, but the observed differences may still have some practical significance.

BIOT-GEERTSMA EQUATIONS

Biot (1956) developed a three-dimensional theory which described the deformation relation between solid and fluid in a porous rock. Geertsma (1961) developed theoretical expressions for the compressional-wave velocity as a function of frequency. Equations 26 and 27 are given below for the limiting cases of zero and infinite frequency. For zero frequency:

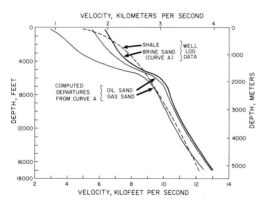

FIG. 37—Velocity versus depth for Miocene in-situ shales and sandstones containing different pore fluids (Gardner et al, 1974).

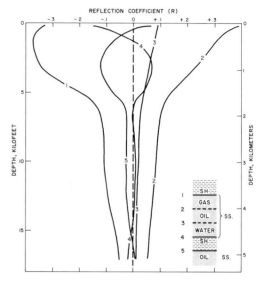

FIG. 38—Reflection coefficient versus depth for Miocene sediments in situ.

$$V_p = \left\{ \frac{1}{\rho_b} \left[\left(\frac{\beta}{c_s} + \frac{4}{3} G_b \right) \right. \right.$$
$$\left. \left. + \frac{(1 - \beta)^2}{(1 - \phi - \beta)c_s + \phi c_f} \right] \right\}^{1/2}, \qquad (26)$$

where c_s = compressibility of rock matrix material, c_b = compressibility of bulk rock material (dry) = $1/(\rho_b V_p^2 - 4/3 G_b)$, $\beta = c_s/c_b$, G_b = shear modulus of dry bulk (frame) material = $V_s^2 \rho_b$, and c_f = compressibility of pore fluid.

Bulk density (ρ_b) for sandstone containing a water-gas mixture is

$$\rho_b = \phi S_w \rho_w + \phi (1 - S_w)\rho_g$$
$$+ (1 - \phi)\rho_s; \qquad (27)$$

where ϕ = fractional porosity, S_w = fractional water saturation, ρ_w = density of water, ρ_s = density of sand grain, and ρ_g = density of gas. For infinite frequency:

$$V_p = \left\{ \left[\left(\frac{\beta}{c_s} + \frac{4}{3} G_b \right) \right. \right.$$
$$\left. + \frac{\frac{\phi \rho_b}{k \rho_f} + (1 - \beta)(1 - \beta - 2\phi/k)}{(1 - \phi - \beta)c_s + \phi c_f} \right]$$

$$\cdot\left.\frac{1}{\rho_b\left(1-\frac{\rho_f}{\rho_b}\frac{\phi}{k}\right)}\right\}^{1/2} ; \qquad (28)$$

where k is the coupling factor between the pore fluid and sand grains. For k = 1, no coupling exists, and for k = ∞, there is perfect coupling. For the case of perfect coupling, Equation 28 reduces to the equation for zero frequency and V_p is independent of frequency.

The pore-fluid density ρ_f for a water-gas mixture is

$$\rho_f = S_w \rho_w + (1 - S_w) \rho_g, \qquad (29)$$

and the pore-fluid compressibility c_f is

$$c_f = S_w c_w + (1 - S_w) c_g. \qquad (30)$$

Domenico (1976) found that theoretical values of V_p computed from the Biot-Geertsma equations were lower than measured values in packings of unconsolidated sand saturated with brine-gas mixtures. The discrepancies between measured and computed compressional-wave velocities at high brine saturations were attributed to uneven microscopic distribution of gas and brine within the porous structure.

OVERBURDEN AND FLUID PRESSURE

The overburden pressure is usually defined as the vertical stress caused by all the material, both solid and fluid, above the formation. An average value is 1 psi for each foot of depth (0.23 kg/sq cm per m of depth), although small departures from this average have been noted. The fluid pressure is usually defined as the pressure exerted by a column of free solution that would be in equilibrium with the formation. The reference to a free solution is significant when dealing with clays or shales with which other pressures such as osmotic, swelling, etc., can be associated. The normal fluid pressure gradient is frequently assumed to be 0.465 psi for each foot of depth (1.07 kg/sq cm per m of depth), although large departures from this value occur in high-pressure shales and sandstones.

The effective overburden pressure ΔP on a reservoir is often defined as the difference between the total overburden pressure P_c and the formation fluid pressure P_i. Hence, when both overburden pressure and formation fluid pressure are varied, only the difference between the two has a significant influence on velocity, that is

$$\Delta P = P_c - P_i. \qquad (31)$$

For practical field applications, Equation 31 is probably a close approximation to the true relation in many cases.

However, it has been suggested by theory (Brandt, 1955; Fatt, 1958) and by some laboratory experimental work (Banthia et al, 1965) that the true relation may be of the form

$$\Delta P = P_c - n P_i, \qquad (32)$$

where $n = C_B - C_r / C_B$ = coefficient of internal deformation, C_B = bulk volume compressibility of rock, and C_r = matrix compressibility of rock. The value of n ranges from 0.7 to 1.0 (Fatt, 1958), is different for different rocks, and is also a function of P_c. Some experimental results of Desai et al (1969) suggest that the shear-wave velocity follows the relation in Equation 31, whereas the compressional-wave velocity is a function of both P_c and P_i. It appears that more definitive experimental work is required to evaluate the relations in Equations 31 and (32).

A set of compressional-wave velocity data for an oil-wet water-saturated sample of Berea Sandstone is pertinent to the relations discussed above (Fig. 39). Whenever the effective pressure ΔP is increased, the velocity increases; whenever ΔP remains constant, the velocity remains constant except at low values of confining pressure P_c. The small rise in velocity noted for $\Delta P = 0$, 70, 141, and 211 kg/sq cm at low confining pressure can be interpreted to mean that: (1) n is less than 1.0, (2) the velocity of the pore fluid increased because the pressure on the pore fluid increased, (3) the acoustic coupling between pore fluid and the oil-wet matrix increased when P_i increased, or (4) a combination of all these factors.

It generally is believed that if a core is taken from a particular depth and subjected in the laboratory to the effective pressure corresponding to this depth, then the velocity will be the same as that obtained from a continuous velocity log. Equality of the two velocities is not self-evident because the measurements are made on different samples of greatly different configuration with different equipment, and the core is subjected to considerable stresses and may be damaged during drilling and recovery operations. Figure 40 shows a comparison which was made for fairly shallow depth, and the agreement between the sonic log data and the core data is quite satisfactory.

At depths greater than about 12,000 ft (3,658 m), or in high-velocity formations, the variation

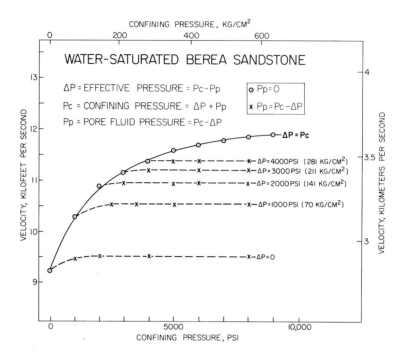

FIG. 39—Velocities through water-saturated oil-wet sandstone sample under confining and effective pressures (Wyllie et al, 1958).

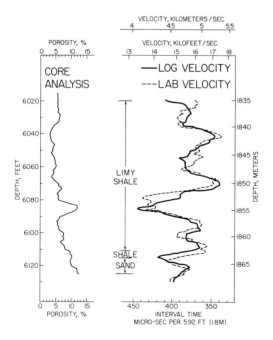

FIG. 40—Comparison between velocities measured in
laboratory on recovered cores and sonic log of cored
interval.

of velocity with pressure often can be ignored.
The velocity for cores can be measured at any
high pressure, to approximate the velocity of
deep formations. However, at shallow depths,
or in low-velocity formations, it is advisable to
duplicate downhole conditions when obtaining
data from cores.

Hottmann and Johnson (1965) presented pertinent data for velocity measured in shales and
the corresponding fluid pressure and depth.
When there was no excess fluid pressure, they
found that the interval travel time, Δt in microseconds per foot, decreased with depth, Z
(in feet), according to the formula

$$Z = A - B \log_e \Delta t, \qquad (33)$$

where A = 82,776 and B = 15,695. They also
gave data from wells that penetrated zones with
abnormally high fluid pressure. All of these data
can be correlated by the equation

$$\left[\frac{P_o - P_F}{\alpha - \beta} \right]^{1/3} Z^{2/3} = A - B \log_e \Delta t, \qquad (34)$$

where P_o = overburden pressure, P_F = fluid

pressure, α = normal overburden pressure gradient, and β = normal fluid pressure gradient. For a normally pressured section, Equation 34 reduces to Equation 33. One interesting feature of Equation 34 is that both the pressure difference $P_o - P_F$ and the depth Z are present. The factor $Z^{2/3}$ may be interpreted as the effect of increased consolidation with depth. For sandstones, the effect of consolidation outweighs the effect of pressure.

Definitions (after Stuart, 1970)

Hydropressures—P_H are the pore-fluid pressures generated by the effective weight on overlying waters plus back pressure of outflowing waters. Hydropressure reservoirs leak.

$$P_H = M_w \, g \, Z + hbp, \qquad (35)$$

where $\rho wo = M_w g$ or weighted average density of overlying waters (psig/ft, or kPa/m)[4], Z = effective hydrostatic column (ft or m), M_w = mass of overlying waters, g = gravity of earth, and hbp = hydropressure back pressure.

The hydropressure gradient P_{hg} (psig/ft, or kPa/m) is

$$P_{hg} = P_h / d, \qquad (36)$$

where d = depth of observation (ft or m).

Geopressures—P_g are those pore-fluid pressures that are generated by a pressure source or mechanism greater than hydropressures. P_g approaches the weight of the overburden. Geopressured reservoirs are considered to be sealed, or they leak at a slower rate than required to maintain hydropressure.

$$P_g = P_h + (P_s - P_h) + f(TP), \qquad (37)$$

where P_h = hydropressure (psig, or kPa)[5], P_s = pressure source which exceeds hydropressure (psig or kPa), and f(TP) = function of thermal expansion due to earth's temperatures and compressibility due to geostatic pressure (see Stuart, 1970, for details).

Geostatic Pressure—(weight of overburden) W_b is

$$W_b = [\rho m(1 - \phi r)Z_r]$$
$$+ \rho w \, \phi w \, Zw + \rho sw \, Zml, \qquad (38)$$

where ρm = weighted average of grain density

[4] 1 kPa/m = 22.6206 psi/ft
[5] 1 kPa = 6.894757 psi

(psig/ft or kPa/m), ρw = weighted average of pore-fluid density (psig/ft or kPa/m), ρsw = weighted average of sea water density (psig/ft or kPa/m), ϕr = weighted average of rock porosity (fraction), ϕw = weighted average of porosity of water-filled rock column (fraction), Z_r = column of rock (ft or m). Below ground level onshore and below mud line offshore, Zw = column of water in rock (ft or m) above depth of observation, and Zml = depth of mud line below mean sea level (ft or m); Zml is zero onshore. The geostatic gradient is

$$W_{bg} = W_b / d. \qquad (39)$$

FLUID SATURATION EFFECTS ON ROCK PROPERTIES

The results of some laboratory studies were reported by Gregory (1976a, b). The purpose of that work was to measure and examine the influence of fluid saturation on the velocity and density of several types of sedimentary rocks under different environments. Of particular interest is the effect of two-phase (water-gas and oil-gas) saturations on velocities, reflection coefficients, and elastic moduli. These relations are important because brine, oil, and gas commonly share the available pore space in the upper part of gas-capped reservoirs. Results of elastic-wave propagation studies in the laboratory, such as those reported here, can serve as a guide to the interpretation of sonic well logs and seismic field data. It is recognized that wide differences exist between the frequencies employed in this work (1 MHz) and those used in sonic logging (10 to 30 KHz) and in seismic prospecting (<200 Hz); however, there appears to be no significant dispersion of velocity with frequency in consolidated sedimentary rocks over this frequency range (McDonal et al, 1958).

In principle, the seismic reflection method should be sensitive to factors which influence any of the three key parameters: velocity, bulk density, and attenuation. The chief factors influencing the key parameters are porosity, mineral composition, intergranular elastic behavior, and properties of the formation fluid. These factors depend on overburden pressure, formation fluid pressure, temperature, microcracks, and age. Attenuation and its dependence on frequency is not considered in this work.

The suite of 20 test samples contained sandstones and carbonates ranging in porosity from 4.5 to 41% and in age from Pliocene to Early Devonian. Eleven of the rocks were classified as sandstones, three were limestones, five were chalks, and one was tripolite. Samples were cut into right cylinders of 2-in. (5.08 cm) diameter and lengths ranging from 0.5 to 1.5 in. (1.27 to 3.81 cm). The ends were ground flat with a magnetic-bed reciprocating grinder and were parallel within 0.001 in. (.0025 cm). Samples were oven-dried at 100°C for at least 12 hours, cooled to room temperature, and weighed to the nearest 0.001 g. Physical dimensions were measured with a dial caliper to a resolution of 0.001 in. (.0025 cm). Porosities of specimens were determined by weighing the samples before and after saturation.

Compressional (P-wave) and shear (S-wave) velocities were measured at room temperature on samples enclosed in rubber jackets and subjected to confining pressures applied by hydraulic fluid uniformly in the lateral and axial directions over the external surfaces. The range of pressures was 0 to 10,000 psi (703 kg/sq cm). Transit times of P and S waves were generated and received by barium titanate transducer discs which operated at a fundamental resonant frequency of 1 MHz. S waves of the same frequency were formed by a mode-conversion method. The experimental techniques were described by Gregory (1967).

EFFECTS OF FLUID SATURATION AND PRESSURE ON BULK DENSITY

The relation between bulk density and porosity for the suite of air-saturated (dry) and water-saturated sedimentary rocks was reasonably linear (Fig. 41). Solid lines drawn through the experimental data were derived from the relation

$$\rho = \phi\tilde{\rho} + (1 - \phi)\hat{\rho}, \qquad (40)$$

where ρ is the bulk density of a rock with fractional porosity ϕ, and superscripts $\sim$ and $\wedge$ refer to liquid and grain densities, respectively. Extrapolation of bulk density to zero porosity indicates that the grain density of most of the sandstones and chalks is about 2.65 g/cc.

Increases in the bulk density P_0 at atmospheric pressure caused by hydrostatic pressure applied isothermally were computed from P-wave transit time T_p and S-wave transit time T_s to give a new density ρ_1, using the relation

$$S = 1 + \frac{1 + \Delta}{3L_0^2 \rho_0} \int_0^P (T_p^{-2}$$
$$- 4/3\ T_s^{-2})^{-1}\ dp, \qquad (41)$$

where

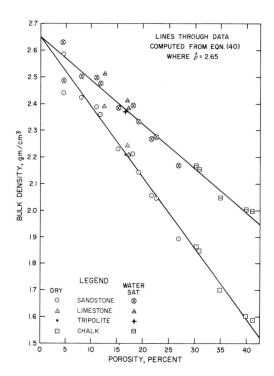

FIG. 41—Bulk density versus porosity for dry and water-saturated suite of sedimentary rocks (Gregory, 1976a).

$$\rho_1 = S^3 \rho_0.$$

Equation 41 was derived for isotropic solids by Cook (1957). Sample length L_0 measured at atmospheric pressure is also required for the computation and $\Delta \simeq 0.01$. Changes in density were less than 1% in the pressure range 0 to 10,000 psi (0 to 703 kg/sq cm). Some results are shown for Boise sandstone in Figure 42.

Effect of Fluid Saturation on Compressional (P-wave) Velocity

The difference ΔV_p between velocities of gas-saturated rocks ($S_w = 0$) and water-saturated rocks ($S_w = 100$) is unpredictable when measurements are made near atmospheric pressure. A wide scatter of points is obtained when the ratio $V_p(S_w = 0)/V_p(S_w = 100)$ is plotted against porosity for a variety of sandstone and carbonate rocks at pressures less than 100 psi (7 kg/sq cm). As the pressure increases, however, a definite trend develops showing a large saturation effect at low porosities and a smaller effect at high porosities. Figure 43 illustrates the pattern of data obtained at a pressure of 10,000 psi (703 kg/sq cm); the velocity ratio rarely exceeds 1.0.

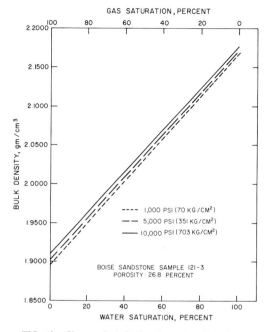

FIG. 42—Changes in bulk density as function of pressure and fluid saturation; computed from Equation 41 (Gregory, 1976a).

The effects of fluid saturation at $S_w = 0$ and $S_w = 100$ are more easily understood by examining generalized velocity curves for rocks saturated by mixtures of water and air and tested under constant pressure (Fig. 44). The shapes of the curves differ for porosities that are low (<10%), medium (10 to 25%), and high (>25%).

Effect of Fluid Saturation on Shear (S-Wave) Velocity

The theory of Biot (1956), based on the assumption that microcracks are negligible, concludes that the shear velocity of dry rocks should always exceed the velocity of the same rocks when fully saturated with liquid, i.e., $V_S(S_w = 0) > V_S(S_w = 100)$. The experimental data support Biot's theory for all porosities at pressures above 9,000 psi or 633 kg/sq cm (Fig. 45). However, as pressures decline, the data for low-porosity rocks begin to depart from theory. Rocks can be divided into two classes then, based on the effects of fluid saturation on S-wave velocities: class I, those with high porosity, above the curve, which agree with theory; and class II, those with lower porosity, below the curve, which disagree with theory.

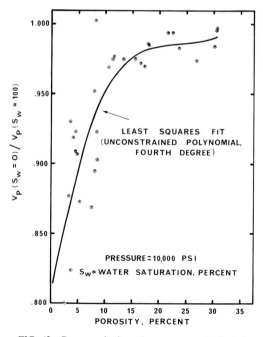

FIG. 43—P-wave velocity ratio versus porosity for fully gas-saturated rocks and fully water-saturated rocks under confining pressure of 10,000 psi (703 kg/sq cm; Gregory, 1976a).

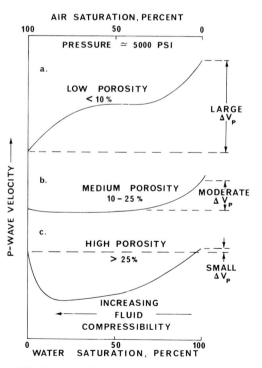

FIG. 44—Generalized curves showing variation of P-wave velocity with full and partial water-air saturations for consolidated sedimentary rocks at pressure of 5,000 psi (351 kg/sq cm; Gregory, 1976a).

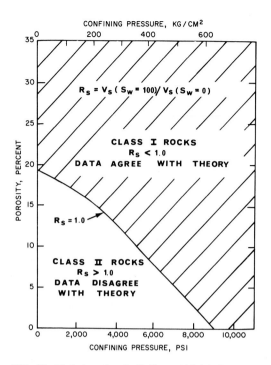

FIG. 45—Variation of ratio $V_s(S_w = 100)/V_s(S_w = 0)$ with porosity and confining pressure (Gregory, 1976a).

Further, the distinction between classes is pressure-dependent, a circumstance suggesting that microcracks in and around the cementing material of low-porosity rocks may account for discrepancies with theory.

Summary of Saturation and Pressure Effects on P-Wave and S-Wave Velocities

Figure 46 shows the range and limits of P-wave and S-wave velocities as a function of porosity for sedimentary rocks under various conditions of liquid-gas saturation and at low and high pressures. As a general rule, the lowest P-wave velocities are recorded for gas-saturated rocks under low pressure, the highest S-wave velocities for gas-saturated rocks under high pressure. Hence, the area of overlapping P- and S-wave velocities in the porosity range 0 to 27% indicates the presence of gas.

Effect of Liquid-gas Saturation on Reflection Coefficients

The reflection coefficient, R, was computed from the relation

$$R = \left| \frac{\rho_2 V_2 - \rho_1 V_1}{\rho_2 V_2 + \rho_1 V_1} \right|; \qquad (42)$$

where $\rho_1 V_1$ is the acoustic impedance of a rock fully saturated by a liquid, and $\rho_2 V_2$ is the acoustic impedance of the same rock containing gas in amounts varying from 0 to 100%. Effectively the above relation is a measure of R at fluid interfaces within the rock. Usually, R is maximum at liquid-gas interfaces or at interfaces between liquids and liquid-gas mixtures of high gas content.

Results of numerous measurements on consolidated sedimentary rocks, ranging in porosity from 4.5 to 41% and in age from Pliocene to Early Devonian, are summarized in Figure 47. The value of R decreases as the pressure increases and as S_w increases. Maximum values of R occur for $S_w = 5$ to 10 and range from 0.1 to 0.21 depending on the pressure. Values of R for shear waves are much lower than those for compressional waves (Gregory, 1976a).

Effects of Fluid Saturation on Dynamic Elastic Moduli

Elastic moduli were computed as a function of pressure and for various fluid saturations. P-wave and S-wave velocities were measured in only one direction through the media; the rocks were assumed to be isotropic. At a constant pressure of 5,000 psi (351 kg/sq cm), Poisson's ratio increases as water saturation increases and as the porosity increases (Fig. 48); the values range from 0.11 to 0.33 for water-saturated rocks and from −0.12 to 0.19 for gas-sa-

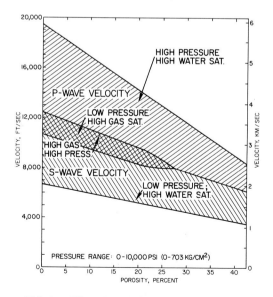

FIG. 46—Effects of fluid saturation and confining pressure on P-wave and S-wave velocities for suite of consolidated sedimentary rocks (Gregory, 1976a).

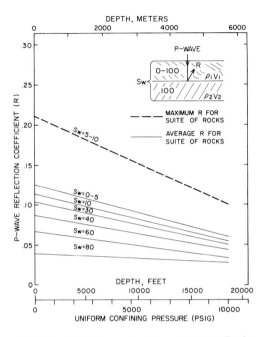

FIG. 47—Effect of gas saturation on P-wave reflection coefficient for suite of sedimentary rocks under effective uniform confining pressure (Gregory, 1976b).

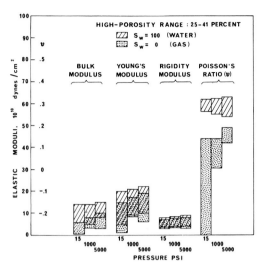

FIG. 49—Effects of gas and water saturation on elastic moduli of high-porosity sedimentary rocks at confining pressures of 15, 1,000, and 5,000 psi (1.05, 70.3, and 351.5 kg/sq cm, respectively; Gregory, 1976a).

turated rocks. The combination of high gas saturation, high porosity, and low pressure produce the largest negative values (Fig. 49).

Usually, the bulk moduli increase as liquid

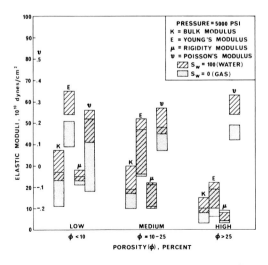

FIG. 48—Comparison of elastic moduli showing effect of porosity and fluid saturation on consolidated sedimentary rocks under pressure of 5,000 psi (351 kg/sq cm; Gregory, 1976a).

saturation increases and as porosity decreases (Fig. 48); lowest values are recorded for high-porosity rocks containing large amounts of gas (Fig. 49). The bulk moduli of this suite of sedimentary rocks range from 1.0×10^{10} dynes/sq cm for high-porosity chalks under low pressure and saturated with gas to 35×10^{10} dynes/sq cm for low-porosity sandstones under high pressure and saturated with water. Quartz, by comparison, has a bulk modulus of 37.9×10^{10} dynes/sq cm (White, 1965).

Diagnostic Potential of Elastic Moduli

Laboratory experiments show that the elastic behavior of a gas-filled rock is different from that of a liquid-filled rock. The differences can best be described in terms of the elastic moduli, computed from measured P- and S-wave velocities and bulk densities. As a rule, the moduli decrease as the porosity increases (Fig. 48); Poisson's ratio is an exception to the rule. Although some overlapping occurs in the values of elastic moduli for gas- and water-saturated rocks, the values for gas are consistently and markedly low. It is significant that certain moduli are singularly different for gas and water saturations, i.e., Young's modulus E for low porosity and Poisson's ratio v for high-porosity rocks. Generally, an increase in pressure causes the moduli to increase.

Poisson's ratio is useful for distinguishing between gas- and water-saturated rocks when

44 A. R. Gregory

the bulk density is unknown and cannot be estimated. Assuming that laboratory results can be extrapolated to the subsurface, the pronounced reduction in elastic moduli caused by gas may become diagnostically significant in seismic work, i.e., when both P- and S-wave velocities are available. Some encouraging field tests of systems which generate and detect SH-wave energy have been conducted in recent years (Erickson et al, 1968; Geyer and Martner, 1969). The potential usefulness of elastic moduli in seismic exploration may rekindle interest in utilizing shear-wave reflection methods in the field.

Significance of V_p/V_s Ratios

When both P-wave and S-wave velocity information is available from field measurements, the simple ratio V_p/V_s may indicate whether the formations are consolidated or unconsolidated and if gas or oil is present.

Gardner and Harris (1968) showed that V_p/V_s ratios greater than 2.0 were characteristic of unconsolidated sands, and values less than 2.0 indicated either a well-consolidated rock or the presence of gas in an unconsolidated sandstone.

The results reported here not only confirm that V_p/V_s ratios are less than 2.0 in consolidated rocks, but also establish limits of variation for a wide range of porosities and differentiate

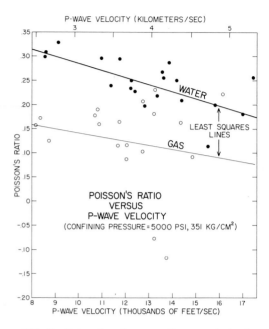

FIG. 51—Poisson's ratio versus P-wave velocity for gas- and water-saturated sedimentary rocks under moderate pressure.

between water and gas saturations. Increasing the gas saturation was found to decrease V_p/V_s by 3 to 30% in consolidated rocks. The plots in Figure 50 show that the ratios also decrease with decreasing porosity and range from 1.42 to 1.96 for water-saturated rocks and from 1.30 to 1.66 for gas-saturated rocks. The area of overlapping data represents an ambiguous zone which includes values common to both gas-saturated and water-saturated rocks. Pressure was found to reduce V_p/V_s ratios somewhat, but the effect was not always consistent or predictable over a wide range of porosities. Use of the ratio in seismic work, of course, has the additional advantage of not requiring density information.

Elastic moduli, Poisson's ratio, and V_p/V_s ratios may be estimated from P-wave velocity data only when satisfactory correlations can be established in the geologic area of interest. Figure 51 shows the correlation between Poisson's ratio and V_p for the suite of diversified sedimentary rocks tested in this study.

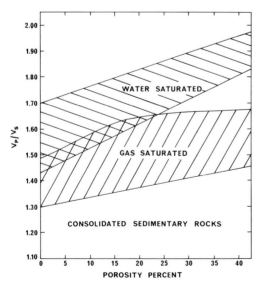

FIG. 50—Variation of V_p/V_s with porosity for water- and gas-saturated sedimentary rocks at confining pressures from 0 to 10,000 psi (0 to 703 kg/sq cm; Gregory, 1976a).

REFERENCES CITED

Afinogenov, Yu. A., 1969, How the liquid permeability of rocks is affected by pressure and temperature: SNIIGIMS, no. 6, p. 34–42. (Translation from Consultants Bureau, 227 West 17th St., N.Y., N.Y., 10011).

Attewell, P. B. and D. Brentnall, 1964, Attenuation measurements on rocks in the frequency range 12 Kc/s to 51 Kc/s and in the temperature range 100°K to 1150°K: 6th Symposium Rock Mechanics, Proc., Rolla, University of Missouri, p. 330-357.

Banthia, B. S., M. S. King, and I. Fatt, 1965, Ultrasonic shear-wave velocities in rocks subjected to simulated overburden pressure and internal pore pressure: Geophysics, v. 30, p. 117-121.

Biot, M. A., 1956, Theory of propagation of elastic waves in a fluid-saturated porous solid, 1. Low-Frequency Range, 2. Higher Frequency Range: Acoust. Soc. America Jour., v. 28, p. 168-191.

Brandt, H., 1955, A study of the speed of sound in porous granular media: Jour. Appl. Mechanics, v. 22, p. 479-486.

Brannan, G. O., and W. D. Von Gonten, 1973, The effect of temperature on the formation resistivity factor of porous media: 14th Ann. Logging Symp., Trans. (SPWLA), Lafayette, La., p. U1-U17.

Casse, Francis J. and Henry J. Ramey, Jr., 1976, The effect of temperature and confining pressure on single phase flow in consolidated rocks: Soc. Petroleum Engineers AIME, Paper 5877, 46th Ann. Calif. Regional Mtg. (Long Beach).

Clark, K. K., 1964, Reduction of fracture pressure of rocks by intensive bore-hole heating: Master's thesis, Univ. of California, Berkeley.

Cook, R. K., 1957, Variation of elastic constants and static strains with hydrostatic pressure: a method for calculation from ultrasonic measurements; Acoust. Soc. America Jour., v. 29, p. 445-449.

Desai, K. P., D. P. Helander, and E. J. Moore, 1969, Sequential measurement of compressional and shear velocities of rock samples under triaxial pressure: Society Petroleum Engineers Jour., v. 9, p. 378-394.

Domenico, S. N., 1976, Effect of brine-gas mixture velocity in an unconsolidated sand reservoir: Geophysics, v. 41, p. 882-894.

Erickson, E. L., D. E. Miller, and K. H. Waters, 1968, Shear-wave recording using continuous signal methods part II—later experimentation: Geophysics, v. 33, p. 240-254.

Fatt, I., 1958, Compressibility of sandstones at low to moderate pressures: AAPG Bull., v. 42, 1924-1957.

Faust, L. Y., 1951, Seismic velocity as a function of depth and geologic time: Geophysics, v. 16, p. 192-206.

Gardner, G. H. F., and M. H. Harris, 1968, Velocity and attenuation of elastic waves in sands: 9th Ann. Logging Symposium, Trans. (SPWLA), p. M1-M19.

—— L. W. Gardner, and A. R. Gregory, 1974, Formation velocity and density—the diagnostic basics of stratigraphic traps: Geophysics, v. 39, p. 770-780.

Gassmann, F., 1951, Ueber die elastizität Poröser Medien: Natur. Ges. Zürich, Vierteljahrssch. 96, p. 1-23.

Geertsma, J., 1961, Velocity log interpretation: the effect of rock bulk compressibility: Soc. Petroleum Engineers AIME Trans., v. 222, p. 235-253.

Geyer, R. L., and S. T. Martner, 1969, SH waves from explosive sources: Geophysics, v. 34, p. 893-905.

Gladwin, M. and F. D. Stacey, 1974, Anelastic degradation of acoustic pulses in rock: Physics Earth and Planetary Interiors, v. 8, p. 332-336.

Gregory, A. R., 1962, Shear wave velocity measurement of sedimentary rock samples under compression: 5th Symposium on Rock Mechanics, Proc., University of Minnesota, p. 439-467.

—— 1967, Mode conversion technique employed in shear wave velocity studies of rock samples under axial and uniform compression: Soc. Petroleum Engineers Jour., v. 240, p. 136-148.

—— 1976a, Fluid saturation effects on dynamic elastic properties of sedimentary rocks: Geophysics, v. 41, p. 895-921.

—— 1976b, Dynamic elastic rock properties key to pore-fluid identity: Oil and Gas Jour., v. 74, no. 43, p. 130-138.

—— and A. L. Podio, 1970, Dual-mode ultrasonic apparatus for measuring compressional and shear-wave velocities of rock samples: IEEE Trans., Sonics & Ultrasonics, v. SU-17, p. 77-85.

Griggs, D., F. J. Turner, and H. C. Heard, 1960, Deformation of rocks at 500°-800°C: Geol. Soc. America Mem. 79, p. 39-104.

Helander, D. P., and J. M. Campbell, 1966, The effect of pore configuration, pressure, and temperature on rock resistivity: 7th Ann. Logging Symposium, Trans., (SPWLA), p. W1-W29.

Hottmann, C. E., and R. K. Johnson, 1965, Estimation of formation pressures from log-derived shale properties: Jour. Petroleum Technology, v. 17, p. 717-722.

Hughes, D. S., and J. H. Cross, 1951, Elastic wave velocities in rocks at high pressures and temperatures: Geophysics, v. 16, p. 577-593.

Ide, J. M., 1936, Comparison of statically and dynamically determined Young's modulus of rocks: Natl. Acad. Sci. Proc., v. 22, p. 82-91.

King, M. S., 1969, Static and dynamic moduli of rocks under pressure: 11th Symposium on Rock Mechanics (Berkeley, Calif.) 35 p.

Kissell, F. N., 1972, Effect of temperature variation on internal friction in rocks: Jour. Geophys. Research, v. 77, p. 1420-1423.

Lobree, D. T., 1968, Measurement of the compressibilities of reservoir type rocks at elevated temperatures: Thesis, University of California, Berkeley, 78 p.

Matthews, C. S., and D. G. Russell, 1967, Pressure buildup and flow tests in wells: Soc. Petroleum Engineers AIME, Monograph series, v. 1, 167 pages.

McDonal, F. J. et al, 1958, Attenuation of shear and compressional waves in Pierre Shale: Geophysics, v. 23, p. 421-439.

McLatchie, A. S., R. A. Hemstock, and J. W. Young, 1958, The effective compressibility of reservoir rock and its effect on permeability: AIME, Trans. v. 213, p. 386-388.

Mobarek, S. A. M., 1971, The effect of temperature on wave velocities in porous rocks: Master's thesis, University of California, Berkeley, 83 p.

Morgans, W. T. A., and N. B. Terry, 1957, Measurement of the static and dynamic elastic moduli of coal: British Jour. Appl. Physics, v. 8, p. 201-219.

Onodera, T. F., 1962, Dynamic investigation of foundation rocks in situ: 5th Symposium on Rock Mechanics, Proc., University of Minnesota, p. 517-533.

Panek, L. A., 1965, Testing techniques for rock mechanics: Am. Soc. Testing and Materials Spec. Tech. Publ. 402, p. 106-132.

Papadakis, E. P., 1966, Ultrasonic diffraction loss and phase change in anisotropic materials, Acous. Soc. America Jour., v. 40, p. 863-876.

Podio, Augusto L., 1968, Experimental determination of the dynamic elastic properties of anistropic rocks—ultrasonic pulse method: Ph.D. Thesis, Univ. Texas, Austin, p. 8-30.

Ramana, Y. V., and M. V. M. S. Rao, 1974, Q by pulse broadening in rocks under pressure: Physics Earth and Planetary Interiors, v. 8, p. 337-341.

Reuss, A., 1929, Berechnung der fleissgrenze von mischkristallen auf grund der plastizitats bedingung fur einkrisalle: Zeitshr. Angew. Math. Mech., v. 9, p. 49-58.

Sanyal, S. K., S. S. Marsden, Jr., and H. J. Ramey, Jr., 1972, The effect of temperature on electrical resistivity of porous media: 13th Ann. Logging Symposium, Trans., (SPWLA), Tulsa, Okla., p. 11-I35.

—— —— ——, 1974, Effect of temperature on petrophysical properties of rocks: Soc. Petroleum Engineers AIME Paper 4898, 49th Ann. Fall Mtg. (Houston), 23 p.

Savage, J. C., 1966, Thermoelastic attenuation of elastic waves by cracks: Jour. Geophys. Research, v. 71, p. 3929-3938.

Schlumberger log interpretation, principles, 1972, v. 1, p. 39.

Shumway, George, 1958, Sound velocity vs. temperature in water-saturated sediments: Geophysics, v. 23, p. 494-505.

Simmons, Gene, and W. R. Brace, 1965, Comparison of static and dynamic measurements of compressibility of rocks: Jour. Geophys. Research v. 70, p. 5619-5656.

Somerton, W. H., and M. A. Selim, 1961, Additional thermal data for porous rocks—thermal expansion and heat of reaction: Soc. Petroleum Engineers Jour., p. 249.

—— M. M. Mehta, and G. W. Dean, 1965, Thermal alteration of sandstones: Soc. Petroleum Engineers AIME Trans., v. 234, p. 589-593.

Stuart, Charles A., 1970, Geopressures, supplement to the proceedings of the second symposium on abnormal subsurface pressure: Baton Rouge, Louisiana State University, 121 p.

Sutherland, R. B., 1962, Some dynamic and static properties of rock: 5th Symposium on Rock Mechanics, Proc., University of Minnesota, p. 473-491.

Voigt, W. 1928, Lehrbuch der Kristallphysik, (mit Ausschluss der Kristalloptik): B. G. Teubner, Leipzig, 978 p.

Volarovich, M. P., and A. S. Gurvich, 1957, Investigation of dynamic moduli of elasticity for rocks in relation to temperature: Bull. Acad. Sci. USSR (Geophys. Ser., English Transl.), No. 4, p. 1-9.

—— et al, 1969, The correlation between attenuation decrements and the elastic moduli of rock: Physics Solid Earth (Engl. Ed.), no. 12, p. 741-746.

Von Gonten, W. D., and B. K. Choudhary, 1969, The effect of pressure and temperature on pore volume compressibility: Soc. Petroleum Engineers AIME Paper 2526, 44th Ann. Fall Mtg. (Denver).

Weinbrandt, R. M., and H. J. Ramey, Jr., 1972, The effect of temperature on relative permeability of consolidated Rocks: Soc. Petroleum Engineers AIME Paper 4142, 47th Ann. Fall. Mtg. (San Antonio).

White, J. E., 1965, Seismic waves: radiation transmission and attenuation: New York, McGraw Hill, 302 p.

Wyllie, M. R. J., A. R. Gregory, and L. W. Gardner, 1956, Elastic wave velocities in heterogeneous and porous media: Geophysics, v. 21, p. 41-70.

—— —— and G. H. F. Gardner, 1958, An experimental investigation of factors affecting elastic wave velocities in porous media: Geophysics, v. 28, p. 459-493.

—— G. H. F. Gardner, and A. R. Gregory, 1962, Studies of elastic wave attenuation in porous media: Geophysics, v. 27, p. 569-589.

section 2:
application of seismic reflection configuration to stratigraphic interpretation

Seismic Stratigraphy and Global Changes of Sea Level[1]

P. R. VAIL ET AL

P. R. Vail, R. M. Mitchum, Jr., R. G. Todd, and J. M. Widmier
Exxon Production Research Co.
Houston, Texas, 77001.

S. Thompson, III
New Mexico Bureau of Mines and Mineral Resources
Socorro, New Mexico, 87801.

J. B. Sangree
Esso Exploration, Inc.
Walton-on-Thames, Surrey, England, KT12 2QL.

J. N. Bubb
Exxon Production Malaysia, Inc.
Kuala Lumpur, Malaysia.

W. G. Hatlelid
Imperial Oil, Ltd.
Calgary, Alberta, Canada.

[1]Manuscript received January 6, 1977, accepted June 13, 1977.
Note—The following paper, consisting of 11 parts, is a systematic treatment of the subject of seismic stratigraphy and its relation to global changes of sea level. The presentation format was designed and prepared by P. R. Vail and R. M. Mitchum, Jr.

Seismic Stratigraphy and Global Changes of Sea Level, Part 1: Overview[1]

P. R. VAIL and R. M. MITCHUM, JR.[2]

Seismic stratigraphy is basically a geologic approach to the stratigraphic interpretation of seismic data. The unique properties of seismic reflections allow the direct application of geologic concepts based on physical stratigraphy. Primary seismic reflections are generated by physical surfaces in the rocks, consisting mainly of stratal (bedding) surfaces and unconformities with velocity-density contrasts. Therefore, primary seismic reflections parallel stratal surfaces and unconformities. Whereas all the rocks above a stratal or uniformity surface are younger than those below it, the resulting seismic section is a record of the chronostratigraphic (time-stratigraphic) depositional and structural patterns and not a record of the time-transgressive lithostratigraphy (rock-stratigraphy).

Because seismic reflections follow chronostratigraphic correlations, it is not only possible to interpret postdepositional structural deformation, but also it is possible to make the following types of stratigraphic interpretations from the geometry of seismic reflection correlation patterns: (1) geologic time correlations, (2) definition of genetic depositional units, (3) thickness and depositional environment of genetic units, (4) paleobathymetry, (5) burial history, (6) relief and topography on unconformities, and (7) paleogeography and geologic history when combined with geologic data. However, one limiting factor is that lithofacies and rock type *can not* be determined directly from the geometry of reflection correlation patterns.

To accomplish the geologic objectives just listed, we recommend the following three-step interpretational procedure: (1) seismic sequence analysis; (2) seismic facies analysis; and (3) analysis of relative changes of sea level.

Seismic sequence analysis is based on the identification of stratigraphic units composed of a relatively conformable succession of genetically related strata termed *depositional sequences.* The upper and lower boundaries of depositional sequences are unconformities or their correlative conformities. The time interval represented by strata of a given sequence may differ from place to place, but the range is confined to synchronous limits marked by ages of the sequence boundaries where they become conformities. Depositional sequence boundaries are recognized on seismic data by identifying reflections caused by lateral terminations of strata termed *onlap, downlap, toplap,* and *truncation.* The depositional sequences, because they consist of genetically related strata having chronostratigraphic significance, provide an ideal framework for stratigraphic analysis.

Analysis of seismic facies is the delineation and interpretation of reflection geometry, continuity, amplitude, frequency, and interval velocity, as well as the external form and associations of seismic facies units within the framework of depositional sequences. Where the seismic facies are described and mapped, an interpretation of the sedimentary processes and environmental settings allows the interpreter to predict the lithology of the seismic facies.

Analysis of relative changes of sea level consists of constructing chronostratigraphic correlation charts and charts of cycles of relative changes of sea level on a regional basis and comparing them with global data. Similarities of the regional cycles to the global cycles are significant in seismic stratigraphic analysis because they introduce a dimension of predictability into stratigraphy, allowing more accurate prediction of age, time of unconformities, paleoenvironments, and lithofacies. Differences between regional and global curves indicate times of local structuring or errors in interpretation.

In addition to this overview, this part of *Memoir* 26 consists of a group of nine related articles and a glossary. In these articles, we separate the purely geologic concepts from the seismic applications of these concepts. For example, the first three articles (parts 2, 3, and 4) describe three geologic concepts based on the depositional sequence: (1) the depositional sequence as a basic stratigraphic unit; (2) the determination of relative changes of sea level from analysis of coastal onlap, and (3) global cycles of relative changes of sea level. These articles present a fundamental system of stratigraphy independent of seismic interpretation.

The next six articles describe application of these geologic concepts to seismic interpretation. Part 5 explains the fundamental relation of seis-

[1]Manuscript received, January 6, 1977; accepted, June 13, 1977.

[2]Exxon Production Research Co., Houston, Texas, 77001.

mic reflections to stratal surfaces and unconformities, and emphasizes their chronostratigraphic properties. Part 6 describes and interprets significant patterns of reflection terminations and configurations observable on seismic sections, and Part 7 summarizes a procedure for seismic stratigraphic analysis. Part 8 is a case history which serves as an example for the documentation of regional cycles of relative change of sea level—in this case, Jurassic-age strata of offshore West Africa and the Gulf of Mexico. Parts 9 and 10 describe seismic facies analysis in clastic and carbonate strata, respectively; and Part 11 is a glossary of some commonly-used seismic stratigraphic terms.

A recurring theme is the interdependence of seismic data with all other forms of geologic information, such as chronostratigraphic and environmental information from paleontology, lithologic data from cores and wireline logs, and regional information from outcrop and literature. Seismic data form one more tool for geologic investigation. Geologic concepts of depositional sequences, interregional unconformities, and global changes of sea level are all subjects of active research by many, long predating our seismic stratigraphic contributions. We welcome comments on further refinements or revisions of our ideas in this ongoing area of geologic research.

This series of papers is a culmination of many years' work by a large group of Exxon explorationists in research and operations, and the authors wish to acknowledge and thank them collectively for their many contributions and support.

Seismic Stratigraphy and Global Changes of Sea Level, Part 2: The Depositional Sequence as a Basic Unit for Stratigraphic Analysis [1]

R. M. MITCHUM, JR., P. R. VAIL, [2] and S. THOMPSON, III [3]

Abstract A depositional sequence is a stratigraphic unit composed of a relatively conformable succession of genetically related strata and bounded at its top and base by unconformities or their correlative conformities. This concept of a "sequence" is modified from Sloss. A depositional sequence is determined by a single objective criterion, the physical relations of the strata themselves. The combination of objective determination of sequence boundaries and the systematic patterns of deposition of the genetically related strata within sequences makes the sequence concept a fundamental and extremely practical basis for the interpretation of stratigraphy and depositional facies. Because distribution and facies of many sequences are controlled by global changes of sea level, sequences also provide an ideal basis for establishing comprehensive stratigraphic frameworks on regional or global scales.

A depositional sequence is chronostratigraphically significant because it was deposited during a given interval of geologic time limited by ages of the sequence boundaries where the boundaries are conformities; however the age range of the strata within the sequence may differ from place to place where the boundaries are unconformities. The hiatus along the unconformable part of a sequence boundary generally is variable in duration. The hiatus along the conformable part is not measureable, and the surface is practically synchronous. Stratal surfaces within a sequence are essentially synchronous in terms of geologic time.

Depositional sequences may range in thickness from hundreds of meters to a few centimeters. Sequences of different magnitudes may be recognized on seismic sections, well-log sections, and surface outcrops.

To define and correlate a depositional sequence accurately, the sequence boundaries must be defined and traced precisely. Usually the boundaries are defined at unconformities and traced to their correlative conformities. Discordance of strata is the main criterion used in the determination of sequence boundaries, and the type of discordant relation is the best indicator of whether an unconformity results from erosion or nondeposition. Onlap, downlap, and toplap indicate nondepositional hiatuses; truncation indicates an erosional hiatus unless the truncation is a result of structural disruption.

Examples of depositional sequences are presented on well-log and seismic sections. Both examples depend primarily on correlation of physical stratigraphic surfaces for identification of the unconformities bounding the sequences, and on biostratigraphic zonation for determination of the geologic ages of the sequences.

INTRODUCTION

This paper defines the depositional sequence, describes its chronostratigraphic significance and scale, discusses the significance of unconformities and their correlative conformities as sequence boundaries, discusses the relation of strata to sequence boundaries, and gives examples of depositional sequences on well-log and seismic sections. A following paper (Part 3, Vail et al, this volume) describes how relative changes of sea level may be recognized from onlap of coastal deposits within a depositional sequence, and how charts showing cycles of relative changes of sea level in given regions may be constructed. A third paper (Part 4, Vail et al, this volume) presents a chart of global cycles of relative change of sea level and describes how the chart was derived by comparison of regional curves from many areas around the globe. The global cycles appear to have controlled the general distribution of the major sequences deposited in marine and coastal environments. Therefore, with global cycle charts, the ages, distributions, and facies of the depositional sequences may be better predicted before drilling in areas where seismic coverage is available.

CONCEPT OF DEPOSITIONAL SEQUENCE

Definition

A *depositional sequence* is a stratigraphic unit composed of a relatively conformable succession of genetically related strata and bounded at its top and base by unconformities or their correlative conformities. Figure 1 illustrates the basic concepts of a depositional sequence.

Because it is determined by a single objective criterion—the physical relations of the strata themselves—the depositional sequence is useful in establishing a comprehensive stratigraphic framework. It is not primarily dependent on determinations of rock types, fossils, depositional processes, or other criteria that generally are subjective and varied within a given sequence.

Unconformities that bound depositional sequences are observable discordances in a given stratigraphic section that show evidence of erosion or nondeposition with obvious stratal terminations, but in places they may be traced into less obvious paraconformities recognized by biostratigraphy or other methods. A more complete dis-

[1] Manuscript received, January 6, 1977; accepted, June 13, 1977.

[2] Exxon Production Research Company, Houston, Texas 77001.

[3] New Mexico Bureau of Mines and Mineral Resources, Socorro, New Mexico 87801.

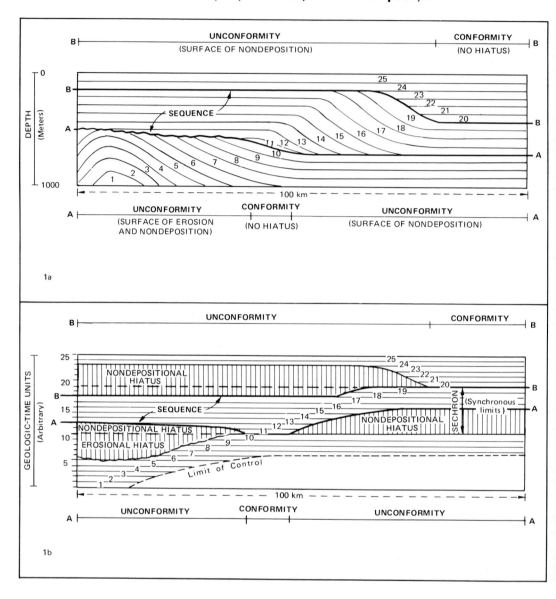

FIG. 1—Basic concepts of depositional sequence. A depositional sequence is a stratigraphic unit composed of relatively conformable successions of genetically related strata and bounded at its top and base by unconformities or their correlative conformities.

A. Generalized stratigraphic section of a sequence. Boundaries defined by surfaces A and B which pass laterally from unconformities to correlative conformities. Individual units of strata 1 through 25 are traced by following stratification surfaces, and assumed conformable where successive strata are present. Where units of strata are missing, hiatuses are evident.

B. Generalized chronostratigraphic section of a sequence. Stratigraphic relations shown in **A** are replotted here in chronostratigraphic section (geologic time is the ordinate). Geologic-time ranges of all individual units of strata given as equal. Geologic-time range of sequence between surfaces A and B varies from place to place, but variation is confined within synchronous limits. These limits determined by those parts of sequence boundaries which are conformities. Here, limits occur at beginning of unit 11 and end of unit 19. A sechron is defined as maximum geologic-time range of a sequence.)

cussion of depositional sequence boundaries is given in the next section.

This concept of a *sequence* is modified from Sloss (1963, p. 93), who stated: "Stratigraphic sequences are rock-stratigraphic units of higher rank than group, megagroup, or supergroup, traceable over major areas of a continent and bounded by unconformities of interregional scope." He recognized (p. 109) that such sequences have chronostratigraphic significance.

In comparison to Sloss's sequences, our depositional sequences are an order of magnitude smaller. However, they may be bounded not only by interregional unconformities, but also by equivalent conformities, and are traceable over major areas of continents and ocean basins. Distinct groups of superposed depositional sequences, called supersequences, are recognized; they are of the same order of magnitude as Sloss' original sequences. Some supersequences are practically the same as his named sequences.

We seriously considered using the term "synthem" instead of "sequence." A synthem is defined as an unconformity-bound unit (Chang, 1975; Hedberg, 1976, p. 92). This term has merit in that it has no previous connotations and avoids applying a specific definition to a term that already has a general meaning. However, we continue the use of the term "sequence" for the following reasons: (1) it has a long-term similar usage and familiarity due to the works of Sloss; (2) a sequence is more than an unconformity-bound unit, because we recognize it even where its boundaries are conformities; (3) a depositional sequence often represents a predictable succession (or sequence) of rocks deposited during a regional or global cycle of relative change of sea level; and (4) some of our supersequences are practically the same as some of Sloss' named sequences in North America, and are based on similar criteria.

Chronostratigraphic Significance

A depositional sequence is chronostratigraphically significant because it was deposited during a given interval of geologic time limited by the ages of the sequence boundaries where they are conformities, although the age range of the strata within the sequence may differ from place to place where the boundaries are unconformities. Two types of chronostratigraphic surfaces are related to sequences: (1) unconformities and their correlative conformities forming sequence boundaries, and (2) stratal (bedding) surfaces within sequences. These surfaces are illustrated diagrammatically in Figure 1 (see Part 5, Vail et al, this volume).

The hiatus represented by the unconformable part of a sequence boundary generally is variable. It may range from about a million years to hundreds of millions of years; however, the unconformity is chronostratigraphically significant because, in general, rocks above an unconformity are everywhere younger than those below it. The conformable part of a sequence boundary is practically synchronous because the hiatus is not measurable; the time span generally is less than a million years. The physical surfaces that separate groups of strata or individual beds and laminae within a sequence are essentially synchronous (Part 5, Vail et al, this volume). Some may form in instants of geologic time as discussed by Campbell (1967).

Boundaries of lithostratigraphic units such as formations and lithofacies units may be parallel to stratal surfaces, or in the case of diachronous lithostratigraphic boundaries, may cross stratal surfaces (see Campbell 1967, 1971, and 1973; Part 5, Vail et al, this volume). Because these gradational formations and lithofacies boundaries are diachronous, they practically are useless in seismic chronostratigraphic analyses, although they may be significant in seismic facies analyses.

Hedberg (1976, p. 92) recognized that unconformity-bounded units (synthems) have great significance in chronostratigraphy because unconformities frequently serve as guides to the approximate placement of chronostratigraphic boundaries. However, he indicated that such units cannot be chronostratigraphic units because they are not bounded by synchronous (isochronous) surfaces. In contrast to those of a synthem, the boundaries of a depositional sequence are conformable and, therefore, synchronous in many places. A depositional sequence has chronostratigraphic significance because all the rocks of the sequence were deposited during the interval of geologic time defined by the ages of the sequence boundaries where they are conformities.

A depositional sequence may have more significance in geologic history than a unit bounded only by synchronous surfaces that are chosen arbitrarily. A sequence represents a genetic unit that was deposited during a single episodic event, whereas the arbitrarily chosen unit may span two or more incomplete portions of genetic depositional units, and therefore not accurately portray the depositional history.

A *sechron* (from *se*-quence and *chron* time) is defined as the total interval of geologic time during which a sequence is deposited. As shown on Figure 1, a sechron is determined as the time interval between the upper and lower sequence boundaries where they are conformities. Because the conformities at the base and the top of a depositional sequence may lie in different areas, it may not be possible to date the entire sechron at a single locality. Where the lower and/or upper boundaries of a depositional sequence are unconformities, partial sechrons would be determined that may differ in duration from place to place. Further aspects of the chronostratigraphic significance of depositional sequences are discussed in the sections on sequence boundaries and the relations of the strata to the boundaries.

MAGNITUDE OF DEPOSITIONAL SEQUENCES

A depositional sequence generally is tens to hundreds of meters thick, although the range may be from thousands of meters down to a few millimeters. In general, the smaller scale sequences can be correlated only very short distances. Also, there are practical limits imposed by the resolution limits of the correlation tool (seismic sections, well-log correlations, or outcrops).

Depositional sequences can be identified on seismic sections (seismic sequences can be correlated only to the nearest reflection cycle). At the present resolution of good data, a seismic reflection represents a unit of strata with a minimum thickness of several tens of meters. If such strata are continuous, sequences can be traced by seismic-cycle correlation for tens or hundreds of kilometers in a dip direction across a basin, and even farther in a strike direction. Seismic reflections tend to parallel stratification surfaces, rather than the gross boundaries of lithologic units that may cut across stratification surfaces (Part 5, Vail et al, this volume). Thus a seismic reflection correlation approaches a chronostratigraphic correlation, and the result provides a level of geologic synchrony that is commensurate with the scale of strata represented. However, a reflection may terminate on a seismic section owing to a thinning of the unit of strata represented, and this unit may continue below the resolution of the seismic tool. Where most pre-Quaternary seismic sequences, have been dated by biostratigraphy from well samples, they generally compare in duration to standard chronostratigraphic units of about stage or series rank, and cover a geologic-time span on the order of 1 to 10 million years (Part 4, Vail et al, this volume).

Depositional sequences determined by correlation with closely spaced well logs may be smaller in scale than sequences on seismic sections. Not only are the larger scale (seismic) sequences recognized, but sequences of intermediate size may also be detected within them. With fine electric or radioactivity tools, stratigraphic units may

be differentiated on logs down to a thickness of a meter or even less. Because stratification surfaces generate both seismic reflections and well-log responses, both these tools may be used to make accurate chronostratigraphic correlations. However, log-marker correlation can be made to a more precise level of geologic synchrony because unconformities involving smaller units of strata can be recognized. Generally, the smaller sequences detected by log correlation are at a lower order of magnitude than many standard chronostratigraphic units. The major disadvantage of well-log correlation is the lack of continuous correlation between wells, although closely spaced well control can be nearly as effective as seismic correlation.

The most detailed chronostratigraphic correlation can be reached with stratum correlation in cores or good exposures of surface outcrops, where individual beds or even laminae can be traced (Campbell, 1967). Control usually limits drastically the areal extent over which stratal correlations can be made. However, within the smallest stratigraphic units recognized with log-marker units, one may see terminations of bedding against log-marker unit boundaries and terminations of laminae against bedding surfaces. Thus, at least two additional levels of more precise geologic synchrony are indicated by these smaller unconformities (or diastems).

BOUNDARIES OF DEPOSITIONAL SEQUENCES

To define and correlate a depositional sequence accurately, the sequence boundaries must be defined and traced. Usually, the boundaries are defined at unconformities based on discordant relations of strata to unconformities. Also, the geologic ages of the rocks above and below the unconformity are determined to provide a measure of the hiatus at a given location. As a particular sequence boundary is traced laterally, the strata may become concordant, but enough of a hiatus may still be evident to continue the designation of an unconformity (a paraconformity). Finally, the boundary may be traced between concordant strata to a place where there is no evidence of a hiatus, and the unconformity has been traced into its correlative conformity(Fig. 1).

Unconformities and Conformities

An *unconformity* is a surface of erosion or nondeposition that separates younger strata from older rocks and represents a significant hiatus (at least a correlatable part of a geochronologic unit is not represented by strata). A *conformity* is a surface that separates younger strata from older rocks, but along which there is no physical evidence of erosion or nondeposition, and no significant hiatus is indicated.

Dunbar and Rodgers (1957, Fig. 57) illustrated the conventional classification of unconformities and defined the familiar terms *nonconformity, angular unconformity, disconformity,* and *paraconformity.* That classification places more emphasis on the angularity or parallelism of the strata above and the strata below the unconformity than on the relation of strata to the unconformity itself. Although this criterion gives some indication of the degree of folding that took place prior to the deposition of the strata above, it is of little value in chronostratigraphy, so we rarely use the conventional classification of unconformities.

Hiatuses

A *hiatus* is the total interval of geologic time that is not represented by strata at a specific position along a stratigraphic surface. If the hiatus encompasses a significant interval of geologic time, the stratigraphic surface is an unconformity (Fig. 1).

To place a sequence within the standard chronostratigraphic framework, some measurement of the hiatuses along the unconformable sequence boundaries must be made. Also, the conformities that mark the maximum geologic-time range (sechron) of a sequence are determined only where the hiatuses along the sequence boundaries are shown to have decreased to an insignificant interval of geologic time.

Ideally, the hiatus is measured quantitatively by some radiometric method. In practice, the hiatus commonly is measured in qualitative units (such as periods, epochs, or faunal zones) by biostratigraphy, paleomagnetic reversal correlations, or some other method. These units are then correlated to a radiometric scale to make a best estimate of the age in millions of years.

The concept of the magnitude of a hiatus along a sequence boundary is analogous to the concept of the magnitude of a sequence. The order of magnitudes depends on the degree of resolution of the tool used. Whereas a seismic sequence commonly encompasses a geologic-time span of a few million years, its absence suggests a hiatus of that magnitude. A well-log sequence that is smaller than a seismic sequence may be used to detect a hiatus ranging in magnitude from a million years to hundreds of thousands of years. Although a very small-scale sequence of bedsets or laminasets can be used to determine a hiatus of very short duration, it commonly falls below the practical limit of what may be considered a significant hiatus, and so it would not be used to define an unconformity.

Hiatuses may be attributable either to erosion or to nondeposition of strata, or to both. As shown on Figure 1 the distinction is based on whether the strata in a depositional sequence terminate against a boundary as a result of erosional truncation or by lapout.

RELATIONS OF STRATA TO SEQUENCE BOUNDARIES

Figure 2 illustrates concordant and discordant relations of strata to boundaries of depositional sequences. These relations are based on the parallelism, or the lack of it, between the strata and the boundary surface itself. If the strata both above and below a surface are concordant (that is, essentially parallel to it), then there is no physical evidence of an unconformity along that part of the surface. On the other hand, if either the strata above or the strata below a surface are discordant (if they terminate against it), then there is physical evidence of an unconformity (or structural disruption).

Concordant relations may be seen at the upper or lower boundary of a depositional sequence. The concordance may be recognized as parallelism of a stratum to an initially horizontal, inclined, or uneven surface. At the base of a sequence, the concordance may be expressed as parallel draping over a bottom irregularity (Fig. 2).

Discordance is the main physical criterion used in the determination of sequence boundaries. The type of discordant relation is the best indicator of whether an unconformity results from erosion or nondeposition. The direction of progressive termination from older to younger strata above the unconformity is the direction of increasing nondepositional hiatus along the unconformity.

The type of discordance is based on the manner in which strata terminate against the unconformable boundary of a depositional sequence (or a structural boundary). *Lapout* is the lateral termination of a stratum at its original depositional limit. *Truncation* is the lateral termination of a stratum as a result of being cut off from its original depositional limit. More specific types of discordance and their significance in stratigraphic analysis are given below. They are determined with greater confidence where several strata within the sequence show a systematic pattern of discordance along a given surface.

Baselap: Onlap, Downlap

Baselap is lapout at the lower boundary of a depositional sequence. Two important types are recognized. *Onlap* is baselap in which an initially horizontal stratum laps out against an initially inclined surface, or in which an initially inclined stratum laps out updip against a surface of greater initial inclination. *Downlap* is baselap in which an initially inclined stratum terminates downdip against an initially horizontal or inclined surface.

Onlap or downlap usually can be readily identified. However, later structural movement may necessitate the reconstruction of depositional surfaces. In areas of great structural complication, the discrimination between onlap and downlap may be practically impossible, and the worker may be able to determine only that the strata are in a baselap relation.

The diagrammatic illustrations in Figure 2 are two-dimensional representations, and the assumption generally may be made that the sections are in the dip direction, and that strike sections at right angles would show essentially horizontal traces of strata. However, a two-dimensional section may show an *apparent* onlap and a section at right angles to it could show *true* downlap. If two sections intersecting at right angles both show apparent onlap, then true onlap is the likely relation.

The regional positions of *proximal onlap* (onlap in the direction of the source of sediment supply) and *distal downlap* (downlap in a direction away from source of sediment supply) commonly mark the lateral beginning and the lateral ending of deposition of a given stratum (or unit; Fig. 1). In narrow or regionally confined depositional sites, *distal onlap* (onlap in a direction away from the source of sediment supply) is commonly encountered, for example against the opposite side of a basin. In some depositional sites, the pattern of distal onlap may be controlled more by local bottom irregularities than by the proximal or distal relations to sediment supply.

Onlap and downlap are indicators of nondepositional hiatuses (Fig. 1) rather than erosional hiatuses. Successive terminations of strata at their depositional limits along the initial depositional surface produce an increasing nondepositional hiatus in the direction of onlap or downlap.

Toplap

Toplap is lapout at the upper boundary of a depositional sequence. Initially inclined strata,

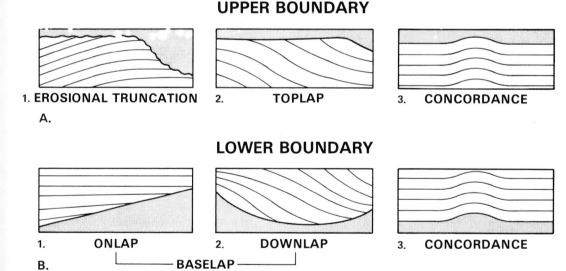

FIG. 2—Relations of strata to boundaries of depositional sequences:
 A. Relations of strata to upper boundary of a sequence. **A1.** Erosional truncation: strata at top of given sequence terminate against upper boundary mainly as result of erosion (e.g., tilted strata terminating against overlying horizontal erosion surface, or horizontal strata terminating against later channel surface). **A2.** Toplap: initially inclined strata at top of given sequence terminate against upper boundary mainly as result of nondeposition (e.g., foreset strata terminating against overlying horizontal surface at base-level equilibrium where no erosion or deposition took place). **A3.** Top-concordance: relation in which strata at top of given sequence do not terminate against upper boundary.
 B. Relations of strata to lower boundary surface of a sequence. **B1.** Onlap: at base of sequence initially horizontal strata terminate progressively against initially inclined surface, or initially inclined strata terminate updip progressively against surface of greater initial inclination. **B2.** Downlap: at base of sequence initially inclined strata terminate downdip progressively against initially horizontal or inclined surface (e.g., initially inclined strata terminating against underlying initially horizontal surface). **B3.** Base-concordance: strata at base of sequence do not terminate against lower boundary.

such as foreset beds and clinoforms, may show this relation. The lateral terminations updip may taper and approach the upper boundary asymptotically. On seismic sections, the resolution may be such that reflections appear to terminate abruptly against the upper surface at a high angle.

Toplap is evidence of a nondepositional hiatus. It results from a depositional base level (such as sea level) being too low to permit the strata to extend farther updip. During the development of toplap, sedimentary bypassing and possibly minor erosion occurs above base level while prograding strata are deposited below base level. Although toplap commonly is associated with shallow marine deposits, such as deltaic complexes, it may also occur in deep-marine deposits (such as fans) where depositional base level is controlled by turbidity currents and other deepwater processes.

Truncation: Erosional and Structural

Erosional truncation is the lateral termination of a stratum by erosion. It occurs at the upper boundary of a depositional sequence; and it may extend over a wide area or be confined to a channel. Strata tilted by structural movement commonly have their updip limits truncated by subaerial or submarine erosion, and the reconstruction to the original depositional limits of onlap or toplap may be a difficult choice. In some cases, the distinction of toplap and erosional truncation may be difficult, but in the latter relation the strata tend to maintain parallelism as they terminate abruptly against the upper boundary rather than taper to it. Such truncation is evidence of an erosional hiatus.

Structural truncation is the lateral termination of a stratum by structural disruption. Such truncation is most easily recognized where it cuts across the strata within a sequence or a group of sequences. The disruption may be produced by faulting, gravity sliding, salt flowage, or igneous intrusion. The distinction between erosional and structural truncation may be difficult, but it should be made before further stratigraphic analyses are attempted. Although structural truncation may produce a discordant relation of strata to a sequence boundary, such disruption has minor, if any, regional chronostratigraphic significance with respect to unconformities or hiatuses.

EXAMPLES OF DEPOSITIONAL SEQUENCES

Figures 3 and 4 are examples of depositional sequences on well-log and seismic sections, respectively. These examples depend primarily on physical stratigraphy for identification of the unconformities at the sequence boundaries, and on

biostratigraphic zonation for determination of the geologic ages of the sequences.

Well-Log Section

Figure 3 is a subsurface cross section from the Western Canada basin of northeastern British Columbia prepared by G. T. MacCallum and provided by Imperial Oil, Limited. The stratigraphic relations are determined by detailed log-marker correlation, including the discordances which are used to determine sequence boundaries. The following features are associated with each sequence: (1) sequence J with erosional truncation at its top; (2) sequence SR-1 with onlap at its base and concordant strata at its top; (3) sequences SR-2, 3, and 4 with downlapping strata at their bases and subtle toplap above; (4) sequence SR-5 with concordant strata at its base and top; and (5) sequence SR-6 similar to SR-4.

Each sequence has an internal stratification pattern that is genetically significant. For example, the marine Jurassic sequence (J) is made up of even, parallel strata. The Lower Cretaceous Gething Formation (G) is a nonmarine sequence of discontinuous sandstones and shales which are difficult to correlate on the electric logs. The thin basal sequence of the Spirit River Formation (SR-1) is a littoral deposit of sandstone and shale which tends to fill in the lows of the unconformity on top of the Gething (G). In contrast, the overlying sequences of the Spirit River (SR-2 to 6) consist of sandstones and shales which exhibit marked progradation into deep water. Note how the undaform sandstones at the tops of these sequences exhibit toplap.

Seismic Section

Figure 4 is a seismic section from offshore northwestern Africa. It trends parallel with depositional dip and encompasses units ranging in age from Triassic to Quaternary as projected from surface geology and well control. The physical stratigraphic relations have been worked out by means of detailed correlation of seismic reflections. The same principles of interpretation used with the well-log example apply equally well to this seismic example. However, on the seismic data, larger-scale features are recognized more easily, and continuity of reflection cycles provides a degree of correlation reliability not easily obtained with logs, especially where wells are far apart. Moreover, the fact that much of the physical stratigraphy can be interpreted before drilling emphasizes the advantage of the seismic approach.

In this particular example, the seismic section has been given the following stratigraphic interpretation. The oldest unit correlated is the Trias-

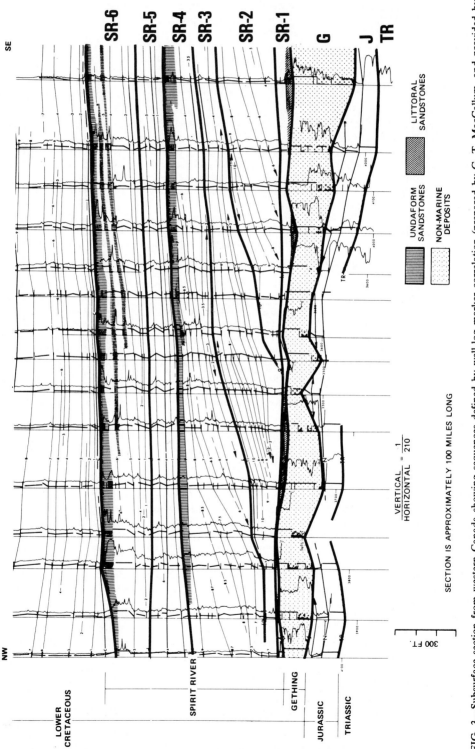

FIG. 3—Subsurface section from western Canada showing sequences defined by well-log marker correlation (prepared by G. T. MacCallum, and provided by Imperial Oil, Ltd.) Marker correlation of closely spaced well logs displays depositional patterns of Lower Cretaceous strata in northwestern British Columbia. Erosion truncation of Jurassic beds (surface J), onlap of lowermost Spirit River onto Gething (surface G), and progressive downlap and minor onlap of prograding upper Spirit River beds are used to determine several sequence boundaries.

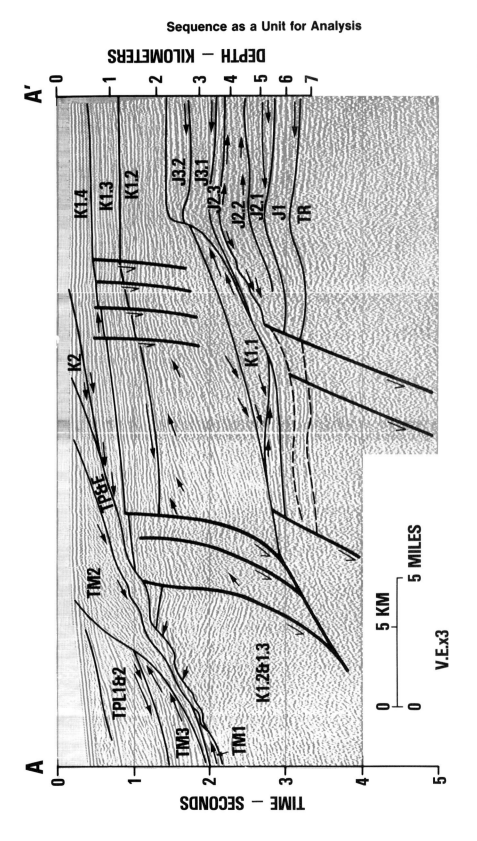

FIG. 4—Seismic section from offshore northwest Africa showing sequences defined by seismic reflections. Systematic reflection termination patterns used to determine downlap, onlap, truncation, and toplap of strata at sequence boundaries. Geologic ages of sequences determined from wells on section and in surrounding area. Age notations explained in text.

sic (TR) shown on the eastern half of the section. (Paleozoic and Precambrian rocks occur in the region but are not differentiated here). Five sequences have been designated within the Jurassic (see Part 8, Vail et al, this volume). The Lower Jurassic (J1) strata onlap the somewhat irregular unconformity at the base of this sequence. Low-angle downlap characterizes the bases of the Middle Jurassic (J2) sequences. The upper reflections of the lower Middle Jurassic (J2.1) are in a classic undaform-clinoform-fondoform relation (Rich, 1951). The lower Upper Jurassic (J3.1) and the Upper Jurassic–Berriasian (J3.2) sequences are similar to the Middle Jurassic, but their deposition culminates in an impressive carbonate shelf margin with a well-developed reef.

The Valanginian sequence (K1.1) is areally restricted to the basin in front of the shelf and consists of deep-marine clastic deposits with high initial dips at both the onlap and downlap terminations.

A great thickness of Lower Cretaceous shales and deltaic sandstones loaded the platform built during Jurassic time and produced a major system of down-to-basin faults and mobile shale structures. Three sequences recognized by downlap and onlap within the Lower Cretaceous are: Hauterivian to lower Aptian (K1.2), middle to upper Aptian (K1.3), and Albian to lower Cenomanian (K1.4). A thin unit of marine shelf deposits is designated as Upper Cretaceous (K2).

The boundary between the Cretaceous and Tertiary is marked by a sloping surface with submarine erosional truncation. A thin unit of early Tertiary age (Paleocene to lower Oligocene [T1, 2]) laps out basinward. The middle Tertiary sequences (middle to upper Oligocene [TO2, 3]; lower Miocene [TM1]) of deep-marine shales onlap steeply updip to terminate shelfward against lower Tertiary rocks. Upper Tertiary (middle

Miocene [TM2], upper Miocene [TM3], and Pliocene [TPL1, 2]) and Quaternary shales (Q) form a complex set of prograding and onlapping sequences. Probably most of the Albian to pre-middle Oligocene sequences extended originally over the eastern part of the section, but during Oligo-Miocene time they were uplifted and eroded from this shelf area.

Much information can be determined from analysis of sequences on seismic sections, even without well control. However, stratigraphic analyses are best documented where sufficient well control is available to date the sequences and determine their depositional facies.

REFERENCES CITED

Campbell, C. V., 1967, Lamina, laminaset, bed and bedset: Sedimentology, v. 8, p. 7-26.
————1971, Depositional model—Upper Cretaceous Gallup beach shoreline, Ship Rock area, northwestern New Mexico: Jour. Sed. Petrology, v. 41, p. 395-401.
————1973, Offshore equivalents of Upper Cretaceous Gallup beach sandstones, northwestern New Mexico, in J. E. Fasset, ed., Cretaceous and Tertiary rocks of the southern Colorado Plateau: Durango, Colo., Four Corners Geol. Soc., (Cretaceous-Tertiary Mem.), p. 78-84.
Chang, K. H., 1975, Concepts and terms of unconformity-bounded units as formal stratigraphic units of distinct category: Geol. Soc. America Bull., v. 86, p. 1544-1552.
Dunbar, C. O., and John Rodgers, 1957, Principles of stratigraphy: New York, John Wiley and Sons, 356 p.
Hedberg, H. D., ed., 1976, International stratigraphic guide: New York, John Wiley and Sons, 200 p.
Rich, J. L., 1951, Three critical environments of deposition and criteria for recognition of rocks deposited in each of them: Geol. Soc. America Bull., v. 62, p. 1-20.
Sloss, L. L., 1963, Sequences in the cratonic interior of North America: Geol. Soc. America Bull., v. 74, p. 93-114.

Seismic Stratigraphy and Global Changes of Sea Level, Part 3: Relative Changes of Sea Level from Coastal Onlap [1]

P. R. VAIL, R. M. MITCHUM, JR.,[2] and S. THOMPSON, III[3]

Abstract Relative changes of sea level can be determined from the onlap of coastal deposits in maritime sequences. The durations and magnitudes of these changes can be used to construct charts showing cycles of the relative rises and falls of sea level. Such charts summarize the history of the fluctuations of base level that control the distribution of the sequences and the strata within them.

A relative rise of sea level is indicated by coastal onlap, which is the landward onlap of littoral and/or nonmarine coastal deposits. The vertical component, coastal aggradation, can be used to measure a relative rise, but it should be adjusted for any thickening due to differential basinward subsidence. During a relative rise of sea level, a transgression or regression of the shoreline, and a deepening or shallowing of the sea bottom may take place. A common misconception is that transgression and deepening are synonymous with a relative rise, and that regression and shallowing are synonymous with a relative fall. A relative stillstand is indicated by coastal toplap; intermittent stillstands between rapid rises are characteristic of a cumulative rise. A relative fall of sea level is indicated by a downward shift in coastal onlap from the highest position in a sequence to the lowest position in the overlying sequence. After a major relative fall of sea level, the shelf tends to be bypassed, and the coastal onlap may be restricted to the apex of a fan at the basin margin.

Seismic sections provide the best means of determining the onlap and toplap patterns within the depositional sequences, and well control can provide the determinations of coastal and marine facies. Each cycle is plotted on a chart in chronologic order, dating and measuring the relative rise by increments of coastal aggradation, dating any relative stillstands by the duration of coastal toplap, and dating and measuring the relative fall by the downward shift of coastal onlap. Seismic examples illustrate the procedures and some of the problems encountered.

INTRODUCTION

Relative changes of sea level can be determined from the onlap of coastal deposits in depositional sequences. Relative stillstands can be determined from coastal toplap. As shown in Part 2 (Vail et al, this volume), seismic sections provide the best means of recognizing onlap and toplap patterns within the depositional sequences. Well control can provide data for the distinction between coastal and marine facies within the sequences.

Determinations of the durations and magnitudes of relative changes of sea level and the durations of the relative stillstands are needed to construct charts showing cycles of relative rises, stillstands, and falls of sea level. Such charts summarize the history of the fluctuations of sea level,

which is the effective base level during the deposition of most maritime sequences and during subsequent erosion. These fluctuations control the distribution of the sequences and the strata within them, and the extent of the unconformities and correlative conformities along the sequence boundaries.

This study examines concepts of relative changes of sea level, some of the depositional patterns related to these changes, and how regional charts showing cycles of relative rise, stillstand, and fall of sea level may be constructed. The following paper (Part 4, Vail et al, this volume) will show how charts of regional cycles are used to determine global cycles.

CYCLES OF RELATIVE CHANGE OF SEA LEVEL

A *relative change of sea level* is defined as an apparent rise or fall of sea level with respect to the land surface. Either sea level itself, or the land surface, or both in combination may rise or fall during a relative change. A relative change may be operative on a local, regional, or global scale. This paper will work with conceptual models showing relative changes and stillstands on a regional scale. Simultaneous relative changes in three or more widely spaced regions around the globe are interpreted as global changes of sea level (Part 4, Vail et al, this volume).

A *cycle of relative change of sea level* is defined as an interval of time during which a relative rise and fall of sea level takes place. A cycle may be recognized on a local, regional, or global scale, but this paper will deal with the regional ones. Most regional cycles are eventually determined to be global, but even those that are not are useful in regional stratigraphic studies.

A cycle of relative rise and fall of sea level typically consists of a gradual relative rise, a period of stillstand, and a rapid relative fall of sea level (Fig. 1). In detail, the gradual cumulative rise consists of a number of smaller scale rapid rises and stillstands. Such a small-scale event is a *paracycle*, defined as a relative rise and stillstand of

[1]Manuscript received, January 6, 1977; accepted, June 13, 1977.

[2]Exxon Production Research Co., Houston, Texas 77001.

[3]New Mexico Bureau of Mines and Mineral Resources, Socorro, New Mexico 87801.

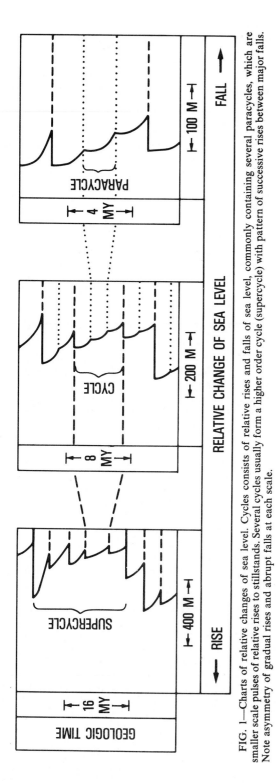

FIG. 1—Charts of relative changes of sea level. Cycles consists of relative rises and falls of sea level, commonly containing several paracycles, which are smaller scale pulses of relative rises to stillstands. Several cycles usually form a higher order cycle (supercycle) with pattern of successive rises between major falls. Note asymmetry of gradual rises and abrupt falls at each scale.

sea level, followed by another relative rise with no significant fall intervening. These small-scale events commonly are not detected with seismic data, but are more readily recognized with data observed in outcrops, cores, and well logs.

Commonly, a set of several regional or global cycles will form a distinctive pattern consisting of successive rises to higher relative positions of sea level, followed by one or more major relative falls to a lower position. This set forms a cycle of higher order (Part 4, Vail et al, this volume). Such a set of cycles is termed a supercycle.

A hierarchy of supercycle, cycle, and paracycle reflects relative changes of sea level of different orders of magnitude. As seen in Figure 1, a supercycle contains several cycles, and a cycle may contain several paracycles. These relations are shown on a regional or global cycle curve.

RELATIONS OF DEPOSITIONAL SEQUENCES TO CYCLES AND PARACYCLES OF RELATIVE CHANGES OF SEA LEVEL

Our basic operational stratigraphic unit is the depositional sequence, defined as a stratigraphic unit composed of a relatively conformable succession of genetically related strata and bounded at its top and base by unconformities or their correlative conformities (Part 2, Vail et al, this volume). If a sufficient sediment supply is available, one or more depositional sequences are deposited during one cycle of relative rise and fall of sea level.

If a cycle contains a continuous relative rise to stillstand of sea level, only one sequence is likely to be deposited during that time. The abrupt fall at the end of the cycle tends to produce an unconformity that will separate the sequence from the overlying one of the next cycle.

If a cycle contains two or more paracycles, at least two sequences are likely to be deposited during the cycle. The boundary between the sequences would be marked most commonly by downlap of the overlying sequence, although toplap of the underlying sequence may be present.

Two or more sequences may be deposited during a cycle or a paracycle. After a rapid rise of sea level, a surface of non-deposition may be developed before the progradational deposits of the stillstand are laid down. The surface should be marked by downlap of the overlying progradational deposits. Frazier (1974) recognized such surfaces in defining depositional episodes during the Pleistocene of the Gulf of Mexico. Each sequence of transgressive sandstones is overlain by a sequence of upward coarsening, progradational strata.

The smaller scale sequences deposited during paracycles or in shorter pulses during development of delta lobes may be too thin to be recognized with seismic data alone; outcrops and/or close well control with adequate cores may be needed. Stratal terminations are commonly subtle and depositional in nature.

In the construction of charts of relative changes of sea level, a simple one-to-one relation of depositional sequences to sea-level cycles should not necessarily be expected. Each sequence is a building block in the regional stratigraphic framework; and comprehensive analyses of the strata, their facies, and their ages are needed to determine if one sequence or a set of them represents one cycle of relative change of sea level.

In general, the greater the sea-level fall the easier it is to recognize sequence boundaries by onlap, downlap, and truncation. If structuring occurs, erosion of the tilted strata during a succeeding lowstand commonly causes spectacular angular unconformities.

Maritime and Hinterland Sequences

Distribution of the strata and facies is controlled directly by relative changes in sea level in some depositional sequences and indirectly or not at all in others. The primary consideration is whether a sequence was deposited in a maritime or hinterland environment.

A *maritime sequence* is a depositional sequence that consists of genetically related coastal and/or marine deposits. The coastal facies of a maritime sequence especially is controlled by the position of sea level as a base level. Shallow marine facies are partially controlled by sea level, but the deep marine facies are not directly controlled by it. Therefore, cyclical changes in the relative position of sea level exert a major control on the landward extent of maritime depositional sequences.

A *hinterland sequence* is one that consists entirely of nonmarine deposits laid down at a site interior to the coastal area, where depositional mechanisms are controlled indirectly or not at all by the position of sea level. Although our experience with hinterland sequences is limited, they seem to be deposited independently of the maritime sequences, and therefore are omitted from the discussion of relative changes of sea level.

INDICATORS OF RELATIVE CHANGES OF SEA LEVEL

The most reliable stratigraphic indicators of relative changes of sea level are the depositional limits of onlap and toplap (Part 2, Vail et al, this

volume) within the coastal facies of maritime sequences. With adequate paleobathymetric control, marine facies may be used; however, the deep-marine control generally is not adequate. Other methods may be employed to help measure relative changes of sea level, but there are pitfalls in using alone such phenomena as transgression/regression of the shoreline and deepening/shallowing of the sea bottom.

The next sections discuss some basic concepts dealing with the stratigraphic indicators used to determine relative changes of sea level in maritime sequences. For purposes of illustration, diagrams show deposits of terrigenous clastics, but the models apply also to carbonates and other rock types. Parts of these concepts are modified from the works of Weller (1960, p. 498-501), van Andel and Curray (1960), Curray (1964), and others. Pitman (1977) discussed interrelations of eustatic changes of sea level, tectonic movements, and rates of sediment supply.

Relative Rise of Sea Level

A *relative rise of sea level* is an apparent sea level rise with respect to the underlying initial depositional surface (Fig. 2) and is indicated by coastal onlap. It may result from (1) sea level actually rising while the underlying initial surface of deposition subsides, remains stationary, or rises at a slower rate; (2) sea level remaining stationary while the initial surface of deposition subsides; or (3) sea level falling while the initial surface of deposition subsides at a faster rate. Additionally, *coastal onlap* is the progressive landward onlap of littoral and/or nonmarine coastal deposits in a given maritime sequence. Coastal deposits may be determined by paleoecology or sedimentology.

During a relative rise of sea level, where the sedimentary supply is sufficient, coastal deposits progressively onlap the underlying initial surface of deposition. The process is unable to build much above sea level, which approximates effective depositional base level. Without the rise of effective base level, the depositional site would be unable to accommodate the sediment, and each increment of coastal deposition would be terminated laterally before it could onlap the depositional surface.

A relative rise of sea level can be measured most accurately where littoral deposits (those laid down between low and high tide) onlap the underlying depositional surface. However, nonmarine coastal deposits (those laid down on the coastal plain above high tide) most commonly onlap the surface; they may build a few meters above sea level, introducing a small error in the

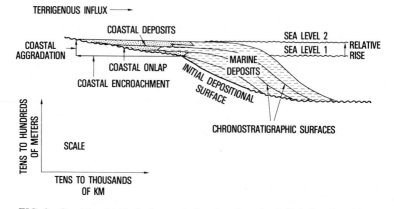

FIG. 2—Coastal onlap indicates a relative rise of sea level. Relative rise of base level allows coastal deposits of a maritime sequence to aggrade and onlap initial depositional surface.

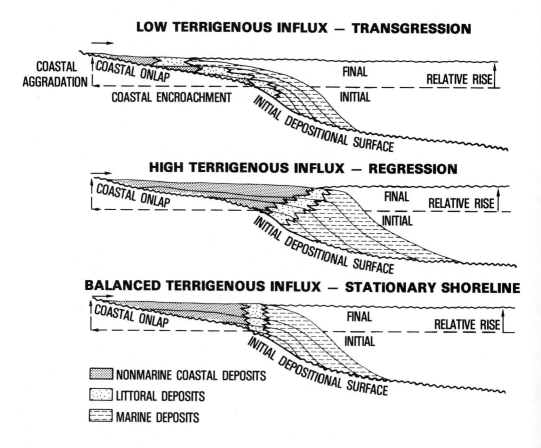

FIG. 3—Transgression, regression, and coastal onlap during relative rise of sea level. Rate of terrigenous influx determines whether transgression, regression, or stationary shoreline is produced during relative rise of sea level.

estimate of the relative rise. On coastal plains that are hundreds of kilometers wide, the error may be several tens of meters.

A relative rise of sea level may be measured on a stratigraphic cross section provided that structural deformation is not too intense. A seismic section is excellent for this purpose. Either the vertical or the horizontal components of coastal onlap can be used, and are termed coastal aggradation and coastal encroachment, respectively (Fig. 2). However, measurements of coastal aggradation should be adjusted for any distortion due to basinward thickening of strata that may have resulted from differential basinward subsidence. Likewise, measurements of coastal encroachment need to be adjusted for variations in the slope of the underlying initial depositional surface that would distort measurements of sea level rise. Other variables, such as compaction, may need to be considered in individual cases.

The onlapping coastal deposits of a particular sequence may have been removed by erosion at a given locality. In some cases the missing strata may be restored to a projection of the underlying initial deposition surface, sometimes with the help of isolated erosional remnants, but generally it is better to search the region for a section in which the coastal onlap is preserved.

Where a relative rise of sea level is more rapid than the rate of deposition, the result may be on-lap of marine strata (marine onlap) instead of coastal onlap, and paleobathymetric control will be needed to help measure the relative rise. Assuming that structural movements are not complicated, the amount of the relative rise may be approximated by measuring the vertical component of marine onlap (marine aggradation) if the paleobathymetry remains constant. If not, the amount of rise may be estimated by determining the marine aggradation plus any amount of deepening, or minus a lesser amount of shallowing. Because paleobathymetric measurements are given in intervals of several hundreds of feet, the measurement of relative rise with marine onlap is only an approximation and should be checked against measurements of coastal onlap of the same unit in other areas.

During a relative rise of sea level, a transgression or regression of the shoreline, and deepening or shallowing of the sea bottom, may take place. Marine transgression and regression during a relative rise of sea level are illustrated in Figures 3 and 4. A transgression of the shoreline is indicated by landward migration of the littoral facies in a given stratigraphic unit, and a regression is indicated by seaward migration of the littoral facies. Instead of transgression or regression, the shoreline may be stationary. Similarly, a deepening of the sea bottom (Fig. 4) is indicated by evidence of increasing water depth, and a shallowing

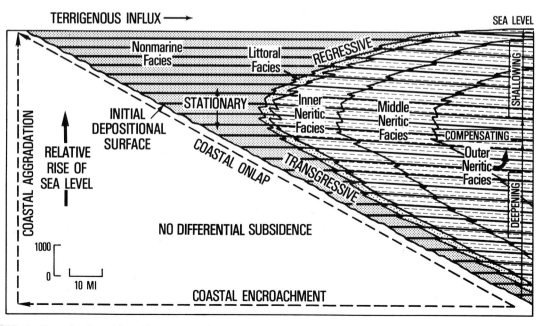

FIG. 4—Coastal onlap with marine transgression and regression. During relative rise of sea level, littoral facies may be transgressive, stationary, or regressive, and neritic facies may be deepening, compensating, or shallowing.

of the sea bottom is indicated by evidence of decreasing water depth. Instead of deepening or shallowing, the sea bottom may be compensating. Any of the above may take place during a relative rise of sea level. A common misconception is that transgression and deepening are synonymous with a relative rise, and that regression and shallowing are synonymous with a relative fall of sea level.

Although it is true that either transgression or deepening may indicate at least a part of a relative rise of sea level, neither can be assumed to indicate the entire rise. A transgression may be terminated by an increase in terrigenous clastic supply to produce a stationary shoreline or a regression while the relative rise of sea level continues (Grabau, 1924, p. 724; Weller, 1960, p. 500; van Andel and Curray, 1960; Curray, 1964). A bathymetric deepening may also be terminated by an increase in sediment supply to produce a shallowing sea bottom while the relative rise continues. Moreover, the distribution of sediment may be such that the littoral facies is regressive while a starved marine section is deepening during a relative rise. Because regression and shallowing can occur during a relative rise, stillstand, or fall of sea level, they cannot be indicative of any one of them. However, regression is most common during a relative rise or stillstand.

On a seismic section through the coastal facies of a maritime sequence, a relative rise of sea level can be recognized by onlapping reflections. Transgressions and regressions of the shoreline are recognized with more difficulty by lateral changes in reflection characteristics, such as amplitude, frequency, and wave form, indicating changes from coastal to marine facies. This point is illustrated in the following example.

Northwestern Africa Example

On a seismic section from the continental margin of northwestern Africa (Fig. 5), a major relative rise of sea level is indicated by coastal onlap. Also, marine transgression and regression are indicated by landward and seaward migration of the littoral facies during this rise. A large sedimentary wedge, Triassic through Tertiary in age, thickens seaward into the Atlantic. Boundaries of depositional sequences (shown by the heavy black lines) have been interpreted from the onlapping and downlapping reflection terminations (shown by arrows). Paleontologic control in wells indicates the two major sequences to be Lower and Upper Cretaceous. Coastal and marine deposits are also identified from well control. Individual seismic reflectors can be traced from interpreted marine deposits on the left into coastal

deposits on the right and finally to the point of onlap against the underlying unconformity. Shown on the section are an Early Cretaceous transgression, stationary shoreline, and regression, followed by a Late Cretaceous transgression and regression. Both these cycles of transgression occurred during a relative rise of sea level documented by continuous coastal onlap during Early Cretaceous and much of Late Cretaceous time.

Relative Stillstand of Sea Level

A *relative stillstand of sea level* is an apparently constant position of sea level with respect to the underlying initial surface of deposition, and is indicated by coastal toplap. It may result if both sea level and the underlying initial suurface of deposition actually remain stationary, or if both rise or fall at the same rate.

During a relative stillstand of sea level, where the sedimentary supply is sufficient, deposition in the coastal environment is hindered in any attempt to build above the effective base level, and the strata are prevented from onlapping the initial depositional surface (Fig. 6). The result is coastal toplap—the toplap (Part 2, Vail et al, this volume) of coastal deposits in a depositional sequence. Each unit of strata laps out in a landward direction at the top of the unit, but the successive terminations lie progressively seaward. Toplap can be recognized on seismic sections, and the evidence of coastal deposits can be determined from paleoecology or sedimentology using well data.

If a relative stillstand occurs after a rise that is more rapid than the rate of deposition, the result may again be marine onlap. In such a case, the stillstand may be reflected in the paleobathymetry as a shallowing at the same rate as the marine aggradation, again assuming no differential structural movements or other complications.

A cumulative relative rise of sea level that occurs over several million years commonly is characterized by shorter pulses of sea-level rise alternating with intervals of stillstand. In general, the rapid rises are more frequent and are of greater magnitude in the early part of the cumulative rise, and the stillstands are more frequent and last longer in the later part of the cumulative rise. This gradual diminishing of the rapid rises slows down the rate of the cumulative rise with time.

The cyclic pulses consisting of alternations of rapid rises and stillstands are called paracycles (Fig. 1). They are recognized frequently as smaller scale depositional sequences on detailed well-log or outcrop sections, but commonly are too small to be recognized on seismic sections.

Minor surfaces displaying toplap may be produced by periods of rapid excess deposition of

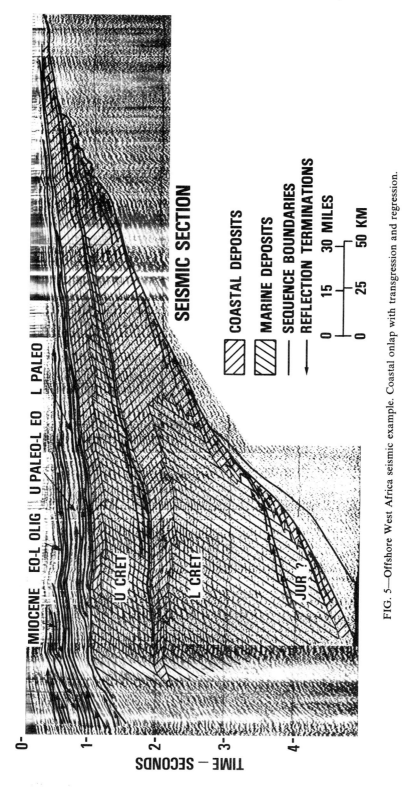

FIG. 5—Offshore West Africa seismic example. Coastal onlap with transgression and regression.

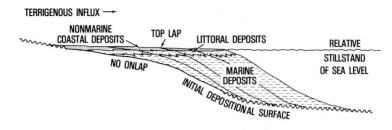

FIG. 6—Coastal toplap indicates relative stillstand of sea level. With no relative rise of base level, nonmarine coastal and/or littoral deposits cannot aggrade, so no onlap is produced; instead, by-passing produces toplap.

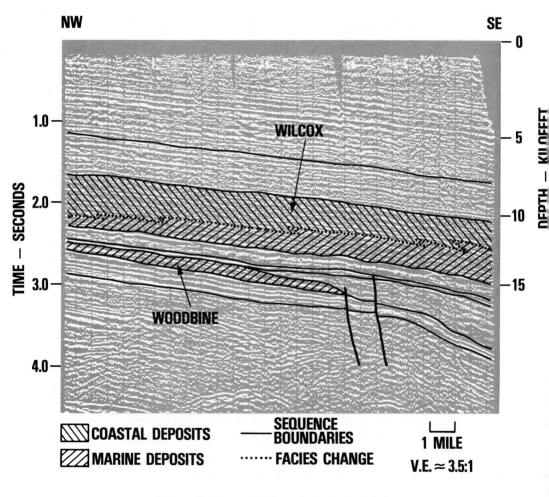

FIG. 7—East Texas seismic example of coastal toplap.

clastic sediments during a relative rise. These indicators of relative stillstand may be localized to areas of abnormally high depositional rates, such as in deltaic lobes. These local paracycles may or may not be related to more regional or global stillstands.

Texas Gulf Coast Example

Two examples of relative stillstand marked by coastal toplap are shown on the seismic section of Figure 7. These are from the Woodbine and Wilcox stratigraphic units of eastern Texas. Coastal and marine deposits were identified from wells.

The Woodbine (middle Cenomanian) forms a depositional sequence that consists mostly of shale in this area and lies basinward from the famous East Texas field. High-amplitude, slightly divergent, continuous reflectors mark the top and the base of the Woodbine sequence. A series of more steeply dipping reflections between these two strong reflections forms an oblique progradational configuration indicating depositional clinoforms (Part 6, Vail et al, this volume). These steeply dipping reflectors terminate by toplap immediately beneath the reflection marking the base of the overlying sequence. Thin deltaic sandstones at this toplap position produce gas in the area (Part 5, Vail et al, this volume). This Woodbine example closely resembles the diagrammatic example in Figure 6.

The Wilcox (Paleocene-Eocene) shows a variation of this toplap pattern. Reflections above and below the Wilcox are parallel. Within the Wilcox, a number of reflections dip at a greater angle than the bounding reflectors, indicating prograding clinoforms. To the northwest (left), reflections from the oblique prograding clinoforms show toplap below a stratigraphic horizon in the lower third of the Wilcox. To the southeast (basinward), the toplap pattern occurs at progressively higher stratigraphic levels. This pattern is interpreted as an alternation of relative stillstands and relative rises of sea level.

Relative Fall of Sea Level

A *relative fall of sea level* is an apparent fall of sea level with respect to the underlying initial surface of deposition, indicated by a downward shift of coastal onlap. It may result: (1) if sea level actually falls while the initial surface of deposition rises, remains stationary, or subsides at a slower rate; (2) if sea level remains stationary while the surface is rising; or (3) if sea level rises while the surface is rising at a faster rate.

A *downward shift of coastal onlap* is a shift downslope and seaward from the highest position of coastal onlap in a given maritime sequence to the lowest position of coastal onlap in the overlying sequence. In Figure 8a, the downward shift occurs between the highest coastal onlap of unit 5 in sequence A and the lowest coastal onlap of unit 6 in sequence B. The patterns of onlap indicate a relative rise of sea level during deposition of sequence A, then an abrupt relative fall to the position of unit 6 in sequence B, followed by another rise during deposition of sequence B.

In the seismic examples we have studied, fall of sea level is indicated by such an abrupt shift, and appears to occur as one event, rather than a succession of events. Where accurately dated, each fall is "rapid," occurring within a million years or less.

A stratification pattern, presented earlier by Weller (1960, Fig. 189B) and reproduced in Figure 8b, shows a series of units prograding at successively lower levels during a relative gradual fall of sea level. An example is the lower Gallup Formation in New Mexico, where several small falls in a short period of time produce a pattern similar to that of Figure 8b. A cross section showing these features in the lower Gallup is shown in Campbell (1977). These features are commonly beyond seismic resolution. In addition, the areas where we have seismic and well control are commonly areas of thick sedimentary deposits. In such areas, regional subsidence may proceed at a faster rate than any gradual fall of sea level and a relative rise is produced. Therefore, an actual fall in sea level, especially a gradual one, might not be detected by using coastal onlap as a criterion.

To measure a relative fall of sea level, the initial difference in elevation is determined between the highest coastal onlap in the underlying sequence (unit 5 in sequence A, Fig. 8a) and the lowest coastal onlap in the overlying sequence (unit 6 in sequence B). Difficulties encountered in the actual measurement are: (1) the underlying unit commonly is eroded during a relative fall, and if so, it must be approximated by restoration; (2) differential basinward subsidence that took place during the fall and the ensuing rise must be corrected; and (3) marine (instead of coastal) onlap is commonly encountered in the oldest beds of the overlying unit and accurate paleobathymetric determinations are needed to determine a relative fall of sea level.

After a relative fall, marine onlap again may result if the relative rise is more rapid than the rate of deposition. Paleobathymetric shallowing in deep marine strata may be helpful in measuring actual falls of sea level, where there are no structural complications.

a) DOWNWARD SHIFT IN COASTAL ONLAP INDICATES RAPID FALL

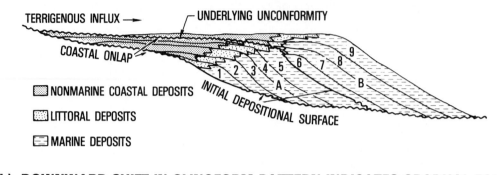

b) DOWNWARD SHIFT IN CLINOFORM PATTERN INDICATES GRADUAL FALL

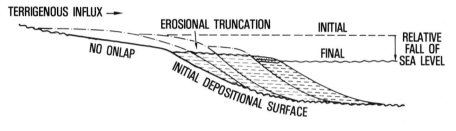

FIG. 8—Downward shift of coastal onlap indicates relative fall of sea level. With relative fall of base level, erosion is likely: deposition is resumed with coastal onlap during subsequent rise. (a) Downward shift in coastal onlap indicates rapid fall observed in all cases studied so far. (b) Downward shift in clinoform pattern (after Weller, 1960), indicates gradual fall; but has not been observed on seismic data.

Because a major fall is so difficult to measure accurately, it rarely can be plotted quantitatively with any degree of confidence.

After a major relative fall of sea level, the shelf tends to be bypassed, and the area of coastal onlap may be restricted to the apex of a fan at the basin margin. The block diagrams in Figure 9 show differences in stratal patterns during periods of highstand and lowstand of sea level. A marked change in the depositional pattern of deep-marine strata commonly occurs as a result of a major relative fall of sea level.

Figure 9a illustrates an idealized pattern developed during a highstand of sea level when shallow seas cover much of the shelf. Deposition occurs in clinoform lobes that prograde across the shallow shelf; with sufficient sediment supply, the progradation continues into deep water as shown in this illustration. Coarse clastics tend to be trapped on the shelf, and finer material is transported to the toes of the clinoforms.

Figure 9b illustrates the depositional pattern after a major fall of sea level below the shelf edge. The shelf is exposed to subaerial erosion, and rivers tend to bypass the shelf and deposit directly onto the slope. During the ensuing rise of sea lev-

el, any coastal onlap occurs near the sediment source; marine onlap may be produced if the sediments are funnelled through a submarine canyon. Most of the onlap commonly occurs in marine deposits at the proximal edge of a submarine fan, as shown in Figure 9b. This type of marine onlap may extend to very deep water, where detailed bathymetric control is needed to calculate the magnitude of relative changes of sea level. However, the overall pattern is diagnostic of a major relative fall of sea level and is common along continental margins and in deep-marine basins.

San Joaquin Basin (California) Example

A seismic section from San Joaquin basin, California (Fig. 10), shows the upper Tertiary section divided into nine depositional sequences; the depositional environments are identified generally. In the middle Miocene sequence, consisting of the lower part of the Fruitvale formation, a prominent shelf-edge is evident, and the sequence is thickest on the shelfward side to the east. The sequence thins basinward along well-developed clinoforms. This sequence is very similar to the basinward part of sequence A in Fig. 8a.

The upper Miocene sequence consists of the upper Fruitvale–McLure shale and Santa Margarita Sandstone. It closely resembles sequence B (in Fig. 8a) as it onlaps against the underlying middle Miocene sequence; and it laps out completely at the middle Miocene shelf edge. The sequence thins as it progrades into the basin. Well control indicates marine onlap in the lower part of the sequence. The upper part of the sequence is composed of the Santa Margarita Sandstone which was deposited in a shallow-marine to deltaic environment. Therefore, coastal onlap is demonstrated in the upper part of the sequence.

Such coastal onlap of a sequence restricted to the basin indicates a major relative fall of sea level. Because this sequence overlies a widespread shelf-type sequence, and because there is no significant difference in age between them, the fall must have been relatively rapid.

North Sea Example

Several examples of a shift in depositional patterns with relative falls of sea level are shown in Figures 11 and 12. These are overlapping seismic sections showing the Tertiary of the North Sea.

The two sections together are more than 160 km long, and vertical exaggeration is about 20 to 1. Depositional sequences are dated and the environments of the strata are determined from well control. Shelf sediments pass into basinal deposits from right to left.

The sequences that are widespread on the shelf are interpreted as highstand deposits, and the sequences that lap out against the slope or lower part of the shelf are interpreted as lowstand deposits. Highstand sequences are thickest near the shelf edge and extend long distances across the shelf to the right. The overall depositional pattern is that of progradation and downlap in a basinward direction. Prominent highstand sequences were deposited in late Paleocene–early Eocene, early Oligocene, middle Miocene, and early Pliocene times, as shown on the chart of relative change of sea level in Figures 11 and 12.

In contrast, lowstand sequences are thickest in the basin to the left and thin rapidly shelfward by marine onlap. Lowstand sequences were deposited in middle Paleocene, early middle Eocene, middle and late Oligocene and early Miocene, late Miocene, and late Pliocene times. The low-

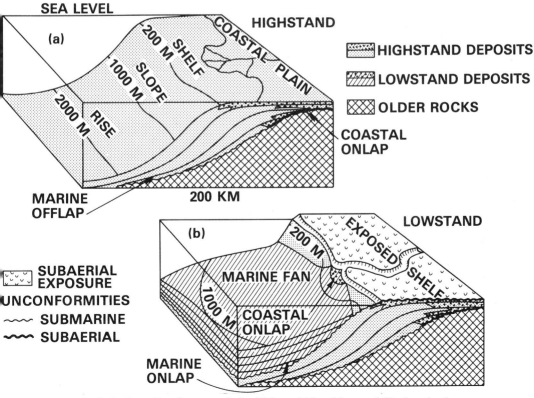

FIG. 9—Depositional patterns during highstand (**a**) and lowstand (**b**) of sea level.

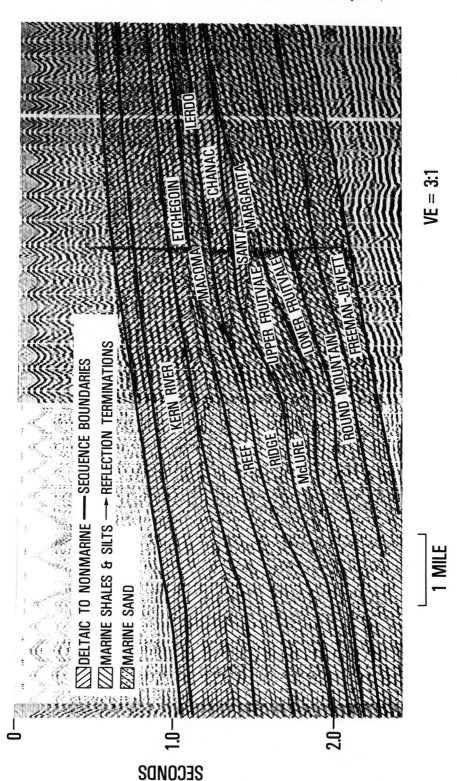

FIG. 10—San Joaquin basin (California) seismic example of downward shift in coastal onlap.

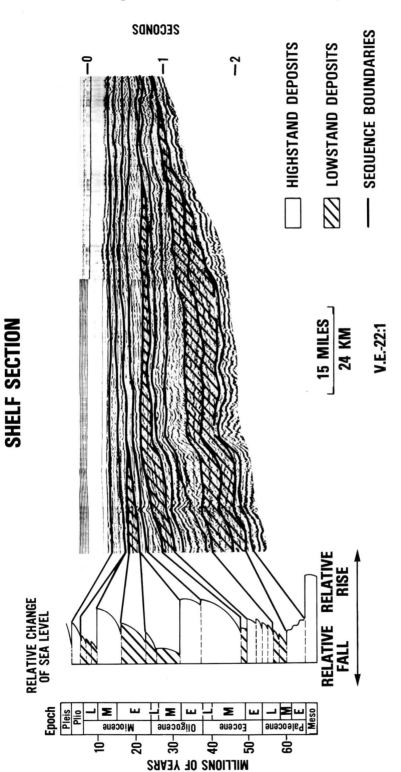

FIG. 11—North Sea Tertiary shelf seismic example of depositional patterns during highstands and lowstands of sea level.

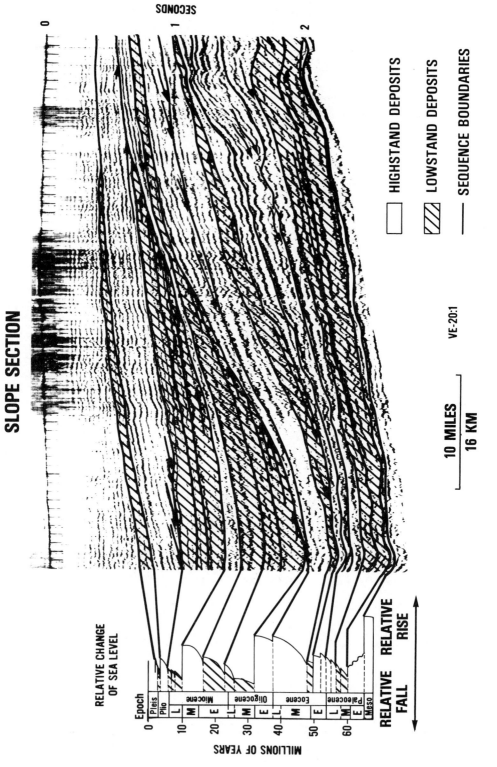

FIG. 12—North Sea Tertiary slope seismic example of depositional patterns during highstands and lowstands of sea level.

stand sequences of late-middle and late Oligocene age are missing over much of north-central Europe. Pleistocene units are not discussed here.

CONSTRUCTION OF REGIONAL CURVES OF RELATIVE CHANGES OF SEA LEVEL

The preceding sections showed how to recognize cycles of relative rise, stillstand, and fall of sea level, using coastal onlap, toplap, and other stratigraphic criteria. This section describes how to plot curves of these cycles for a given region. A *region,* in the sense used here, can be an entire oceanic domain, a continental margin, an inland sea, or a strip of coast. The main requirement is that the sequences of strata had some original physical continuity, so that the regional charts are based on strata that have a definite relation in both time and space.

A sea-level curve may be compared with other regional data by plotting it on a chronostratigraphic correlation chart for the region (Part 7, Vail et al, this volume). Such a combination shows the relations of sea-level changes to geologic age, distribution of depositional sequences, unconformities, facies and environment, and other information.

The next paper of this suite (Part 4, Vail et al) will show how to compare several sea-level curves from different regions and produce a global chart of relative changes of sea level. The global chart may be used in an undrilled region to predict age and general depositional characteristics of sequences. As data are acquired, the regional chart can be improved, and comparisons with the global chart may show regional anomalies.

Procedure

The procedure for constructing a sea-level curve will be illustrated diagrammatically and also with a seismic example from northwestern Africa. The curve is plotted with respect to geologic time and shows the relative rises, stillstands, and falls of sea level determined in the region. A regional grid of stratigraphic cross sections is needed to draw the curve; these sections should include the shelf edge and show the most complete record of coastal onlap in the region. Seismic sections are excellent for this purpose if they have good reflection quality and sufficient well control, and if structural deformation has not been too complicated.

Figure 13 diagrammatically illustrates the steps in constructing a sea-level curve. The first step is to analyze the maritime sequences (A to E in Fig. 13a). This step includes determinations of sequence boundaries, ages, areal distributions, and the presence of coastal onlap and toplap. On seismic sections the reflections representing strata within each sequence are traced to the proximal depositional limits of onlap or toplap. Environmental control is added to distinguish coastal and marine facies, and available age control is used to determine the geologic-time range of each sequence. The same procedures may be followed using subsurface well logs or surface sections, as long as the stratification surfaces are traced accurately.

The second step is to construct a chronostratigraphic correlation chart (Fig. 13b) of the sequences (Part 2, Vail et al, this volume). This chart plots the information shown on the stratigraphic cross section (Fig. 13a) against geologic time. The geologic-time ranges and relative areal distributions of the sequences are shown on this type of display.

After determining and plotting the ages of the depositional sequences, the third step is to identify the cycles of relative rise and fall of sea level, to measure the magnitudes of the rises and falls, and to plot them and the stillstands with respect to geologic time (Fig. 13c). Coastal aggradation is the best measure of a relative rise, and a downward shift of coastal onlap is the best measure of a relative fall of sea level. Coastal toplap indicates a relative stillstand. Where the onlap is in marine deposits, the paleobathymetry may sometimes be used to determine a relative change of sea level. If the record of the onlap is obscure, or has been removed by erosion, it is preferable to search for more complete sections elsewhere in the region.

In this example, relative rises and falls are determined with measurements of coastal aggradation, the vertical component of coastal onlap (Fig. 13a). Each increment of coastal aggradation is plotted on a geochronologic chart (Fig. 13c). Beginning with the oldest unit, sequence A, the first increment of coastal aggradation is 100 m, and the time interval is from 26 to 24 m.y. before present. Similarly, the successive increments of 150, 100, and 50 m are determined to give a total of 600 m of coastal aggradation during the deposition of sequence A. The measurements of coastal aggradation are made as closely as possible to the underlying unconformity to minimize the effect of differential basinward subsidence.

The unconformity at the top of sequence A shows erosional truncation of some units. However, the youngest coastal onlap in sequence A is dated 17-18 m.y., approximately the same as the oldest coastal onlap in sequence B. Evidence for a significant stillstand is missing. The relative fall of sea level at the end of sequence A occurred in less than one million years. The magnitude of the fall, from the highest coastal onlap in sequence A to

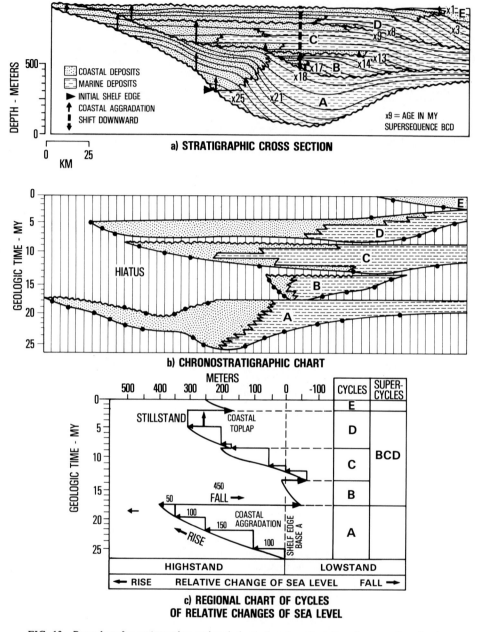

FIG. 13—Procedure for constructing regional chart of cycles of relative changes of sea level.

the lowest coastal onlap in sequence B, is measured as 450 m (heavy dashed line, Fig. 13a). Because the stratal surfaces are parallel over the area of measurement in sequences B, C, and D, no correction for differential basin subsidence is necessary.

Repeating the steps, the cycles of relative changes of sea level are plotted on Figure 13c

from onlap patterns of sequences B, C, D, and E on Figure 13a. In the upper part of sequence D, coastal toplap indicates a relative stillstand of sea level.

In this example, each cycle of relative change of sea level is asymmetrical, with a slow rise to stillstand and a rapid fall (Fig. 13c). This asymmetry of cycles has been observed in the investigations to date. The cycles as a group also show a pattern of asymmetry. Cycle A represents a highstand, cycles B, C, and D begin with a lowstand and gradually rise to a highstand, and cycle E rapidly falls to a lowstand. A higher order cycle (supercycle) BCD is recognized by the progressive rise and fall within the asymmetrical pattern.

A *highstand* is defined as the interval of time when sea level is above the shelf edge, and a *lowstand* as the interval of time when sea level is below the shelf edge. A *comparative* lowstand may be recognized as when sea level is at its lowest position on the shelf during the deposition of a series of sequences, or at its lowest position but when a shelf edge is not evident. Knowledge of these times can be helpful in predicting sedimentary events. For example, the highstands are the most likely times for trapping terrigenous clastics in deltas on the shelf, and the lowstands are the most likely times that the clastics will be funneled through submarine canyons or other notches in the shelf edge and deposited in submarine fans in the basin.

As shown in Fig. 13c, the times when sea level is at the shelf edge may be plotted as arrows on the curves. Then the highstands are to the left of the arrows and the lowstands are to the right. In preparing a group of regional cycle charts, we orient them with the relative rises plotted toward the left (as in Fig. 13c) to correspond to the general convention of showing landward on the left. However, in a given region, the relative rises of sea level are plotted to correspond to the landward direction on the stratigraphic section. Thus a section with landward on the right may be used to construct a chart with relative rises plotted toward the right, as in Figures 11 and 12. When charts for several regions are compared, they all should be oriented the same way.

Together, the stratigraphic section (Fig. 13a), the chronostratigraphic correlation chart (Fig. 13b), and the sea-level cycle curve (Fig. 13c) summarize the relations between sequences, correlation, and cycles of relative changes in sea level. Such a combination provides a summary of the geology within a regional stratigraphic framework. The regional cycles of relative change of sea level shown on the geochronologic chart can be compared with curves of global relative changes of sea level (Part 3, Vail et al, this volume).

In many regional seismic stratigraphic studies, practical problems such as the distribution and quality of data, erosion, or structural displacement of stratigraphic units, make a complete analysis of coastal onlap impractical or impossible. In such areas, a partial curve may be plotted for the cycles represented by data, and the remainder of the regional curve may be inferred from the global cycle chart.

Northwestern Africa Example

Figures 14 and 15 illustrate the procedure for constructing a curve of relative changes of sea level from seismic data for an offshore region in northwestern Africa. Figure 14 shows the same seismic section used in Figure 2, but the environmental interpretation was deleted and the measurement of coastal aggradation was added. From well control, we know the geologic age, and that the onlapping strata were deposited in a coastal environment. Figure 15 shows the chronostratigraphic correlation chart and the chart of relative changes of sea level.

The first step in constructing the curve is to determine the magnitudes of the relative rises from coastal aggradation. On the seismic section (Fig. 14), note where the onlapping reflections in a given sequence are parallel, and where they begin to diverge as a result of differential basinward subsidence. Measure increments of coastal aggradation where the reflections are parallel to avoid adjustments where they diverge. Beginning at the base of the sequence, measure the first increment as the vertical component of coastal onlap from the base of the sequence up to the highest parallel reflection. Then trace this reflection laterally to its point of onlap and measure the next increment at that point. Repeat the process until the top of the sequence is reached. Sum the lengths of the vertical components; seismic times must be converted to depth—the summation gives the total coastal aggradation and thus the total relative rise in sea level.

On this section, we measured 1,100 m of coastal aggradation by the incremental method. Note that the total thickness of the Cretaceous (measured where the reflections are divergent on the left side of the section) is approximately 5,000 m and thus is much greater than the aggradation because of differential basinward subsidence. Incremental measurements of aggradation are made near the point of onlap to avoid such problems with stratal thickening.

In the next step, relative falls are determined from the downward shift in coastal onlap. A prominent downward shift occurs at the Tertiary-Cretaceous boundary between the highstand position of the youngest Upper Cretaceous coastal strata and the oldest coastal onlap of the overlying Paleocene strata. Unfortunately the highest coastal onlap of the Cretaceous strata is not seen on the seismic section. However, much of the relative fall can be determined indirectly by measuring the subsequent rise in early Tertiary time. Differential subsidence occurring during the fall; or

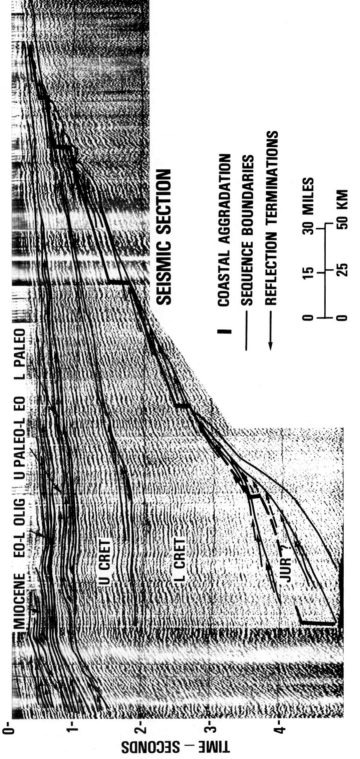

FIG. 14—Offshore West Africa example of seismic stratigraphic interpretation.

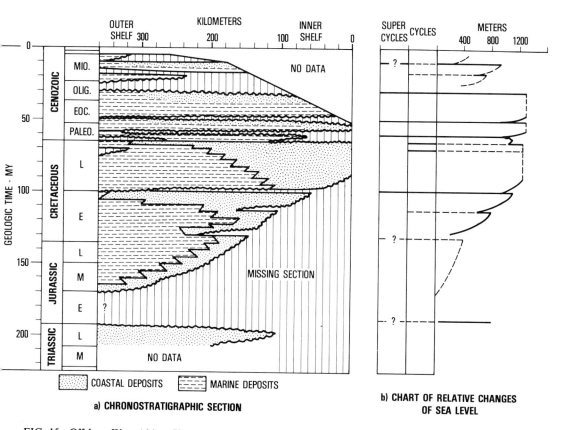

FIG. 15—Offshore West Africa. Chronostratigraphic section (**a**) and chart of relative changes of sea level (**b**).

before the rise, would make the subsequent rise greater than than the fall. The estimates of the relative falls are shown on Fig. 15b.

Such complications are encountered so often that accurate measurements of relative falls are rarely obtained. Nevertheless, the best approximations are given on the sea level curves so the magnitudes of relative change of sea level can be expressed to some degree in quantitative terms. Additional data and further research on this problem may yield more accurate results.

The combination of the chronostratigraphic section (Fig. 15a) and the chart of relative changes of sea level (Fig. 15b) provides a summary of the geologic history of the area based on the seismic section (Fig. 14).

REFERENCES CITED

Campbell, C. V., 1977, Depositional model of beach shoreline—Gallup Sandstone (Upper Cretaceous), northwestern New Mexico: New Mexico Bur. Mines and Mineral Resources Circular (in press).

Curray, J. R., 1964, Transgressions and regressions, *in* R. L., Miller, ed., Papers in marine geology (Shepard commemorative vol.): New York, Macmillan Co., p. 175-203.

Frazier, D. E., 1974, Depositional episodes—their relationship to the Quaternary stratigraphic framework in the northwestern portion of the Gulf basin: Texas Univ. Bur. Econ. Geology Geol. Circ. 74-1, 28 p.

Grabau, A. W., 1924, Principles of stratigraphy: New York, D. G. Seiler, 1,185 p.

Pitman, W. C., III, 1977, Relationship between sea level changes and stratigraphic sequences: Geol. Soc. America Bull., (in press).

van Andel, T. H., and J. R. Curray, 1960, Regional aspects of modern sedimentation in northern Gulf of Mexico and similar basins, and paleogeographic significance, *in* F. P. Shepard, F. B. Phleger, and T. H. van Andel, eds., Recent sediments, northwest Gulf of Mexico: AAPG, p. 345-364.

Weller, J. M., 1960, Stratigraphic principles and practices: New York, Harper and Brothers, 725 p.

Seismic Stratigraphy and Global Changes of Sea Level, Part 4: Global Cycles of Relative Changes of Sea Level. [1]

P. R. VAIL, R. M. MITCHUM, JR.,[2] and S. THOMPSON, III[3]

Abstract Cycles of relative change of sea level on a global scale are evident throughout Phanerozoic time. The evidence is based on the facts that many regional cycles determined on different continental margins are simultaneous, and that the relative magnitudes of the changes generally are similar. Because global cycles are records of geotectonic, glacial, and other large-scale processes, they reflect major events of Phanerozoic history.

A global cycle of relative change of sea level is an interval of geologic time during which a relative rise and fall of mean sea level takes place on a global scale. A global cycle may be determined from a modal average of correlative regional cycles derived from seismic stratigraphic studies.

On a global cycle curve for Phanerozoic time, three major orders of cycles are superimposed on the sea-level curve. Cycles of first, second, and third order have durations of 200 to 300 million, 10 to 80 million, and 1 to 10 million years, respectively. Two cycles of the first order, over 14 of the second order, and approximately 80 of the third order are present in the Phanerozoic, not counting late Paleozoic cyclothems. Third-order cycles for the pre-Jurassic and Cretaceous are not shown. Sea-level changes from Cambrian through Early Triassic are not as well documented globally as are those from Late Triassic through Holocene.

Relative changes of sea level from Late Triassic to the present are reasonably well documented with respect to the ages, durations, and relative amplitudes of the second- and third-order cycles, but the amplitudes of the eustatic changes of sea level are only approximations. Our best estimate is that sea level reached a high point near the end of the Campanian (Late Cretaceous) about 350 m above present sea level, and had low points during the Early Jurassic, middle Oligocene, and late Miocene about 150, 250, and 200 m, respectively, below present sea level.

Interregional unconformities are related to cycles of global highstands and lowstands of sea level, as are the facies and general patterns of distribution of many depositional sequences. Geotectonic and glacial phenomena are the most likely causes of the sea-level cycles.

Major applications of the global cycle chart include (1) improved stratigraphic and structural analyses within a basin, (2) estimation of the geologic age of strata prior to drilling, and (3) development of a global system of geochronology.

INTRODUCTION

Cycles of relative change of sea level on a global scale are evident throughout Phanerozoic time. The evidence is based on the fact that many regional cycles determined on different continental margins are simultaneous and that the relative magnitudes of the changes generally are similar. Concepts and methods of determination of rela-tive changes of sea level and regional cycles were given previously (Part 3, Vail et al, this volume). In this paper are presented charts of global cycles, the methods for constructing the charts from a modal average of correlative regional cycles based on seismic stratigraphy, and our estimates of the actual magnitudes of the sea-level changes.

Because the global cycles are records of geotectonic, glacial, and other large-scale processes, they reflect major events of Phanerozoic history. The timing and relative importance of these events are indicated by charts of the cycles. Such a composite record offers a means of subdividing Phanerozoic time into significant geochronologic units based on a single criterion.

Fairbridge (1961) summarized the historical development of concepts of sea-level change on a global scale, including the classic works of Haug (1900), Suess (1906), Stille (1924), Grabau (1940), Umbgrove (1942), Kuenen (1940, 1954, 1955), Arkell (1956), and others. These pioneer investigations laid the foundation for later work including ours. However, some developments have confused "transgressions and regressions" of the shoreline with "rises and falls" of sea level. Grabau (1924) recognized this problem. The charts we present in this paper show relative and eustatic rises and falls of sea level on a global scale, and differ from charts that show transgressions and regressions of the shoreline.

GLOBAL CYCLES

Figures 1 through 3 are charts of relative changes of sea level on a global scale. The vertical axis of each chart is scaled in millions of years (Ma, after Van Hinte, 1976 a, b), with standard periods and epochs plotted alongside. The horizontal axis shows relative positions of sea level and is scaled from 1.0 to 0.0, with 1.0 being the maximum relative highstand (65 Ma) and 0.0 being the minimum relative lowstand (30 Ma). Relative rises of sea level are plotted toward the left, and relative falls toward the right. The present position of sea level is extended through

[1]Manuscript received, January 6, 1977; accepted, June 13, 1977.

[2]Exxon Production Research Co., Houston, Texas 77001.

[3]New Mexico Bureau of Mines and Mineral Resources, Socorro, New Mexico 87801.

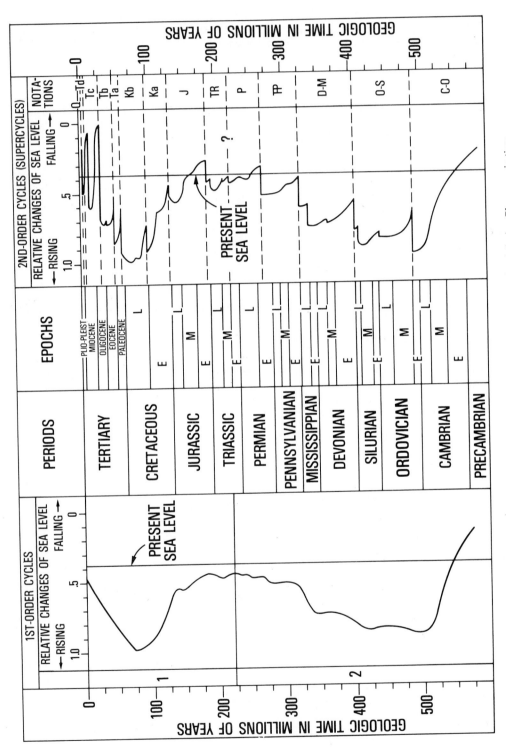

FIG. 1—First- and second-order global cycles of relative change of sea level during Phanerozoic time.

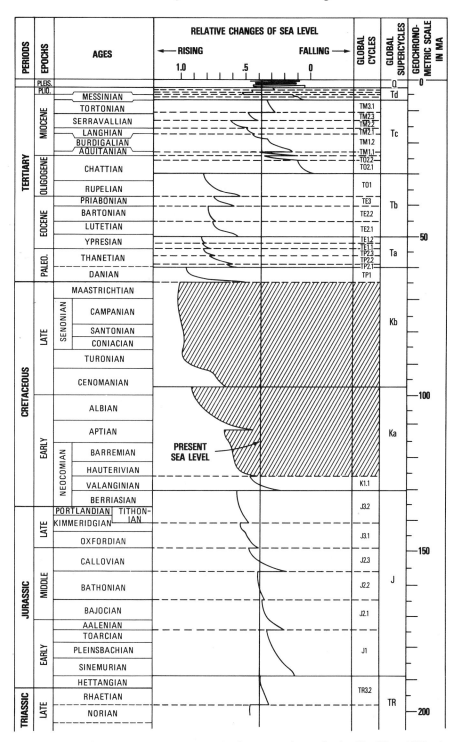

Jurassic-Cretaceous time scale after Van Hinte 1976 a, b

FIG. 2—Global cycles of relative change of sea level during Jurassic-Tertiary time. Cretaceous cycles (hatchured area) have not been released for publication.

Phanerozoic time as a vertical reference line, although it rarely can be related to measurement of ancient changes of sea level.

A *global cycle* of relative change of sea level is an interval of geologic time during which a relative rise and fall of sea level takes place on a global scale. Intermittent stillstands (and therefore paracycles) may occur in any part of the cycle, but tend to predominate after the major part of the rise has taken place and before the fall begins.

The global cycle charts (Figs. 1-3) show cycles of three orders of magnitude. The older of the two first-order cycles (Fig. 1) occured from Precambrian to Early Triassic, with a duration of over 300 million years; the younger first-order cycle occurred from middle Triassic to the present within a duration of about 225 million years. The durations of the 14 second-order cycles (Fig. 1) range from 10 to 80 million years. Over 80 third-order cycles (Figs. 2, 3), not including late Paleozoic cyclothems, have durations that range from approximately 1 to 10 million years. In this paper we do not show the third-order cycles for the pre-Jurassic and Cretaceous periods. Pre-Jurassic cycles are not included because documentation comes mainly from North America with only limited data from other continents. Cretaceous cycles have not been released for publication.

Trends of rise and fall of sea level reveal a marked asymmetry in the second- and third-order cycles. The relative rise generally is gradual and the relative fall generally is abrupt. In the first-order cycles, the cumulative falls tend to be more gradual and the curves are relatively symmetrical. Although the ages and durations of the first-, second-, and third-order cycles are fairly accurate, the amplitudes of the relative changes are only approximations.

Figure 1 shows first- and second-order cycles of the entire Phanerozoic. No distinct boundary occurs between the two first-order cycles, but the best dividing point appears to be between Early and Middle Triassic.

Figure 2 is a chart of relative changes of sea level on a global scale during Jurassic-Triassic time. The second- and third-order cycles are more clearly shown at this expanded time scale. The horizontal scale is the same as that of Figure 1 (see Part 7, Vail et al, this volume).

Figure 3 is a chart of the second- and third-order cycles during Tertiary and Quaternary time. A scale expanded from that of Figure 2 is needed to show the third-order cycles during these times; however, even at this scale all the cycles of glacial events in late Quaternary time are not clearly shown. The ages and durations of the Cenozoic third-order cycles have the best documentation, based mainly on zones of planktonic

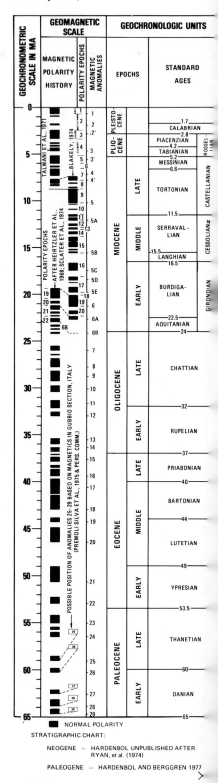

■ NORMAL POLARITY

STRATIGRAPHIC CHART:

NEOGENE — HARDENBOL UNPUBLISHED AFTER RYAN, et al. (1974)

PALEOGENE — HARDENBOL AND BERGGREN 1977

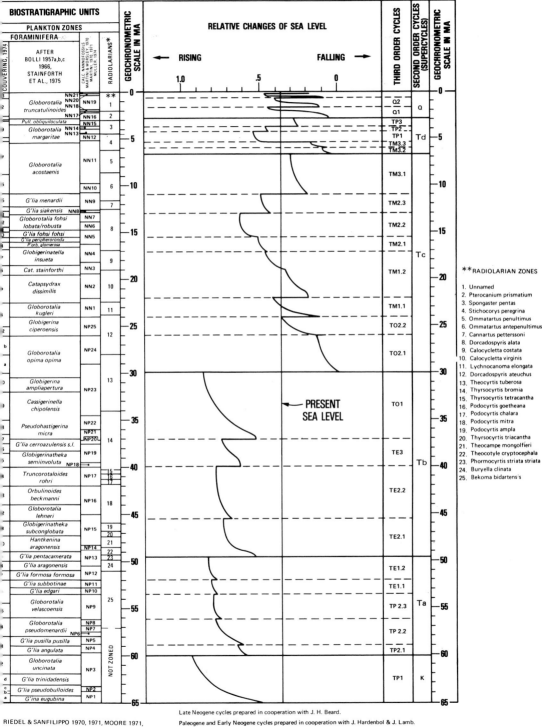

Late Neogene cycles prepared in cooperation with J. H. Beard.

RIEDEL & SANFILIPPO 1970, 1971, MOORE 1971,   Paleogene and Early Neogene cycles prepared in cooperation with J. Hardenbol & J. Lamb.
FOREMAN 1973, THEYER & HAMMOND 1974

Notations: (e.g. $TE^{1.1}$) may be used to designate sequences and supersequences on seismic sections, etc.

G. 3—Global cycles of relative change of sea level during Cenozoic time. Basic references for the stratigraphic part of the chart are Hardenbol (unpublished after Ryan et al, 1974) and Hardenbol and Berggren (1977).

FIG. 4—Location of regional studies of seismic stratigraphy used in construction of global cycle chart for Phanerozoic time.

Foraminifera that have been tied to the geochronometric scale. Amplitudes of the relative changes are determined mainly from seismic stratigraphic analysis of grids of seismic sections tied to well data.

On the right side of the global cycle charts (Figs. 1-3), are columns containing notations that are useful for identifying stratigraphic units on seismic and stratigraphic sections. We identified depositional sequences according to their ages and their relations with cycles of relative changes of sea level. For example, the supersequence corresponding to the Jurassic supercycle is identified as J. It in turn is subdivided into J1, J2, or J3 corresponding to the Early, Middle, or Late Jurassic epoch in which the sequence occurs. Where more than one cycle of sea-level change occurs within a given epoch, such as the Middle Jurassic (J2), we used the notation J2.1 and J2.2, depending on the number of cycles. If we wanted to identify more than one sequence within a single cycle, we used the notation J2.1A, J2.1B, etc.

It also is important to identify sequence boundaries, especially those that are unconformities. This is because a given unconformity may truncate strata with a wide range of ages and also may be onlapped by strata of different ages. We identified an unconformity as pre- the oldest overlying sequence at the point where the surface

approaches conformity. For example, an unconformity that becomes conformable at the base of the Callovian would be identified as pre-Callovian (pre-J2.3).

CONSTRUCTION OF GLOBAL CYCLE CHART

Figure 4 is a world map that shows the general locations of the regional grids of seismic data used in the construction of the global cycle charts (Figs. 1-3). The seismic sections, supplemented by well control and other geologic data, provided the regional stratigraphic framework needed to measure and date the relative changes of sea level. These studies of seismic stratigraphy are drawn from all continents except Antarctica, and provide a representative worldwide sample for Jurassic and younger cycles. However, data are concentrated in areas of petroleum exploration where sedimentary sections are relatively thick and subsidence rates are high, with resultant higher rates of relative rise than would be recognized in thinner sections. Pre-Jurassic cycles are determined primarily on the evidence from North America, with supporting data from other continents.

The global cycle chart is simply a modal average of correlative regional cycles from many areas around the world. The construction of regional cycle charts is explained in Part 3 (Vail et al, this volume). A global cycle can be approximated with a modal average of three or more correlative regional cycles from different continents. The more continents represented, the greater the accuracy of the resulting chart. Some obviously provincial effects such as orogenic deformation, excessive tectonic subsidence, or excessive sedimentary loading may distort the average amplitudes of sea level changes and should not be used without adjustment. Ages of the cycles in these regions generally correspond to the global pattern.

Figure 5 is an example of how four regional cycles are correlated and averaged to construct global cycles (column 5). In this example Cenozoic cycles of specific regions on four continents are used. Although many regional differences in the four curves are obvious, ages and durations of the cycles generally are correlative and amplitudes of the relative changes generally are similar. The global cycle curve is not based on these four regional curves alone, but from those of other areas shown in Figure 4.

The main bases of comparison are the ages of the major relative falls of sea level. For example, the pre-late Miocene fall (10.8 Ma) occurs on all three charts where data are present and is indicated by a major unconformity in all these regions. A pre-middle late Oligocene fall (30 Ma) can be recognized on all charts. The latest early Eocene fall (49 Ma) and the mid-Paleocene fall (60 Ma) are recognized in all three regions where Eocene and Paleocene deposits are found. After major falls are correlated and charted, ages of individual cycles are charted. These are quite similar in all four areas where data are available. Some differences in age relations may be explained by local differences in paleontologic techniques of age dating. Some cycles are not recognized with seismic data because the stratigraphic section is too thin, such as within the Tertiary of northwestern Africa.

Determination of average amplitudes of cycles is the least quantitative step in this procedure. With the exception of the Gippsland basin, shapes of the curves relating to the first- and second-order cycles generally show an overall fall with large fluctuations at supercycle boundaries. The Gippsland basin curve shows an anomalous overall rise probably related to the geotectonic history of Australia. Amplitudes of third-order cycles are charted, giving greatest weight to regions where most complete data are available. Although the first-order cycle of the Gippsland basin is anomalous, the third-order cycles fit the global pattern well.

Accuracy of the global cycle chart depends on the quality and quantity of the regional charts

that are used to construct it. On the charts in this report (Figs. 1-3), the ages, durations, and relative amplitude of the global cycles represent a relatively high level of accuracy, but measurement of the actual amplitudes is still a major problem. Direct measurement is made difficult by: (1) a sometimes wide difference range in thickness of coastal onlap for the same global cycle from various regions; (2) practical difficulties in making complete regional onlap analyses such as lack of control, erosion of critical portions of coastal onlap, or structural complications; (3) necessity of inferring coastal onlap from other facies relations where onlap for portions of the curve can not be measured directly; and (4) difficulties in measuring relative falls of sea level from seaward shifts in coastal onlap. For these reasons, the horizontal scale on the global charts showing amplitude of relative rises and falls is not calibrated in meters, but with a relative scale normalized on the maximum range of sea level positions of the curve. The highest position of sea level, occurring at the end of the Cretaceous (65 Ma), is set at 1.0 and the lowest position, at the mid-Oligocene (30 Ma), is set at 0.0. In the example cited (Fig. 5), each regional curve has been normalized according to this pattern. Where the regional curves do not include the Late Cretaceous high or the Oligocene low, they are normalized by making the best fit of the overlapping portions of the curves. If a given Upper Cretaceous or mid-Oligocene regional cycle has an anomalous magnitude, the regional curve is normalized by making a best fit to the other curves with a less distorted portion of the curve.

ESTIMATION OF EUSTATIC CHANGES OF SEA LEVEL

Global cycle curves (Figs. 1-3) summarize relative changes of sea level as described above. However, these curves include large-scale subsidence that must be discounted to determine the eustatic curve. An estimate of the true eustatic curve (Fig. 6c) has been made for Jurassic to Holocene time by calibrating the global cycle curve (Fig. 2) with the work of Pitman (in press), Sleep (1976), and J. H. Beard (personal commun.). These authors calculated quantitative values for the position of sea level during parts of Cretaceous, Tertiary, and late Neogene time. Their results conform to our preliminary estimates of eustatic change.

Pitman (in press) presented a curve of sea level change from Late Cretaceous to late Miocene time (Fig. 6a). His calculations, based only on rates of seafloor spreading and resultant volumes of midocean ridges, show a cumulative fall from

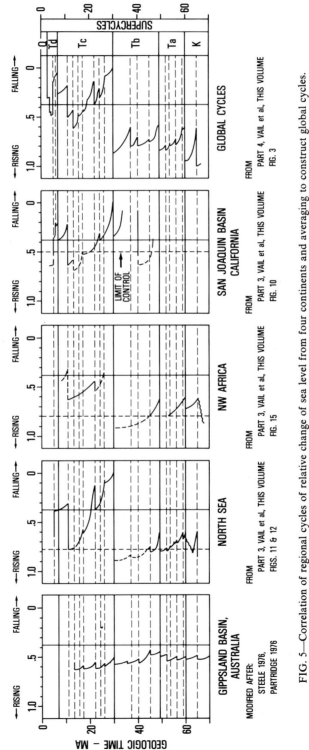

FIG. 5—Correlation of regional cycles of relative change of sea level from four continents and averaging to construct global cycles.

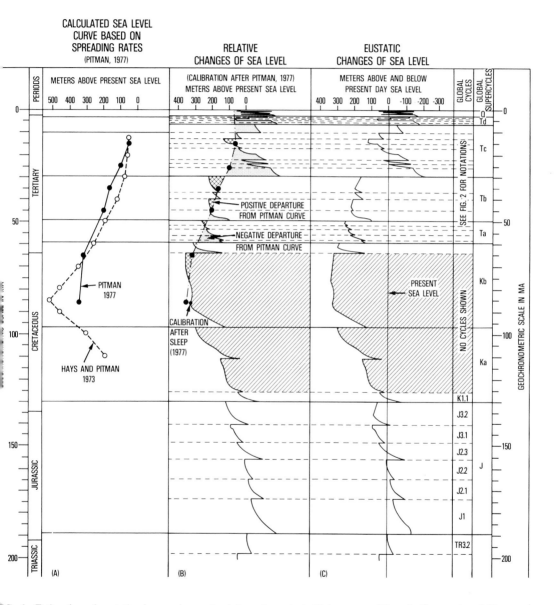

FIG. 6—Estimation of eustatic changes in sea level from Jurassic to Holocene: **a.** Pitman's (in press) and Hays and Pitman's (1973) calculated sea-level curves based on rates of seafloor spreading and volumes of midocean ridges; **b.** Pitman's (in press) curve from (a.) overlain on global curve of relative changes of sea level; and **c.** best estimate of eustatic changes of sea level, calibrated from Pitman's (in press) curve.

350 to 60 m above the present position of sea level; he explained the remaining 60 m as water retained in present ice caps. His curve matches closely the Tertiary part of our Triassic-Holocene first-order cycle curve (Fig. 1). Sleep (1976) suggested sea level in the late Turonian to be 300 m above the present, based on work in the Precambrian shield area of Minnesota. J. H. Beard (personal commun.) estimated late Neogene eustatic changes by relating them to known Pleistocene eustatic changes using paleontology and seismic sequence analysis. This work provided a calibration for the youngest part of our chart.

Pitman's curve and our first-order cycle curve are very similar. However, our second- and third-order cycles show significant departures, both positive (higher) and negative (lower), from Pitman's curve (Fig. 6b). Where our curve is higher than Pitman's, his curve indicates that an overall gradual eustatic fall of sea level may be in progress, but our analysis of onlap indicates a relative rise. This discrepancy may be due to the fact that in our study areas with relatively thick sedimentary columns, subsidence occurs at a greater relative rate than sea level falls, producing a relative rise of sea level (see Part 3, Vail et al, this volume). During such a relative highstand, Pitman's curve (showing a gradual fall) is considered more representative of eustatic change than is ours.

The negative departures of our second- and third-order cycle curves from Pitman's curve are rapid falls of sea level probably caused by glacial withdrawals of water not evaluated by Pitman (see later section on causes of global cycles). Subsequent deglaciation produces a rise in sea level that diminishes in rate until the rises reach the position of the eustatic fall shown on Pitman's curve. As discussed later, glaciation is not documented in the geologic record prior to Oligocene within our last first-order cycle, so additional evidence of glaciation or of some other cause is needed to explain some of the rapid changes, especially the falls, of the second- and third-order cycles in the Early Tertiary and Mesozoic.

Although not covered by Pitman's curve, a general first-order rise from Early Jurassic to Late Cretaceous is indicated from our curve. If glaciation or some other factor produced the rapid second- and third-order falls, the subsequent rises should not cross the first-order curve. Therefore, no gradual eustatic falls are indicated except possibly during the Tithonian and Berriasian.

Our best estimate of true eustatic changes from Jurassic to Holocene is shown on Fig. 6c. Amplitudes of the changes are calibrated in meters with respect to present sea level, based on our curves,

those of Pitman (in press), and data from Sleep (1976) and J. G. Beard (personal commun.). A more accurate curve for the Triassic through Early Cretaceous segment can be made when the first-order rise of sea level for that part of the curve has been calculated from rates of seafloor spreading or from some other long-term cause.

GLOBAL HIGHSTANDS AND LOWSTANDS AND MAJOR INTERREGIONAL UNCONFORMITIES

Table 1 lists the major global highstands and lowstands (or comparative lowstands) of sea level during Phanerozoic time. They are separated by major global falls of sea level. These falls are associated with major interregional unconformities.

A *global highstand* is an interval of geologic time during which the position of sea level is above the shelf edge in most regions of the world. Conversely, a *global lowstand* is an interval during which sea level is below the shelf edge in most regions. A *comparative lowstand* occurs when sea level is at its lowest position on the shelf between periods of highstand.

Global highstands of sea level (Part 3, Vail et al, Fig. 9, this volume) are characterized by widespread shallow marine to nonmarine deposits on the shelves and "starved" basins. If the supply of terrigenous sediment is abundant, delta lobes may prograde over the shelf edge into deep water. Global lowstands are characterized by erosion and nondeposition on the shelves, and deposition of deep marine fans in the basins. After a major fall of sea level to a global lowstand, a major interregional unconformity commonly is developed by subaerial erosion and nondeposition on shelves and basin margins, and by periods of nondeposition or shifts in depositional patterns in the deep-water parts of the basin.

The two first-order global cycles of sea-level change (Fig. 1) may be generally described in terms of global lowstands and highstands. During the older first-order cycle, sea level rose from a lowstand in late Precambrian time to a highstand during a long interval from Late Cambrian to Mississippian, and gradually fell to a lowstand which was at a broad minimum through Permian and Early Triassic time. In the cycle from Triassic to the present, cumulative rises built to a highstand peak in Late Cretaceous followed by cumulative falls to lowstand with many fluctuations.

CAUSES OF GLOBAL CYCLES

According to Fairbridge (1961), a eustatic change of sea level on a global scale may be produced by a change in the volume of sea water, by a change in the shape of the ocean basins, or by a combination of both. A change in the volume of seawater may be produced by glaciation and deglaciation, or by additions of juvenile water from magmatic sources (Rubey, 1951, p. 1137-1138), volcanos, or hot springs (Egyed, 1960). A change in the shape of the ocean basins may be produced by geotectonic mechanisms or sedimentary filling of the basins.

Among these factors, only geotectonic mechanisms appear to be of sufficient duration and magnitude to account for the first-order and most of the second-order cycles. Glaciation and deglaciation probably account for many third-order cycles and some of the second-order cycles, especially those in the late Neogene. Other unidentified causes may produce the rapid changes evident in second- and third-order cycles, or may work in combination with geotectonics and/or glaciation to accentuate or diminish the changes. Pitman (in press) discussed the effect of abrupt changes in rates of seafloor spreading on onlap patterns along continental shelves.

Changes in volume or elevation of midocean ridges, which are related to changes in the rate of

seafloor spreading, appear to produce significant changes in the shape of ocean basins (Hallam, 1963, 1969, 1971; Menard, 1964; Russell, 1968; Wise, 1972, 1974; Flemming and Roberts, 1973; Rona, 1973a, b; Hays and Pitman, 1973; and Rona and Wise, 1974; Pitman, in press). Volumetric changes along subduction zones are more difficult to quantify and evaluate and have not been treated quantitatively to our knowledge. Carey (1976) suggested that earth expansion may be the major cause for the rapid changes.

Pitman (in press) stated that, except for glacial effects, volumetric change in the midocean ridges related to change in rate of seafloor spreading is potentially the fastest and volumetrically the most significant way to change sea level. According to his calculations, sea level has fallen steadily but at varying rates from the Late Cretaceous, due to a contraction in size of oceanic ridges related to decreasing rates of seafloor spreading. At the same time, passive continental margins of the Atlantic and other ocean basins have subsided tectonically at decreasing rates, following a pre-

Table 1. Global Highstands and Lowstands of Sea Level and Associated Major Interregional Unconformities During Phanerozoic Time.

SEA LEVEL HIGHSTANDS	MAJOR GLOBAL SEA LEVEL FALLS	SEA LEVEL LOWSTANDS
	PRE-LATE PLIOCENE & PRE-PLEISTOCENE (3.8 & 2.8 MA)	LATE PLIOCENE - EARLY PLEISTOCENE
EARLY & MIDDLE PLIOCENE		
	PRE-LATE MIOCENE & PRE-MESSINIAN (10.8 & 6.6 MA)	LATE MIOCENE
MIDDLE MIOCENE		
	PRE-MIDDLE LATE OLIGOCENE (30 MA)	MIDDLE LATE OLIGOCENE
LATE MIDDLE EOCENE & EARLY OLIGOCENE		
	PRE-MIDDLE EOCENE (49 MA)	EARLY MIDDLE EOCENE
LATE PALEOCENE - EARLY EOCENE		
	PRE-LATE PALEOCENE (60 MA)	MID-PALEOCENE
CAMPANIAN & TURONIAN		
	PRE-MIDDLE CENOMANIAN (98 MA)	MID-CENOMANIAN
ALBIAN - EARLIEST CENOMANIAN		
	PRE-VALANGINIAN (132 MA)	VALANGINIAN
EARLY KIMMERIDGIAN		
	PRE-SINEMURIAN (190 MA)	SINEMURIAN
NORIAN & MIDDLE GUADALUPIAN		
	PRE-MIDDLE LEONARDIAN (270 MA)	MID-LEONARDIAN
WOLFCAMPIAN & EARLIEST LEONARDIAN		
	PRE-PENNSYLVANIAN (324 MA)	EARLY PENNSYLVANIAN
OSAGIAN & EARLIEST MERAMECIAN		
	PRE-DEVONIAN (406 MA)	EARLY DEVONIAN
MIDDLE SILURIAN		
	PRE-MIDDLE ORDOVICIAN (490 MA)	EARLY MIDDLE ORDOVICIAN
LATE CAMBRIAN & EARLY ORDOVICIAN		
		EARLY CAMBRIAN & LATEST PRECAMBRIAN

dictable thermal cooling curve (Sclater et al, 1971). The general model can be used to explain the cumulative first-order fall since the Late Cretaceous.

There may be a general correspondence of times of orogenic movement and volcanism with times of second-order sea-level highstands (Fig. 1). In general, high rates of seafloor spreading should be associated with relatively shallow ocean floors, highstand flooding of continental and basin margins, and greater subduction. Increased subduction should tend to produce increased volcanism and orogeny from continent-continent collisions. Such orogenic episodes should have fairly long terms of occurrence associated with durations of second-order highstands. However, pronounced angular unconformities associated with rapid falls of sea level can give impressions of short periods of orogeny. If short-term orogenies can be documented, they may be related to third-order cycles.

First-order cycles show a possible overall relationship to patterns of seafloor spreading rates and orogeny. For example, rifting and continental pull-apart dominate times of major sea-level rise in Cambrian and Jurassic–Early Cretaceous, and orogenies dominate intervals of falling sea level within each first-order cycle.

Many workers have noted unconformities which occur simultaneously in many regions of the globe (Stille, 1924; Arkell, 1956; Sloss, 1963, 1972; Gussow, 1963; Vail and Wilbur, 1966; Moore et al, 1974; and Dennison and Head, 1975). Most of these unconformities are coincident with major relative falls of sea level on our charts (Figs. 1-3). The unconformities that correspond to major sea-level falls at the end of the second-order cycles are shown on Table 1.

We concur with Sleep (1976) that unconformities caused by major eustatic changes of sea level modify the history of subsidence within cratonic basins and continental margins. Where thermal contraction of the lithosphere controls basin subsidence, basins should continue to subside even during times of eustatic lowstands. Sleep calculated that significant unconformities in the geologic record could be produced by eustatic falls of sea level, with amplitudes of those presented on Figure 6c, even in rapidly subsiding basins. Sleep's work supports our contention that interregional unconformities are not primarily due to uplift of the continental interior or continental margins, but are primarily caused by erosion or non-deposition during eustatic changes in sea level.

Glaciation and deglaciation are the only well understood causal mechanisms that occur at the relatively rapid rates of third-order cycles (Pit-

man, in press). Rates of geotectonic mechanisms related to seafloor spreading are too slow. Glaciation has been documented in the Pleistocene, late and early Miocene, and to some degree in the late Oligocene, but there is no evidence of glaciation at the times of many other lowstands. Other evidence for climatic changes, such as from oxygen isotopes (Savin, in press; Savin and Douglas, in press; Fischer and Arthur, in press) and other faunal studies (Haq and Lohmann, 1976) show that lowstands generally represent climatically cool conditions, and highstands represent climatically warm conditions.

Other evidences of cyclicity which correlate in general with global cycles include frequency of unconformities in deep sea cores and cycles of faunal diversity (Fischer and Arthur, in press), and changes of calcite compensation depth (van Andel, 1975, in press).

In summary, the cause for the first-order and some second-order cycles may be related to geotectonic mechanisms. Some of the second- and third-order cycles can be explained by glaciation. The empirically observed rapid falls of sea level at the ends of the third-order cycles remain unexplained where evidence for glaciation is not known.

APPLICATIONS

Major applications of global cycles fall into three categories: (1) improved stratigraphic and structural analyses incorporating the effects of sea-level changes, (2) estimation of geologic age ahead of the drill, and (3) development of a global system of geochronology.

In regional stratigraphic studies, after analysis of seismic sequences and regional sea-level changes is begun (Parts 2, 3, 7, Vail et al, this volume) comparison of regional and global sea level curves can aid prediction of age of sequences for which control is lacking and fill gaps in regional sea-level curves. Correlation of regional curves with times of unconformities, lowstands, and highstands on global curves aids prediction of depositional facies and distribution of sequences (Parts 3, 9, 10, Vail et al, this volume). Moreover, departures of the regional curve from the global curve indicate anomalous regional effects such as tectonic subsidence or uplift.

Estimation of geologic age of strata prior to drilling is a seismic-stratigraphic technique commonly applied in areas of sparse or no well control. Where wells are present and biostratigraphic zones are determined, they can be tied to seismic sequences for accurate age dating throughout the area of the seismic grid. If there is no well control within the grid, geologic age can be inferred by

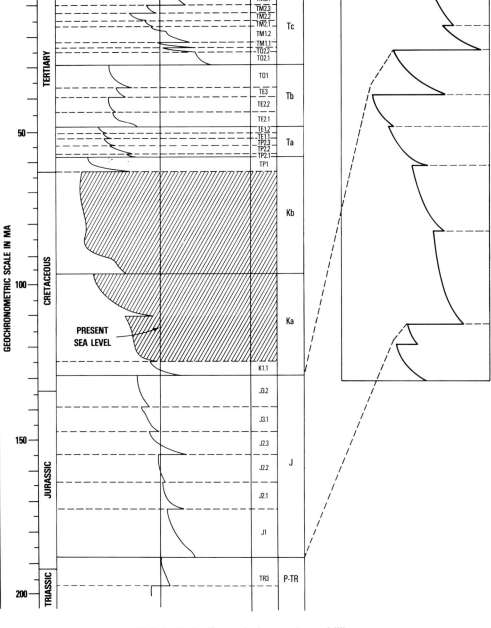

FIG. 7—Estimating geologic age prior to drilling.

building a regional chart of relative changes of sea level from seismic data and matching it with the global chart (Fig. 7). Accuracy can be improved with information from outcrops or distant wells that help to establish the general age of strata that are known to be present in the basin.

One of the greatest potential applications of the global cycle chart is its use as an instrument of geochronology. Global cycles are geochronologic units defined by a single criterion—the global change in the relative position of sea level through time. Determination of these cycles is dependent on a synthesis of data from many branches of geology. As seen on the Phanerozoic chart (Fig. 1), the boundaries of the global cycles in several cases do not match the standard epoch and period boundaries, but several of the standard boundaries have been placed arbitrarily and remain controversial. Using global cycles with their natural and significant boundaries, an international system of geochronology can be developed on a rational basis. If geologists combine their efforts to prepare more accurate charts of regional cycles, and use them to improve the global chart, it can become a more accurate and meaningful standard for Phanerozoic time.

REFERENCES CITED

Arkell, W. J., 1956, Jurassic geology of the world: London, Oliver and Boyd Ltd., 806 p.

Carey, S. W., 1976, The expanding earth—developments in geotectonics, part 10: Amsterdam, Elsevier, 470 p.

Dennison, J. M., and T. W. Head, 1975, Sea level variations interpreted from the Appalachian basin Silurian and Devonian: Am. Jour. Sci., v. 275, p. 1089-1120.

Egyed, L., 1960, On the origin and constitution of the upper part of the earth's mantle: Geol. Rundschau, v. 50, p. 251-258.

Fairbridge, R. W., 1961, Eustatic changes in sea level, in L. H. Ahrens, et al, eds., Physics and chemistry of the earth: London, Pergamon Press, v. 4, p. 99-185.

Fischer, A. G., and M. A. Arthur, (in press), Secular variations in the pelagic realm, in H. E. Cook, and P. Enos, eds., Basinal carbonate sediments: SEPM Spec. Pub. no. 25.

Flemming, N. C., and D. G. Roberts, 1973, Tectono-eustatic changes in sea level and sea floor spreading: Nature, v. 243, p. 19-22.

Foreman, H. P., 1973, Radiolaria of Leg 10 with systematics and ranges for the families *Amphipyndacidae, Artostrobiidae* and *Theoperidae*: deep sea drilling project, leg 10, in Initial reports of the DSDP, v. 10: Washington (U.S. Govt. Printing Office) p. 407-474.

Grabau, A. W., 1924, Principles of stratigraphy: New York, A. G. Seiler, 1,185 p.

——— 1940, The rhythm of the ages: Peking, Henri Vetch Pub., 56 p.

Gussow, W. C., 1963, Metastacy, in D. C. Mungan, ed., Polar wandering and continental drift: SEPM Spec. Pub. 10, p. 146-169.

Hallam, A., 1963, Major epeirogenic and eustatic changes since the Cretaceous and their possible relationship to crustal structure: Am. Jour. Sci., v. 261, p. 397-423.

——— 1969, Tectonism and eustacy in the Jurassic: Earth-Sci. Rev., v. 5, p. 45-68.

——— 1971, Mesozoic geology and the opening of the North Atlantic: Jour. Geology, v. 79, p. 129-157.

Haq, B. U., and G. P. Lohmann, 1976, Early Cenozoic calcareous nannoplankton biogeography of the Atlantic Ocean: Marine Micropaleontology, v. 1, no. 2, p. 119-194.

Hardenbol, J., and W. A. Berggren, (1977), A new Paleogene numerical time scale: AAPG Studies Geology, no. 6, in press.

Haug, E., 1900, Les geosynclinaux et les aires continentales: Soc. Geol. France Bull., Ser. 3, v. 28, p. 617-711.

Hays, J. D., and W. C. Pitman, 1973, Lithospheric plate motion, sea level changes, and climatic and ecological consequences: Nature, v. 246, p. 18-22.

Kuenen, Ph. H., 1940, Causes of eustatic movements: 6th Pacific Sci. Cong., Proc., v. 2, p. 833-837, Berkeley, Univ. Calif. Press.

——— 1954, Eustatic changes of sea-level: Geologie en Mijnbouw, v. 16, p. 148-155.

———1955, Sea level and crustal warping; crust of the earth—a symposium: Geol. Soc. America, v. 62, p. 193-204.

Menard, H. W., 1964, Marine geology of the Pacific: New York, McGraw-Hill, 271 p.

Moore, T., 1971, Radiolaria, deep sea drilling project, leg 8, in Initial reports of the DSDP, v. 8: Washington (U. S. Govt. Printing Office) p. 727-748.

Moore, T. C., Jr., et al, 1974, Cenozoic hiatuses in pelagic sediments, in E. Siebold and W. R. Riedel, eds, Marine plankton and sediments; 3rd Plankton Conference (Kiel) Proc.

Partridge, A. D., 1976, The geologic expression of eustacy in the early Tertiary of the Gippsland basin: APEA Jour., v. 16, p. 73-79.

Pitman, W. C., (in press), Relationship between sea level change and stratigraphic sequences: Geol. Soc. America Bull.

Riedel, W. R., and A. Sanfilippo, 1970, Radiolaria, deep sea drilling project leg, Leg 4, in Initial reports of the DSDP, v. 4: Washington (U.S. Govt. Printing Office) p. 503-575.

———1971, Cenozoic radiolaria from the western tropical Pacific, deep sea drilling project, leg 7, in Initial reports of the DSDP, v. 7: Washington (U.S. Govt. Printing Office) p. 1,529-1,672.

Rona, Peter A., 1973a, Relations between rates of sediment accumulation on continental shelves, sea-floor spreading, and eustacy inferred from the central North Atlantic: Geol. Soc. America Bull., v. 84, p. 2851-2872.

———1973b, Worldwide unconformities in marine sediments related to eustatic changes of sea level: Nature Phys.-Sci., v. 244, p. 25-26.

—— and D. U. Wise, 1974, Symposium: global sea level and plate tectonics through time: Geology, v. 2, p. 133-134.

Rubey, W. W., 1951, Geologic history of sea water: Geol. Soc. America Bull., v. 62, p. 1111-1147.

Russell, K. L., 1968, Oceanic ridges and eustatic changes in sea level: Nature, v. 218, p. 861-862.

Savin, S. M., (in press), The history of the earth's surface temperature during the past hundred million years: Annu. Rev. Earth and Planetary Sci., v. 5.

—— and R. Douglas, (in press), Changes in bottom-water temperatures in the Tertiary and its implications: Geology.

Sclater, J. G., R. N. Anderson, and M. L. Bell, 1971, Elevation of ridges and evolution of the central eastern Pacific: Jour. Geophys. Research, v. 76, p. 7888-7916.

Sleep, N. H., 1976, Platform subsidence mechanisms and "eustatic" sea-level changes: Tectonophysics, v. 36, p. 45-56.

Sloss, L. L., 1963, Sequences in the cratonic interior of North America: Geol. Soc. America Bull, v. 74, p. 93-113.

——1972, Synchrony of Phanerozoic sedimentary-tectonic events of the North American craton and the Russian platform. Sect. 6, in Stratigraphie et sedimentologie: 24th Int. Geol. Cong. (Montreal).

Stille, H., 1924 Grundfragen der vergleichenden Tektonik: Berlin, Borntraeger.

Suess, E., 1906, The face of the earth: Oxford, Clarendon Press, v. 2, 556 p.

Theyer, F., and S. R. Hammond, 1974a, Paleomagnetic polarity sequence and radiolarion zones, Brunhes to polarity Epoch 20: Earth and Planetary Sci. Letters, v. 22, p. 307-319.

—— 1974b, Cenozoic magnetic time scale in deep-sea cores—completion of the Neogene: Geology, v. 2, no. 10, p. 487-492.

Umbgrove, J. H. F., 1942, The pulse of the earth: The Hague, Nijhoff 179 p.

Vail, P. R., and R. O. Wilbur, 1966, Onlap, key to worldwide unconformities and depositional cycles (abs.): AAPG Bull., v. 50, p. 638.

van Andel, Tj. H., 1975, Mesozoic-Cenozoic calcite compensation depth and the global distribution of carbonate sediments: Earth and Planetary Sci. Letters, v. 26, p. 187-194.

—— (in press), An eclectic overview of plate tectonics, paleogeography, and paleoceanography, in Historical biogeography, plate tectonics and changing environment: 37th Biology Colloquium, Oregon State Univ. Press.

Van Hinte, J. E., 1976a, A Jurassic time scale: AAPG Bull., v. 60, p. 489-497.

—— 1976b, A Cretaceous time scale: AAPG Bull., v. 60, p. 498-516.

Wise, D. U., 1972, Freeboard of continents through time: Geol. Soc. America Mem. 132, p. 87-100.

—— 1974, Continental margins; freeboard and volumes of continents and oceans through time, in C. A. Burke and C. L. Drake, eds., The geology of continental margins: New York, Springer-Verlag, p. 45-58.

Seismic Stratigraphy and Global Changes of Sea Level, Part 5: Chronostratigraphic Significance of Seismic Reflections [1]

P. R. VAIL, R. G. TODD,[2] and J. B. SANGREE[3]

Abstract Primary seismic reflections follow chrono-stratigraphic (time-stratigraphic) correlation patterns rather than time-transgressive lithostratigraphic (rock-stratigraphic) units. Physical surfaces that cause seismic reflections are primarily stratal surfaces and unconformities with velocity-density contrasts. Stratal surfaces are major bedding surfaces and thus represent ancient depositional surfaces. Unconformities are surfaces of erosion or nondeposition that represent significant chronostratigraphic gaps.

Both stratal surfaces and unconformities have time significance because of the Law of Superposition. In terms of geologic time, reflections from stratal surfaces approximate time-synchronous events, while reflections from unconformities are commonly time-variable. However, unconformity reflections are time-significant because all the strata below the unconformity are older than all the strata above the unconformity.

No physical surface that could generate a reflection parallel with the top of a time-transgressive formation (lithostratigraphic unit) exists in nature. The continuity of the seismic reflection follows the stratal surfaces across the time-transgressive formation boundaries, although reflection character (amplitude, cycle breadth, and waveform) will change as the reflection coefficients and spacing of the stratal surfaces change laterally. Commonly, a given seismic reflection character transgresses reflection continuity as a rock formation transgresses geologic time.

Geologic time correlations made from paleontologic data tie with seismic-reflection correlations even where the latter cross major facies boundaries. In addition, unconformities or their correlative conformities that bound sequences, also commonly bound paleontologic zones especially in the Paleozoic and Mesozoic.

Understanding the chronostratigraphic significance of seismic-reflection correlations and relating them to available well control is essential for stratigraphic trap exploration.

There also are other types of continuous physical surfaces that are locally present in sedimentary rocks. Most significant of these are gas-water, gas-oil, and oil-water fluid contacts and gas hydrate zones. Reflections from these types of physical surfaces cut across the reflections originating from the stratal surfaces, if they are at an angle to each other.

Chronostratigraphic correlations of seismic data with well data are accurate only to ± ½ cycle breadth owing to possible changes of reflection character caused by changes in bed spacing and reflection coefficients.

INTRODUCTION

An understanding of the relation of seismic reflections to lithologic units and geologic time is fundamental to interpreting stratigraphy from seismic data. This report demonstrates that primary reflections on a seismic section show chronostratigraphic correlation patterns rather than the gross lithostratigraphic units. Physical surfaces in the rocks that produce reflections are primarily stratal surfaces and unconformities with velocity-density contrasts. Both types of surfaces have chronostratigraphic significance in the sense that all rocks above the surface are younger than those below. Conversely, there is no continuous physical surface that follows the top of a time-transgressive lithologic unit.

Because seismic reflections approximate chronostratigraphic correlations, it is possible to make the following types of stratigraphic interpretations from the geometry of seismic reflection patterns: (1) postdepositional structural deformation and thickness changes, (2) geologic time correlations, (3) definition of genetic depositional units, (4) depositional environment as revealed by depositional topography, (5) paleobathymetry, (6) burial history, (7) relief and topography on unconformities, and (8) paleogeography and geologic history. However, one limiting factor is that rock type can not be determined directly from reflection correlation patterns.

RELATIONS OF PHYSICAL SURFACES TO SEISMIC REFLECTIONS

Two types of physical surfaces are present in sediments at the time of deposition—stratal surfaces and unconformities. Each type causes seismic reflections if sufficient velocity or density contrast occurs.

Stratal surfaces are the bedding surfaces (Gary et al, 1974) that separate the principal sedimentary strata. They represent periods of nondeposition or change in the depositional sequence. Within a given interval, bedding surfaces may extend from a region where they separate recognizably different strata to a region where the bordering strata are the same type of rock. Thus, at any particular point, bedding surfaces may or may not be readily recognizable (Campbell, 1967). Only those stratal surfaces that have velocity or density contrasts across them will influence the seismic reflections. In most sedimentary sections there are ample variations in velocity and

[1]Manuscript received, January 6, 1977; accepted, June 13, 1977.

[2]Exxon Production Research Co., Houston, Texas 77001.

[3]Esso Exploration, Inc., Walton-on-Thames, Surrey, England, KT12 2QL.

density to produce seismic reflections even where reflections may be low in amplitude. Seismic automatic gain control (AGC) commonly makes these low-amplitude reflections visible on a seismic section.

Unconformities are surfaces of erosion or nondeposition that separate younger strata from older rocks, and represent a significant geologic time gap (Gary et al, 1974). Unconformity surfaces occur commonly at an angle to the underlying or overlying stratal surfaces. They may have either erosional truncation below the unconformity or onlap or downlap above the unconformity, or combinations of the three patterns (Part 2, Vail et al, this volume). Seismic, angular unconformities are expressed in three ways. Lacking significant velocity-density contrast across the unconformity, no unconformity reflection will be generated. However, the unconformity can be located on seismic data by the discordance between the underlying truncated reflections and the overlying onlapping or downlapping reflections. If there is a significant velocity-density contrast across the unconformity, the unconformity will appear as either a *continuous* or a *discontinuous* reflection. The unconformity reflection will be continuous if the reflection coefficient produced by the strata above and below the unconformity is significantly greater than the reflection coefficients of the underlying and overlying stratal surfaces, especially if either the underlying or overlying adjacent strata are concordant. If the underlying and/or overlying stratal surfaces dip with respect to the unconformity and have significant reflection coefficients of approximately the same magnitude as the unconformity surface, the unconformity reflection will go in and out of phase with the reflections from the truncated, onlapping or downlapping stratal surfaces and appear as a discontinuous reflection.

Nonangular unconformities commonly exhibit strong reflections resulting from a characteristic change in velocity or density and constructive interference of reflections from the overlying or underlying parallel strata. A nonangular unconformity surface will not have a reflection associated with it if there is no velocity-density contrast between the overlying and underlying strata. Nonangular unconformities are commonly detected by tracing through a seismic grid from areas where they show angularity. Locally, paleontologic criteria for a chronostratigraphic gap are required to document an unconformity.

CHRONOSTRATIGRAPHIC SIGNIFICANCE OF STRATAL SURFACES AND UNCONFORMITIES

Both stratal surfaces and unconformities have chronostratigraphic significance. Stratal surfaces are geologic-time surfaces because they are former depositional bedding surfaces that were synchronous over the area of their occurrence. We can observe bedding surfaces forming today in beaches, turbidites, varves, ash falls, and fluvial deposits. Comparison of these recent sediments with ancient rocks shows identical bedding features. Therefore, if these surfaces form in hours, days, or even years, they are practical synchronous surfaces representing a point in geologic time. Small-scale patterns of scour, onlap, or downlap commonly exist along many bedding surfaces, indicating variations in time along the bedding surface. However, in terms of geologic time, the chronostratigraphic break is insignificant and the bedding or stratal surface is essentially synchronous.

Unconformity surfaces have chronostratigraphic significance in that all rocks below the unconformity are older than the rocks above the unconformity. Commonly, large-scale onlapping or downlapping strata are present overlying the unconformity surface, or truncated strata are present beneath it. This type of unconformity is *not* time-synchronous, and represents a significant hiatus of variable duration laterally due either to nondeposition or erosion. Thus, most unconformities are time-variable boundaries, but they have chronostratigraphic significance because they separate younger rocks from older rocks.

RELATIONS BETWEEN SEISMIC REFLECTIONS, STRATAL SURFACES, AND LITHOSTRATIGRAPHY

Primary seismic reflections are generated by stratal surfaces which are chronostratigraphic, rather than by boundaries of arbitrarily defined lithostratigraphic units, many of which commonly transgress geologic time. Many lithostratigraphic units cross seismic reflections because they are defined by lithologic content rather than by bedding geometry. In addition, lithostratigraphic units are commonly composited from fragmentary outcrop and/or subsurface data where stratal surfaces can not be traced.

Tertiary Example: South America

Physical Stratigraphy—Five wells are located down the crest of a plunging anticline (Fig. 1). A geologic cross section (Fig. 2), using a regional electric log marker as a datum, illustrates the correlation of physical stratal surfaces from well to well. In correlating the electric logs, a technique of pattern correlation was used which carries a group of stratal interfaces rather than one or two large "kicks" on the electric log. This approach is useful because it stresses the overall stratal geometry of a cross section rather than the position of a few lithostratigraphic "tops" which may be time-transgressive.

Although a number of stratal markers were correlated from well 6 updip to well 1, horizons 15, 10, 8, and an unconformity are specifically identified because they permit a threefold subdivision of the geologic section. The upper interval from 15 to 10 shows similar log configurations for all wells, indicating that the bedding is uniform with no major lateral changes. The interval is all shale in well 6, but becomes slightly more sandy in wells 2 and 1. The middle interval from 10 to 8 also is totally shale in well 6, but becomes increasingly more sandy updip, where well 1 is essentially all sandstone. The lower interval from horizon 8 to the unconformity is dominantly sandstone across the section. It thins appreciably updip because of onlap against the underlying unconformity. It is evident that the sandstone unit identified between horizon 8 and the unconformity in wells 6 and 5 is considerably older than the lower sandstone unit above the unconformity in wells 2 and 1. Therefore, the sandstone in both these units can be characterized collectively as a basal time-transgressive formation.

Velocity Relations—It is necessary to determine lateral velocity variations of the geologic section (Fig. 2) to evaluate the seismic response from the stratigraphy shown. This is accomplished by lateral correlation of the velocities derived from continuous velocity logs (CVLs) for wells 5 and 1, based on detailed E-log correlations (Fig. 3). On Figure 3, correlation lines 15, 10, 8, and the unconformity are indicated along with other stratal surfaces. The diagonal dashed line which extends from the top of the lower sandstone unit in well 5 to the top of the sandstone unit in well 1, represents a lithostratigraphic correlation and clearly crosses the stratal surfaces. There is no physical surface corresponding to the diagonal dashed line.

In general, the velocity derived from CVL data varies with lithologic type. High velocities, indicated by the stipple pattern on Figure 3, range from 3,300 to 4,200 m/sec and correspond to sandstone. Medium velocities range from 2,700 to 3,300 m/sec and are associated with silty beds. Low velocity ranges from 2,200 to 2,700 m/sec, and correlates with shale. A gradual decrease in all the velocities occurs northward, owing to a decrease in depth. The upper interval, between correlation lines 15 and 10, shows slightly higher velocities downdip, probably because of more compaction in a rather uniform shale. The middle interval, between horizons 10 and 8, is characterized by a northward change from shale to sandstone with a corresponding increase in velocity

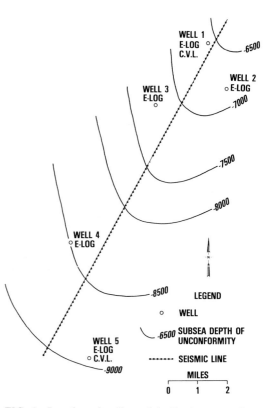

FIG. 1—Location of wells used in Tertiary example, South America.

across the zone of change. The lower interval, between correlation line 8 and the unconformity, is characterized by a high-velocity sandstone that thins and onlaps the unconformity just north of well 3.

Seismic Models— Using the velocity profile developed for Figure 3, a series of synthetic seismic sections was made by convolving the velocity data with an input pulse at 0.6-km intervals across the section. A synthetic seismogram was generated at each position, and, when plotted side by side, the seismograms resemble a conventional wiggly trace seismic section.

Figure 4 is a synthetic seismic section generated by a 90-Hz sine-wave input pulse. The section is a high-frequency presentation and shows the chronostratigraphic relation of the seismic reflections and correlation lines. Correlation lines 15, 10, 8, and the unconformity are marked on the section, as are the locations of wells 5 and 1. In well 6, correlation line 8 corresponds to the top of the sandstone unit. The corresponding reflection pinches out by onlap before it reaches well 1. The top of the sandstone unit present in well 1 corresponds to correlation line 10. This relation indi-

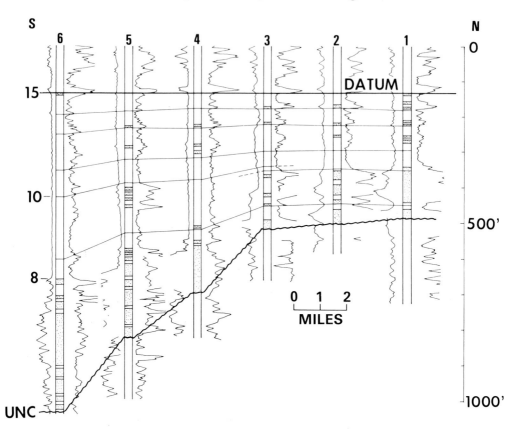

FIG. 2—Geologic cross section showing electric-log correlations, Tertiary example, South America. Stippled pattern in well bore represents sandstone. Key horizons indicated on left.

cates the following: (1) the sandstone unit in well 1 is younger than the sandstone in well 6; (2) reflections generated at the top and within the sandstone unit in well 1 can be traced downdip toward well 6 where they diminish in amplitude in the shale section; and (3) no diagonal reflection (as depicted in Fig. 3) is seen to correspond to the sandstone top. This is because no continuous physical surface exists in nature between the tops of the sandstones in wells 6 and 1, as was diagrammed in Figure 3.

Figure 5 is a synthetic seismic section generated in the same manner as Figure 4, but using a 20-Hz input pulse which is more typical of seismic data in this area. Although resolution is lower, essentially the same observations can be made that were discussed for the synthetic section made with the 90-Hz input pulse. The sandstone unit in well 1 is stratigraphically above, and therefore younger than, the sandstone in well 6.

VDF Seismic Section—Figure 6 is a single-fold variable density film (VDF) magnetic-tape seis-

mic section recorded several years ago. The seismic line is located near wells 1 through 6 (see Fig. 1), and their location is plotted along the top of the section. Plotted near the base of the seismic line is the position of correlation lines 15, 10, 8, and the unconformity. The top of the sandstone unit in wells 5 and 6 equates with horizon 8. The black reflection directly below correlation line 8 terminates updip as a result of onlap between wells 3 and 2. The top of the sandstone unit in well 1 equates with horizon 10. This reflector diminishes in amplitude downdip and is stratigraphically in the shaly section above the sandstone top in wells 5 and 6. As in the synthetic seismic section, no diagonal reflector corresponding to the top of the sandstone is present on the section because it does not exist physically in nature.

Cretaceous Example: San Juan Basin, New Mexico

A geologic section (Fig. 7) and a parallel seismic line displayed in time and depth, respectively

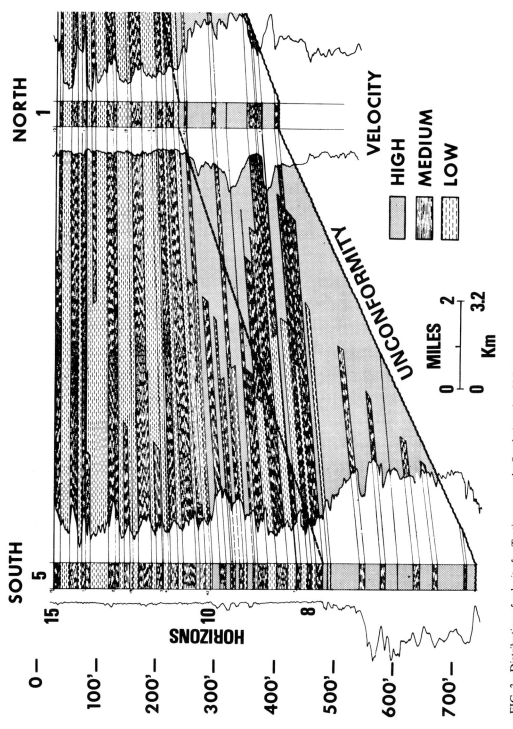

FIG. 3—Distribution of velocity for Tertiary example, South America. Velocity was obtained from CVLs for wells 5 and 1 and distributed laterally using stratal surface correlations.

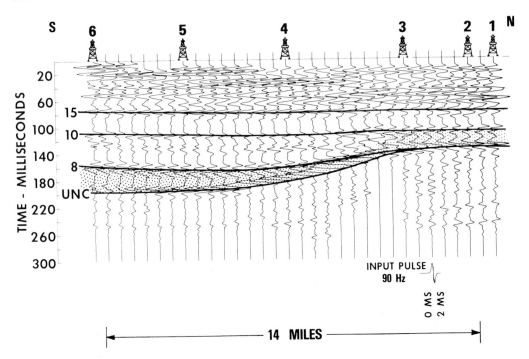

FIG. 4—High frequency synthetic seismic section constructed by inputting 90-Hz sine wave. Stipple pattern represents sandstone.

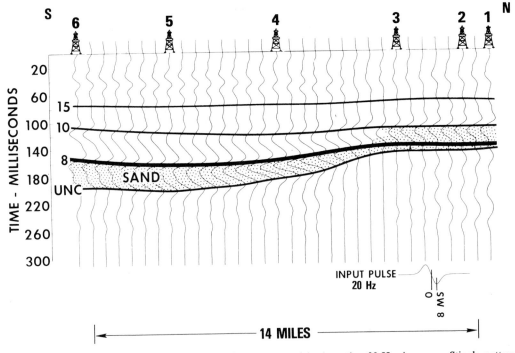

FIG. 5—Normal frequency synthetic seismic section constructed by inputting 20-Hz sine wave. Stipple pattern represents sandstone.

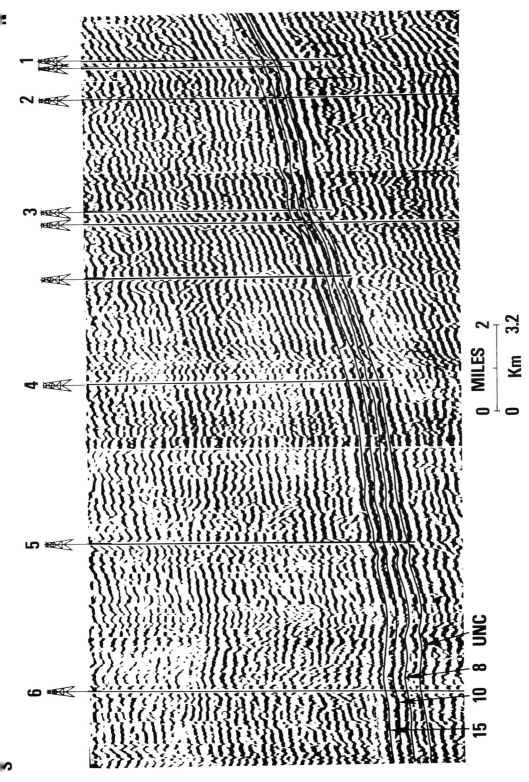

FIG. 6—Single-fold VDF magnetic tape seismic section from a Tertiary basin in South America.

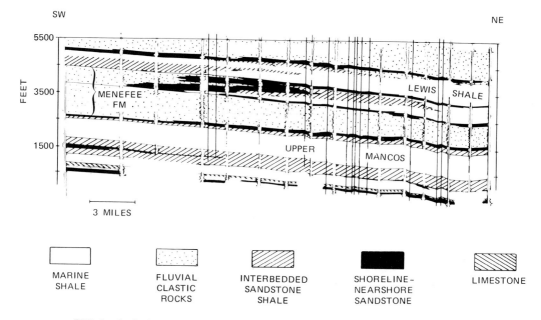

FIG. 7—Geologic section based on electric logs from the San Juan basin, New Mexico.

(Figs. 8, 9), show sediments deposited in shallow-marine to nonmarine environments. The geologic section is about 48 km long and averages almost one well per 1½ km; the wells are very close to the seismic section. The interval of interest is a Cretaceous nonmarine wedge, overlain and underlain by marine shales; the overlying shale is the Lewis and the underlying shale is the upper Mancos. These marine shales intertongue with the nonmarine wedge, and at the transition between the nonmarine and marine units, a series of shoreline sandstones are formed. For example, in the interval between 2,500 ft (750 m) and 3,300 ft (1,000 m) on the right side of the section, marine facies of the Lewis Shale are equivalent to the sandstones of the shoreline and nearshore environment, and finally to the fluvial clastic deposits of the nonmarine wedge. Detailed log marker correlations not included on the section show that individual stratal surfaces cross all the facies patterns.

The seismic section was shot as close as possible to the wells, and seismic time (Fig. 8) has been carefully converted to depth (Fig. 9), using close-spaced velocity analyses along the line. High-amplitude reflections mark the top and base of the nonmarine wedge at the zone of interbedding of marine shales and shoreline and nearshore sands.

On the seismic section, the zone of interest is between the arrows marking the top and base of the Menefee Formation (Figs. 8, 9), in which marine shales to the right pass into nonmarine sedi-

ments to the left. Striking changes are present in the seismic parameters associated with reflections in that interval. On the right, reflection character is one of relatively low amplitude and moderate continuity. Where the shales interfinger with the shoreline sandstones, amplitudes increase sharply because of the interbedding of sandstones and shales and the possible presence of gas in some of these sandstones. Farther left, reflections are present in a discontinuous pattern of both low and high amplitudes, consistent with the discontinuous bedding of the fluvial sediments.

RELATION BETWEEN PALEONTOLOGIC AND SEISMIC-REFLECTION CORRELATION

A lithostratigraphic unit may be time-transgressive and therefore cross seismic reflections. A series of inclined sigmoidal seismic reflections represents essentially time-synchronous stratal surfaces developed across major facies changes, based on paleontologic data obtained above and below the events in question.

Tertiary Example

An early and middle Eocene formation locally exhibits a well-developed sigmoidal reflection pattern on seismic data. The formation is a carbonate unit composed of nummulitic foram concentrations, chalk, and micrite. A large part of the carbonate is in the silt- and clay-sized fraction, which was deposited as a series of northward-dip-

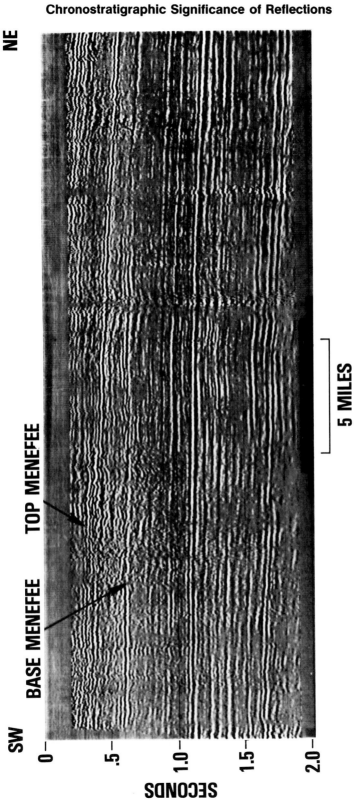

FIG. 8—Seismic section in time which parallels the geologic section (Fig. 7), San Juan basin, New Mexico. Datum elevation is 5,500 ft (1,676 m).

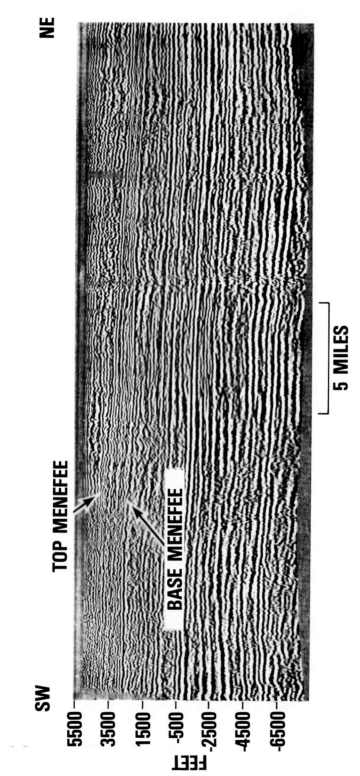

FIG. 9—Seismic section converted to depth, San Juan basin, New Mexio. Datum elevation is 5,500 ft (1,676 m).

ping clinoform beds. Figure 10 is a seismic section oriented north-south and approximately 31 km in length. Three reflections, indicated by the gray bands, are identified on the seismic line. Horizons I and II are sigmoidal reflections of Eocene age strata, whereas the lowest reflection is of Paleocene age strata. Three wells with velocity logs are located on the seismic line and are used to tie the reflections to specific depths in each well.

An interval extending above and below both horizons I and II was selected for study. Figure 11 is a geologic cross section incorporating the three wells noted on the seismic section. The three correlation lines coincide with the three seismic events marked on the seismic section (Fig. 10). Paleoenvironmental interpretations shown on this cross section were derived from examinations of ditch samples in the three wells, and are the result of qualitative determinations and are not intended to represent absolute water depths. Undoubtedly, the middle neritic (shallow) sediments (Fig. 11) were deposited in deeper waters than the inner neritic sediments, but it is not possible to determine the actual depth ranges.

The first seismic event, horizon I (Fig. 10), corresponds to the boundary between the middle Eocene and early Eocene. Two nummulite species are seen in all three wells at this horizon. They are *Nummulites globulus* and *N. nitidus*, both widely recorded as early Eocene index fossils in southern Europe and in northern Africa. Sediments directly below the clinoform surface are interpreted as inner neritic on the basis of faunal elements (biofacies) in the three wells. However, the biofacies variation in the three wells is sufficient to interpret a deepening northward, even though the broad environmental designation of inner neritic does not change. The top of the lower Eocene in well A contains a fauna consisting predominantly of Foraminifera *N. globulus* and *N. nitidus* with other shallow-water genera such as *Asterigerina rotula*, *Operculina* sp., and *Lockhartia hunti*, along with pelecypod and bryozoan fragments. The forams *Quinqueloculina* sp., *Lockhartia hunti*, and larger forms of *Rotalia*, *Eponides*, and *Textularia* are found at the top of the lower Eocene in well B. The fauna in well C at the same stratigraphic position is similar to that in well B, but with an occasional pelagic foram. The biofacies above the stratal surface at horizon I change markedly from well to well, indicating deeper waters to the north toward well C, and therefore a sloping sea bottom.

The second strong seismic event, horizon II, is located within the lower Eocene section. *Nummulites globulus* and *N. nitidus* were recorded in abundance in well A; in well B, only a few of each of these species were present; and in well C only sparse specimens of *N. globulus* were noted. In well C, pelagic forams make up approximately 20% of the faunal assemblages, and benthonic forams *Nonion*, *Textularia*, *Bulimina*, *Cibicides*, and *Uvigerina* are present. These faunal relations indicate progressively deeper water in wells B and C. Horizon II is an inclined depositional surface which is part of a classic progradational pattern downlapping against the underlying limestone.

Both seismic horizons I and II represent synchronous surfaces within initially inclined deposits which pass northward from shallow into deeper water, as demonstrated by paleontologic data. These relations indicate that seismic reflections follow strata surfaces, even where they cross major facies changes.

SIGNIFICANCE OF INTERPRETING CHRONOSTRATIGRAPHY PATTERNS FROM SEISMIC DATA FOR STRATIGRAPHIC TRAP EXPLORATION

Seismic stratigraphic techniques are useful in recognizing and exploiting a stratigraphic-trap play, such as the Woodbine gas trend in Polk and Tyler counties, southeastern Texas. Seismic sequence analysis (Part 6, Vail et al, this volume) permits identification of two prograding wedges, diagrammatically shown in Figure 12, which lie above a Buda carbonate shelf margin of early Late Cretaceous (early Cenomanian) age, and below a flat-lying high-amplitude reflection of the interface between the Austin Chalk and the Eagle Ford Shale. The oldest wedge is the Woodbine of late Cenomanian age which onlaps the frontal edge of the Buda shelf and, in turn, downlaps and ultimately laps out in a seaward direction. Internal reflections display an oblique progradational configuration which terminates at the top of the wedge as a series of toplapping reflections.

The Woodbine reflection pattern represents a complex of deltaic environments, including delta plain, delta front, and prodelta. The oblique progradational pattern is considered sand-prone (Part 9, Vail et al, this volume). The reflection pattern is indicative of energy levels sufficiently high to transport coarse clastics; however, if no source of sand existed for a particular area, none will be present in the sediments. It remains for well control to confirm or refute sand content for specific areas along the Woodbine trend.

The younger Eagle Ford wedge of Late Cretaceous Turonian age displays strong marine onlap along the frontal edge of the Woodbine, then changes to gentle coastal onlap after encroaching beyond the crest of the wedge (Fig. 12). Seaward, the Eagle Ford wedge gradually thickens; near

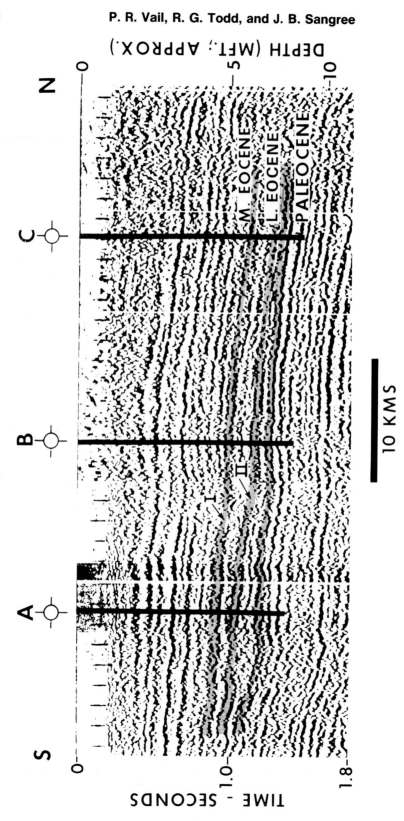

FIG. 10—North-south seismic section showing reflection correlations.

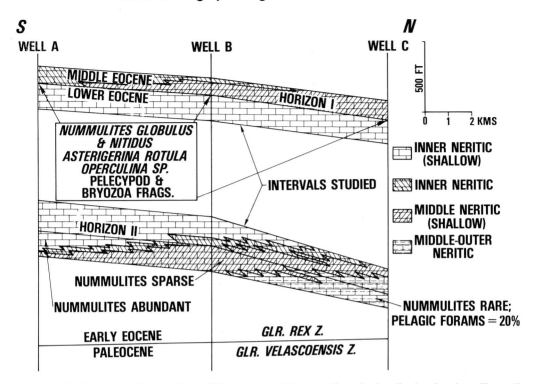

FIG. 11—North-south correlation using well log control and incorporating seismic reflection data from Figure 10.

the end of several seismic lines in the area, the beginning of toplapping reflections is present at the top. Eagle Ford seismic reflections exhibit considerably lower amplitude than the Woodbine, suggesting shale. Well penetrations are sparse, but tend to confirm this interpretation.

The seismic line in Figure 13 crosses several producing Woodbine gas fields and is located in central Polk County, Texas. The Woodbine prograding clinoforms (shown on Fig. 12) are quite evident, as is the strong Eagle Ford onlap on the front side of the Woodbine delta. All but the uppermost part of the Woodbine delta laps out against the Buda shelf edge in this area; unfortunately, the position of the Buda shelf margin is not seen on this seismic line but is located about 3

km north of the end of the seismic line. The two wells located on the seismic section, the Shell 1 and 2 Southland Paper Mills (SPM), are dry holes that were drilled as Woodbine tests in the early 1960s. The Shell 1 SPM well contains many thick Woodbine sandstones, but the sandstones carried salt water. The Shell 2 SPM well subsequently was drilled in a more seaward direction and found an overall thicker Woodbine section, but only one, thin, nonproductive sandstone bed. A well drilled a comparable distance north of the Shell 1 well made a gas discovery at the updip stratigraphic pinchout of the sandstones present in the well.

Synthetic seismograms, generated for the two wells, are overlain on the seismic line (Fig. 13).

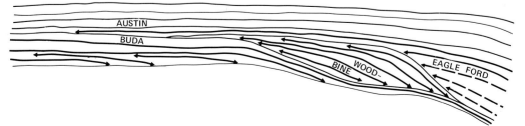

FIG. 12—Diagrammatic geologic cross section of the Woodbine and Eagle Ford wedges, southeast Texas.

P. R. Vail, R. G. Todd, and J. B. Sangree

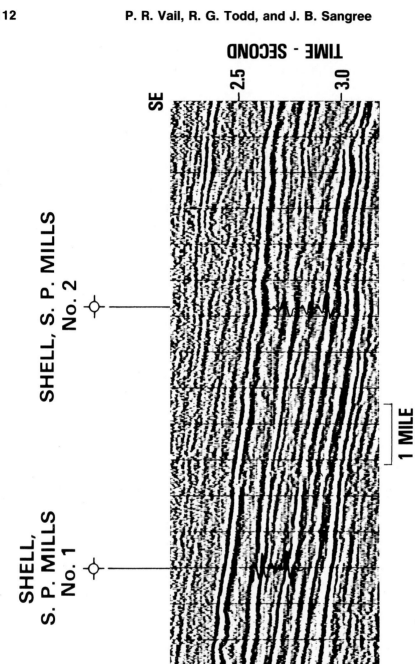

FIG. 13—Blowup of a seismic section showing reflection pattern details of Woodbine deltaic wedge.

INPUT PULSE 15 Hz

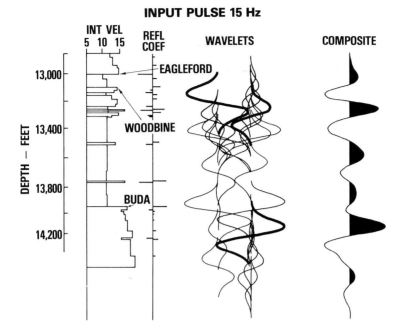

FIG. 14—Synthetic seismogram, Shell 1 Southland Paper Mills, Polk County, Texas. Interval velocity in thousands of feet per second.

The sonic log intervals available are short; however, the ties are good. Figures 14 and 15 are detailed plots of the synthetic seismograms for the two wells. The input pulse used for the synthetic seismogram is a 15-Hz first Gaussian derivative wavelet, essentially a sine wave. The detailed synthetic plots display from left to right: (1) well depth; (2) interval velocities in thousands of feet per sec, with stratigraphic tops and lithologies noted; (3) a plot of positive and negative reflection coefficients; (4) negative and positive wavelets individually displayed; and (5) the composite synthetic seismic trace.

Well-defined reflections are obtained from both synthetic seismograms for the Austin–Eagle Ford interface and from the top of the Buda Limestone (both indicated by heavy lines). Woodbine sandstones are not identifiable on the composite trace for the Shell 1 SPM well (Fig. 14) because of masking by the Eagle Ford wavelet and destructive interference from the top and bottom of individual sandstones. The Shell 2 well (Fig. 15) contains only one, very thin, sandstone bed which also is masked totally by the Eagle Ford wavelet; cancellation by the wavelet coming from the base of the sandstone also occurs.

It is not possible to directly detect hydrocarbons or reservoir sandstones within the producing part of the Woodbine wedge owing to resolution problems. What then can seismic stratigraphic interpretation techniques offer? The overall cycle reflection pattern for the Woodbine suggests a high-energy, probably sand-prone, deltaic environment. Two types of stratigraphic-trap plays emerge from this interpretation. One is an updip pinchout of Woodbine sandstone by onlap against the underlying Buda carbonate shelf margin (Fig. 12), described by Sheriff (1976, Fig. 10). The second play is the possibility of sandstone pinchouts just updip of toplap cycle terminations at the top of the Woodbine wedge. As an interpreter, one needs to remember even though a reflection may terminate on a seismic section, the sandstone bed causing it may continue for some lateral distance, although below seismic resolution. As a result, this second play is even a higher risk but seems to be supported by Woodbine sandstone production data in the area.

Figure 16 is a well-log cross section that parallels the seismic line (Fig. 13). The two wells (Figs. 14, 15) are located on the southeastern half of the cross section; two additional wells are plotted updip which provide control for known production. Sandstone correlations were made using the reflection patterns displayed by the Woodbine as a guide. Treating the log "kicks" as inclined prograding units resulted in correlations that agree with the seismic line (Fig. 13) and also explain the

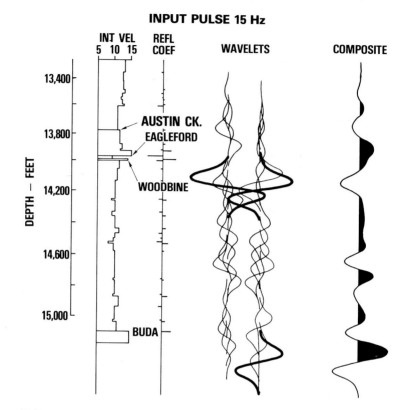

FIG. 15—Synthetic seismogram, Shell 2 Southland Paper Mills, Polk County, Texas. Interval velocity in thousands of feet per second.

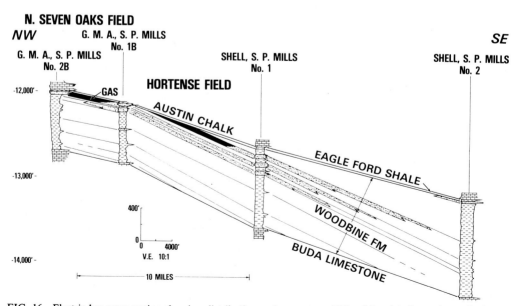

FIG. 16—Electric-log cross section showing distribution and geometry of Woodbine deltaic sandstone beds, Polk County, Texas.

gas production. The uppermost Woodbine sandstone unit in the Shell 1 SPM terminates updip as a result of toplap. This pinchout produces a stratigraphic trap, the Hortense field. This cross section was drawn before the discovery well was drilled, which is the reason the electric log is not included on the section.

Updip from Hortense field is another dry hole (Fig. 16), the G.M.A. 1B Southland Paper Mills (SPM). It was drilled prior to the discovery of Hortense field and provided early evidence for its existence. Updip from the G.M.A. 1B SPM is another producing gas field, North Seven Oaks field. It is also a stratigraphic trap and pinches out updip as a result of toplap. The Woodbine sandstones are not in lateral communication at the top of the deltaic wedge, resulting in a series of stratigraphic traps where individual sandstones lap out. Geologically, this probably is unusual for deltaic deposits, but may be more common in areas where coarse clastic influx is minor and the finer grained clastics dominate.

In summary, seismic data (Fig. 13) not only provide correlation of individual beds but also can be used to closely approximate the position of updip pinchouts.

REFLECTIONS FROM FLUID CONTACTS AND HYDRATE ZONES

In the past several years, attention has been focused on direct hydrocarbon indicators. The realization that impedance contrasts between gas, oil, and water could generate reflections from fluid contacts has modified seismic-exploration techniques, particularly in younger sediments. Another type of hydrocarbon anomaly present in some areas is the gas-hydrate deposit. Again, the reflection is not generated by a stratal surface, but is produced by impedance characteristics of the hydrate itself or associated diagenesis. In many instances, reflections from fluid contacts and gas hydrate zones cut across the reflections originating from stratal surfaces where they are at an angle to each other.

ACCURACY OF CHRONOSTRATIGRAPHIC CORRELATIONS

The accuracy of chronostratigraphic correlations using seismic data is dependent on data quality and resolution. A computer model of several sandstone bodies interbedded with massive shale illustrates the problem. Figure 17 shows a geologic section with overlays of two noise-free synthetic seismic sections made with different simple sine-wave pulses. The reflection lag time has been adjusted so the reflection peaks super-

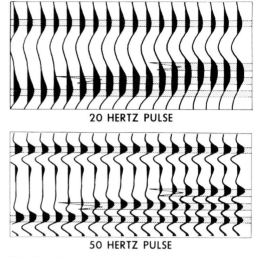

20 HERTZ PULSE

50 HERTZ PULSE

FIG. 17—Comparison of resolving power between 50-Hz and 20-Hz pulse.

impose the appropriate sandstone body to better illustrate the relation of sandstones to reflections.

In the lower part of Figure 17, a high-resolution pulse with a characteristic frequency of 50 Hz reveals the continuity of the upper and lower sandstones, and also indicates the lateral facies change from sandstone to shale of the two middle sandstones, by a loss of amplitude in the shale. The upper part of the figure shows the same geologic model and a synthetic section made (this time) with a low-resolution 20-Hz pulse. The continuity of the upper sandstone is preserved, because this sandstone is separated from the other sandstones. However, the lower three sandstones interfere in the composite waveform to produce a cycle that drifts across the depositional units, thus appearing to violate the principle of parallelism.

This exception is less of a problem than one might think: (1) the step-like geometry of a lateral facies transition must have an unusual degree of symmetry to produce the indicated effect; (2) abrupt shifts of this type may be visible as phase shifts, particularly if reflections above and below are continuous and parallel; and (3) improvements in seismic resolution continually push the limit of the thinnest bed that can be resolved by a complete seismic cycle, although the problem of resolving thin beds at great depths remains a severe limitation of the seismic system. Our experience indicates chronostratigraphic correlations from seismic data are accurate only to ± ½ wavelength due to possible changes of reflection character caused by changes in bed spacing and reflection coefficients.

PITFALLS OF CHRONOSTRATIGRAPHIC CORRELATIONS

Not all reflections on a seismic section are generated by bedding interfaces. Coherent noise patterns, such as multiples, ringing, steep dip events, diffractions, and sideswipe are examples of events which, if not recognized, may mislead the interpreter because they are not primary reflections or migrated to their proper positions.

Resolution and correlation of individual reflections commonly becomes a problem in highly "squeezed" seismic data, to produce a large vertical exaggeration, especially in high-frequency, shallow-penetration, arcer data. One example is high-frequency oceanographic data gathered over long distances in deep water. Correlation of individual reflections for long distances in such data is difficult, and the tendency is to correlate "bundles" of high-amplitude events or "transparent" zones as markers in the section. Faunal data from boreholes show that these zones may be time-transgressive. Detailed reflection correlation in some of these examples shows that individual reflections are commonly transgressed by reflection character changes, particularly amplitudes, or are lost in the "transparent" zones. As a result, the true chronostratigraphy shown by individual reflections is masked by time-transgressive reflection character changes.

REFERENCES CITED

Campbell, C. V., 1967, Lamina, laminaset, bed, and bedset: Sedimentology, v. 8, p. 7-26.

Gary, M., R. McAfee, Jr., and C. L. Wolf, 1974, Glossary of geology: Falls Church, Va., Am. Geol. Inst., 805 p.

Sheriff, R. E., 1976, Inferring stratigraphy from seismic data: AAPG Bull., v. 60, p. 528-542.

Seismic Stratigraphy and Global Changes of Sea Level, Part 6: Stratigraphic Interpretation of Seismic Reflection Patterns in Depositional Sequences [1]

R. M. MITCHUM, JR., P. R. VAIL, [2] and J. B. SANGREE [3]

Abstract Seismic stratigraphy is the study of stratigraphy and depositional facies as interpreted from seismic data. Seismic reflection terminations and configurations are interpreted as stratification patterns, and are used for recognition and correlation of depositional sequences, interpretation of depositional environment, and estimation of lithofacies.

Seismic sequence analysis subdivides the seismic section into packages of concordant reflections, which are separated by surfaces of discontinuity defined by systematic reflection terminations. These packages of concordant reflections (seismic sequences) are interpreted as depositional sequences consisting of genetically related strata and bounded at their top and base by unconformities or their correlative conformities. Reflection terminations interpreted as stratal terminations include erosional truncation, toplap, onlap, and downlap.

Seismic facies analysis interprets environmental setting and lithofacies from seismic data. Seismic facies units are groups of seismic reflections whose parameters (configuration, amplitude, continuity, frequency, and interval velocity) differ from adjacent groups. After seismic facies units are recognized, their limits defined, and areal associations mapped, they are interpreted to express certain stratification, lithologic, and depositional features of the deposits that generated the reflections within the units. Major groups of reflection configurations include parallel, subparallel, divergent, prograding, chaotic, and reflection-free patterns. Prograding configurations may be subdivided into sigmoid, oblique, complex sigmoid-oblique, shingled, and hummocky clinoform configurations. External forms of seismic facies units include sheet, sheet drape, wedge, bank, lens, mound, and fill forms. Seismic facies units are interpreted in terms of the depositional environments, the energy of the depositing medium, and the potential lithologic content of the strata generating the seismic facies reflection pattern.

INTRODUCTION

Preceding papers (Parts 1 through 5, Vail et al, this volume) discuss the concept of sequences as a tool for stratigraphic analysis. The depositional sequence is an objectively defined stratigraphic unit composed of genetically related strata and bounded by unconformities and their correlative conformities. Deposition of most major sequences is related to cycles of regional and global changes of sea level.

The reflection seismic method is the most effective tool for applying sequence concepts, although depositional sequences may also be observed on well-log sections, outcrops, and cores. Seismic reflections are composites of the individual reflections generated by surfaces separating

strata of differing acoustical properties. For this reason, the reflections tend to parallel stratal surfaces and to have the same chronostratigraphic significance as stratal surfaces. Therefore it is possible to make chronostratigraphic correlations using seismic reflection patterns.

The present paper describes our approach to seismic stratigraphy, describes the stratification patterns interpreted from seismic reflection terminations and configurations, and summarizes their use for recognition and correlation of depositional sequences, interpretation of depositional environment, and estimation of lithofacies. Our approach involves: (1) seismic sequence analysis—subdividing the seismic section into packages of concordant reflections separated by surfaces of discontinuity, and interpreting them as depositional sequences; and (2) seismic facies analysis—analyzing the configuration, continuity, amplitude, frequency, and interval velocity of seismic reflection patterns within seismic sequences. These patterns are interpreted in terms of environmental setting and estimates of lithology.

SEISMIC SEQUENCES—DEFINED BY REFLECTION TERMINATIONS

A seismic sequence is a depositional sequence (see Part 2, Vail et al, this volume) identified on a seismic section. It is a relatively conformable succession of reflections on a seismic section, interpreted as genetically related strata; this succession is bounded at its top and base by surfaces of discontinuity marked by reflection terminations and interpreted as unconformities or their correlative conformities. Seismic sequences have all the properties of depositional sequences subject only to the condition that these properties may be recognized and interpreted from the seismic reflection data. This paper will discuss only the specific subjects of recognition and interpretation of depositional sequences from seismic data.

Reflection terminations are the principal criteria for recognition of seismic sequence bound-

[1]Manuscript received, January 6, 1977; accepted June 13, 1977.

[2]Exxon Production Research Co., Houston, Texas, 77001.

[3]Esso Exploration, Inc., Walton-on-Thames, Surrey, England, KT12 2QL.

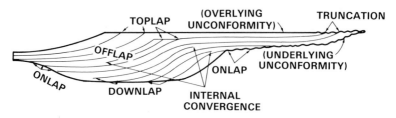

FIG. 1—Seismic stratigraphic reflection terminations within idealized seismic sequence.

aries. The types of reflection terminations are based on the types of stratal terminations that were defined in Part 2, Vail et al, (this volume). These types (listed in Table 1, first column) are illustrated diagrammatically on Figure 1, and on the seismic sections (Figs. 2; 3).

Top-discordant relations include *erosional truncation* and *toplap.* Erosional truncation implies the deposition of strata and their subsequent removal along an unconformity surface. Interpretation of reflection terminations as erosional truncation may be straightforward or quite subjective, depending on the angularity of the reflections to the erosional surface. In some instances, the erosional surface itself may produce a seismic reflection; elsewhere there is no reflection from the surface and only the systematic terminations of underlying reflections may define the surface. In general, however, erosional truncation is the most reliable top-discordant criterion for a sequence boundary.

Toplap is the termination of reflections interpreted as strata against an overlying surface as a result of nondeposition (sedimentary bypassing) and only minor erosion. In practice, many depositional boundaries marked by toplap are found to be rather local in extent, and in many cases can not be correlated regionally. For this reason, minor occurrences of toplap are commonly included within mapped depositional sequences and at their upper boundaries.

Base-discordant relations include seismic *onlap* and *downlap* (Figs. 1; 3). Onlap is a relation in which seismic reflections are interpreted as ini-

Table 1. Geologic Interpretation of Seismic Facies Parameters.

REFLECTION TERMINATIONS (AT SEQUENCE BOUNDARIES)	REFLECTION CONFIGURATIONS (WITHIN SEQUENCES)		EXTERNAL FORMS (OF SEQUENCES AND SEISMIC FACIES UNITS)
LAPOUT	PRINCIPAL STRATAL CONFIGURATION		
BASELAP	PARALLEL		SHEET
ONLAP	SUBPARALLEL		SHEET DRAPE
DOWNLAP	DIVERGENT		WEDGE
TOPLAP	PROGRADING CLINOFORMS		BANK
TRUNCATION	SIGMOID		LENS
EROSIONAL	OBLIQUE		MOUND
STRUCTURAL	COMPLEX SIGMOID-OBLIQUE		FILL
CONCORDANCE	SHINGLED		
(NO TERMINATION)	HUMMOCKY CLINOFORM		
	CHAOTIC		
	REFLECTION-FREE		
	MODIFYING TERMS		
	EVEN	HUMMOCKY	
	WAVY	LENTICULAR	
	REGULAR	DISRUPTED	
	IRREGULAR	CONTORTED	
	UNIFORM		
	VARIABLE		

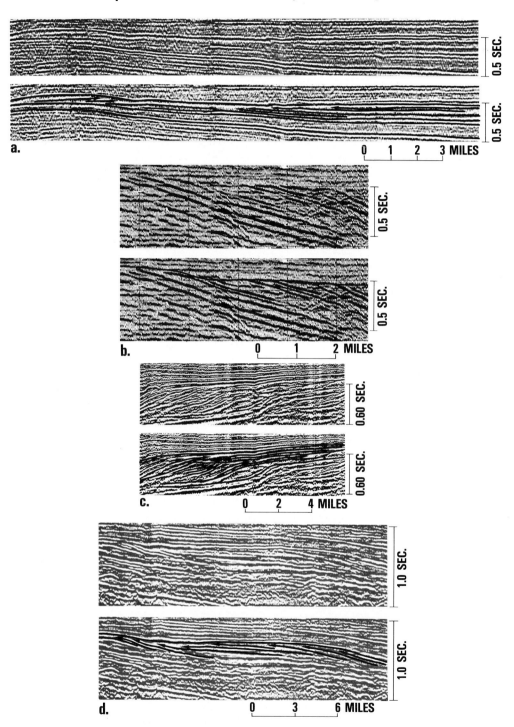

FIG. 2—Top-discordant seismic reflection patterns: **a** and **b** are erosional truncation; **c** and **d** are toplap. Second section of each pair shows interpretation.

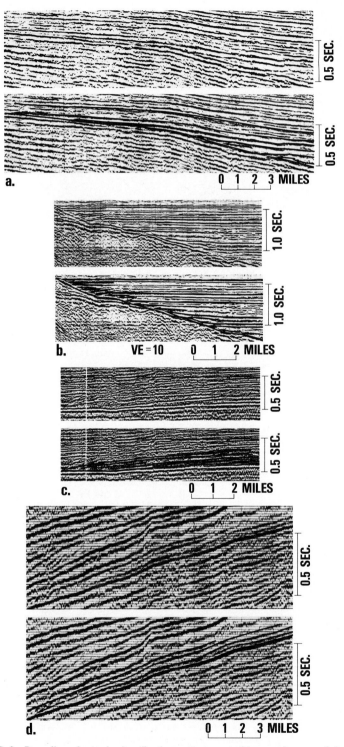

FIG. 3—Base-discordant seismic reflection patterns: **a** and **b** are onlap; **c** and **d** are downlap. Second of each pair is interpreted.

tially horizontal strata terminating progressively against an initially inclined surface, or as initially inclined strata terminating progressively updip against a surface of greater inclination.

Downlap is a relation in which seismic reflections are interpreted as initially inclined strata terminating downdip against an initially inclined or horizontal surface. If onlap can not be distinguished from downlap because of subsequent deformation, the more inclusive term *baselap* may be necessary.

A given two-dimensional seismic section may show apparent onlap (see Part 2, Vail et al, this volume), although the true configuration in intersecting sections may be downlap. If two sections intersecting at right angles both show apparent onlap, then true onlap is the likely relation. True onlap is the most reliable base-discordant criterion for a sequence boundary.

On seismic sections, apparent downlap may occur where reflections representing units of inclined or tangential strata terminate downdip, but the strata themselves actually flatten and continue as units which are so thin that they are below the resolution of the seismic tool. In some cases where control is available to test this possibility, true downlap terminations are indicated. In other examples, the thin, flattened, horizontal "toes" of the downlapping strata apparently accumulate and produce one or more horizontal seismic reflections, on which succeeding reflections terminate by downlap. This produces a new surface of discordance (usually local in extent) one or more cycles above the original surface. In most cases such minor occurrences of downlap are not mappable regionally and are included within the mapped depositional sequence rather than as a new sequence.

Nonsystematic reflection terminations within sequences, due to thinning of strata to below seismic resolution (internal convergence), should not be confused with terminations along sequence boundaries. Offlap (Fig. 1) is a term commonly used by seismic interpreters for reflection patterns from strata prograding into basins.

Seismic expression of sequence boundaries varies considerably depending on the velocity-density contrasts within the beds above and below the unconformity, and across the unconformity surface itself. Lacking significant velocity-density contrast across the unconformity, no unconformity reflection will be generated. However, angular unconformities can be located on seismic data by the discordance between the underlying truncated reflections and the overlying onlapping or downlapping reflections. If there is significant velocity-density contrast across the unconformity, the unconformity will appear as either a continuous or a discontinuous reflection. The unconformity reflection will be continuous if the reflection coefficient of the unconformity is significantly greater than the reflection coefficients of the underlying and overlying beds (especially if either the underlying or overlying adjacent strata are concordant). If the underlying and overlying beds dip with respect to the unconformity and have significant reflection coefficients of approximately the same magnitude as the unconformity, the unconformity reflection will go in and out of phase with the reflections from the truncated, onlapping or downlapping strata, and appear as a discontinuous reflection.

Nonangular unconformities commonly exhibit strong reflections from a characteristic change in velocity or density and constructive interference of reflections from the overlying or underlying parallel strata. A nonangular unconformity will not have a reflection associated with it if there is no velocity-density contrast. Nonangular unconformities are commonly detected by tracing them through a seismic grid from areas where they show angularity. In some cases paleontologic criteria for a chronostratigraphic gap are required to document a nonangular unconformity.

A strong reflection along an erosional surface may produce a "follow-cycle" beneath the principal reflection. The "follow-cycle" may mask underlying reflections so that they appear to terminate against it, rather than against the principal reflection. The true sequence boundary should be drawn above the principal reflection. In many cases, onlap by the overlying sequence will show the correct position of the unconformity

SEISMIC FACIES—DEFINED BY REFLECTION CONFIGURATION

Seismic Facies Parameters

After seismic sequences are defined, environment and lithofacies within the sequences are interpreted from seismic and geologic data. Seismic facies analysis is the description and geologic interpretation of seismic reflection parameters, including configuration, continuity, amplitude, frequency, and interval velocity. Each parameter provides considerable information on the geology of the subsurface (Table 2). Reflection configuration reveals the gross stratification patterns from which depositional processes, erosion, and paleotopography can be interpreted. In addition, fluid contact reflections (flat spots) commonly are identifiable. Reflection continuity is closely associated with continuity of strata; continuous reflections suggest widespread, uniformly stratified deposits. Reflection amplitude contains informa-

Table 2. Seismic Reflection Parameters Used in Seismic Stratigraphy,
and Their Geologic Significance.

SEISMIC FACIES PARAMETERS	GEOLOGIC INTERPRETATION
REFLECTION CONFIGURATION	• BEDDING PATTERNS • DEPOSITIONAL PROCESSES • EROSION AND PALEOTOPOGRAPHY • FLUID CONTACTS
REFLECTION CONTINUITY	• BEDDING CONTINUITY • DEPOSITIONAL PROCESSES
REFLECTION AMPLITUDE	• VELOCITY-DENSITY CONTRAST • BED SPACING • FLUID CONTENT
REFLECTION FREQUENCY	• BED THICKNESS • FLUID CONTENT
INTERVAL VELOCITY	• ESTIMATION OF LITHOLOGY • ESTIMATION OF POROSITY • FLUID CONTENT
EXTERNAL FORM & AREAL ASSOCIATION OF SEISMIC FACIES UNITS	• GROSS DEPOSITIONAL ENVIRONMENT • SEDIMENT SOURCE • GEOLOGIC SETTING

tion on the velocity and density contrasts of individual interfaces and their spacing. It is used to predict lateral bedding changes and hydrocarbon occurrences. Frequency is a characteristic of the nature of the seismic pulse, but it is also related to such geologic factors as the spacing of reflectors or lateral changes in interval velocity, as associated with gas occurrence. Grouping of these parameters into mappable seismic facies units allows their interpretation in terms of depositional environment, sediment source, and geologic setting. Seismic reflection configuration is the most obvious and directly analyzed seismic parameter. Stratal configuration is interpreted from seismic reflection configuration, and refers to the geometric patterns and relations of strata within a stratigraphic unit. These are commonly indicative of depositional setting and processes, and later structural movement. Analysis and mapping of stratal configuration from seismic reflection configuration are emphasized here as the first steps in a complete seismic facies analysis.

Seismic Facies Units

Seismic facies units are mappable, three-dimensional seismic units composed of groups of reflections whose parameters differ from those of adjacent facies units. Where the internal reflection parameters, the external form, and the three-dimensional associations of these seismic facies units are delineated, the units can then be interpreted in terms of environmental setting, depositional processes, and estimates of lithology. This interpretation is always done within the stratigraphic framework of the depositional sequences previously analyzed.

External Form and Reflection Configuration

The overall geometry of a seismic facies unit consists of the external form and the internal reflection configuration of the unit (Table 1, second and third columns). Both must be described to understand the geometric interrelation and depositional setting of the facies units. However, initial analysis always starts in the two-dimensional mode of a single seismic section, and these apparent configurations are later corroborated in a three-dimensional grid of seismic sections. Single sections obviously may cut stratal geometry at any angle. However, for purposes of present discussion, diagrammatic sections illustrating reflection configurations are assumed to be parallel with sedimentary dip unless otherwise indicated.

Table 1 lists principal reflection configurations and external forms of seismic facies units. Within a given external form, one or several internal reflection configurations may occur; also, a variety of related external forms may be classified as one type. For example, mounds range widely in shape, and fills occur in many types of depressions. These variations are discussed and illustrated. The external boundaries of seismic facies units may be identified by the termination of a series of seismic reflections against a common reflection, by a conformable reflection that bounds a particular configuration, or by an arbitrary boundary within a sequence across a gradational change in continuity, amplitude, frequency, or interval velocity.

A practical limitation imposed on the seismic interpreter is the relative size and location of his seismic sections with respect to the limits of seismic facies units and depositional sequences. The lateral extent of some seismic facies units may exceed seismic coverage so lateral boundary effects play only a minor role in analysis, and only the internal reflection configuration is available. Some large wedge or sheet units with parallel to subparallel reflections are thus classified. Small seismic facies units, such as mounds or fill structures, are more easily defined using more obvious lateral boundary effects and internal reflection patterns.

Depositional Energy Concept

Where seismic facies units are described and mapped, they are interpreted first in terms of sedimentary processes, environmental setting, and depositional energy of the environment; then in terms of the lithologic potential of the seismic facies. If, for example, the interpreted environmental setting of a clastic facies unit has sufficiently high energy to transport and deposit significant quantities of sand, then the facies can be considered sand-prone. Conversely, if depositional energies are low and insufficient to develop significant sand accumulations, then the facies is considered shale-prone. Unfortunately, many specific examples of high-energy sand-prone facies will prove to consist entirely of shale and silts because no source of sand was available. Thus, the criteria of high and low energy apply only to the potential for sand deposition. Other techniques, for example calibration of interval velocity for sand prediction, must be used to interpret the actual presence of bedded sand.

Types of Reflection Configuration Patterns

Some significant reflection configuration patterns are listed on Table 1, and are diagrammatically illustrated in Figures 4-11. Description and interpretation of reflection configurations begin with simple patterns and continue to the more complex. Variations within configurations commonly can be described with modifying terms

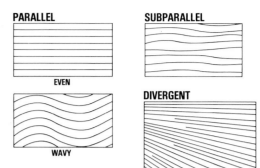

FIG. 4—Parallel, subparallel, and divergent seismic reflection configurations.

such as those shown on Table 1 and Figure 13. Obviously this is not a complete listing, and names and types of configurations should be modified to meet particular needs if necessary.

Parallel and Subparallel—These reflection configurations are shown on Figure 4, and seismic examples are shown on Figures 5A, B. Lateral limits of the seismic facies unit are not included. Modifying terms such as "even" or "wavy" describe part of the diagram or example. Parallel configurations may occur in several external forms, but are probably most common in sheet, sheet drape, and fill units. Subdivisions of this simple configuration are based on variations in other seismic parameters such as amplitude, continuity, or cycle breadth. This pattern suggests uniform rates of deposition on a uniformly subsiding shelf or stable basin plain setting.

Divergent—This reflection configuration (Fig. 4; 5C, D) is characterized by a wedge-shaped unit in which most of the lateral thickening is accomplished by thickening of individual reflection cycles within the unit, rather than by onlap, toplap, or erosion at the base or top. Nonsystematic lateral terminations of seismic reflections occur commonly within the wedge in the direction of convergence (Fig. 1). These terminations are probably due to progressive thinning of strata to below the resolution of the seismic tool. Divergent configurations suggest lateral variations in the rate of deposition, or progressive tilting of the depositional surface.

Prograding Reflection Configurations—Several more complex reflection configurations occur (Fig. 6), interpreted as strata in which significant deposition is due to lateral outbuilding or prograding. Sigmoid, oblique, complex, shingled, and hummocky progradational patterns form through progressive lateral development of gently sloping depositional surfaces, called *clinoforms*. The cli-

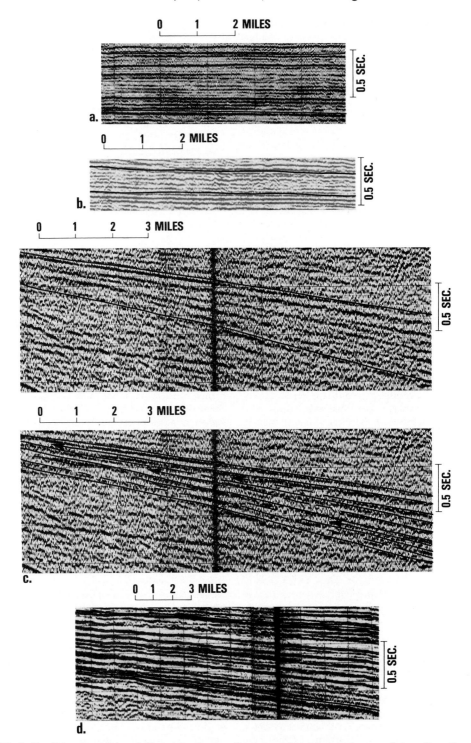

FIG. 5—Parallel, subparallel, and divergent seismic reflection configurations: **a** is a parallel configuration with good continuity and high to medium amplitude; **b** is subparallel configuration with good to fair continuity and high to medium amplitude; **c** and **d** are divergent configurations, with thickening of individual reflection cycles in direction of divergence. Nonsystematic reflection terminations occur in direction of thinning.

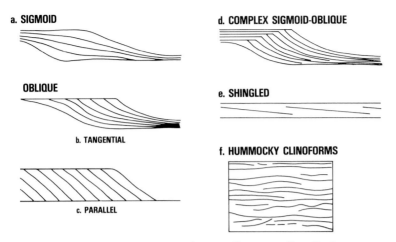

FIG. 6—Seismic reflection patterns interpreted as prograding clinoforms.

noform surface is one of the most common depositional features. Differences in prograding clinoform patterns result in large part from variations in rate of deposition and water depth. A variety of environmental settings is possible. If the upper part of the pattern was deposited in shallow water, and the lower part gently slopes into deeper water, the configuration can be divided into upper, middle, and lower zones, corresponding to Rich's (1951) undaform, clinoform, and fondoform topographic environments, respectively. Prograding clinoforms occur in units with many external shapes, including all those included in Table 1.

A *sigmoid* progradational configuration (Figs. 6A; 7A) is a prograding clinoform pattern formed by superposed sigmoid (S-shaped) reflections interpreted as strata with thin, gently dipping upper and lower segments, and thicker, more steeply dipping middle segments. The upper (topset) segments of the strata have horizontal or very low angles of dip and are concordant with the upper surface of the facies unit. The thicker middle (foreset) segments form lenses superposed to allow successively younger lenses to be displaced laterally in a depositionally downdip direction, forming overall outbuilding or prograding patterns. Depositional angles are quite low (usually less than 1°). The lower (bottomset) segments of the strata approach the lower surface of the facies unit at very low angles, and the seismic reflections show real or apparent downlap terminations as the strata terminate or become too thin to be recognized on seismic sections. On sections parallel with depositional strike, reflections indicate that strata are commonly parallel and concordant with unit boundaries.

The most distinctive feature of the sigmoid reflection configuration is the interpreted parallelism and concordance of the upper stratal (topset) segments, suggesting a degree of continued upbuilding (aggradation) of the upper segments coincident with prograding of the middle segments. This configuration implies relatively low sediment supply, relatively rapid basin subsidence, and/or rapid rise in sea level to allow deposition and preservation of the topset units. A relatively low-energy sedimentary regime is interpreted (Part 8, Vail et al, this volume).

An *oblique* progradational reflection configuration (Figs. 6B, C; 7B, C) is interpreted as a prograding clinoform pattern consisting ideally of a number of relatively steep-dipping strata terminating updip by toplap at or near a nearly flat upper surface, and downdip by downlap against the lower surface of the facies unit. Successively younger foreset segments of strata build almost entirely laterally in a depositionally downdip direction. They may pass laterally into thinner bottomset segments, or terminate abruptly at the lower surface at a relatively high angle. They build out from a relatively constant upper surface characterized by lack of topset strata and by pronounced toplap terminations of foreset strata. Depositional dips are characteristically higher than in the sigmoid configuration, and may approach 10°.

In the *tangential oblique* progradational pattern, dips decrease gradually in lower portions of the foreset strata, forming concave-upward strata which pass into gently dipping bottomset strata. Seismic reflections terminate tangentially against the lower boundary of the facies unit by real or apparent downlap as the strata from which they

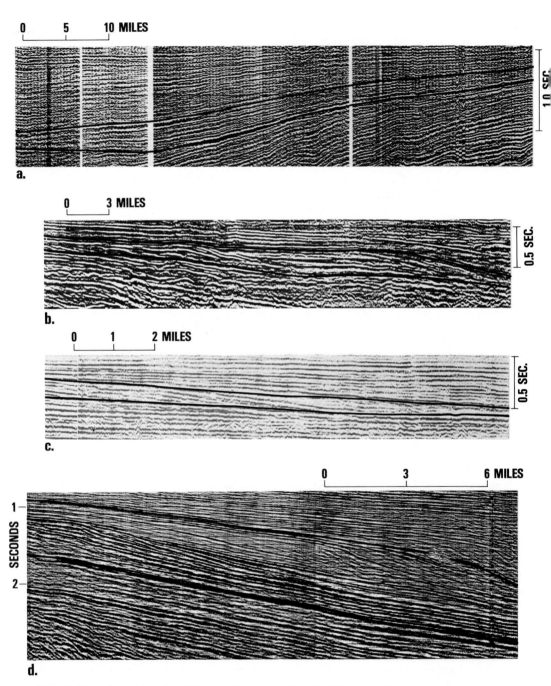

FIG. 7—Examples of sigmoid, oblique, and complex sigmoid-oblique seismic reflection configurations: **a** is sigmoid, **b** is mostly tangential oblique with some sigmoid, **c** is mostly parallel oblique, **d** is complex sigmoid-oblique.

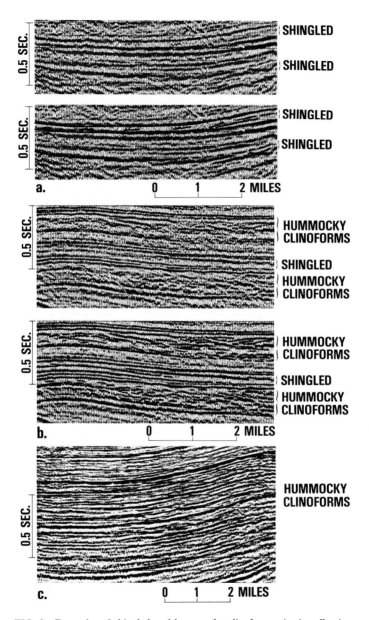

FIG. 8—Examples of shingled and hummocky clinoform seismic reflection configurations: **a** is a shingled configuration; **b** is hummocky clinoform configuration
with minor shingling; **c** is hummocky clinoform configuration. Both configurations
are interpreted as strata deposited in small clinoforms with relief approaching, or at,
the point of seismic resolution. Clinoforms of **a** and **b** are slightly larger than those
of **c** with correspondingly better resolution. Second sections of pairs **a** and **b** shows
interpretation.

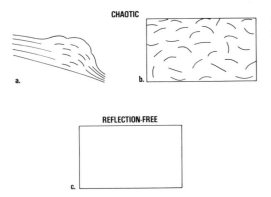

FIG. 9—Diagrams of chaotic and reflection-free seismic reflection patterns. **a** represents a chaotic pattern which may be interpreted as original stratal features still recognizable after penecontemporaneous deformation; in **b** reflections may not be interpreted in any recognizable stratal pattern; **c** represents a reflection-free area.

are derived terminate or become thinner downdip.

In the *parallel oblique* progradational pattern, the relatively steep-dipping parallel foreset strata terminate downdip at a high angle by downlap against the lower surface. In sections parallel with depositional strike, reflections in these seismic facies units may range from parallel to low-angle oblique or sigmoid progradational, possibly with small channel-fill configurations.

The oblique progradational configuration implies depositional conditions with some combination of relatively high sediment supply, slow to no basin subsidence, and a stillstand of sea level to allow rapid basin infill and sedimentary bypass or scour of the upper depositional surface. A relatively high-energy sedimentary regime is indicated (Part 8, Vail et al, this volume).

A *complex sigmoid-oblique* progradational reflection configuration (Figs. 6D; 7D) is a prograding clinoform pattern consisting of a combination of variably alternating sigmoid and oblique progradational reflection configurations within a single seismic facies unit. The upper (topset) segment of the facies unit is characterized by a complex alternation of horizontal sigmoid topset reflections and segments of oblique configuration with toplap terminations. This variability implies strata with a history of alternating upbuilding and depositional bypass in the topset segment, within a high-energy depositional regime. In other respects this configuration is similar to the sigmoid configuration.

This reflection configuration illustrates short segments of toplap within a seismic sequence

rather than at its upper boundary. The short segments of toplap indicate a number of smaller scale depositional sequences whose boundaries are below seismic resolution except where toplap is prominent. These smaller scale units are commonly interpreted as discrete lobes of a prograding depositional unit.

A *shingled* progradational reflection configuration (Figs. 6E; 8A, B) is a thin prograding seismic pattern, commonly with parallel upper and lower boundaries, and with gently dipping parallel oblique internal reflectors that terminate by apparent toplap and downlap. Successive oblique internal reflectors within the unit show little overlap with each other. The overall pattern resembles that of the parallel oblique progradational configuration, except that the thickness of the unit is just at the point of seismic resolution of the oblique beds. In some thin units, the internal reflector is a series of discontinuous events whose obliquity is only suggested. Shingled seismic configurations are most common in seismic facies units interpreted as depositional units prograding into shallow water.

A *hummocky clinoform* reflection configuration (Figs. 6F; 8B, C) consists of irregular discontinuous subparallel reflection segments forming a practically random hummocky pattern marked by nonsystematic reflection terminations and splits. Relief on the hummocks is low, approaching the limits of seismic resolution. This pattern commonly grades laterally into larger, better defined clinoform patterns, and upward into parallel reflections. The reflection pattern is generally interpreted as strata forming small, interfingering clinoform lobes building into shallow water in a prodelta or inter-deltaic position.

Chaotic Reflection Configuration—Chaotic patterns (Figs. 9; 10A, B) are discontinuous, discordant reflections suggesting a disordered arrangement of reflection surfaces. They are interpreted either as strata deposited in a variable, relatively high-energy setting, or as initially continuous strata which have been deformed so as to disrupt continuity. Some reflection patterns (Figs. 9A; 10A) may be interpreted as original stratal features still recognizable after penecontemporaneous deformation. Other patterns (Figs. 9B; 10B) are so disordered that reflections throughout a significant part of a sequence may not be interpreted in any recognizable pattern of stratal configuration. Penecontemporaneous slump structures, cut-and-fill channel complexes, and highly faulted, folded, or contorted zones may have chaotic seismic expression.

Reflection-Free Areas—Homogeneous, nonstratified, highly contorted, or steeply dipping geo-

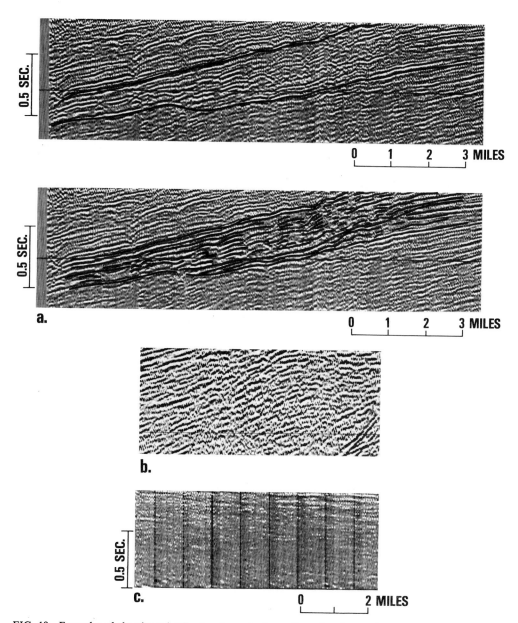

FIG. 10—Examples of chaotic and reflection-free seismic configuration. In **a** reflections may be interpreted as contorted stratal surfaces; in **b** no stratal patterns may be reliably interpreted; **c** is largely reflection-free, where no or very few reflections occur in seismically homogeneous shale.

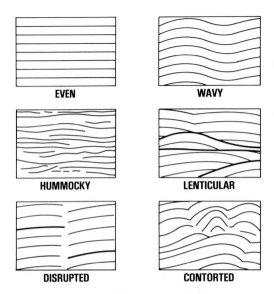

EVEN

WAVY

HUMMOCKY

LENTICULAR

DISRUPTED

CONTORTED

FIG. 11—Some modifying seismic reflection configurations.

logic units may be expressed as essentially reflection-free areas on seismic data (Figs. 9C; 10C). For example, some large igneous masses, salt features, or thick seismically homogeneous shales or sandstones could appear reflection-free.

Modifying Terms—Minor variations in basic patterns of reflection configurations may be described by common modifying terms, some of which are listed on Table 1 and illustrated on Figure 11. Such terms as wavy, even, hummocky, lenticular, disrupted, contorted, regular, irregular, uniform, and variable are self-explanatory.

Types of External Forms of Seismic Facies Units

An understanding of three-dimensional external forms and areal associations of seismic facies units is important in their analysis. Table 1 and Figure 12 show important external forms. Some of these, such as mounds and fills, can be divided into subtypes, depending on origin, internal reflection configuration, and modifications of external forms.

Sheets, wedges, and banks (Fig. 12) may be large, and are the most common shelf seismic facies units. A variety of parallel, divergent, and prograding patterns makes up the internal reflection configuration within these units. Sheet drapes commonly consist of parallel reflections interpreted as strata draped over underlying topography in a pattern suggesting uniform, low-energy, deep-marine deposition independent of

bottom relief. Lenses may occur in many seismic facies associations, but are most common as the external form of prograding clinoform seismic facies units. Mounds and fills are groups of seismic forms derived from strata with diverse origins, forming prominences or filling depressions on depositional surfaces.

Mounds are reflection configurations interpreted as strata-forming elevations or prominences, rising above the general level of the surrounding strata. Most mounds are topographic buildups resulting from either clastic or volcanic depositional processes, or organic growth. They are generally small enough that their external limits can be defined on a grid of seismic sections, and are characterized by onlap or downlap of overlying strata which fill around the mounds. Because of diverse origins, mounds may have diverse external shapes and internal stratal configurations. A descriptive subdivision, purely on the basis of internal configuration and external geometry, should be considered only a preliminary step in genetic interpretation of the mound. Deep sea fans, lobes, slump masses, some deepsea current and contourite deposits, carbonate buildups and reefs, and volcanic piles could have mounded two-dimensional configurations.

Figure 13 is a diagram of mound types, showing two-dimensional external form and internal reflection configuration of several genetic groups of mounds. Vail et al, Part 8 (this volume), discusses in more detail most of the clastic mounds, and Vail et al, Part 9 (this volume), gives details of carbonate buildups. One configuration, the migrating wave (Figs. 13; 14), is not common, but is sufficiently distinctive to be described further. It is a number of superposed wave-shaped reflections, each of which is progressively offset laterally from the preceding reflection, and is interpreted as a series of migrating sediment waves moving across a horizontal surface. These waves have been observed with crests up to 6 mi (9.6 km) apart, and with vertical relief of 300 ft (91 m) between crests and troughs. Where vertical upbuilding is significant, the pattern of successive waves resembles that of "climbing ripples," a sedimentary structure in sandstones, scaled in inches. Where vertical upbuilding is less prominent, deposition of successive strata is accompanied by some scour of the "up-current" part of the wave, giving a dunelike form to the wave. This configuration appears to be formed by current action in deep water. Damuth (1975, p. 28, 36) illustrated and described this configuration.

In many instances, mound size is so small that individual mounds can not clearly be defined or

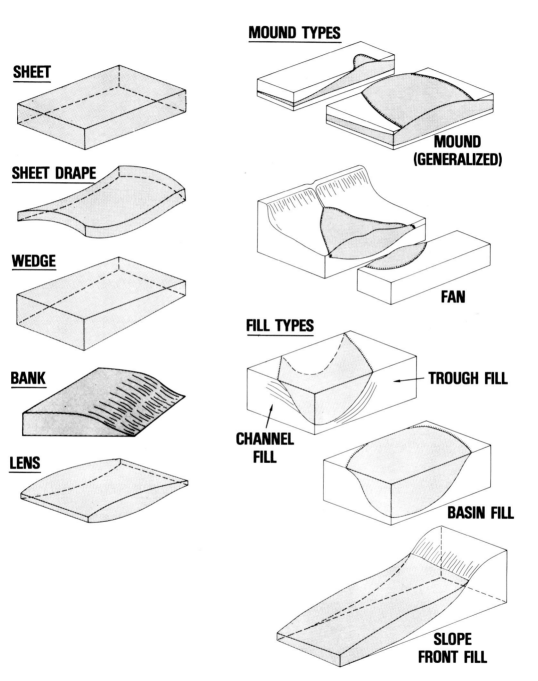

FIG. 12—External forms of some seismic facies units.

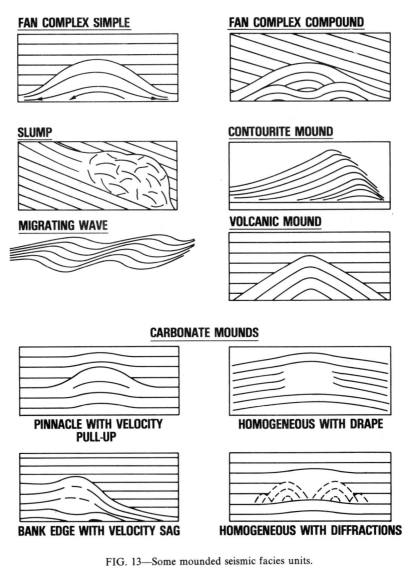

FIG. 13—Some mounded seismic facies units.

FIG. 14—Example of migrating-wave seismic reflection configuration.

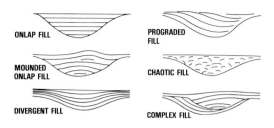

FIG. 15—Some fill seismic facies units.

mapped seismically. A "hummocky" or "mounded" reflection pattern is the common seismic expression of groups of these small features.

Fill reflection patterns are interpreted as strata filling negative-relief features in the underlying strata. Underlying reflections may show either erosional truncation or concordance along the basal surface of the fill unit. Fill units may be classified by external form (channel fill, trough fill, basin fill, or slope-front fill; Fig. 12). They also show a variety of internal reflection configurations, some of which are illustrated diagrammatically in Fig. 15.

Fill patterns represent structures which may have a variety of origins such as erosional channels, canyon fills, structural-trough fills, fans, slumps, and others. Large fill structures may be mapped as separate entities, but as size and clarity of definition diminish, these features may be grouped into complexes or treated as subordinate features of larger seismic facies units.

CONCLUSIONS

Interpretation of stratigraphy from seismic reflection data can be rewarding, either in frontier areas of sparse well control or in data-rich provinces. The seismic identification and interpretation of depositional sequences (seismic sequence analysis) provides basic stratigraphic units which are both objectively defined seismically and, to a degree, predictable because of their relation to global cycles of relative change of sea level.

Recognition and mapping of seismic facies units within the seismic sequences, based on objectively defined seismic parameters such as reflection configuration, continuity, amplitude, frequency, and interval velocity, are called seismic facies analysis. Seismic facies units may be interpreted in terms of environmental setting, depositional processes, and estimates of lithology. A combination of these two approaches affords a powerful tool called seismic stratigraphy.

REFERENCES CITED

Damuth, J. E., 1975, Echo character of the western equatorial Atlantic floor and its relationship to the dispersal and distribution of terrestrial sediments: Marine Geology, v. 18, p. 17-45.

Rich, J. L., 1951, Three critical environments of deposition and criteria for recognition of rocks deposited in each of them: Geol. Soc. America Bull., v. 62, p. 1-20.

Seismic Stratigraphy and Global Changes of Sea Level, Part 7: Seismic Stratigraphic Interpretation Procedure [1]

R. M. MITCHUM, JR., and P. R. VAIL [2]

Abstract A generalized procedure for making regional stratigraphic studies using seismic data involves analysis of seismic sequences and seismic facies, which are interpreted from terminations and configurations of seismic reflections generated by sedimentary strata. Generalized steps in the procedure include (1) recognition, correlation, and age determination of seismic sequences; (2) recognition, mapping, and interpretation of seismic facies; and (3) regional analysis of relative changes of sea level.

INTRODUCTION

This paper discusses a generalized procedure for making stratigraphic studies using seismic data. Previous papers in this volume have discussed aspects of the sequence concept and seismic stratigraphy. The procedure discussed here represents one simple approach to the application of these concepts in a given regional analysis.

Depositional sequences have two characteristic properties which make them ideal for interpretation on seismic data. First, they may be defined objectively using terminations of reflections along surfaces of discontinuity, interpreted as stratal terminations along unconformable sequences boundaries (Part 2, Vail et al, this volume). Second, deposition of most major sequences appears related to global changes of sea level (Part 4, Vail et al, this volume). This relation allows a degree of predictability of age and depositional characteristics based on comparison with global cycle charts. The procedure is designed to take advantage of these properties.

Our approach to seismic stratigraphy involves (1) seismic sequence analysis: subdividing the seismic section into sequences which are the seismic expression of depositional sequences—stratigraphic units of relatively conformable, genetically related strata bounded by unconformities or their correlative conformities; (2) seismic facies analysis: analyzing the configurations of reflections interpreted as strata within depositional sequences to determine environmental setting and to estimate lithology; and (3) sea level analysis: analysis of regional relative changes of sea level for comparison with global data.

PROCEDURE

Procedure for interpreting stratigraphy from seismic data consists of three generalized stages.

Step 1: Seismic Sequence Analysis—Recognition, Correlation, Age Determination

Seismic sequences first need to be defined by recognizing surfaces of discontinuity from reflection terminations. To be objective, seismic sequences are delineated and boundaries established on seismic sections before ages are determined from well control. This approach contrasts with conventional methods in which mapped seismic reflections are picked by reference to age or formation tops in wells and may have little relation to sequence boundaries.

Surfaces of discontinuity are recognized by interpreting systematic patterns of reflection terminations along the surfaces, which are catagorized as onlap, downlap, toplap, or truncation (Part 6, Vail et al, this volume). Sequence boundaries are then extended over the complete section, including parts where reflections are concordant with boundaries.

Delineation of sequence boundaries is repeated on other intersecting lines in the grid of seismic data until the boundaries have been correlated and loop-tied throughout the grid. This process tends to verify the regional extent of major discontinuity surfaces and indicates other similar surfaces as local in nature.

Complete correlation through the seismic grid should produce a three-dimensional framework of successively stratified seismic sequences, separated by surfaces of discontinuity. Each seismic sequence represents a depositional sequence with its own particular regional distribution, stratal geometry, thickness pattern, and geologic history. This state of analysis may be reached without reference to any information other than the seismic sections or without any outside knowledge of age or facies. Normally, however, some outside information is available and is incorporated into the study.

Age determination of sequence is based on well and outcrop information and on seismic prediction. Dating the physically defined seismic sequences allows them to be tied into the standard

[1]Manuscript received, January 6, 1977; accepted June 13, 1977.

[2]Exxon Production Research Co., Houston, Texas 77001.

135

chronostratigraphic framework. Once the sequences are dated using biostratigraphic, radiometric, or other types of information, other well data such as lithology, depositional facies, environmental interpretations, fluid content, and velocity information may be applied to an appropriate degree.

If there is no well information for part or all of the stratigraphic framework, then age commonly is projected and predicted on the basis of information from outside the area, sequence characteristics, and use of the global cycle chart (Part 4, Vail et al, this volume). Accuracy of age predictions varies, but coupled with knowledge of the regional geologic setting, predictions have been quite successful.

Although seismic stratigraphy may be effectively applied to any area, it commonly has been used where well and outcrop information is minimal. One advantage of the sequence framework is that well or outcrop information anywhere in the seismic grid is optimized because it can apply to the complete grid.

Figure 1 is an example of a seismic section from offshore West Africa, repeated from Part 2 (Vail et al, this volume). On this section, seismic sequences have been identified and their ages designated to the degree possible with available well control. Some ages are projected, based on the global cycle chart (Parts 4 and 8, Vail et al, this volume).

Step 2: Seismic Facies Analysis—Recognition, Mapping and Interpretation

Seismic facies analysis (Parts 6, 9, and 10, Vail et al, this volume) is the analysis of reflection configuration and other seismic parameters within the seismic-sequence correlation framework. These parameters, including reflection configuration, amplitude, continuity, frequency, and interval velocity, are interpreted to express certain gross lithologic, stratification, and depositional features of the sediments generating the cycles. The parameters are mapped as seismic facies units, which are three-dimensional groups of reflections whose elements differ from those of adjacent units. These units are always interpreted within the chronostratigraphic connotation of the seismic sequence framework.

There are several alternate approaches to seismic facies analysis, depending on the parameters analyzed and the purpose of the analysis. Although our discussion is limited to visual inspection of these parameters, especially seismic reflection configuration, more quantitative approaches are available through use of computer techniques.

Reflection configuration commonly is interpreted in terms of environmental setting, depositional processes, and estimates of lithology of the strata involved. Although there is no unique relation between types of reflection configuration and specific lithologies, the integration of these seismic reflection elements with all available nonseismic data affords a reasonable prediction of depositional environment and lithology.

Several steps are involved in visual analysis of reflection configuration. First is the recognition and differentiation of seismic facies units within each sequence on all the seismic sections in the areal grid to be mapped. Next is the transfer of seismic facies descriptions from the seismic sections to a map for each sequence. The simplest approach is to divide each sequence on the seismic section into segments representing the lateral extent of various seismic facies units. These segments are then transferred to a shotpoint map with appropriate descriptions or names of seismic facies units. The limits of the units on successive seismic sections are connected on the map to outline their distribution. Thicknesses of facies units also may be mapped.

Included in the basic seismic facies information are (1) the relation of reflections to upper and lower sequence boundaries (segments of onlap, downlap, toplap, truncation, and concordance; and direction of onlap and downlap); and (2) the dominant types of reflection configuration (segments of parallel, divergent, sigmoid, or oblique configurations). In many instances one aspect of the seismic facies proves more significant than others, and a separate map is prepared to portray that single aspect. For example, dip direction of downlapping reflections is important in interpreting sediment transport direction and possible source areas.

After distribution and thickness of seismic facies units are mapped, this information is combined with the map distribution of any other diagnostic seismic parameters such as interval velocity or localized amplitude anomalies. Nonseismic information such as well and outcrop data also is integrated with seismic facies distribution.

Next is the interpretation of the seismic facies map in terms of environmental setting such as shelf, slope, basin position, marine or nonmarine environments, paleobathymetric water depths, energy of the depositional medium, transport direction, or other depositional aspects.

Estimates of lithology are based on all possible data, including the above environmental interpretations. These estimates may range from gross in-

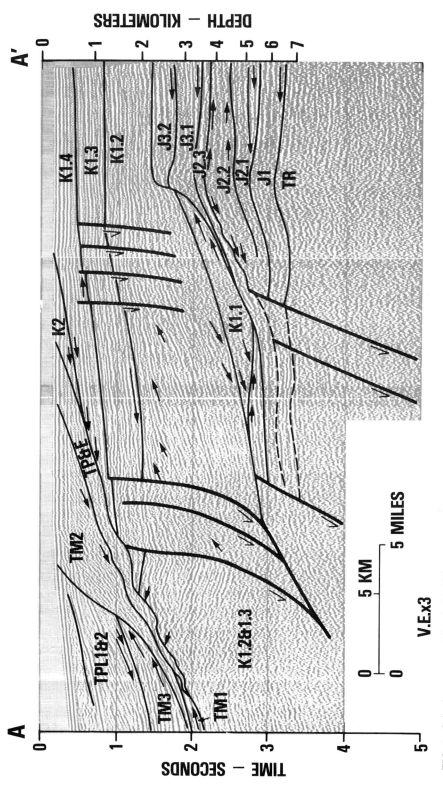

FIG. 1—Seismic sequences, Line A-A' offshore western Africa (map location Figs. 4, 5). Sequence boundaries shown by black lines. Arrows mark cycle terminations of onlap, downlap, and truncation which provide criteria for recognition of sequence boundaries. Local occurrences of toplap also shown by arrows. Ages of sequences designated by symbols in Part 4, Vail et al, this volume (Fig. 2).

terpretations of clastic versus carbonate sedimentary regimes to fine-tuned predictions of lithologic percentages, depending on the purpose of the study, types of data, size of area, and other factors. Lithologic predictions are commonly based on a combination of parameters that are interpreted in terms of overall depositional setting and energy potential of the depositional medium, and on parameters such as interval velocity that are direct measures of certain physical properties of the strata involved. Calibration of the latter parameters with known data in the area greatly enhances the prediction of specific lithologies.

Another level of interpretation is possible, dealing with more general subjects such as stratigraphic and structural history of the area, tectonostratigraphic relations, hydrocarbon source-reservoir migration, or relations to regional and global changes of sea level. (Many of the relations discussed in Parts 3 and 4, Vail et al, this volume, were observed in the course of this type of seismic stratigraphic investigation.)

Step 3: Sea Level Analysis—Chronostratigraphic Charts, Regional and Global Curves of Relative Changes of Sea Level

Because most major depositional sequences in a given region appear related in their ages, distribution, and general depositional characteristics to global changes of sea level, these properties are to some extent predictable through comparison of regional sea level curves to global curves. Information necessary for preparation of regional curves of relative change of sea level includes (1) age and duration of depositional sequences; (2) relative nature of distribution of sequences (restricted to basin and slope or widespread on shelf); (3) nature and measurement of coastal onlap of each sequence. (The procedure for preparation of regional curves is explained in Part 3, Vail et al, this volume; especially Fig. 13).

In many specific regional study areas, a complete analysis of sea level changes is impractical or impossible because of practical problems such as the distribution and quality of data, erosion of coastal onlap, or structural displacement of stratigraphic units. However, no matter how incomplete the onlap data may be, a partial analysis should be made using whatever information is available. Where onlap data are missing, other indirect or only supportive data, such as occurrence of abrupt marine transgressions, abrupt changes from prograding shelf patterns to deep marine fans, units showing persistent marine onlap or downlap, thickness of leached zones in carbonate rocks, and other general facies relations, can be

supplementary aids in building a chart of relative changes of sea level.

Comparison of regional and global sea level curves helps predict ages of sequences for which no other age data exist; and forecasts generalized depositional characteristics of sequences related to cycles of relative rise and fall of sea level. For example, sequences deposited during times of sea-level lowstands and highstands commonly have quite different regional distributions and sedimentary characteristics (Part 3, Vail et al, this volume). Differences between regional and global sea level curves may indicate times of local structuring of sufficient magnitude to change the regional curve.

A useful means of summarizing stratigraphic information for sea level analysis is a chronostratigraphic correlation chart in which geologic time is plotted as the vertical scale and distance across the area of interest is plotted as the horizontal scale. An example (Fig. 2) was prepared for the region of the seismic section of Figure 1. This chart summarizes a wide range of information on: (1) relations of sequences to bounding unconformities, showing areas of onlap, downlap, toplap, and truncation; (2) correlation of sequences to standard geochronologic units such as ages and epochs; (3) hiatal gaps along unconformities; (4) the relation and correlation of named lithostratigraphic units such as groups and formations within the sequences (not shown on Fig. 2); (5) distribution of facies and environment; and (6) well and outcrop control. The chart also aids in the selection of units for stratigraphic mapping and for structural analysis.

Figure 2 also includes the regional curve of relative change of sea level, derived mainly from the type of information shown on the chronostratigraphic correlation chart. Construction of the Jurassic–Early Cretaceous part of the curve and its comparison to the global curve is described in more detail in Part 8 (Vail et al, this volume). Similarities in the regional and global curves for the Jurassic–Lower Cretaceous sequences indicate a normal pattern of relative changes of sea level during this time. In contrast, the Late Cretaceous–Tertiary curve is anomalous, indicating significant regional uplift in Oligo-Miocene time.

EXAMPLE OF STRATIGRAPHIC INTERPRETATION OF SEISMIC REFLECTION PATTERNS

An area off the coast of western Africa is used as a simplified example to illustrate the seismic stratigraphic analysis procedure described above. Figure 1 is an example of subdivision of a seismic section into seismic sequences using the criteria of

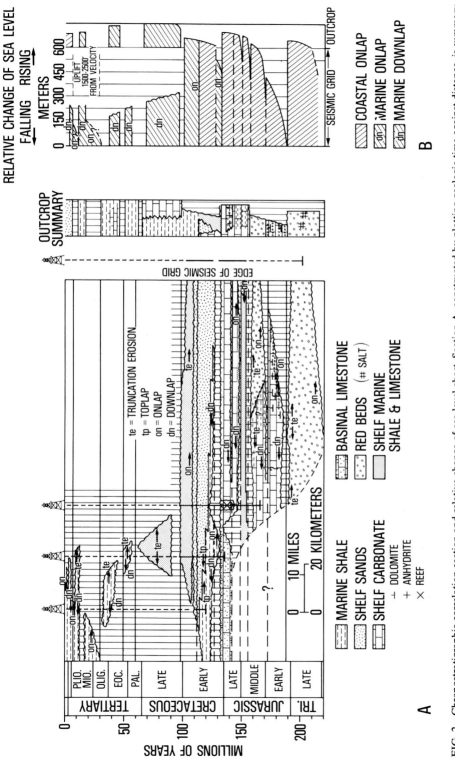

FIG. 2.—Chronostratigraphic correlation section and relative change of sea level chart. Section A, constructed by plotting geologic time against distance, is summary of relations of sequences to bounding unconformities, correlation to geochronologic units, hiatal gaps along unconformities, facies and environment, and well and outcrop control. Chart B shows regional relative changes of sea level based on measurements of coastal aggradation and other information summarized on section A.

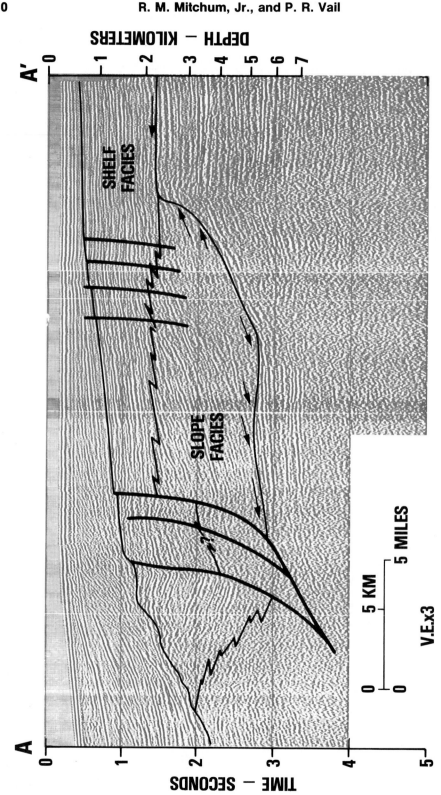

FIG. 3.—Generalized seismic facies, Line A-A', offshore western Africa (map location Figs. 4, 5). Two generalized seismic facies interpreted. Strata deposited in shelf-depositional environment prograde right to left over deposits of slope environment.

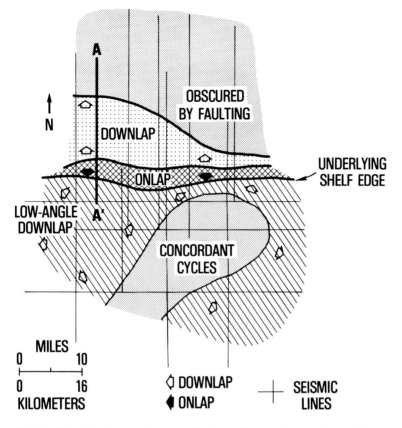

FIG. 4—Seismic facies map showing reflection patterns at lower surface of Lower Cretaceous Valanginian through Aptian sequences, offshore western Africa.

onlap, downlap, toplap, and truncation to recognize sequence boundaries. Using well control and a knowledge of the areal geology, the ages of the sequences are determined as accurately as possible, and the stratigraphy of the section is interpreted. This gross stratigraphy is summarized on the chronostratigraphic chart of Figure 2. On this chart the distribution of sequences by age and geographic extent can be plotted, in addition to their relations to the unconformities at their upper and lower boundaries and their gross lithologies, and other information. Comparison of regional and global sea level curves (Fig. 2) shows close similarities in the Jurassic–Early Cretaceous.

Description of Seismic Section

Figures 1 and 3 are interpretations of seismic section A-A', which is identified on the intersecting grid of seismic lines shown on the map (Figs. 4, 5). Figure 1 designates ages of sequences and Figure 3 shows generalized seismic facies of part of the Lower Cretaceous section. Although this interval includes three mappable sequences (Valanginian, Hauterivian–lower Aptian, and middle–upper Aptian on Fig. 1), these sequences have been grouped for purposes of simplification in this generalized example.

The lower bounding surface of the mapped sequences represents the upper surface of a large Tithonian-Berriasian carbonate shelf which culminates in a major shelf margin reef. The shelf carbonates become thin (to the left) as they pass into slope and basin deposits. This bathymetric distribution has controlled the deposition of the overlying Cretaceous units. The overlying Valanginian sequence (Fig. 1) is at this location a steeply dipping marine wedge that thins updip by apparent onlap, and downlap by downdip against underlying slope and basin deposits. This unit is not present in detectable thickness on the shelf where Hauterivian beds directly overlie Berriasian strata.

Valanginian, Hauterivian, and Aptian sequences display two different generalized seismic facies units interpreted as shelf and slope facies

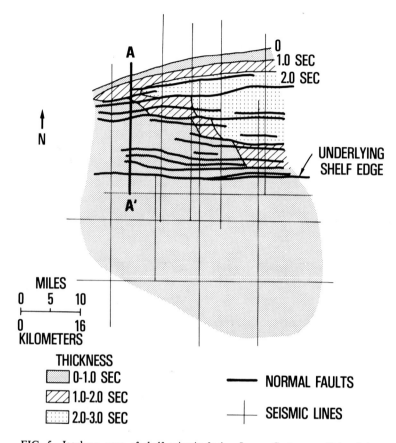

FIG. 5—Isochron map of shelf seismic facies, Lower Cretaceous Valanginian through Aptian sequences, offshore western Africa. Thicknesses measured in two-way travel time.

(Fig. 3). To the right, the reflection pattern of the shelf facies is mostly parallel to slightly divergent, with high-amplitude, continuous cycles. Low-angle downlap dips in a right-to-left direction against the lower bounding surface. Depositional environment of strata generating this seismic facies is interpreted as a nonmarine to shallow marine shelf with high or variable depositional energy. To the left, this seismic facies progrades over the slope seismic facies. The slope facies consists of a complex sigmoid-oblique reflection pattern with fairly low-amplitude, discontinuous character, and prominent high-angle downlap at the base. This seismic facies is interpreted as strata deposited in a deep marine slope environment of low depositional energy.

The contact between the two seismic facies is interpreted as a prograding shelf margin, which becomes progressively younger to the left. Erosional truncation is pronounced along the upper

surface of the mapped sequence to the left, and large growth faults are present to the left. The shelf seismic facies is interpreted as being much thicker on the downthrown side of some major faults whose growth appears contemporaneous with sedimentation.

Seismic Facies Maps

Two seismic facies maps are presented: (1) a seismic reflection pattern map showing distribution and direction of onlap, downlap, and concordance at the base of the mapped sequences (Fig. 4); and (2) an isochron map (thickness measured in seconds of two-way time) of the shelf seismic facies unit (Fig. 5). Although made from different parameters, the two maps support the same environmental and depositional interpretation.

Figure 4 shows the types of seismic reflection relations such as onlap, downlap, and concordance, along the lower surface of the Lower Cretaceous mapped unit. The shelf area lies to the south and the basin area to the north of the underlying prominent shelf edge of the Tithonian-Berriasian carbonate sequence. Seismic reflections in the basin area to the north are predominantly downlap, dipping north onto the deep water basin floor. Strong apparent onlap defines the steep slope produced by the underlying Tithonian-Berriasian carbonate shelf edge. These relations are obscured by faulting over much of the area.

The seismic facies pattern on the platform to the south shows a series of arrows indicating low-angle downlap radiating from an elongate central area where reflections are mostly concordant to the lower surface. This pattern of downlap and concordance is interpreted as a lobe of parallel-bedded strata, built in the central area from a southern source, and prograding from the center, whereas the deltaic lobe grew. As the basin filled, the parallel-bedded shelf facies overrode the slope facies and prograded northward across it (Fig. 3).

Figure 5 shows the isochron map (two-way time thickness) of the shelf seismic facies unit. Thickness may be measured in two-way time (isochron) or converted to feet (isopach) if there are sufficient velocity data. The map shows a uniform thickness of less than 1.0 sec over the platform area and most of the western basin area, but a pronounced thickening to greater than 2.0 sec occurs in the eastern basin area. Thickening seems to occur along growth faults, which are most common in the area of thick shelf facies. This relation suggests that sedimentary loading may be one control of the faulting.

Comparison of the two maps suggests an areal relation of the direction of elongation of the area of concordant reflections on Figure 4 and the thick shelf facies on Figure 5. This relation could be interpreted as occurring in a deltaic feature prograding northeastward across the platform and into the eastern half of the basin.

Seismic Stratigraphy and Global Changes of Sea Level, Part 8: Identification of Upper Triassic, Jurassic, and Lower Cretaceous Seismic Sequences in Gulf of Mexico and Offshore West Africa [1]

R. G. TODD AND R. M. MITCHUM, JR.[2]

Abstract Seismic stratigraphic techniques permit identification of Upper Triassic, Jurassic, and Lower Cretaceous sequences in strata from the North American Gulf Coast and West Africa. Several distinct sequences are remarkably persistent from the Florida panhandle, around the perimeter of the Gulf Coast and into northern Mexico, a distance of more than 1,500 mi (2,400 km). Their identification requires the integration of seismic data with lithologic, environmental-facies, biostratigraphic, radiometric, and well-log information. A comparison with strata of comparable age offshore West Africa indicates the same sequences can be recognized there. The sequences in North America and Africa are interpreted to be controlled by global changes of sea level because they occupy the same time-stratigraphic positions and display coastal onlap patterns similar to those previously recognized elsewhere.

Gulf Coast seismic sequences and the formations present within them are: (1) Upper Triassic and Lower Jurassic (Hettangian?)—Eagle Mills Formation; (2) Lower Jurassic (Sinemurian through Toarcian)—known only from southern Mexico and not identified in the study area; (3) Middle Jurassic (Bajocian-Bathonian)—Werner anhydrite–Louann Salt interval; (4) upper Middle Jurassic (Callovian)—Norphlet Formation; (5) Upper Jurassic (Oxfordian-Kimmeridgian)—Smackover-Buckner-Gilmer (Haynesville) Formations; (6) Upper Jurassic–Lower Cretaceous (Tithonian-Berriasian)—most of the Cotton Valley Group; and (7) Lower Cretaceous (Valanginian)—a restricted sedimentary wedge seen only in a basinward position.

West African seismic sequences are: (1) Upper Triassic and Lower Jurassic (Hettangian?); (2) Lower Jurassic (Sinemurian through Toarcian); (3) Middle Jurassic (Bajocian); (4) Middle Jurassic (Bathonian); (5) Middle Jurassic (Callovian); (6) Upper Jurassic (Oxfordian-Kimmeridgian); (7) Upper Jurassic–Lower Cretaceous (Tithonian-Berriasian); and (8) Lower Cretaceous (Valanginian).

INTRODUCTION

Seismic stratigraphic techniques (Part 7, Vail et al, this volume) permit identification of Upper Triassic, Jurassic, and Lower Cretaceous sequences (Parts 3 and 4, Vail et al, this volume) in strata from the North American Gulf Coast and from West Africa. Our intent is to document a group of related Jurassic–Lower Cretaceous sequences which represent one supersequence; we interpret these sequences to be caused by changes of sea level on a global scale because they occupy the same time-stratigraphic positions and display coastal onlap patterns similar to those recognized elsewhere.

Although time stratigraphy can be worked out by means of electric-log correlations, seismic sections can provide a more rapid and (commonly) more accurate way of doing so. Improved accuracy results because a grid of seismic sections, although having considerably lower resolution than electric logs, provides a continuous profile of stratigraphy. However, electric logs must be correlated from well to well to depict the stratigraphy.

NORTHERN GULF OF MEXICO

The subsurface Jurassic trend (Fig. 1) can be traced from the Florida panhandle around the perimeter of the Gulf Coast into northern Mexico, a distance of over 1,500 mi (2,400 km). Although facies changes are locally pronounced, a number of distinct sequences are identified along this trend by integrating seismic data with lithologic, environmental-facies, biostratigraphic, radiometric, and well-log information. The Gulf Coast Jurassic trend within the United States is restricted to the subsurface; however, numerous excellent outcrops in northern Mexico provide valuable lithologic and biostratigraphic information which can be applied to our understanding of correlative subsurface Jurassic units along the northern Gulf Coast.

Identification of Sequences

Seismic line A (Fig. 2) is on the west side of the East Texas basin. To the left, in an updip direction, is a well-developed, down-to-basin fault; a low-relief salt swell is present on the right side of the section. Four prominent sequences are identified. Well 1, near the center of the section, ties the identified seismic sequences to existing subsurface stratigraphy. Two blow-ups of the seismic section (Figs. 3, 4) illustrate the criteria used to identify the sequences. The seismic sequences and their contained stratigraphic units in ascending order are:

[1]Manuscript received, January 6, 1977; accepted, June 13, 1977.

[2]Exxon Production Research Company, Houston, Texas 77001.

145

R. G. Todd and R. M. Mitchum, Jr.

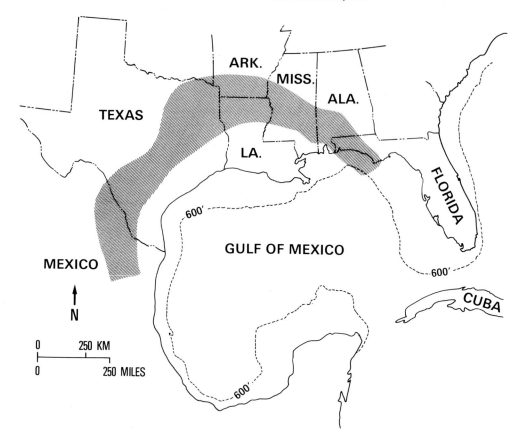

FIG. 1—Subsurface Jurassic trend for northern Gulf of Mexico.

1. A sequence consisting of the Eagle Mills Formation (EM) defined by erosional truncation in an updip position.

2. The Werner anhydrite–Louann Salt sequence (L-W) which displays several cycles of onlap against the older Eagle Mills sequence updip, and possibly onlaps Upper Paleozoic rocks farther downdip. In addition, the Eagle Mills sequence shows evidence of erosional truncation. The salt section on seismic line A is less widespread landward than the Eagle Mills Formation and nearly laps out against it.

3. A sequence which contains the Smackover, Buckner, and Gilmer Formations (S-B-G) [3] and

shows gentle onlap against the underlying Louann-Werner sequence. The Norphlet Formation, which is between the Louann Salt and Smackover Formation, probably represents a discrete sequence but is thin in this area and is below seismic resolution. The Smackover onlap is subdued; however several basal terminating cycles are visible on seismic line A. The sharp lithologic break between Smackover marine carbonate rock and underlying nonmarine Norphlet sandstone and siltstone also helps support this lower sequence boundary. A prominent seismic sequence boundary occurs at the top of the Gilmer Limestone as a result of strong onlap by black shale of the Bossier Formation onto the underlying Gilmer Limestone.

4. Finally, a sequence containing the Cotton Valley Group (CV) overlies the Smackover-Buckner-Gilmer sequence. The onlap of Bossier shale against the Gilmer Limestone surface defines the lower boundary. The upper boundary for the Cotton Valley sequence, although representing a

[3]Recently, Forgotson and Forgotson (1976, p. 1119) proposed the name Gilmer Limestone for the "limestone section underlying the Bossier (shale) Formation and overlying either the Buckner Formation or the Smackover Formation." Gilmer limestone would replace the informal terms *Cotton Valley limestone* or *Haynesville limestone* commonly used in East Texas. The Gilmer Limestone is coeval with the Haynesville Formation of Arkansas and Louisiana, a terrigenous sandstone and shale unit.

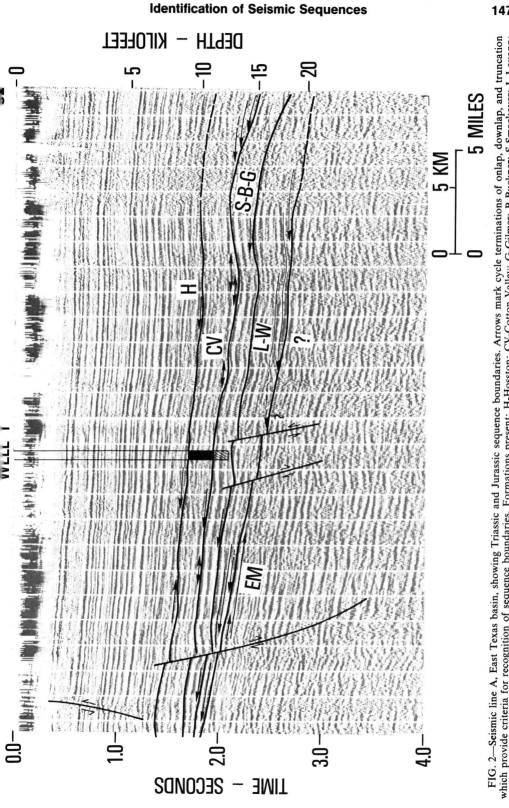

FIG. 2.—Seismic line A, East Texas basin, showing Triassic and Jurassic sequence boundaries. Arrows mark cycle terminations of onlap, downlap, and truncation which provide criteria for recognition of sequence boundaries. Formations present: H-Hosston; CV-Cotton Valley; G-Gilmer; B-Buckner; S-Smackover; L-Louann; W-Werner; EM-Eagle Mills.

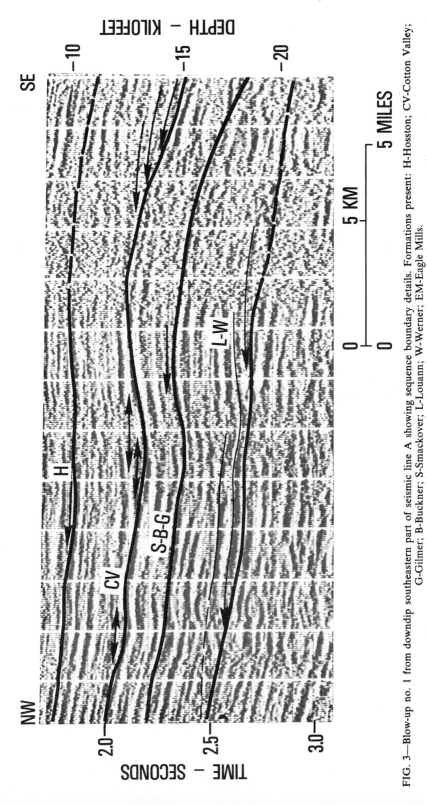

FIG. 3—Blow-up no. 1 from downdip southeastern part of seismic line A showing sequence boundary details. Formations present: H-Hosston; CV-Cotton Valley; G-Gilmer; B-Buckner; S-Smackover; L-Louann; W-Werner; EM-Eagle Mills.

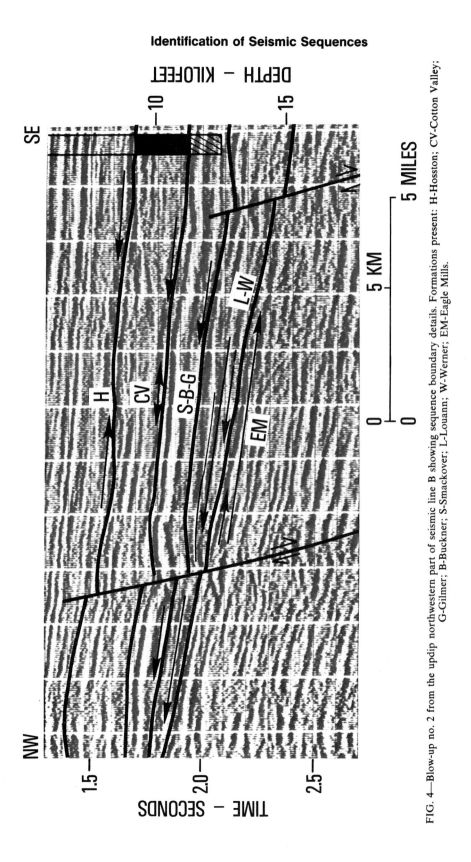

FIG. 4—Blow-up no. 2 from the updip northwestern part of seismic line B showing sequence boundary details. Formations present: H-Hosston; CV-Cotton Valley; G-Gilmer; B-Buckner; S-Smackover; L-Louann; W-Werner; EM-Eagle Mills.

major unconformity in the Gulf Coast, commonly is difficult to document seismically because overlying Hosston sandstones (H) lie on Cotton Valley sandstones. Hosston onlap can be seen on the southeast half of the section and some irregularity, visible on the northwest end of the section, probably is associated with erosional truncation of the Cotton Valley section.

East Texas Stratigraphy

Figure 5 is a generalized cross section across the East Texas basin using seismic, well log, and core and cuttings information. It shows distribution of the Jurassic units and their lithofacies. The heavy black lines represent sequence boundaries. On the right side of Figure 5 is a sea level cycle chart, showing regional relative changes of sea level for each sequence.

Eagle Mills Formation—At the base of the cross section is the Eagle Mills Formation which directly underlies the Jurassic supersequence. The Eagle Mills is a red-bed section of shale, siltstone, sandstone, and conglomerate. Intrusive igneous sills and dikes are common.

Louann–Werner Formation—The Louann Salt and underlying Werner anhydrite occur at the base of the Jurassic supersequence. The Louann-Werner interval onlaps the older Eagle Mills section and ultimately laps out against it in a landward direction, as shown diagrammatically on Figure 5. The Louann is represented mainly by salt. The Werner is probably a marginal equivalent of the Louann Salt and consists mostly of horizontal-laminated and cross-laminated anhydrite, but coarse sandstone and conglomerate locally underlie the anhydrite.

Norphlet Formation—The Norphlet Formation overlies the Louann Salt and represents a comparatively thin interval of red and pink sandstone and siltstone. Farther east in Mississippi, the Norphlet is more than 1,000 ft (305 m) thick and contains numerous coarse conglomerate beds. Regionally in the northern Gulf Coast, the Norphlet oversteps the Louann landward.

Smackover–Buckner–Gilmer Formations—The Smackover Formation onlaps the Norphlet and contains three major facies: (1) a basal laminated micrite which formed in an intertidal to slightly subtidal environment (this facies commonly laps out in a landward direction); (2) an overlying pelletal micrite facies which formed in a shallow subtidal setting; and (3) at the top, a complex of high-energy coated grains (hardened pellets are present in an updip position, higher energy oolites occur farther seaward, and finally, a mixed-coated grain facies of hardened pellets, oolites, and algal pisolites are formed farthest seaward). Dolo-

mitization of coated grains forms most of the Smackover reservoirs in East Texas. The Smackover oversteps the Norphlet updip and overlies Eagle Mills or Upper Paleozoic deposits.

The Buckner Formation overlies the Smackover and is in part correlative with it; it consists of interbedded red beds and nodular anhydrite which originated in a supratidal (sabkha) setting landward of the Smackover intertidal oolite environments. Locally, salt and thin limestone beds occur. Brines derived from Buckner evaporites dolomitized the underlying Smackover carbonate grainstones, creating important secondary porosity. Seaward the Buckner red beds and anhydrite change facies into limestone containing a combination of oolites, algal pisolites, and hardened pellets.

The Gilmer Limestone in East Texas is almost entirely limestone which has transgressed across Buckner supratidal red beds and evaporites. The correlative formation in Arkansas and Louisiana is the Haynesville Formation composed mainly of sandstone and shale. A high terrigenous clastic influx in eastern Louisiana and Mississippi diminishes toward East Texas which permits carbonate sediments to form there. The rather abrupt vertical change in lithology from red beds into limestone in East Texas suggests a sequence boundary at the Buckner-Gilmer contact, but we have not been able to verify it seismically. Detailed well-log correlations in an updip position indicate an erosional unconformity between the Buckner and Gilmer. The landward peak on the regional sea level cycle chart (Fig. 5) opposite the Gilmer Formation represents the rapid shift we see regionally, based on facies analysis, from supratidal red beds and evaporites of the Buckner to intertidal-subtidal carbonates of the Gilmer. This landward shift of the cycle chart which apparently lacks a preceding relative lowstand of sea level is termed a paracycle. (Parts 3, 4, Vail et al, this volume, for a more complete discussion.)

Cotton Valley Group—The Cotton Valley Group represents the uppermost sequence of the Jurassic supersequence. The Hosston Formation of Early Cretaceous age directly overlies it. The Bossier, Schuler, and Knowles Formations comprise the Cotton Valley Group. The Schuler Formation is a nearshore to continental sandstone, conglomerate, and shale unit; the Bossier Formation represents its offshore deeper water equivalent. The Bossier Formation strongly onlaps the Gilmer Limestone surface and displays obvious cycle terminations against it (Figs. 2-4) thereby defining the lower sequence boundary. The upper several hundred feet of the Cotton Valley Group, in a seaward position, consists of low energy mi-

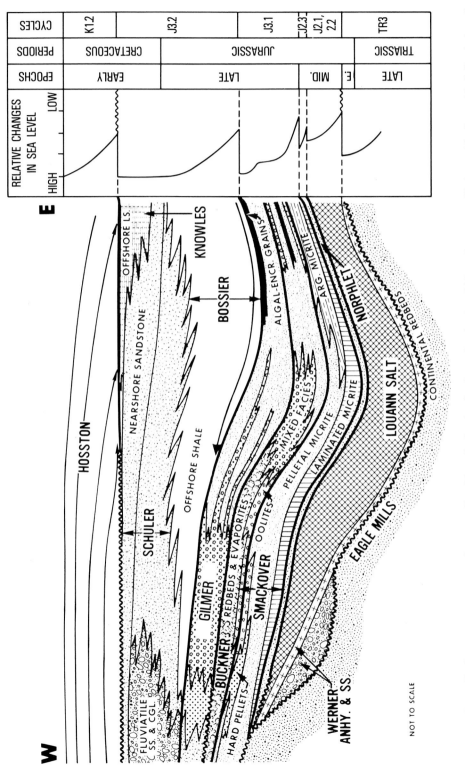

FIG. 5—Generalized west to east cross section across East Texas basin, showing formational units within each sequence (designated by heavy black lines). A cycle chart summarizes regional relative changes of sea level for each sequence. Cycles are not plotted to a linear time scale; cycle notations e.g., J2.1, correspond to Figure 2 in Part 4 (Vail et al, this volume).

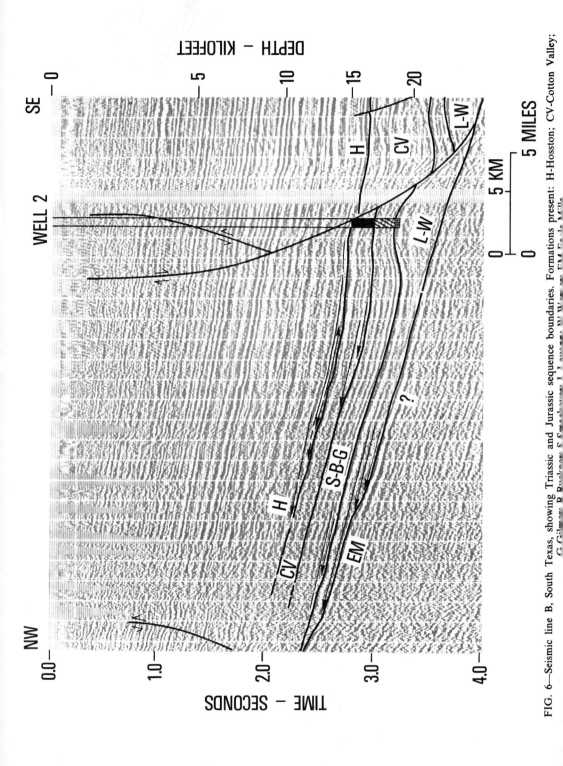

FIG. 6—Seismic line B, South Texas, showing Triassic and Jurassic sequence boundaries. Formations present: H-Hosston; CV-Cotton Valley; G Gilmer; P Bosbara, S Smackover, L Louann, W Werner, EM Eagle Mills.

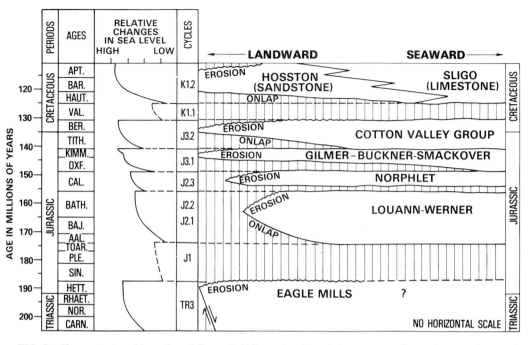

FIG. 7—Chronostratigraphic section of Texas Gulf Coast showing relative amounts of coastal encroachment of Triassic, Jurassic, and Lower Cretaceous sequences. A cycle chart summarizes regional relative changes of sea level for each sequence; cycle notations, e.g. J2.1, correspond to Figure 2 in Part 4 (Vail et al, this volume).

critic limestone and some high energy grainstones interbedded with shale. This dominantly limestone section has been named the Knowles Limestone by Mann and Thomas (1964, p. 145), replacing the term Cotton Valley "B" limestone of informal subsurface usage. The Knowles Limestone becomes progressively thicker seaward whereas the influx of terrigenous clastics diminished farther offshore.

South Texas Seismic Sequences

Seismic line B (Fig. 6) is in South Texas nearly 300 mi (480 km) southwest of line A; it is oriented northwest-southeast. As in the East Texas example (Figs. 2-4), the same four sequences are identifiable. Well 2, near the southeast end of line B, penetrated the Louann Salt and ties the seismic sequences to subsurface stratigraphy. The seismic sequences in ascending order are: (1) an older Eagle Mills sequence which underlies the Louann Salt and Werner anhydrite; (2) a Werner anhydrite–Louann Salt sequence which sharply onlaps older Eagle Mills or Late Paleozoic rocks (several cycles of onlap are evident [Fig. 6] and the salt eventually laps out updip); (3) a Smackover-Buckner-Gilmer sequence onlaps the Louann Salt surface. As in the East Texas example, the Norphlet Formation is below seismic resolution and

may be absent. Basal Smackover coastal deposits progressively onlap against the Louann Salt surface, and finally overlap it and lie directly on Eagle Mills and/or Late Paleozoic rocks in an updip direction; (4) a Cotton Valley sequence composed of the Schuler, Bossier, and Knowles Formations. The Bossier Formation onlaps the Gilmer Limestone but the onlap is not as strong as in the East Texas example. A sequence boundary is seen at the top of the Cotton Valley Group where the Hosston Formation downlaps against the Knowles Limestone section seaward and onlaps the Cotton Valley surface landward. Farther south, Hosston terrigenous clastics change facies into a limestone (Sligo Formation). In South Texas, planktonic microfossils recovered from the limestone indicate a substantial time break at the sequence boundary between the Sligo and Cotton Valley.

Geologic Age of Texas Gulf Coast Sequences

As was discussed in Part 3 (Vail et al, this volume), a chronostratigraphic section differs from a stratigraphic section in having the vertical dimension represent absolute geologic time rather than thickness. We believe it provides valuable information by illustrating how interpreted sequences fit into an area/time framework. Figure 7 is a chronostratigraphic section which shows how the identified Texas Gulf Coast seismic sequences are distributed in time and space. Geologic time is plotted vertically on the right side of the section, and the area distribution of the various sequences is shown horizontally with the landward direction to the left. Sequences which lap out landward, such as the Louann-Werner or the Smackover-Buckner-Gilmer, provide information about sea level position. In these two examples, coastal onlap is documented by well data and seismic reflection patterns. Adjacent to the chronostratigraphic section, sea-level position is summarized on a cycle chart which shows relative changes in sea level for the Texas Gulf Coast region. A column adjacent to the regional cycle chart contains notations which identify specific stratigraphic units (Part 4, Vail et al, this volume).

The geologic ages of the Cotton Valley Group, Gilmer, Buckner, Smackover, and Eagle Mills Formations are reasonably well-established. The age of the Louann, Werner, and Norphlet Formations was interpreted by using the global cycle charts (Part 4, Vail et al, this volume, Fig. 2). Age dating of the identified sequences was accomplished by using three methods—radiometric, biostratigraphic, and by comparison with global cycle charts.

Late Triassic–Early Jurassic (Hettangian?)

The Eagle Mills Formation contains plant macrofossils which represent a Late Triassic to possibly Early Jurassic age (Scott, Hayes, and Fietz, 1961). It is intruded also by basaltic igneous rocks, several of which are radiometrically dated as forming 195, 197 (Baldwin and Adams, 1971), and 180 to 200 m.y. Basaltic intrusions suggest that the Eagle Mills is largely Late Triassic; plant macrofossils suggest that the Eagle Mills could be as young as Early Jurassic at its top. We assign the Eagle Mills sequence to a Late Triassic–Early Jurassic (Hettangian?) age based on the paleontology and radiometric dates and our Mesozoic global chart of relative changes of sea level.

Early Jurassic

Our experience on a worldwide basis indicates that Lower Jurassic sequences are areally restricted (Part 4, Vail et al, this volume, Fig. 2). The relatively widespread distribution of the Louann-Werner and Norphlet in the northern Gulf Coast suggests they fit better into the Middle Jurassic. Conversely, if we assume the Louann-Werner and Norphlet to be Early Jurassic in age, then we must explain the absence of Middle Jurassic rocks in the Northern Gulf of Mexico; worldwide experience with global cycles indicates Middle Jurassic rocks to be compara-

tively widespread and common in their occurrence.

Early Jurassic marine rocks in the gulf region are known only from southern Mexico (Erben, 1956a). Their age, based on ammonites, is Sinemurian–early Pliensbachian. In this same area correlative beds contain coal deposits (Erben, 1956b). The restricted areal distribution of these beds and their association with coals, which probably formed in a warm, temperate climate, suggests the Early Jurassic was not a favorable time for deposition of Louann and Werner evaporites. Although not conclusive, the lack of Hettangian-aged rocks in this Lower Jurassic section suggests that the base of the Jurassic supercycle begins with the Sinemurian.

Middle Jurassic

Paleontologic support for a Middle Jurassic age assignment for the Louann–Werner evaporite section is provided by Kirkland and Gerhard (1971). They identified a Middle or Late Jurassic palynomorph assemblage from calcite cap rock associated with a salt diapir, Challenger Knoll, in the central Gulf of Mexico. Kirkland and Gerhard inferred the cap rock to be derived from the underlying bedded salt and believe the salt is correlative with the Louann Salt of the northern Gulf of Mexico. Recently Watkins, Worzel, and Ladd (1976) reviewed evidence available for age-dating the Louann and concluded it as Middle(?) Jurassic in age.

The Norphlet sandstone is unfossilerous. It is below seismic resolution in East Texas; however, in central Mississippi where it is thickest, it downlaps against the Louann Salt surface, thus defining its lower boundary. Throughout the northern Gulf, marine Smackover carbonate rocks onlap and transgress nonmarine Norphlet clastic rocks, thereby defining the upper sequence surface. Based on global cycle charts, the Norphlet is probably late Middle Jurassic (Callovian) in age.

Early Late Jurassic

The Smackover Formation was dated by Imlay (1943) as late Oxfordian, based on ammonites found in wells in northern Louisiana. Limestones equivalent to the Buckner Formation also contain ammonites dated as early Kimmeridgian by Imlay. Ammonites found in wells in Louisiana and Texas indicate the Haynesville Formation and the coeval Gilmer Limestone are late Kimmeridgian in age.

Late Jurassic–Early Cretaceous

The top of the Gilmer Limestone marks an important sequence boundary. A coccolith zone becomes extinct in the Bossier (shale) Formation directly overlying the Gilmer Limestone. The top of this coccolith zone, which is dated as early Tithonian, laps out against the Gilmer surface. The top of the Jurassic occurs within the Cotton Valley section; it is not a sequence boundary, but can be approximated using coccolith control.

The top of the Cotton Valley Group has been considered the top of the Jurassic in the Gulf Coast for many years. It is certainly an excellent sequence boundary. Nannoconids, which are planktonic microfossils, occur within the Knowles Limestone and indicate that the upper part of the Cotton Valley is not Late Jurassic but Early Cretaceous (Berriasian) in age, which agrees with the coccolith data.

As previously discussed, Hosston terrigenous clastic rocks change facies into carbonate rocks (Sligo Formation) downdip. In South Texas, Early Cretaceous (Hauterivian) nannoconids are found in the carbonate rocks directly overlying the Knowles Limestone which contains Berriasian microfossils. This suggest Valanginian rocks may be missing due to nondeposition or erosion. Several seismic sections in Central and East Texas, but farther downdip, display a restricted sequence lying between the Knowles Limestone and the overlying Sligo (limestone) Formation; although paleontologic confirmation is not available, we believe this wedge represents the missing Valanginian section.

OFFSHORE WEST AFRICA

A number of sequences are identified in Upper Triassic, Jurassic, and Lower Cretaceous deposits on seismic sections offshore West Africa (Part 2, Vail et al, this volume). The fundamental pattern of sequences and supersequences is similar to that observed in the Gulf of Mexico. The sequences are identified seismically, dated and described lithologically from well information, and related to regional cycles of relative changes of sea level.

Identification of Sequences

Two major sequence boundaries are especially evident on seismic line A (Fig. 8). The older one, located just below three seconds in the center of the seismic section, separates a Triassic supersequence from a Jurassic supersequence. This boundary is characterized by strong basal onlap in a landward direction by the Jurassic sequences against the underlying Triassic supersequence. The uppermost sequence in the Triassic supersequence is relatively extensive in a landward direction. It was traced onshore into a well containing a red-bed section with interbedded basic igneous intrusions that have been radiometrically dated as Late Triassic (Carnian, 215 m.y.) in age. Above the basic intrusions, but still within the red-bed

sequence, are sediments which are believed to be Early Jurassic. Ager (1974, p. 26) in his discussion of Mesozoic sediments in northwest Africa noted that the age of the red beds ranges into the Jurassic. Brachiopods occurring in dolomite beds above the Triassic salt pans indicate a marine transgression of probably Early Jurassic (Sinemurian) age. As in the dating of the Eagle Mills Formation in the North American Gulf Coast the evidence is not strong, but does suggest that the uppermost part of the Triassic supercycle incorporates Early Jurassic (Hettangian) sediments. The beginning of the Jurassic supercycle in West Africa is marked by a marine transgression and includes Early Jurassic (Sinemurian) rocks as its base.

The younger major boundary, between the Jurassic and Cretaceous supersequences, is defined on seismic line C (Fig. 9). It is depositional in origin and consists of a carbonate shelf margin which exhibits steep topography. A restricted shale wedge strongly onlaps this steep carbonate shelf margin and downlaps against correlative fondaform beds in a basinward direction. The restricted sequence is interpreted to represent a large seaward shift of marine clastic deposits which resulted from a rapid relative lowering of sea level (Part 3, Vail et al, this volume). This restriction defines the end of the Jurassic supersequence and the beginning of the Cretaceous supersequence. Line A (Fig. 8) has the details of this sequence boundary obscured as a result of faulting in the vicinity of the shelf margin.

The Jurassic supersequence consists of six seismic sequences (Figs. 8, 9) which are defined by minor unconformities and occur between the two major unconformities described above.

Lower Jurassic

The Jurassic supersequence contains at its base a restricted sequence (J1) which overlies and laps out landward against an interpreted Triassic section (TR3; Fig. 8). We interpret the Jurassic onlap to be coastal in nature and to represent a major seaward shift of coastal onlap as compared with that of the underlying widespread Upper Triassic–Lower Jurassic (Hettangian?) sequence (TR3). Although this restricted sequence is not penetrated by wells in the study area, it is interpreted to be of Early Jurassic (Sinemurian through Toarcian) age because of its stratigraphic position and also because of its restricted distribution which agrees with global cycle charts (Part 4, Vail et al, this volume, Fig. 2).

A low-relief Lower Jurassic carbonate shelf margin is present on seismic lines A (Fig. 8) and C (Fig. 9). Line C contains on its eastern end only the distal fondaform and clinoform parts; the car-

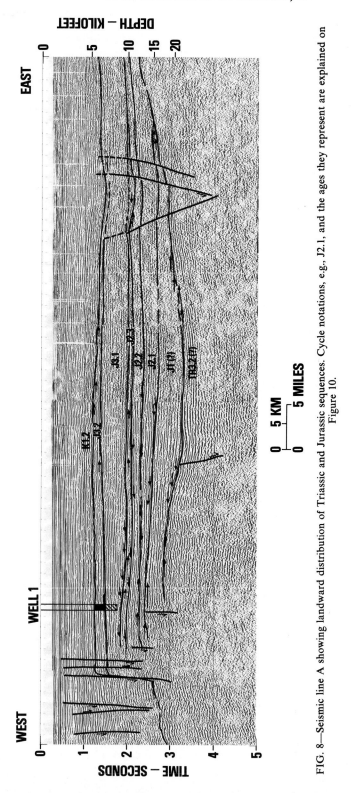

FIG. 8—Seismic line A showing landward distribution of Triassic and Jurassic sequences. Cycle notations, e.g., J2.1, and the ages they represent are explained on Figure 10.

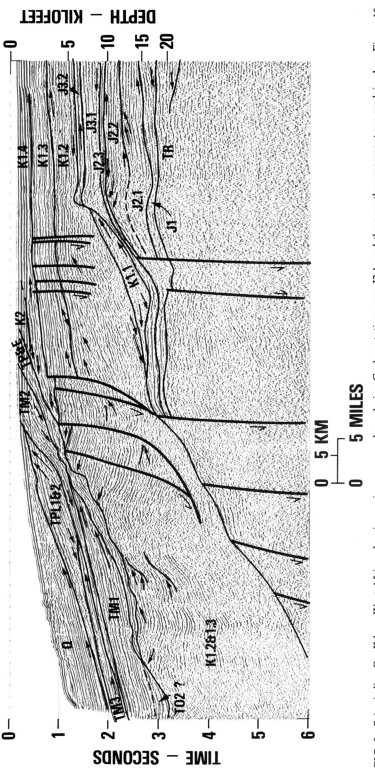

FIG. 9—Seismic line C, offshore West Africa, showing major sequence boundaries. Cycle notations, e.g. J2.1, and the ages they represent are explained on Figure 10.

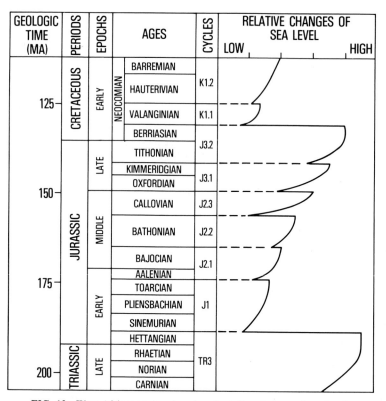

FIG. 10—West African regional cycles of relative changes of sea level.

bonate-shelf edge is inferred to occur east of the end of the seismic section. Based on seismic lines A, B, and C, (Part 2, Vail et al, this volume), water depth seaward of the Lower Jurassic shelf edge probably was not great.

Middle Jurassic

Overlying the Lower Jurassic are three seismic sequences assigned to the Middle Jurassic (J2). The oldest onlaps the Lower Jurassic and oversteps it landward, but also laps out against the underlying Triassic. The remaining two sequences are each distributed successively farther landward than the previous one, but they also lap out against the more widespread Triassic surface. The uppermost sequence was penetrated by a well which recovered fossils of Middle Jurassic (Callovian) age. The two older sequences have not been dated. Using a combination of well control and global cycle charts (Part 4, Vail et al, this volume, Fig. 2), the three sequences are interpreted to be of Middle Jurassic Bajocian, Bathonian, and Callovian age (Fig. 10).

Bajocian (J2.1)—The lower boundary of this sequence is characterized by onlap in a landward

direction as seen on lines A (Fig. 8) and C (Fig. 9). A low-relief shelf margin occurs on the eastern end of line C. It probably consists of high-energy carbonate sediments based on its oblique progradational pattern and toplapping upper reflections.

Bathonian (J2.2)—This sequence shows strong onlap against the older Bajocian shelf margin (Fig. 9) and then only gentle onlap farther landward. The regressive pattern of shelf margins shown by the Lower Jurassic and the Middle Jurassic (Bajocian) continues through the Bathonian in that its shelf edge is located several kilometers seaward of the preceding Bajocian margin. Well penetrations indicate that this shelf edge is composed of carbonate rocks.

The upper boundary for the Bathonian sequence is typified by erosional truncation; this is evident on all three seismic lines, particularly line C (Fig. 9). This erosional truncation at the top of the Bathonian correlates with a substantial drop in sea level.

Callovian (J2.3)—Overlying the truncated Bathonian surface is a thin shale section represented on the seismic sections by one cycle. East of line C (Fig. 11), this cycle diverges to many cycles and forms a prograding shelf margin with clinoform topography. The clinoforms exhibit toplap and are overlain by two flat-lying cycles. This seismic facies pattern suggests a prograding deltaic complex of flat-lying sand-prone delta-front or delta-plain deposits overlying shale-prone prodelta sediments. A well, drilled landward of this shelf margin, indicates it likely is composed of terrigenous sandstones and shales.

A deep-water origin for the thin shale unit present on line C is verified by well control. The shale contains a foram assemblage indicative of upper bathyal water depths. Fossils found in the shale are Callovian in age. Based on global sea level cycle charts, this seismic sequence is dated as Callovian.

Upper Jurassic–Lowermost Cretaceous

Two sequences of Late Jurassic and Early Cretaceous (J3) age which comprise the upper part of the Jurassic supercycle (Fig. 10) are found on West African offshore seismic lines. These sequences have been penetrated by several wells and their geologic age is documented.

Oxfordian–Kimmeridgian (J3.1)—As a result of the depositional topography created by the preceding Callovian shelf margin (Fig. 11), this sequence displays downlap against the underlying Callovian sequence on the east end of seismic lines B (Part 2, Vail et al, this volume, Fig. 4) and C (Fig. 9). Data quality is not as good on line A (Fig. 8), but the shelf edge of the Callovian is seen together with seaward downlap and landward onlap by the Oxfordian-Kimmeridgian sequence against it.

Landward, the Oxfordian-Kimmeridgian sequence displays coastal onlap against the Middle Jurassic (Callovian) and eventually oversteps it, indicating that the Oxfordian-Kimmeridgian has a more widespread distribution. Well 3, located

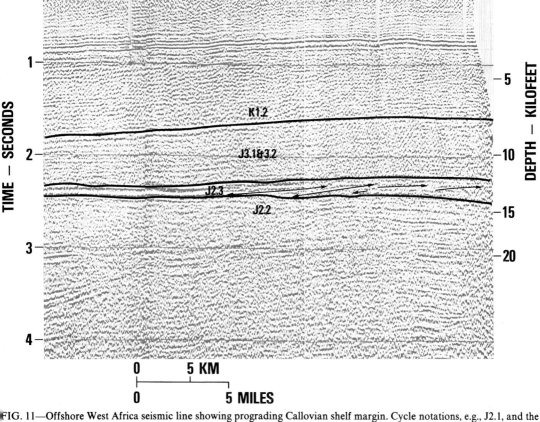

FIG. 11—Offshore West Africa seismic line showing prograding Callovian shelf margin. Cycle notations, e.g., J2.1, and the ages they represent are explained on Figure 10.

on line B (Part 2, Vail et al, this volume, Fig. 4), and well 1, located on line A (Fig. 8) both recovered fossils of Late Jurassic (Oxfordian-Kimmeridgian) age from this sequence. Strong landward encroachment by marine sediments of early Late Jurassic age has been noted along the West African coast (Ager, 1974, p. 30), the North Sea (Williams et al, 1975, p. 1589), the Gulf of Mexico (Imlay, 1943), and elsewhere in the world (Hallam, 1971, p. 142).

Seaward a carbonate shelf margin occurs. Of particular interest is the progressive increase in shelf to basin topography throughout the Jurassic. Corresponding to the increase in water depth is a noticeable thinning or "starving" of the carbonates basinward (Adams et al, 1951).

Tithonian-Berriasian (J3.2)—Wells 2 and 3 of line B (Part 2, Vail et al, this volume, Fig. 4), and well 1 of line A (Fig. 8) penetrated this sequence and found fossils of Late Jurassic (Tithonian) and Early Cretaceous (Berriasian) age (Fig. 10). This sequence, which marks the top of the Jurassic supersequence, incorporates sediments of earliest Cretaceous (Berriasian) age.

The Upper Jurassic–lowermost Cretaceous sequence marks the end of carbonate sedimentation for this area; and displays the greatest depositional topography of all the prograding Jurassic carbonate shelf margins. Cores taken in well 1 (Fig. 8) contain reef-building corals, oolites, and calcareous algae at the top of the unit. Because carbonate rocks on the shelf margin formed at or near sea level, an estimation of water depth seaward of the shelf margin can be made. At the location of well 2 (Part 2, Vail et al, this volume, Fig. 4), the vertical relief between the shelf edge and its correlative basinal position is about 0.7 sec one-way seismic travel time. Using an interval velocity of 12,000 ft/sec (3,658 m/sec) for the intervening shale and carbonate section and multiplying this value by the vertical relief in time, we estimate a water depth of 8,400 ft (2,560 m). This estimate does not take into consideration sediment compaction or postdepositional structural movements; however, for this particular example, they probably were negligible. Substantial shelf to basin relief is supported by well 2 which penetrated a carbonate section of pelagic lime muds containing abundant deep-water planktonic organisms such as calpionellids, nannoconids, calcisphaerulids, and ammonites.

The top of the Jurassic on line B is known in well 2 (based on the calpionellids and nannoconids found in limestone cores), and clearly falls below the top of the sequence boundary. Above this sequence boundary is a deep-water shale section which represents the beginning of a siliciclastic Cretaceous supersequence (K1). (See Part 7, Vail et al, this volume, for a discussion of Lower Cretaceous sedimentation).

Lower Cretaceous (Valanginian)

An important restricted sequence, the Valanginian (K1.1) (Fig. 10), lies directly in front of the Upper Jurassic–Lower Cretaceous carbonate platform and represents an initial restricted sequence of an ultimately more widespread Cretaceous supersequence. This restricted wedge is present on line B (Part 2, Vail et al, this volume, Fig. 4) and displays deep-marine onlap against the shelf margin and downlap seaward. The wedge is also evident on line C (Fig. 9), but the upper sequence boundary is not as well defined. Well 2 on line B was drilled into the restricted sequence and documents its age as Early Cretaceous (Valanginian) and the underlying carbonate unit as Late Jurassic (Tithonian)–Early Cretaceous (Berriasian). Well control on top of the carbonate bank, landward of the restricted unit, suggests that the Valanginian section is missing. Microfossils, mainly dinoflagellates, indicate that Early Cretaceous (Hauterivian) terrigenous clastic rocks unconformably overlie carbonate rocks of Berriasian age. Although a relative lowstand of sea level is indicated by these data, the magnitude of the drop can not be estimated from the deep-marine facies on these seismic lines.

Coastal Onlap of West African Jurassic Sequences

Lateral and vertical distribution of the Triassic, Jurassic, and Lower Cretaceous seismic sequences recognized for West Africa seismic lines A, B, and C are summarized on the geologic cross section in Figure 12. A chronostratigraphic section for offshore West Africa (Fig. 13) shows the distribution of these sequences when plotted against absolute geologic time. Regional cycles of relative changes in sea level also are shown. A Late Triassic–Early Jurassic (Hettangian?) sequence marks the top of the previous supercycle. Early Jurassic (Sinemurian through Toarcian) rocks are restricted. Middle Jurassic rocks are considerably more widespread; and, finally, the youngest sequences of Late Jurassic–Early Cretaceous age are the most widespread. Their widespread distribution is especially evident when compared with the extremely restricted Valanginian wedge which marks the beginning of the Cretaceous supercycle.

Referring to the geologic cross section (Fig. 12), the oldest sequence is assigned to the Late Triassic–Early Jurassic (Hettangian?) based on its widespread distribution and the large amount of coastal onlap seen in overlying sequences. As previously discussed, this sequence has been traced into red beds onshore which contain basaltic in-

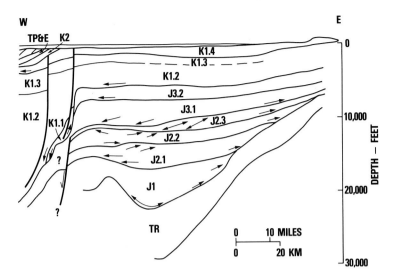

FIG. 12—Generalized geologic cross section for offshore cycle notations, e.g., J2.1, and the ages they represent are explained on Figure 10.

trusions of Late Triassic age and in the area is overlain by marine dolomites containing Early Jurassic (Sinemurian) brachiopods (Ager, 1974, p. 26).

The next overlying sequence has a considerably more restricted distribution. It is dated as Early Jurassic using global cycle charts; however, no wells have penetrated these rocks to confirm the interpretation.

The Middle Jurassic contains three sequences. The Bajocian onlaps the interpreted Lower Jurassic and ultimately lies directly on Triassic in a landward direction. The overlying Bathonian oversteps the Bajocian and is even more widespread landward. The uppermost sequence has been dated as late Middle Jurassic (Callovian) in age.

Above the Middle Jurassic two sequences are evident. The lower one ranges from the Oxfordian to the top of the Kimmeridgian; and the upper sequence includes the Tithonian and Berriasian Stages and ranges from Late Jurassic into the Early Cretaceous. Available well control supports these age assignments.

The restricted wedge, lying just in front of the carbonate platform, is Early Cretaceous (Valanginian) in age (based on well control). This restricted sequence marks the beginning of the Cretaceous supersequence.

COMPARISON OF WEST AFRICA–GULF COAST CYCLES

Plotting the regional cycle charts for the Texas Gulf Coast (Fig. 7) and West Africa (Fig. 13) against absolute geologic time allows comparison

with the Jurassic global cycle chart (Fig. 14). The regional cycle charts from West Africa and the Gulf of Mexico, and similar ones from the north slope of Alaska, the Northwest Shelf of Australia, the Middle East, and western Europe, ultimately have been normalized to remove local tectonic effects, then combined to arrive at our estimate of the global cycle chart for the Jurassic.

The Late Triassic–Early Jurassic (Hettangian?) sequence appears as a widespread unit in West Africa and Gulf of Mexico, and represents a relatively high position of sea level on our cycle chart. Although the Eagle Mills is a red-bed section in the Gulf Coast, we include it on a sea-level chart because the red beds probably formed near sea level and were affected by base-level changes caused by fluctuations in sea level.

The restricted sequence in West Africa which laps out against the underlying Late Triassic–Early Jurassic (Hettangian?) sequence is interpreted to be Early Jurassic in age and to represent a relatively lowstand of sea level. No rocks of a comparable age are known in the northern Gulf of Mexico, although a restricted sequence similar to the West African one is known from southern Mexico (Erben, 1956a).

Middle Jurassic cycles indicate higher relative positions of sea level than Early Jurassic ones. The controversial Louann Salt–Werner anhydrite sequence of the northern Gulf of Mexico is placed by us in the early Middle Jurassic (Bajocian-Bathonian) section based on palynology and its widespread coastal onlap. The Norphlet Formation is generally below seismic resolution but

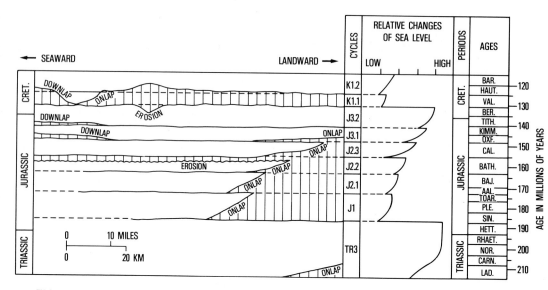

FIG. 13—Chronostratigraphic section offshore West Africa, showing relative amounts of coastal encroachment of Triassic, Jurassic, and Lower Cretaceous sequences. A cycle chart summarizes regional relative changes of sea level for each sequence.

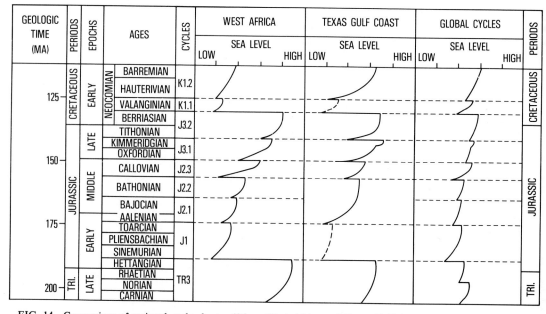

FIG. 14—Comparison of regional cycle charts offshore West Africa and Texas Gulf Coast with global cycle chart.

is dated as Callovian based on its position between the Louann Salt and the Smackover Limestone of well-defined Late Jurassic (Oxfordian) age. In West Africa, Bajocian, Bathonian, and Callovian sequences were recognized which had a similar distribution pattern to those identified in the Gulf Coast.

Sequences of the Late Jurassic (Oxfordian-Kimmeridgian) cycle in West Africa and the Gulf of Mexico display a prominent and widespread marine transgression related to a major relative rise of sea level. The Smackover-Buckner-Gilmer (Haynesville) formations of the northern Gulf of Mexico correspond to this cycle. A Tithonian-Berriasian cycle caps the top of the Jurassic supercycle. It is represented by the Cotton Valley Group of the North American Gulf Coast. Just how much coastal onlap occurred in the Tithonian-Berriasian is difficult to determine in both Africa and North America due to erosion.

A marked lowstand during the Valanginian cycle marks the end of the Jurassic supercycle and the beginning of a Cretaceous supercycle. The restricted Valanginian sequence is well developed on West African seismic data but only can be inferred in the Gulf Coast. On a worldwide basis, there is considerable paleontologic data suggesting that the Berriasian has more affinities with the Jurassic than the overlying Hauterivian and Valanginian (Wiedmann, 1968).

CONCLUSIONS

Seismic-stratigraphic techniques permit recognition of distinct sequences which have time-stratigraphic significance and can be traced reliably over many square miles. We interpret the sequences identified in North America and Africa to be controlled by relative changes of sea level on a global scale because they occupy the same time-stratigraphic positions and display coastal onlap patterns similar to those previously recognized in other basins around the world.

A main difference between the Gulf Coast and offshore West Africa is an interpreted lack of Early Jurassic rocks in the northern Gulf of Mexico. Along the Jurassic trend in the northern Gulf, exploration is conducted mainly along its updip limits and we may not encounter sequences having restricted distributions. A restricted sequence interpreted to represent the Lower Cretaceous (Valanginian) wedge, for example, is found only on seismic lines 40 mi (64 km) downdip from the main Jurassic–Early Cretaceous trend in the gulf.

REFERENCES CITED

Adams, J. E., et al, 1951, Starved Pennsylvanian Midland basin: AAPG Bull., v. 35, p. 2600-2607.

Ager, D. V., 1974, The western High Atlas of Morocco and their significance in the history of the North Atlantic: Geol. Assoc. (London), Proc., v. 85, no. 1, p. 23-41.

Arkell, W. J., 1956, Jurassic geology of the world: London, Oliver and Boyd Ltd., 806 p.

Baldwin, O. D., and J. A. S. Adams, 1971, K^{10}/Ar^{10} ages of the alkalic igneous rocks of the Balcones fault trend of Texas: Texas Jour. Sci., v. 22, no. 2 & 3, p. 223-231.

Erben, H. K., 1956a, Paleogeographic reconstructions for the Lower and Middle Jurassic and for the Callovian of Mexico: 20th Internat. Geol. Cong. (Mexico), Sec. 2, p. 35-41.

———— 1956b, New biostratigraphic correlations in the Jurassic of eastern and south-central Mexico: 20th Internat. Geol. Cong. (Mexico), Sec. 2, p. 43-52.

Forgotson, J. M., and J. M. Forgotson, Jr., 1976, Definition of Gilmer Limestone, Upper Jurassic formation, northeastern Texas: AAPG Bull., v. 60, p. 1119-1123.

Hallam, A., 1971, Mesozoic geology and the opening of the North Atlantic: Jour. Geology, v. 79, no. 2, p. 129-157.

Imlay, R. W., 1943, Jurassic formations of Gulf region: AAPG Bull., v. 27, no. 11, p. 1407-1533.

Kirkland, D. W., and J. E. Gerhard, 1971, Jurassic salt, central Gulf of Mexico, and its temporal relation to circum-Gulf evaporites: AAPG Bull., v. 55, no. 5, p. 680-686.

Mann, C. J., and W. A. Thomas, 1964, Cotton Valley Group (Jurassic) nomenclature Louisiana and Arkansas: Gulf Coast Assoc. Geol. Soc. Trans., v. 14, p. 143-152.

Scott, K. R., W. E. Hayes, and R. P. Fietz, 1961, Geology of the Eagle Mills Formation: Gulf Coast Assoc. of Geol. Soc. Trans., v. 11, p. 1-14.

Watkins, J. S., J. L. Worzel, and J. W. Ladd, 1976, Deep seismic reflection investigation of occurrence of salt in Gulf of Mexico: Texas Univ., Marine Science Inst., Contr. no. 84, 35 p.

Wiedmann, J., 1968, Das problem stratigraphischer grenzziehung und die Jura/Kreide-grenze: Eclogae Geol. Helvetiae, v. 61/2, p. 321-386.

Williams, J. J., D. C. Conner, and K. E. Peterson, 1965, Piper oil field, North Sea—fault-block structure with Upper Jurassic beach/bar reservoir sands: AAPG Bull., v. 58, p. 1585-1601.

Seismic Stratigraphy and Global Changes of Sea Level, Part 9: Seismic Interpretation of Clastic Depositional Facies [1]

J. B. SANGREE[2] and J. M. WIDMIER [3]

Abstract Depositional facies are predictable from seismic data through an orderly approach to the interpretation of seismic reflections. We term this approach "seismic facies analysis." Seismic facies analysis procedures are discussed in detail in Part 6, Vail et al, this volume. Seismic facies types generated by sand-shale strata vary mostly as a function of water depth at the time of deposition. Therefore, a regional environmental framework of shelf, shelf margin, and basin slope and basin floor, provides a useful gross subdivision for classification of clastic seismic facies units.

Shelf environments are characterized by general parallelism of reflections. Changes in reflection amplitude, frequency, continuity, interval velocity, and broad, low-relief mounds are the principal factors in defining seismic facies units.

The shelf-margin and prograded-slope environment typically contains thick marine sediments and has water depths sufficient for the development of complex arrangements of sigmoid and oblique prograding reflection patterns. The basin slope and floor environment includes a variety of deep basin facies as well as nonprograding slope facies and facies that extend from the slope into the basin deep.

INTRODUCTION

Interpretation of sand-shale depositional facies from seismic reflection data is based on a procedure we term *seismic facies analysis*. It involves the delineation and interpretation of reflection geometry, continuity, amplitude, frequency, and interval velocity, as well as the external form and three-dimensional associations of groups of reflections. Each of these seismic reflection parameters contains information of stratigraphic significance (Part 6, Vail et al, this volume).

Where seismic facies patterns are described and mapped, an interpretation of sedimentary processes and environmental facies leads to better lithologic prediction. The approach is useful for both clastic and carbonate rocks, but this report will be limited to a discussion of terrigenous clastic seismic facies.

DEPOSITIONAL SETTINGS

A regional three-fold environmental framework of shelf, shelf margin and prograded slope, and basin slope and floor provides a useful gross subdivision for classification and interpretation of seismic facies units. The shelf environment commonly is composed of marine and nonmarine depositional sequences (Part 2, Vail et al, this volume). Nonmarine sediments in these sequences

are interbedded with, or pass laterally into, marine sediments. In shelf sequences, sea level has been a controlling factor in establishing fluvial base level and the nature of the neritic sediments. Shelf strata characteristically generate mostly parallel reflections. This configuration is, in part, the result of limited seismic resolution; higher seismic resolution may reveal more complicated reflection configurations within these patterns.

The shelf-margin and prograded-slope environment typically contains thick marine sediments and has water depths sufficient for development of complex arrangements of sigmoid and oblique prograding reflection patterns. In this and the following depositional setting, the depositional sequences are easily identifiable and serve as a basic framework for seismic facies analysis.

The basin slope and floor environment includes a variety of deep basin facies, nonprograding slope facies, and facies that extend from the slope to the basin deep. "Deep basin" is used here for any basin, whether inland or along a continental margin, which had water depth sufficient for seismic definition of a slope environment; consequently, water depth at the time of deposition may range from bathyal to abyssal.

We can not present examples for all seismic facies types documented for each environment. Instead, we present examples of one type for each depositional setting.

Seismic characteristics and environmental facies interpretations of clastic seismic facies units are summarized in Table 1. Following sections discuss these units in more detail under headings corresponding to the regional depositional settings of shelf, shelf margin and prograded slope, and basin slope and floor. The use of reflection configuration to interpret high- and low-energy depositional environments and thus to predict sand-prone and shale-prone strata is discussed in Part 6, Vail et al, this volume.

[1]Manuscript received, January 6, 1977; accepted, June 13, 1977.

[2]Esso Exploration, Inc., Walton-on-Thames, Surrey, England, KT12 2QL.

[3]Exxon Production Research Company, Houston, Texas 77001.

Table 1. Seismic Characteristics and Environmental Facies Interpretation of Clastic Seismic Facies Units.

DEPOSITIONAL FRAMEWORK INTERPRETATION	SEISMIC FACIES UNIT	ENVIRONMENTAL FACIES INTERPRETATION	EXTERNAL FORM OF FACIES UNIT	REFLECTION GEOMETRY AT BOUNDARIES	REFLECTION CONFIGURATION: PRINCIPAL INTERNAL CONFIGURATION	CONFIGURATION MODIFIER	LATERAL RELATIONS	OTHER SEISMIC FACIES PARAMETERS: INTERVAL VELOCITY	AMPLITUDE	CONTINUITY	FREQUENCY (CYCLE BREADTH)
SHELF	HIGH-CONTINUITY AND HIGH-AMPLITUDE	TYPICALLY SHALLOW MARINE CLASTICS DEPOSITED PRIMARILY BY WAVE TRANSPORT PROCESSES. COULD ALSO BE FLUVIAL CLASTICS INTERBEDDED WITH WIDESPREAD MARSH DEPOSITS	SHEET OR WEDGE	CONCORDANT AT TOP WITH CONCORDANT TO GENTLE ONLAP OR DOWNLAP AT THE BASE	PARALLEL TO DIVERGENT		MAY GRADE LATERALLY TO ALL OTHER SHELF FACIES OR TO UNDAFORM PORTION OF SLOPE PROGRADATIONAL FACIES		RELATIVELY HIGH, BUT VARIABLE	RELATIVELY CONTINUOUS	VARIABLE, SOME BROAD CYCLES
SHELF	LOW-AMPLITUDE	MARINE CLASTICS DEPOSITED BY LOW-ENERGY TURBIDITY CURRENTS AND BY WAVE TRANSPORT	SHEET OR WEDGE	CONCORDANT AT TOP WITH CONCORDANT TO GENTLE ONLAP OR DOWNLAP AT THE BASE	PARALLEL TO DIVERGENT		MAY GRADE LATERALLY TO SLOPE PRO-GRADATIONAL FACIES OR TO HIGH-CONTINUITY AND HIGH-AMPLITUDE SHELF FACIES		VERY LOW TO LOW	DISCONTINUOUS TO CONTINUOUS	VARIABLE BUT LACKS EXTREMES OF HIGH-CONTINUITY AND HIGH-AMPLITUDE FACIES
		FLUVIAL AND NEARSHORE CLASTICS DEPOSITED BY FLUVIAL AND WAVE TRANSPORT PROCESSES					MAY GRADE LATERALLY TO HIGH-CONTINUITY AND HIGH-AMPLITUDE OR TO LOW CONTINUITY AND VARIABLE AMPLITUDE SHELF FACIES		LOW	DISCONTINUOUS TO MODERATELY CONTINUOUS	
SHELF	LOW-CONTINUITY AND VARIABLE AMPLITUDE	DOMINANTLY NON-MARINE CLASTICS. DEPOSITED BY RIVER CURRENTS AND ASSOCIATED MARGINAL MARINE TRANSPORT PROCESSES:	SHEET OR WEDGE	CONCORDANT AT TOP WITH CONCORDANT TO GENTLE ONLAP OR DOWNLAP AT THE BASE	PARALLEL TO DIVERGENT		COMMONLY GRADES LATERALLY TO HIGH-CONTINUITY AND HIGH-AMPLITUDE FACIES OR TO SAND-PRONE LOW-AMPLITUDE FACIES		LOW TO HIGH: QUITE VARIABLE WITH FREQUENT BURSTS OF HIGH-AMPLITUDE	GENERALLY DISCONTINUOUS TO MODERATELY CONTINUOUS	QUITE VARIABLE
SHELF	MOUNDED	SHELF DELTA COMPLEX	LOW MOUND OR ELONGATE MOUND	CONCORDANT AT TOP WITH GENTLE DOWNLAP AT BASE	MOUNDED TO SIGMOID AND DIVERGENT		COMMONLY GRADES LATERALLY TO HIGH- AND HIGH-AMPLITUDE FACIES OR TO UNDAFORM PORTION OF SLOPE PROGRADATIONAL FACIES		VARIABLE BUT RELATIVELY LOW; BURSTS OF DISCONTINUOUS HIGH AMPLITUDE ARE COMMON	DISCONTINUOUS TO MODERATELY CONTINUOUS	QUITE VARIABLE, SOME NARROW, CYCLES

Table 1. (Continued) Seismic characteristics and environmental facies interpretation of basin-outline seismic reflection facies units.

DEPOSITIONAL FRAMEWORK INTERPRETATION	SEISMIC FACIES UNIT	ENVIRONMENTAL FACIES INTERPRETATION	EXTERNAL FORM OF FACIES UNIT	REFLECTION GEOMETRY AT BOUNDARIES	PRINCIPAL INTERNAL CONFIGURATION	CONFIGURATION MODIFIER	LATERAL RELATIONS	INTERVAL VELOCITY	AMPLITUDE	CONTINUITY	FREQUENCY (CYCLE BREADTH)
SHELF-MARGIN AND PROGRADED-SLOPE	SIGMOID-PROGRADATIONAL	CLAY MUDS DEPOSITED BY LOW ENERGY TURBIDITY CURRENTS AND BY HEMIPELAGIC DEPOSITION FROM LOW-VELOCITY WATER CURRENTS. SHALLOW UNDAFORM PORTIONS MAY INVOLVE WAVE AND EVEN FLUVIAL TRANSPORT PROCESSES.	ELONGATE LENS TO SUBTLE FAN	CONCORDANT AT THE TOP WITH DOWNLAP AT THE BASE	SIGMOID ALONG DEPOSITIONAL DIP AND PARALLEL TO SUBPARALLEL ALONG DEPOSITIONAL STRIKE		MAY GRADE LATERALLY OR VERTICALLY TO OBLIQUE-PROGRADATIONAL FACIES. COMMONLY ONLAPPED BY ONLAP-FILL FACIES. UNDAFORM PART MERGES WITH SHELF FACIES AND FONDOFORM PART MAY GRADE TO SHEET DRAPE FACIES.		GENERALLY MODERATE TO HIGH, RELATIVELY UNIFORM	NORMALLY CONTINUOUS	VARIES PARALLEL TO DIP WITH BROADEST CYCLES ASSOCIATED WITH THICKER BEDS OF THE MIDDLE CLINOFORMING ZONE. CYCLE BREADTH IS UNIFORM ON SECTIONS PARALLEL TO DEPOSITIONAL STRIKE.
SHELF-MARGIN AND PROGRADED-SLOPE	OBLIQUE-PROGRADATIONAL — SUBZONES UNDAFORM	SEDIMENT COMPLEX USUALLY DEPOSITED IN SHELF MARGIN DELTAIC ENVIRONMENT; INCLUDES DELTA PLAIN, DELTA FRONT AND PRODELTA PROCESSES. MAY ALSO BE FORMED IN DEEP WATER ASSOCIATED WITH STRONG BOTTOM CURRENTS.	FAN	CONCORDANT AT TOP IF UNDAFORM CYCLES PRESENT	PARALLEL		COMMONLY MERGES DOWNDIP WITH DEEP BASIN TURBIDITES, MASS TRANSPORT AND HEMIPELAGIC FACIES. FREQUENTLY ONLAPPED BY ONLAPPING-FILL FACIES. MAY GRADE LATERALLY OR VERTICALLY TO SIGMOID-PROGRADATIONAL FACIES. UNDAFORM PORTION MERGES WITH PARALLEL LAYERED SHELF FACIES.		MODERATE TO HIGH	GENERALLY CONTINUOUS	RELATIVELY UNIFORM
	UPPER CLINOFORM			TOPLAP TRUNCATION AT THE TOP	OBLIQUE ALONG DEPOSITIONAL DIP AND PARALLEL OR GENTLY OBLIQUE TO SIGMOID FACIES PARALLEL TO DEPOSITIONAL STRIKE				MODERATE TO HIGH	GENERALLY MODERATELY CONTINUOUS	FAIRLY UNIFORM
	MIDDLE AND LOWER CLINOFORM								VARIABLE GENERALLY LOWER THAN OTHER SUBZONES	DISCONTINUOUS TO MODERATE INCREASES TOWARD LOWER CLINOFORM ZONE	CYCLE BREADTH DECREASES RAPIDLY DOWNDIP AS BEDS THIN
	FONDOFORM			DOWNLAP AT THE BASE					GENERALLY MODERATE TO LOW	CONTINUOUS	RELATIVELY NARROW CYCLE BREADTH DECREASES BASINWARD
BASIN-SLOPE AND BASIN-FLOOR	SHEET-DRAPE	DEEP MARINE HEMIPELAGIC CLAYS AND OOZES	SHEET-DRAPE	CONCORDANT AT TOP AND CONCORDANT OR VERY SLIGHT ONLAP AT BASE	PARALLEL		COMMONLY INTERBEDDED WITH TURBIDITE SANDS & SILTS, GRADES TO GENTLY DIVERGENT FONDOFORM SEDIMENTS OF PROGRADING COMPLEXES		COMMONLY RELATIVELY LOW TO MODERATE	CONTINUOUS	NORMALLY UNIFORMLY NARROW
BASIN-SLOPE AND BASIN-FLOOR	SLOPE-FRONT FILL	DEEP-WATER SEDIMENT COMPLEX COMMONLY RELATED TO SUBMARINE FANS	LARGE FAN	CONCORDANT AT TOP. ONLAPS UPDIP AND DOWNLAPS DOWNDIP	PARALLEL TO SUBPARALLEL		THINS AND GRADES INTO BASIN FLOOR FACIES, COMMONLY PINCHES OUT UPDIP		VARIABLE	VARIABLE	VARIABLE
BASIN-SLOPE AND BASIN-FLOOR	ONLAPPING FILL	RELATIVELY LOW VELOCITY TURBIDITY CURRENT DEPOSITS	BASIN TROUGH CHANNEL AND SLOPE FRONT FILL	ONLAP AT THE BASE AND USUALLY CONCORDANT AT THE TOP			COMMONLY GRADES TO MOUNDED ONLAP OR CHAOTIC FILL FACIES. ALTERNATION WITH OTHER FILL FACIES IS COMMON		VARIABLE	COMMONLY CONTINUOUS	CYCLE BREADTH INCREASES INTO FILL CENTER TRENDS TO BE RELATIVELY NARROW

Table 1 (continued). Seismic Characteristics and Environmental Facies Interpretation of Clastic Seismic Facies Units.

DEPOSITIONAL FRAMEWORK INTERPRETATION	SEISMIC FACIES UNIT	ENVIRONMENTAL FACIES INTERPRETATION	EXTERNAL FORM OF FACIES UNIT	REFLECTION CONFIGURATION				OTHER SEISMIC FACIES PARAMETERS			
				REFLECTION GEOMETRY AT BOUNDARIES	PRINCIPAL INTERNAL CONFIGURATION	CONFIGURATION MODIFIER	LATERAL RELATIONS	INTERVAL VELOCITY	AMPLITUDE	CONTINUITY	FREQUENCY (CYCLE BREADTH)
BASIN SLOPE AND BASIN FLOOR	MOUNDED (FAN COMPLEX)	DEEP-WATER SEDIMENT COMPLEX COMMONLY LOCATED AT MOUTH OF SUBMARINE CANYON, COMPOSED OF TURBIDITES, MASS-MOVEMENT AND HEMI-PELAGIC DEPOSITS ASSOCIATED WITH MAJOR SUBAERIAL DRAINAGE SYSTEMS	FAN	ONLAP OF OVERLYING UNITS DOWNLAP AT BASE	EXTREMELY VARIED COMPLEX MOUNDED		LOCATED NEAR SUBMARINE CANYON COMMONLY GRADES BASINWARD TO SHEET DRAPE FACIES		VARIABLE BUT TENDS TO BE LOW FREQUENTLY DECREASES RAPIDLY WITH INCREASING DEPTH, SUGGESTING HIGH ENERGY ABSORPTION	TENDS TO BE DISCONTINUOUS	HIGHLY VARIABLE
BASIN SLOPE AND BASIN FLOOR	MOUNDED (CONTOURITE)	DEEP-WATER SEDIMENT COMPLEX FORMED BY DEPOSITION FROM DEEP-MARINE CURRENTS POSSIBLY COMPOSED PRIMARILY OF FINE-GRAINED CLASTICS	ELONGATE MOUND	TRUNCATED AND CONCORDANT AT TOP, DOWNLAP AT BASE	ASYMMETRIC MOUNDS		THINS AND GRADES INTO BASIN FLOOR FACIES		VARIABLE	VARIABLE	VARIABLE
BASIN SLOPE AND BASIN FLOOR	MOUNDED ONLAPPING FILL	RELATIVELY HIGH VELOCITY TURBIDITY CURRENT DEPOSITS	MOUNDED, BASIN TROUGH CHANNEL AND SLOPE FRONT FILL	ONLAP AT THE BASE AND CONCORDANT OR EROSIONAL TRUNCATION AT TOP	IRREGULARLY MOUNDED TO PARALLEL		COMMONLY GRADES TO ONLAP FILL OR CHAOTIC FILL FACIES ALTERNATION WITH OTHER FILL FACIES IS COMMON		VARIABLE DECREASES AS CONTINUITY DECREASES	DISCONTINUOUS TO MODERATELY CONTINUOUS GENERALLY LESS THAN NONMOUNDED ONLAP FILL FACIES	CYCLE BREADTH INCREASES IN FILL CENTER
BASIN SLOPE AND BASIN FLOOR	CHAOTIC FILL	GRAVITY MASS TRANSPORT AND HIGH ENERGY TURBIDITY CURRENT SEDIMENTS LITHOLOGY IS FUNCTION OF UPSLOPE SEDIMENT SOURCE	BASIN TROUGH CHANNEL & SLOPE FRONT FILL DEGREE OF MOUNDING VARIABLE, WAVY SUB PARALLEL CHAOTIC PATTERN TENDS TO BE ASSOCIATED WITH SMOOTHER AND LOWER MOUNDS THAN CONTORTED & DISCORDANT CHAOTIC PATTERNS	UNIT ONLAPS AT BASE BUT INDIVIDUAL ONLAP TERMINATIONS ARE RARE BECAUSE OF REFLECTION PATTERN. WHERE PRESERVED, REFLECTION SEGMENTS AT THE TOP MAY SHOW CONCORDANCE OR EROSIONAL TRUNCATION. MASS TRANSPORT GAUGE IS COMMON AT BASE OF CONTORTED DISCORDANT PATTERN	CHAOTIC & CONTORTED	RANGES FROM CONTORTED AND DISCORDANT TO WAVY SUBPARALLEL	MAY GRADE SLOPEWARD TO LOWER & MIDDLE CLINOFORM SUBZONES OF OBLIQUE PROGRADATIONAL FACIES. ALSO MAY LIE DOWNDIP OF PROMINENT DETACHMENT SCARS ALTERNATION WITH OTHER FILL FACIES IS COMMON		RANGES FROM LOW TO HIGH IN CONTORTED DISCORDANT PATTERNS AND IS GENERALLY LOW IN WAVY SUBPARALLEL CHAOTIC FILL PATTERNS	VERY DISCONTINUOUS SHORT SEGMENTS MAY OCCUR IN THE CONTORTED DISCORDANT PATTERNS	VARIABLE IN CONTORTED DISCORDANT FACIES REFLECTING INTERNAL HETEROGENEITY. MORE UNIFORM IN WAVY SUB PARALLEL CHAOTIC FILL PATTERN

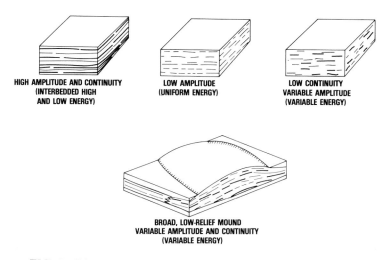

HIGH AMPLITUDE AND CONTINUITY
(INTERBEDDED HIGH
AND LOW ENERGY)

LOW AMPLITUDE
(UNIFORM ENERGY)

LOW CONTINUITY
VARIABLE AMPLITUDE
(VARIABLE ENERGY)

BROAD, LOW-RELIEF MOUND
VARIABLE AMPLITUDE AND CONTINUITY
(VARIABLE ENERGY)

FIG. 1—Diagrammatic illustration of shelf seismic facies types.

SHELF SEISMIC FACIES

Shelf sedimentary environments, as defined here, typically range from neritic to totally non-marine. Sediments deposited in these environments tend to generate parallel to gently divergent reflection configurations having a widespread sheet, or wedge-shaped, external form (Fig. 1; Table 1). One facies, characterized by a broad, low-relief mound composed of gently sigmoid to downlapping reflections, is an exception to these generalizations.

Reflections are normally concordant at the top and range from concordant to gently onlapping and occasionally downlapping at the base. Prediction of depositional energy and sand content in shelf seismic facies units must rely heavily on analysis of variations in reflection amplitude, continuity, cycle breadth, interval velocity, and areal relations with other units. The active transport-deposition modes in this environment include fluvial processes, and wave (surf-zone reworking and longshore drift) and other marine current and flow processes.

High-Continuity and High-Amplitude Facies
(interbedded high- and low-energy deposits)

High continuity of reflections in this facies suggest continuous strata deposited in a relatively widespread and uniform environment, and high reflection amplitudes are interpreted to indicate interbedding of shales with relatively thick sandstones, siltstones, or carbonate rocks. Normally, deposits in this facies consist of neritic marine sediments; however, there are examples in which

fluvial sediments with interbedded, widespread marsh clays and coals also generate high-continuity and high-amplitude reflections. Cycle breadth varies, and it is common to find broad, high-amplitude cycles alternating with cycles having narrow cycle breadth, suggesting vertical variations in bed thickness. Laterally, high-amplitude and high-continuity reflections may grade into any of the other shelf facies or into the undaform (environments above wave base, see Rich, 1951) part of prograded-slope facies.

An example in which this facies contains significant sand is typified by Cretaceous sediments in the San Juan basin. Thirty wells provide a great deal of control for analysis of the Cretaceous part of the San Juan section (Fig. 2). Abundant, thick (50 to 200 ft or 15 to 60 m) nearshore and shoreline sands generate continuous, high-amplitude reflections (Fig. 3) characteristic of this facies. The reflections lie within a pattern of parallel reflections that extends across the entire section and is concordant throughout. Southwestward, sandstones grade into a nonmarine facies characterized by numerous discontinuous sandstones and marsh shales (low-continuity, variable-amplitude seismic facies); toward the northeast, they grade into an open-marine, calcareous shale facies (low-amplitude seismic facies).

In the San Juan basin, the sediments in the high-continuity and high-amplitude facies are beach and shoreface deposits. Consequently, wave transport and deposition processes, including surf-zone reworking and longshore drift, appear dominant in formation of this facies. This section demonstrates how reflections pass from one facies to another laterally.

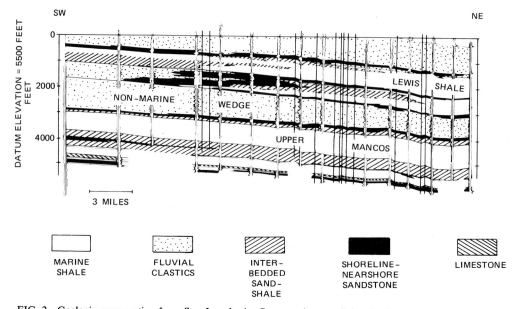

FIG. 2—Geologic cross section from San Juan basin. Cross section parallels seismic line shown on Figure 3.

Low-Amplitude Facies
(uniform-energy deposits, high or low)

Low-amplitude zones on a seismic section indicate either beds too thin to be resolved by seismic methods or a zone of one predominant lithologic type. Consequently, the low-amplitude facies may be either sand-prone (high energy) or shale-prone (low energy). Knowledge of regional geology, lateral seismic facies relations, or other lithologic data is required to identify the correct lithology.

A shale-prone low-amplitude facies typically grades shoreward to a silt- or sand-prone, high-continuity and high-amplitude facies, and basin-ward to a prograded-slope facies. In contrast, a sand-prone low-amplitude facies tend to grade landward to nonmarine low-continuity and variable-amplitude facies, and seaward to high-continuity and high-amplitude marine facies. Of course, there are exceptions to this principle, but it is most important in distinguishing whether a low-amplitude facies consists of massive sandstone or massive shale.

The San Juan basin is a good example of the low-amplitude facies. On the San Juan section (Fig. 3), a vertical and lateral gradation from the low-amplitude shale-prone facies through the high-continuity and high-amplitude interbedded sandstone facies to a low-continuity and variable-amplitude nonmarine sand-prone facies is well illustrated. At the northeast end of the section, the vertical succession is: (1) low-amplitude reflections from the Mancos Formation marine shale, (2) high-amplitude and continuous reflections from the shoreline sandstone, (3) low-continuity, relatively low-amplitude reflections from the non-marine Menefee Formation sandstones, silt-stones, and shales, (4) two high-amplitude and continuous reflections from shoreline sandstones, and (5) relatively low-amplitude reflection from the Lewis Shale. Laterally, the Lewis Shale grades into nonmarine equivalents with a repetition of the succession of seismic facies types (1), (2), and (3) listed above.

Two contrasting types of low-amplitude facies result from contrasting depositional processes. Low-amplitude massive sandstones tend to be nearshore to fluvial sands that are transported and deposited by high-energy fluvial and wave transport processes. Low-amplitude, massive shales tend to be deposited farther offshore or along sand-poor shorelines, and are associated with wave and low-velocity turbid layer and suspension transport processes.

Low-Continuity and Variable-Amplitude Facies
(variable energy)

Sediments deposited by fluvial currents are generally less continuous than sediments deposited under marine conditions. In addition, discontinuous sandstones in many nonmarine environments provide good velocity contrasts with

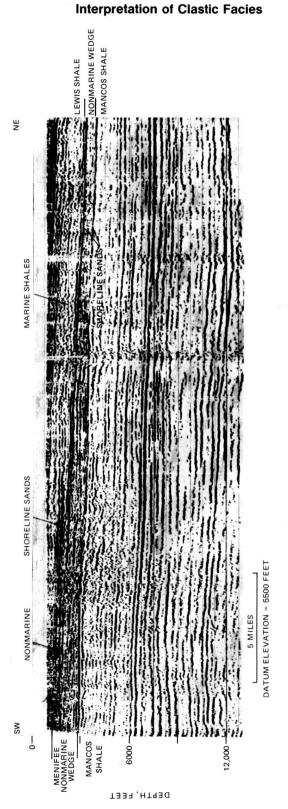

FIG. 3.—Seismic section from San Juan basin. Seismic depth section on Figure 6 is residual amplitude section. In residual amplitude, only amplitudes above selected thresholds are preserved; those lying below thresholds are set equal to zero. Done to emphasize amplitude variations for analysis.

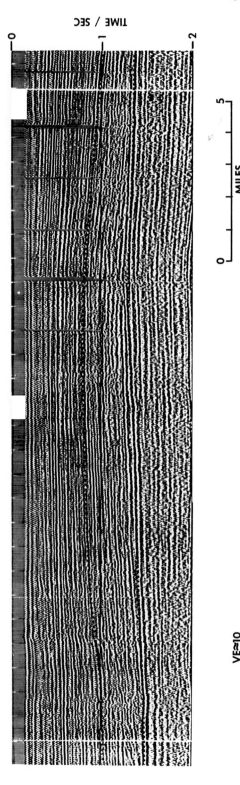

FIG. 4—Broad, low-relief mound seismic facies. Example from offshore Africa, shows mounded external form with reflections concordant at top and downlapping in opposite directions at base.

encasing shales and coals. Consequently, shelf seismic facies characterized by reflections having poor lateral continuity and bursts of high amplitude are typically nonmarine sediment types. Nonmarine sediments may generate high-amplitude and relatively high-continuity reflections, for example where thick coals are interbedded with sandstones or siltstones. However, marine sediments rarely generate the discontinuous high-amplitude events typical of this low-continuity facies without concurrent evidence for large-scale mounding or marine cut-and-fill deposition.

Low-continuity and variable-amplitude seismic facies may occur in widespread sheets or may show prominent wedging with basal onlap where sediment supply is sufficient to fill a rapidly subsiding basin. Reflections are typically parallel in the sheet forms and divergent in the wedge form. Cycle breadth may be variable, probably reflecting complex variations in bed thickness typical of nonmarine deposition. This facies commonly grades seaward into high-continuity and high-amplitude sand-prone facies, and sometimes into low-amplitude facies composed of thick and relatively massive sandstones. Nonmarine sediments in the San Juan basin (Fig. 3) are examples.

Broad, Low-Relief Mound Facies
(variable energy)

This shelf facies, noted at only a few location, is interpreted as reflecting a complex of delta lobes formed on a subsiding shelf. Its key distinguishing feature is a broad, gently mounded external form. Internally, reflections form gently sigmoid to divergent patterns. Cycles are concordant at the top, and downlap gently in an arcuate pattern at the base. Laterally, this unit may grade into any of the parallel shelf facies or into the undaform part of prograded-slope facies. Figure 4 is an example of this type of seismic facies from offshore West Africa.

SHELF-MARGIN AND PROGRADED-SLOPE SEISMIC FACIES

The next major class of clastic seismic facies units is associated with shelf-margin and prograded-slope types of deposition. Two principal facies, defined exclusively on reflection configuration, are recognized in this environment (Table 1; Fig. 5). These are the oblique-progradational facies, and the sigmoid-progradational facies. Both are characterized by downlapping reflections at their base. Downlap represents the outbuilding of sediments from relatively shallow into relatively deep water with thinning of the outer toes of individual beds, commonly below seismic resolution. Ordinarily, the upper parts of these patterns show sediments deposited in fluvial to neritic environments, but examples have been documented where these patterns originate from sediments deposited entirely in bathyal water, presumably through the action of deep-water currents.

Oblique-Progradational Facies
(typically high energy in updip portions)

The distinguishing feature of this facies is a prominent oblique reflection configuration when viewed parallel with depositional dip. Reflections terminate by toplap at or near the upper surface (Fig. 5) and by downlap at the base. Depositional dips may approach 10° and are significantly steeper than dips of the sigmoid-progradational facies. Parallel with depositional strike, reflections may be parallel or may show low-angle oblique or sigmoid patterns. Small channels are commonly related to this configuration, and show best on depositional strike sections. Studies of this facies in delta and delta-front sediments commonly indicate that it is deposited during stillstands or where sea level is slowly rising (if the supply of sediment is sufficient to overwhelm the effects of the relative rise).

Where this seismic patterns occurs on the shelf margin, it is characteristic of fluvial deltas and associated coastal-plain sediments and contains high-energy deposits. The undaform zone (Rich, 1951), corresponding to a delta-plain environment, and the upper part of the clinoform zone, corresponding to a delta-front environment, have good sand potential. In contrast, the lower clinoform and fondoform zones, corresponding to a prodelta environment, are typically shale-prone. Previous studies have documented these relations (Ewing et al, 1963; Lehner, 1969). In some instances, the fondoform part of oblique seismic facies units may contain turbidite deposits, and thus may contain sandstones interbedded with marine shales.

Amplitude, continuity, and cycle breadth vary depending on position in the oblique configuration. Fondoform and undaform reflections have good continuity; fondoform reflections have moderate amplitude, and undaform reflections have moderate or high amplitude. Cycle breadth decreases basinward and may become narrow. Clinoform zone reflections are variable, but show a general decrease in continuity and amplitude from upper clinoform to lower clinoform. Some of the highest amplitudes observed in oblique facies occur in the upper clinoform zone. This probably represents maximum interfingering of shallow-water sandstones and siltstones with delta-front and prodelta clays. Cycle breadth in

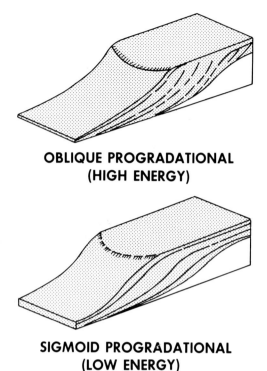

OBLIQUE PROGRADATIONAL
(HIGH ENERGY)

SIGMOID PROGRADATIONAL
(LOW ENERGY)

FIG. 5—Shelf-margin and prograded-slope seismic facies types.

the clinoform zone decreases significantly downdip as beds thin.

In three dimensions, oblique units are frequently fan shaped, and it is not uncommon to find multiple fans forming major sedimentary complexes. Downdip, the facies may be associated with mass transport facies that slumped from the front of the growing complex, or may be onlapped by various fill facies. The oblique facies also may grade laterally and vertically into sigmoid progradational facies.

The upper Miocene in the San Joaquin Valley in California provides a well-documented example of a sandstone-rich oblique facies shown by conventional seismic data. Examination of this section (Fig. 6) shows zones of relatively parallel reflections sandwiching a zone of oblique reflections. Individual reflections terminate against a common upper surface by toplap and against a common lower surface by downlap. The total section thickens from less than 0.2 sec on the left to 0.4 sec in the center. This thinning marks the transition from a thin, dominantly fondoform position to one of thickest development in the oblique clinoform part of the complex. The unda-

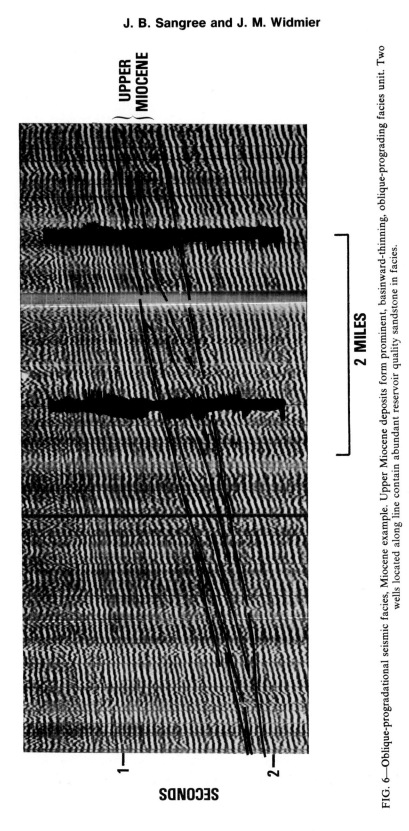

FIG. 6—Oblique-progradational seismic facies, Miocene example. Upper Miocene deposits form prominent, basinward-thinning, oblique-prograding facies unit. Two wells located along line contain abundant reservoir quality sandstone in facies.

A   A'

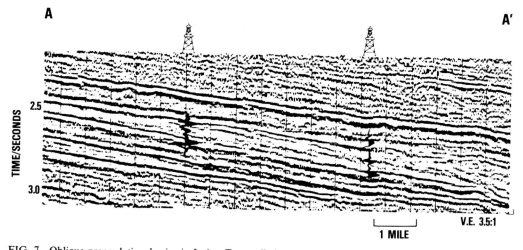

TIME/SECONDS

2.5

3.0

V.E. 3.5:1

1 MILE

FIG. 7—Oblique-progradational seismic facies. Two wells located along line contain only minor amounts sandstone in upper part of oblique-progradational reflection pattern.

form part of the unit (upper right part) occupies two to three parallel cycles.

This example contains an unusually high proportion of sandstone. The sandstone extends from the undaform zone to outer parts of the clinoforms, as is indicated by logs of two wells occurring along the line (Fig. 6). Units above and below are shaly, and the upper unit shows prominent onlap onto the clinoform front of the upper Miocene deposits.

Figure 7 presents another example of an oblique-progradational pattern. In this figure, the oblique pattern is recognized from reflections that are inclined more steeply in a downdip direction than the reflections above or below. The downlap pattern of reflection terminations is visible at the base, and a series of reflections terminate at the top by toplap. Although we interpret this as a high-energy environment, two wells shown on Figure 7 contain only thin deltaic sandstones at the top of the unit. The thinness of these sandstones suggest that a large source of coarse clastics was not available for transportation into this depositional setting.

Sigmoid-Progradational Facies
(low energy)

Sigmoid-progradational units are characterized by gentle sigmoid (S-shaped) reflections along depositional dip. The reflections downlap at the base and are concordant with the top of the unit (Fig. 5). On seismic lines parallel with depositional strike, these reflections are commonly parallel and concordant with unit boundaries. Sediments of this facies are deposited on the slope

along continental margins with undaform parts extending onto shelf-margin areas. Considerable upbuilding produces thick undaform deposits, which require some combination of a rising eustatic sea level or a subsiding land level. Fine-grained clastics dominate this facies and are likely deposited from low-energy turbidity currents as hemipelagic deposits. Undaform sedimentation may involve wave or even fluvial transport processes, and provides some possibility of coarser clastic sediment deposition in the undaform environment.

Typically, sigmoid reflections have moderate to high amplitude and high continuity. Cycle breadth, although uniform on sections parallel with depositional strike, varies parallel with dip, with the broadest cycles correlative with the thicker beds of the middle clinoform zone.

Most commonly, a sigmoid-progradational facies forms a lens, elongated parallel with depositional strike. In addition to frequent association with the oblique facies, this unit is commonly onlapped by various chaotic- and onlapping-fill facies. Undaform reflections merge with parallel shelf facies types, and fondoform reflections may grade into sheet-drape facies of the basin floor. The sigmoid-progradational facies is illustrated in Figure 8.

BASIN-SLOPE AND BASIN FLOOR SEISMIC FACIES

The groups of seismic facies units (Fig. 9) that dominate the basin slope and floor are: (1) sheet-drape seismic facies, (2) slope-front fill seismic facies, (3) onlapping-fill seismic facies, (4) mound-

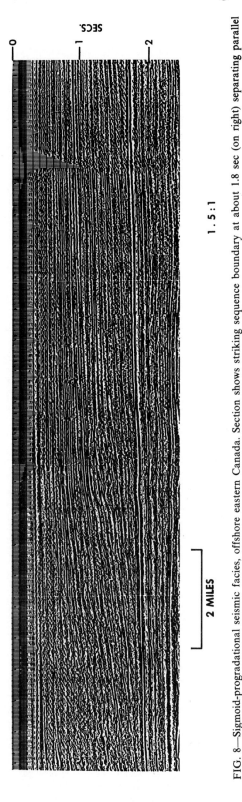

FIG. 8—Sigmoid-progradational seismic facies, offshore eastern Canada. Section shows striking sequence boundary at about 1.8 sec (on right) separating parallel reflections below from sigmoid-prograding reflections above.

ed-fan seismic facies, (5) mounded-contourite seismic facies, (6) mounded onlapping-fill seismic facies, and (7) chaotic-fill facies. Documentation is sketchy for some facies types, and this classification will be refined and probably extended in the next several years.

These facies commonly overlap from the basin floor onto the slope environment. For example, submarine fan complexes characteristically extend from the slope at their apex, out into the basinal environment. Chaotic and onlapping-fill seismic facies units are deposited in topographic lows on the slope and basin floor, including the basin-floor plain, local basins, channels or troughs, or areas of prominent topographic flattening in the configuration of the slope. These facies are most likely deposited by high-density turbidity current and mass transport processes. Presence of displaced, relatively shallow-water fauna is common in cores of all types of chaotic facies units.

Two characteristics most indicative of high depositional energy and consequently of sand-prone units appear to be the increasing irregularity of reflection pattern and character, and the mounding of external form. Hence fan complexes, mounded onlapping, and chaotic units are interpreted as higher energy deposits. The final control of lithology is the sediment source area—no process can transport sand if sand is not present.

Sheet-Drape Facies
(low energy)

Parallel reflections of this seismic facies drape over contemporaneous topography with only gradual changes in thickness or reflection character (Fig. 9) and suggest uniform deposition independent of bottom relief. This pattern is strongly indicative of deep-marine hemipelagic clays and oozes with almost no potential for sand development.

This facies forms relatively thin, widespread sheets of parallel reflections, normally concordant at top and base. However, it will (on occasion) show slight onlap of contemporaneous topography at the base. Continuity is very high, and cycle breadth commonly is uniformly narrow. Amplitude varies but commonly is relatively low because of the uniform lithology of the unit. Hemipelagic clays and oozes of the parallel sheet-drape facies are frequently found interbedded with turbidite and mass movement sandstones, siltstones, and clays, and gradually may pinch out laterally, or may grade to gently divergent fondoform sediments of prograding units. Figure 10 shows an example of this facies.

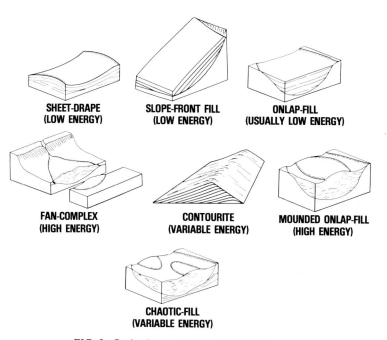

SHEET-DRAPE
(LOW ENERGY)

SLOPE-FRONT FILL
(LOW ENERGY)

ONLAP-FILL
(USUALLY LOW ENERGY)

FAN-COMPLEX
(HIGH ENERGY)

CONTOURITE
(VARIABLE ENERGY)

MOUNDED ONLAP-FILL
(HIGH ENERGY)

CHAOTIC-FILL
(VARIABLE ENERGY)

FIG. 9—Basin slope and floor seismic facies types.

Slope-Front Fill Facies
(low energy)

On dip sections, this common deep-water seismic facies characteristically shows parallel to subparallel reflections sloping seaward with pronounced updip onlap, and downdip downlap. Although data are limited in the strike direction this facies seems to show dips indicative of large deep-water sedimentary fans (Fig. 9). Where studied in three dimensions, updip onlap on dip sections is only apparent onlap—the sediments are in fact sloping downward and away from the apex of the sedimentary fan. A diagrammatic example is shown (Fig. 9, Part 3, Vail et al, this volume) and a seismic example is represented by the Valanginian interval (Fig. 1, Part 7, Vail et al, this volume).

Sediments of this facies are typically deep marine fine-grained clays and silts that were carried downslope from the apex of the sedimentary cone. The more parallel the seismic reflections, the more likely the section will be fine grained.

Onlapping-Fill Seismic Facies
(predominantly low energy)

This low-energy facies does not show the mounding characteristic of mounded onlap- and chaotic-fill facies, and its reflections tend to have a more uniform, parallel to gently divergent pattern with high continuity and variable amplitude. Cycle breadth also tends to be more uniform and relatively narrow, with a slight broadening into the topographic low. Vertical and lateral gradation and interbedding with other fill facies are common for this unit.

As in mounded-fill facies, the onlapping nature and tendency to fill lows suggest that the onlapping-fill facies is deposited by gravity-controlled flows along the bottom. Parallel patterns of continuous reflections created by widespread parallel strata are probably the expression of deposits formed by relatively low velocity turbidity currents interbedded with hemipelagic and pelagic deposits. These patterns can be expected to indicate clay and silt. Thin sandy units below seismic resolution may occur interbedded with the shales and silts more typical of this seismic facies; presumably, such thin sandstones result from intermittent high-energy turbidity current transport from a sand-rich source area. Figure 11 is an example of this seismic facies.

Mounded Seismic Facies
(variable energy)

Part 6, (Vail et al, this volume), illustrates the patterns for a number of different mounded seismic facies. The most prominent deep-water mounds are fan complexes, contourite mounds, and slumps.

J. B. Sangree and J. M. Widmier

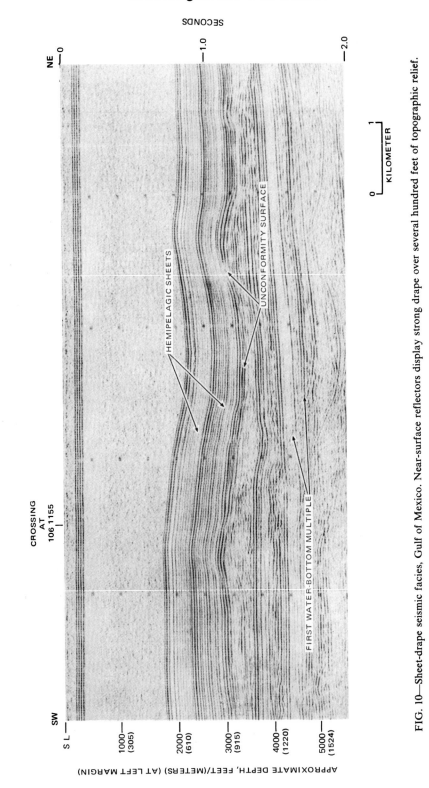

FIG. 10—Sheet-drape seismic facies, Gulf of Mexico. Near-surface reflectors display strong drape over several hundred feet of topographic relief.

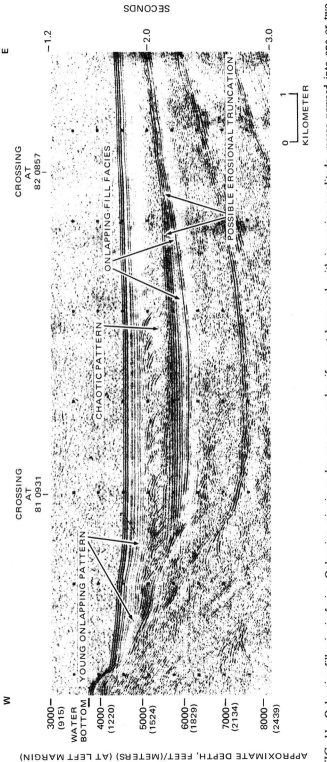

FIG. 11—Onlapping-fill seismic facies. Onlapping seismic cycles are even and uniform at base, and, with increasing amplitude, merge upward into one or two less-continuous cycles. Layered onlapping pattern gives way to chaotic pattern.

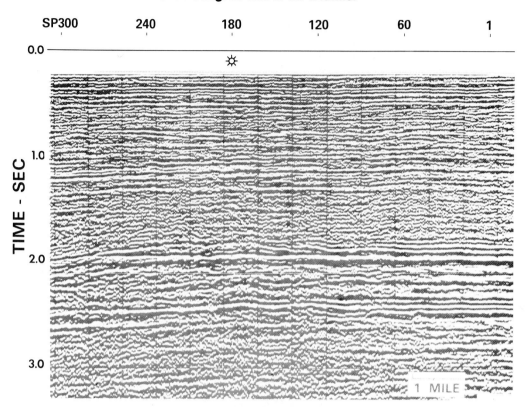

FIG. 12—Fan-complex seismic facies, Tertiary example. Mounded external form of fan is well-developed on left at about 1.8 sec. Internal patterns obscured by fluid contact just above 2.0 sec.

Fan-complex, mounded seismic facies are characterized by their three-dimensional fan shape and internal reflection configurations, types of which include parallel, divergent, chaotic, and a tendency to be reflection-free where the mound is very thick. Further, large fan complexes may incorporate individual chaotic and onlapping-fill units. Any additional generalizations seem premature. Fans consist of large, complex piles of sediments created in the basin-slope and basin-floor environment by gravity transport of sediments through submarine canyons. Fan complexes deposited by gravity transport are distinguished by their fan-shaped external form, complex internal patterns, and common association with major subaerial and submarine drainage systems. These fans may contain good reservoir sandstones, depending on the nature of the source area and on the distance and processes of sediment transport. A seismic example of a fan is shown on Figure 12.

Contourite mounded seismic facies units are commonly elongate features resembling dunes.

Asymmetric mounded internal reflection patterns indicate scouring and redeposition by largely unidirectional current flow. Reflection continuity within single units is good, but scouring, slumping, and superposition of several units may reduce continuity and regularity of reflection patterns. Large depositional mounds which have recently been recognized as commonly occuring on continental slopes and ocean basin floors have these seismic facies charistics. The mounded features are thought to have been formed by deposition from major bottom-flowing oceanic currents, and their sediments are termed *contourites* (Heezen et al, 1966; Schneider et al, 1967). Although this facies may have been deposited under high energy conditions, there is little evidence of sandstones in this environment.

Mounded Onlapping-Fill Facies
(high energy)

Mounded onlapping-fill seismic facies units are characterized by a mounded form and prominent onlap away from topographic lows. Internal re-

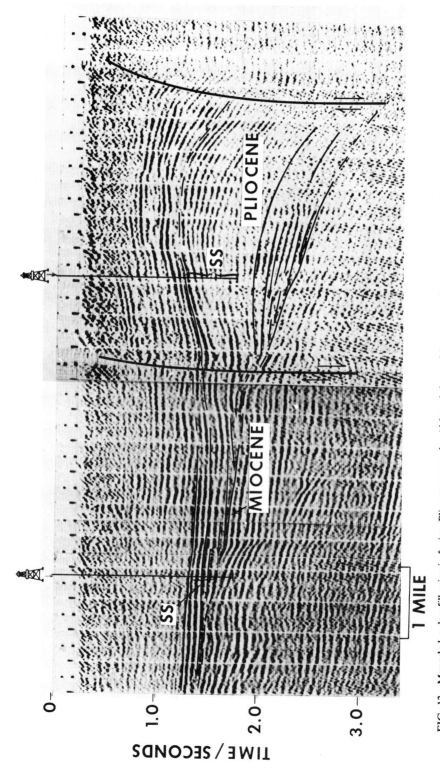

FIG. 13—Mounded onlap-fill seismic facies, Pliocene example. Although located in structurally complex area, sandstone-rich mound is believed to be depositional in origin.

J. B. Sangree and J. M. Widmier

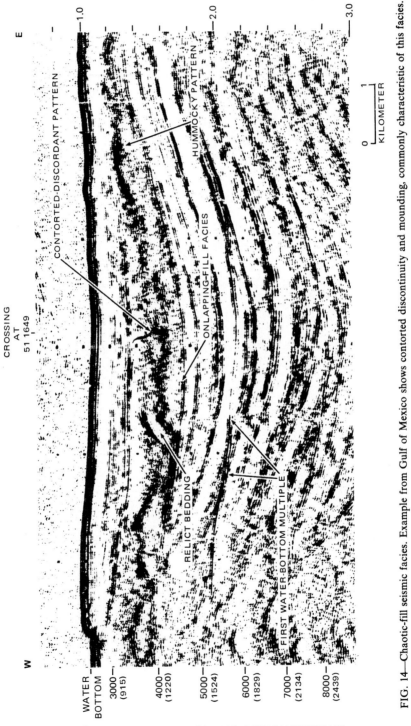

FIG. 14—Chaotic-fill seismic facies. Example from Gulf of Mexico shows contorted discontinuity and mounding, commonly characteristic of this facies.

flections are irregularly parallel to divergent and and tend to be discontinuous. Reflection amplitude is variable; amplitude tends to decrease as continuity decreases. Normally, cycle breadth increases toward the center of the topographic low.

This facies is commonly intermediate to (and laterally gradational with) both the chaotic-fill facies and the onlapping-fill facies, and may also show common vertical interbedding with these two facies. Individual units of this facies range from a few tens of feet to over 1,000 ft (310 m) and, where interbedded with chaotic- and onlapping-fill facies, attain thicknesses of several thousands of feet.

The onlapping nature of this facies, together with its tendency to selectively fill lows in the depositional topography, suggests deposition by gravity-controlled flow along the sea bottom (Hersey, 1965). Where the seismic pattern is formed by discontinuous reflections that show an increased rate of thickening into topographic lows, the deposits probably contain discontinuous beds and reflect a less uniform and possibly high-velocity mode of transportation and deposition. All things considered, turbidity current flow seems the most likely origin for this facies type. Discontinuous and irregular patterns with minor mounding probably represent deposits formed by higher velocity turbidity currents, possibly capable of transporting sands if sands are available at the source.

A large Pliocene turbidite fan that was deposited in offshore California is primarily formed of onlapping mounds (Fig. 13). These mounds are interbedded with parallel-layered units of relatively uniform thickness which are interpreted as pelagic shales and turbidites deposited during intermittent times of low turbidite influx. Two wells drilled on this line contain about 40 to 50% sand in the mounded interval. Subsequent structural uplift has altered the original depositional form; however, careful inspection reveals the internal mounding of the unit.

Chaotic-Fill Facies
(variable energy)

Chaotic-fill seismic facies units are characterized by a mounded external form, by location in topographic lows, and by an internal pattern of contorted and discordant to wavy subparallel reflections (Fig. 14). Mass-transport slump and creep and high-energy turbidity current processes are thought to be responsible for transportation and deposition of this facies. Composition of slump deposits is dependent on the type of material at the source area. Commonly, these sedi-

ments have not been sufficiently winnowed to produce clean sands in the slump deposits.

Chaotic units are thick in topographic lows and onlap from these lows, although the lack of coherent reflections obscures onlap of individual cycles. The top of the unit is commonly hummocky. Thickness of these units vary. Chaotic zones 300 m or more thick are common, and some exceed 200 m in thickness in the Pleistocene of the Gulf Coast.

Contorted internal reflections of chaotic-fill units originate from bent and folded beds that have retained some coherence during downslope *en masse* transportation. (Chaotic patterns also occur in complex and stacked cut-and-fill submarine fan sediments.) Diffractions originating from these segmented masses are common.

Interpretation of a mass-transport origin for a chaotic-fill unit is based on evidence for broken and contorted beds, evidence for updip detachment scars and erosional gouge below the unit, and maps that indicate the unit has lost section upslope and gained section downdip.

Reflection amplitude, continuity, and frequency seem to reflect the extent of homogenization in mass-transport types of chaotic units. Where homogenization is incomplete and reflection contorted, amplitude ranges from low to high and reflects the original sediment velocity contrasts. In contrast, amplitude is low in the wavy subparallel pattern. Figure 14 is an example of this facies.

FUTURE OF SEISMIC FACIES INTERPRETATION

There are four areas in which major improvement in seismic facies analysis is expected. First, improved knowledge of the types and genesis of major nonmarine and marine sedimentary facies and better seismic documentation for these facies are needed. Second, routine computer procedures to model, store, and map a variety of seismic facies data promise to be useful. Third, quantification of seismic parameters such as amplitude, frequency, polarity, and continuity holds advantages for objective prediction of facies. And finally, close-spaced interval velocity has great capability for the prediction of lithology, especially if it is used in conjuction with reflection pattern methods of seismic facies analysis.

REFERENCES CITED

Ewing, J., X. Le Pichon, and M. Ewing, 1963, Upper stratification of Hudson apron region: Jour. Geophys. Research, v. 68, p. 6303-6316.

Heezen, B. C., C. D. Hollister, and W. F. Ruddiman, 1966, Shaping of the continental rise by deep geostrophic contour currents: Science, v. 152, p. 502-508.

Hersey, J. B., 1965, Sediment ponding in the deep sea: Geol. Soc. America Bull., v. 76, p. 1251-1260.

Lehner, P., 1969, Salt tectonics and Pleistocene sediments on continental slope of northern Gulf of Mexi-co: AAPG Bull., v. 53, p. 2431-2479.

Rich, J. L., 1951, Three critical environments of deposition and criteria for recognition of rocks deposited in each of them: Geol. Soc. America Bull., v. 62, p. 1-20.

Schneider, E. D., et al, 1967, Further evidence of contour currents in the western North Atlantic: Earth and Planetary Sci. Letters, v. 2, p. 351-359.

Seismic Stratigraphy and Global Changes of Sea Level, Part 10: Seismic Recognition of Carbonate Buildups[1]

J. N. BUBB[2] and W. G. HATLELID[3]

Abstract Carbonate buildups, including reefs and banks, are ideally suited for stratigraphic interpretation from reflection seismic data because of pronounced differences in depositional or bedding characteristics between the buildups and enveloping strata. Geophysical criteria that allow recognition of buildups can be either direct—those seismic parameters that directly outline buildups such as reflections from the boundaries of the buildups, onlap of overlying cycles, or seismic facies changes between the buildups and enveloping beds; or indirect—those seismic parameters that indirectly outline or indicate the presence of buildups such as drape, velocity anomalies, and spurious events. Use of basin architecture is an additional indirect, but generally geologic, line of evidence to infer locations of buildups. All available geologic and geophysical data should be used; the techniques of seismic stratigraphic and seismic facies analysis provide the framework for this interpretation.

INTRODUCTION

This paper provides seismic interpreters with: (1) criteria for recognizing carbonate buildups on seismic sections, and (2) several examples of the seismic expression of documented carbonate buildups for comparison with seismic data from their own areas of interest.

Carbonate buildups, including reefs and banks, form important and prolific hydrocarbon reservoirs in many operating areas of the world, particularly in the United States, Canada, North Africa, Mexico, southeastern Asia and the Middle East. Their recognition and proper interpretation are important because of variations in reservoir characteristics of strata within and associated with the buildups, and because structural closure on prospects is commonly due to the topography generated during deposition of the buildups. Seismic stratigraphic interpretation of carbonate buildups is enhanced by depositional topography and contrasts in lithology, interval velocity, density, and bedding characteristics between the buildups and enveloping strata.

The seismic interpretation procedure recommended is that outlined in Part 6 (Vail et al, this volume). Emphasis is placed here on visual interpretation of reflection configuration and on other seismic parameters such as amplitude, frequency, continuity, and interval velocity. Modern computer processing also aids in presenting more sophisticated graphic displays of these parameters.

REEFS, BANKS, AND BUILDUPS

The term *carbonate buildup* as used here, is a general term for all sedimentary carbonate deposits that form positive bathymetric features. This inclusive term is used because seismic data do not easily differentiate between the deposits conventionally described as reefs and banks by many authors following Lowenstam (1950), Nelson et al (1967), and Klement (1967). The term *bank*, a descriptive term with genetic implications, denotes a bathymetrically positive sediment accumulation formed by the gregareous growth of organisms which cause and contribute to sediment deposition but do not form a rigid structure. The term *reef* is used for bathymetrically positive rigid structures formed by sedentary, intergrowing organisms. A reef is commonly a bioherm (Cumings, 1932), that is, a mound or lens-shaped feature of organic origin which is lithologically discordant with surrounding deposits. A bank can be a bioherm as above, or a biostrome, which is a layer or bed of coarse skeletal remains which grades into surrounding deposits of different lithology. Although we recommend the use of the terms reef and bank where possible, we have found that the subdivision of carbonate buildups given below is also very useful where reefs and banks can not be separated seismically.

For purposes of seismic analysis, a wide variety of carbonate buildups may be grouped into four major types illustrated in Figure 1: barrier buildups are linear, with relatively deep water on both sides during deposition; pinnacle buildups are roughly equidimensional and were surrounded by deep water during deposition; shelf-margin buildups are linear, with deep water on one side, and shallow water on the other; and patch buildups form in shallow water, either in close proximity to shelf margins, or over broad, shallow seas. Careful analysis of an appropriate grid of seismic data is required to define the shapes and depositional environments of these features.

[1]Manuscript received, January 6, 1977; accepted, June 13, 1977.

[2]Exxon Production Malaysia, Inc., Kuala Lumpur, Malaysia.

[3]Imperial Oil, Ltd., Calgary, Canada.

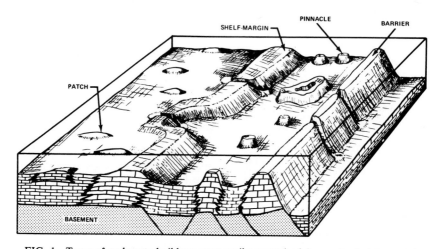

FIG. 1—Types of carbonate buildups most easily recognized from seismic interpretation. Conventional classification of reefs and banks, although preferred, is not easily applicable to seismic data.

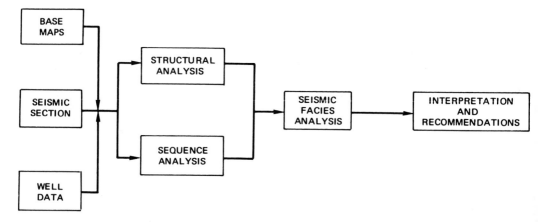

FIG. 2—Diagrammatic procedure for stratigraphic interpretation of seismic data. Three general steps include (1) conversion of all geologic data to a form compatible with seismic data, (2) seismic sequence and structural analysis, and (3) seismic facies analysis, interpretation.

Carbonate buildups as identified from seismic data commonly are composed of a variety of carbonate microfacies. For instance, a feature may have fore-reef facies changing laterally updip to synchronous reef talus and in-situ reef facies which may, in turn, grade to various back reef, lagoonal, or tidal-flat facies. Each facies has its own characteristic carbonate compositions, grain types, textures, and sedimentary structures. Resolution of most of these carbonate microfacies is beyond the limits of most conventional seismic data.

PROCEDURE FOR SEISMIC RECOGNITION OF CARBONATE BUILDUPS

Procedures recommended for the interpretation of carbonate buildups from a grid of seismic data follow the seismic stratigraphic techniques described in Part 6 (Vail et al, this volume). Basically, this interpretation procedure has three general steps (Fig. 2). As a preliminary step, all available geologic data are converted to a format compatable for interpretation with the seismic data. This usually requires construction of various time-linear well logs and overlays of paleontologic data, paleobathymetry, and other well information. Available outcrop data should be projected into the line of the seismic section.

Next, depositional sequences (Part 2, Vail et al, this volume) are interpreted from seismic data through study of the systematic patterns of cycle terminations. The sequences are correlated with well data, their ages determined, and their boundaries extended through seismic grids to complete the stratigraphic and structural framework.

Where the sequence framework is established, the third step is the recognition of seismic facies units within sequences. Seismic parameters are objectively defined, mapped, and correlated to well data where possible, and they are interpreted in terms of depositional processes, environments, and possible lithology.

GEOPHYSICAL CRITERIA FOR RECOGNIZING CARBONATE BUILDUPS

The criteria for recognizing carbonate buildups on seismic sections include seismic parameters that directly outline the buildup, and those that indirectly outline the buildup or infer its presence. Figure 3 diagrammatically illustrates these criteria; each of the diagrams was taken from an actual example.

Direct Criteria

Boundary Outline—Commonly, reflection configurations directly define the boundary of the buildup. These include reflections from the top and sides of the depositional feature, and onlap of overlying reflections onto the buildup. Depositional topography must be sufficiently great for these criteria to be evident on the reflection seismic record.

Seismic Facies Change—Changes may occur in amplitude, frequency, or continuity of reflections from within the buildup, or between the buildup and the laterally adjacent time-synchronous or younger onlapping reflections. Such changes would result where differences in characteristics of bedding continuity, density and/or velocity exist between strata within the buildup, or between the buildup and the strata enclosing the buildup.

Indirect Criteria

Drape—Drape commonly occurs in reflections overlying the buildup because of differential compaction of strata in the buildup and the enveloping strata. This phenomenon is generally most pronounced where a strong contrast exists in lithology of buildup and off-buildup sediments, such as with a limestone buildup surrounded by shale. The effects of drape generally die out stratigraphically upwards.

Velocity Anomalies—A pronounced velocity contrast commonly exists between the buildup and adjacent strata, resulting in differences in seismic travel time through these strata. For instance, reflections from strata beneath a limestone buildup with a higher velocity than laterally adjacent shales would be "pulled up" in time, compared to reflection time from the same strata beneath the shales (see Fig 13). Similarly, reflections below a buildup with slower interval velocities than those of the surrounding strata would be "pulled down" below the buildup (see Fig. 5). The amount of the velocity anomaly is directly related to the contrast of interval velocity between buildup and laterally adjacent strata, and the thickness of the buildup or strata that have the contrasting velocities. Carefully constructed isochron maps of time-stratigraphic units containing buildups are useful exploration tools where these velocity contrasts occur. Such maps were extensively used by industry in Devonian reef exploration in Canada.

Spurious Events—The edges of the buildup commonly are marked by termination of surrounding beds or abrupt changes in internal bedding geometry. These edges can be sites for development of diffractions or odd events. Mapping of such seismic events may offer a clue to the presence and distribution of otherwise hard-to-see carbonate buildups.

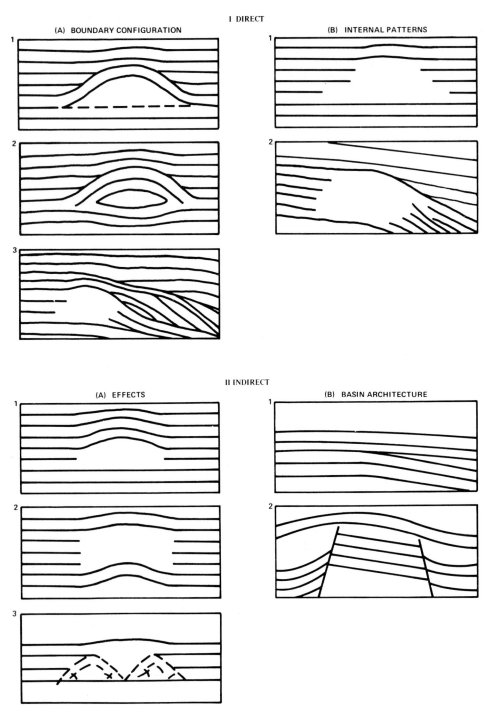

FIG. 3—Seismic criteria for recognizing carbonate buildups. Criteria for directly outlining buildups include reflections from top and sides of buildups and onlap of overlying reflections onto buildups (**I-A**); and patterns of seismic facies change between buildup and enclosing strata (**I-B**). Criteria that indirectly outline or infer presence of buildup include drape, velocity anomalies, and spurious events (**II-A**), and determination of optimum basin positions for buildups (**II-B**).

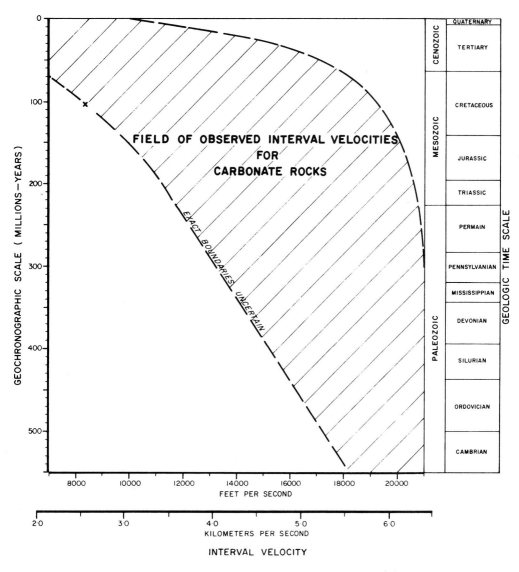

FIG. 4—Field of observed interval velocities from well data for carbonate rocks plotted against geologic age.

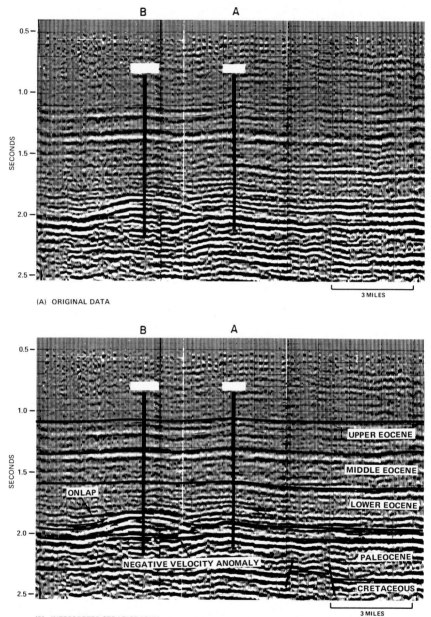

FIG. 5—North Africa (12-fold CDP thumper data). This high-quality seismic line shows anomaly interpreted as pinnacle reef by (1) a reflection outlining tops and sides of buildup, (2) three cycles of onlap, (3) drape in overlying beds, and (4) negative velocity anomaly (off-reef micritic limestones and shales have higher interval velocity than porous, lightly cemented, Tertiary reef carbonates). Overall aspect of this pattern is so-called "eye effect".

Location **A** was first to be drilled, based on poorer seismic data, on closure at Eocene level above Paleocene reef unit. Well encountered about 60 m of gas-filled pay in Paleocene and was abandoned; acreage in area was subsequently dropped. Another company picked up acreage, obtained high-quality seismic data that showed Paleocene reef, and drilled well **B** as discovery well. Well encountered about 300 m of porous algal-foraminiferal and coralline limestone. Pay section was 293 m thick. Oil flowed on test at rate of more than 40,000 bbl/day. Estimated recoverable reserves in this field are approximately 1.5 billion bbl of oil.

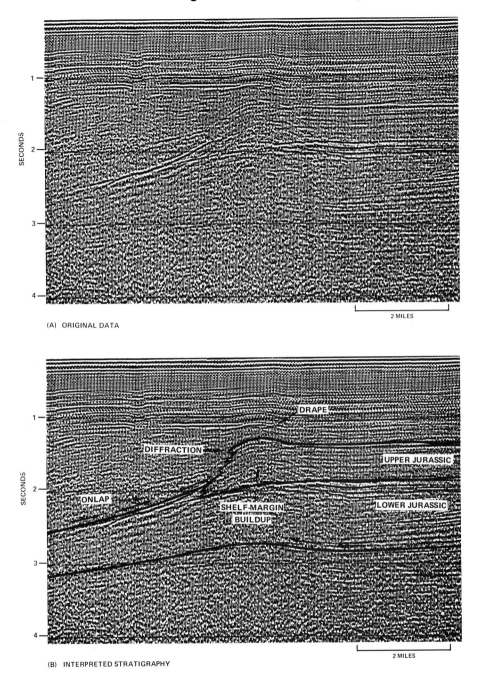

(A) ORIGINAL DATA

(B) INTERPRETED STRATIGRAPHY

FIG. 6—Offshore West Africa (12-fold CDP Aquapulse® data). Shelf-margin carbonate buildup can be seen by (1) reflection from top and front of buildup, (2) onlap of cycles onto buildup, (3) change from continuous, parallel reflectors into discontinuous reflectors, (4) numerous diffractions, (5) drape over buildup, and (6) abrupt change in dip of reflectors.

Wells encountered series of Mesozoic shelf-margin buildups along eastern Atlantic continental margin off Africa. Buildup displayed on this line is interpreted as Late Jurassic.

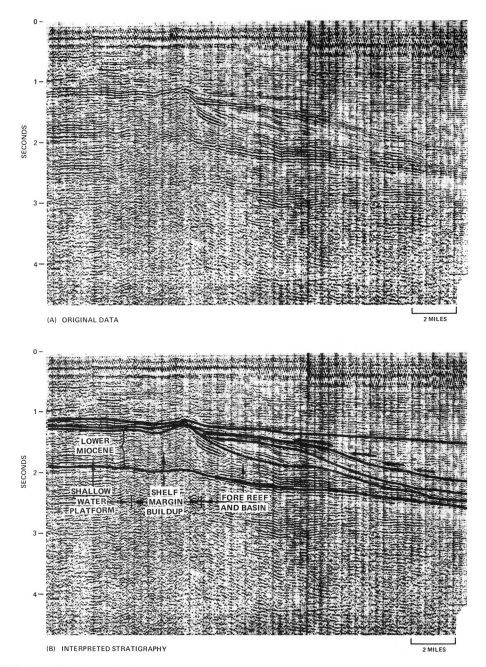

(A)   ORIGINAL DATA   2 MILES

(B)   INTERPRETED STRATIGRAPHY   2 MILES

FIG. 7—Gulf of Papua (6-fold CDP dynamite data). Shelf-margin reef is interpreted on basis of (1) abrupt changes in slope of reflectors at shelf edge, (2) onlap of slope and basin units onto shelf edge, and (3) seismic facies pattern change from zone of more continuous, parallel reflectors (back reef–lagoon) to discontinuous, nearly reflection-free zone (reef facies) to thinner zone of dipping, convergent cycles (fore reef). Interesting prograding lens of convergent to sigmoid cycles, either younger or in part synchronous with reef, has partially infilled basin seaward of shelf-margin buildup. This lens, untested by wells, is interpreted to be shale and/or micritic limestone.

Early Miocene reefs, both shelf-margin and pinnacle types (some with gas) have been encountered in Papua basin. Each prospective area should be considered in terms of (1) closure, (2) seal (sometimes shelf-margin reef has porous facies over it), and (3) freshwater flushing.

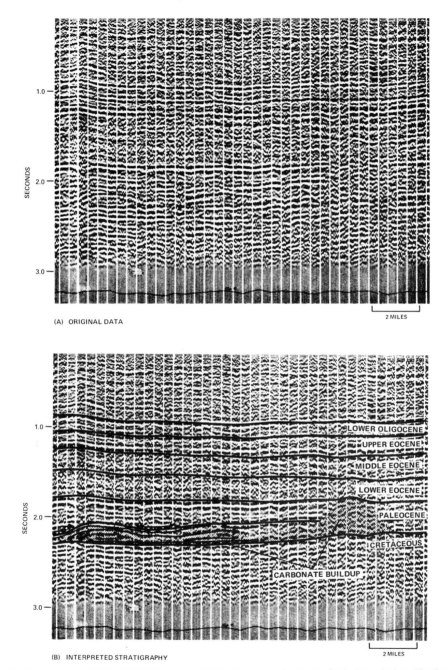

(A) ORIGINAL DATA

2 MILES

(B) INTERPRETED STRATIGRAPHY

2 MILES

FIG. 8—North Africa (12-fold CDP thumper data). Carbonate buildups, both low platformlike banks and pinnacles, are interpreted on basis of (1) seismic facies change from continuous parallel reflectors to mainly reflection-free to very discontinuous reflection zone, and (2) one to two cycles of onlap of overlying units onto buildups (extreme left of section).

Well south of line confirmed buildup seen on right. It encountered shallow-water carbonates of Paleocene age in part of basin normally having shale and deep-water limestone. Only a thin gas pay, apparently associated with small amount of drape, was encountered. Trapping problem was lack of seal. A sheetlike porous carbonate grainstone extends over this part of basin near top of Paleocene sequence and intersects buildup to right. This contact may have allowed drainage of hydrocarbons from buildup tested in well.

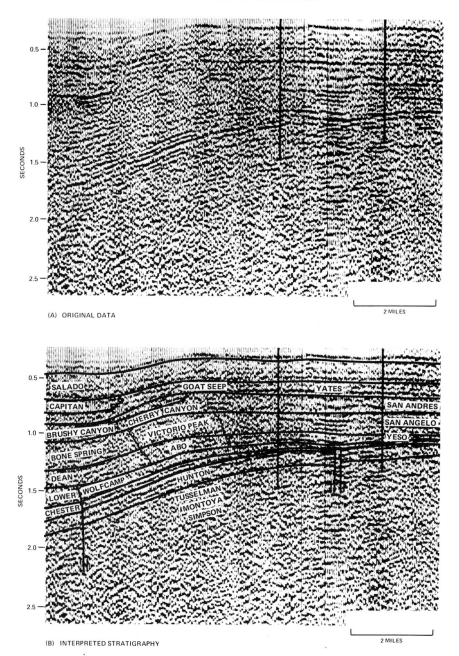

FIG. 9—Central basin platform, Lea County, New Mexico (12-fold CDP Vibroseis® data). Shelf-margin carbonate-bank buildup on this line is indicated by (1) abrupt change in dip at shelf edge, and (2) seismic facies change from high-amplitude, continuous reflections to low-amplitude to nearly reflection-free zone at shelf edge.

Leonardian and Guadalupian shelf-margin banks, composed mainly of dolomitized skeletal limestones of the Abo, Victorio Peak, Goat Seep, Getaway, and Capitan Formations, are documented by wells in this part of Permian basin. Basinward of shallow-water banks are siltstones, shales, and micritic limestones of Dean, Bone Spring, Brushy Canyon, and Cherry Canyon Formations; shelfward of banks are thin-bedded, dolomitized micritic and dolomitized algal-laminated limestones and sandstones of the Yates, Seven Rivers, Queen, Grayburg, San Andres, San Angelo, and Yeso Formations.

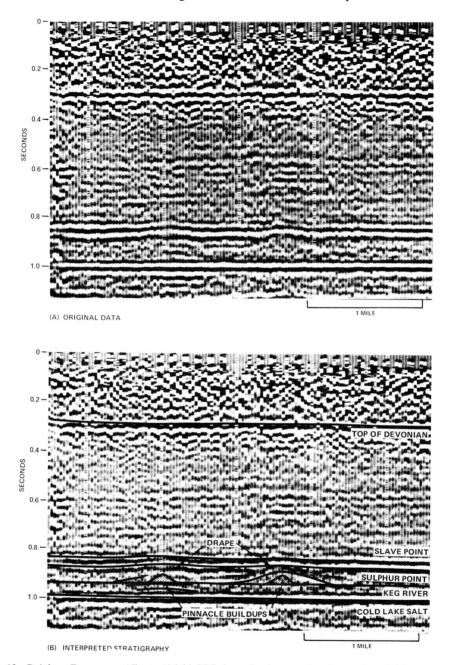

(A) ORIGINAL DATA

1 MILE

(B) INTERPRETED STRATIGRAPHY

1 MILE

FIG. 10—Rainbow-Zama area, Alberta (6-fold CDP dynamite data, section datumed on Cold Lake salt reflection—a near-basement horizon). Presence of two small pinnacle reefs is indicated by small amount of drape in reflectors above reefs and by one cycle of onlap (?).

Buildups indicated on section are Middle Devonian Keg River Formation pinnacle reefs. Drape seen in overlying Sulphur Point and Slave Point Formations is probably due both to differential compaction and to subsequent removal of Muskeg salt deposited in interreef areas. Recoverable reserves in Rainbow-Zama area are estimated at 800 million bbl of oil. Pool size is variable; some are smaller than 8 ha., others are larger than 2 sq km, but average is 20 to 30 ha. Pay thickness is variable.

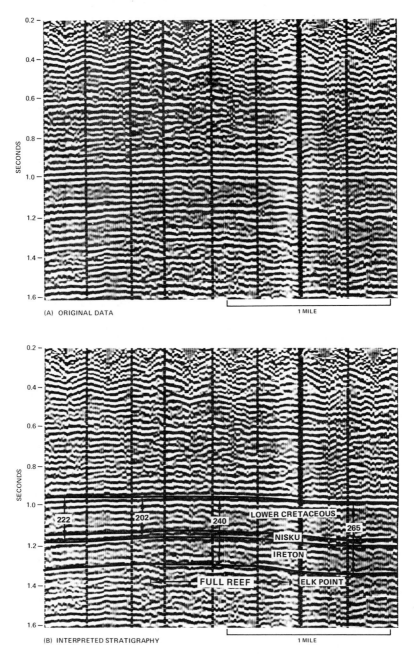

FIG. 11—Golden Spike, Alberta (singlefold conventional seismic data). Carbonate buildup indicated by (1) drape, and (2) velocity anomaly. Drape effect due to differential compaction shown by about 23 msec of thinning in Lower Cretaceous to Ireton isochron. The 30 msec of thinning in Lower Cretaceous to Elk Point isochron, which encompasses reef, is caused by faster velocity in reef limestones than in laterally adjacent shales.

Golden Spike (discovered 1949) is areally small (10 sq km) pinnacle reef, with more than 300 million bbl of oil in place. It formed on Cooking Lake bank seaward of Leduc-Rimbey barrier reef trend and is surrounded by shales of Ireton and Duvernay formations.

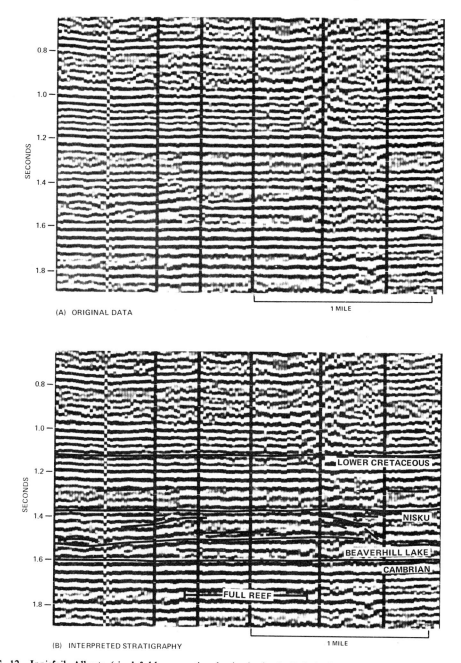

(A) ORIGINAL DATA

1 MILE

(B) INTERPRETED STRATIGRAPHY

1 MILE

FIG. 12—Innisfail, Alberta (singlefold conventional seismic data). Only indications of carbonate buildup are vague, odd dips and diffractions. After reef was discovered, mapping of spurious events produced good outline of reef which was useful in development stages of drilling. Method was not successful in predicting reef prospects in adjacent areas.

Innisfail field, a Leduc reef on west side of Bashaw complex in southwest Alberta, is in an area where off-reef Ireton has high lime content. Hence, no measurable velocity anomaly or drape effect is found. Innisfail field (discovered 1957), produced about 20 million bbl of oil up to December 1968; total reserves near 75 million bbl.

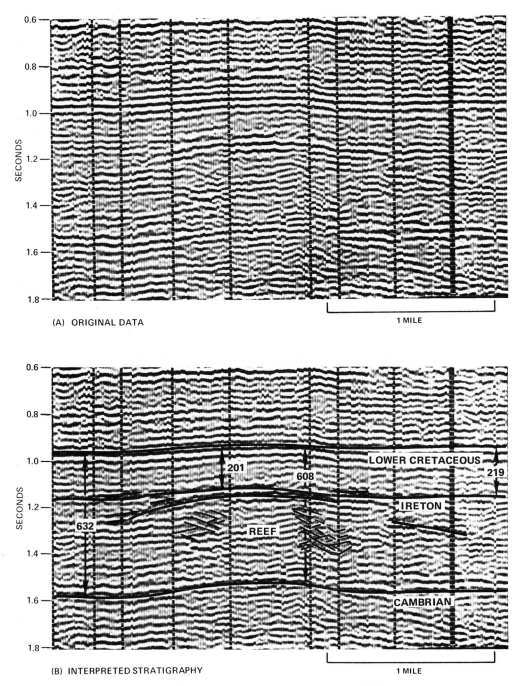

FIG. 13—Yekau Lake, Alberta (singlefold conventional seismic data). This buildup is defined by (1) about 18 msec drape measured in Lower Cretaceous to Ireton interval, (2) about 25 msec velocity anomaly measured in Lower Cretaceous to Cambrian isochron, and (3) spurious events, probably diffractions, near edges of buildup.

Yekau Lake reef is small Leduc reef within Leduc-Rimbey barrier trend. Off-reef Ireton-Duvernay in this area is mainly shale, giving rise to pronounced velocity anomaly and drape effects. Field size is slightly more than 2 sq km, pay thickness averages 6.5 m, and recoverable reserves estimated at 3.6 million bbl.

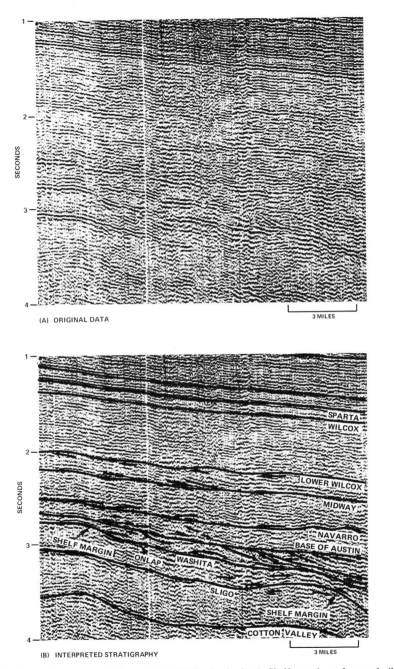

FIG. 14—Washita reef trend, south Texas (6-fold CDP seismic data). Shelf-margin carbonate buildup inferred adjacent to hinge line as indicated by (1) abrupt thinning in Washita-Sligo interval and (2) onlap of about two cycles onto Washita edge. Sligo shelf margin also suggested on right-hand (south) side of section.

Wells encountered Washita-Fredericksburg rudistid reefs and banks that pass seaward into dense deeper water carbonate mud. Exploration problem is that of updip closure. Back-reef and lagoonal sediments behind shelf-margin buildups are commonly porous, and may have permitted hydrocarbons to escape.

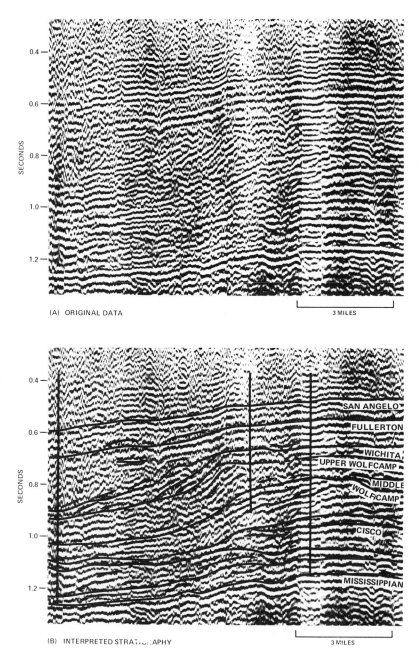

(A) ORIGINAL DATA   ⊢————————⊣
                                              3 MILES

(B) INTERPRETED STRATIGRAPHY   ⊢————————⊣
                                              3 MILES

FIG. 15—Eastern shelf area, Permian basin, Crosby County, Texas (singlefold conventional seismic data). Series of stacked carbonate-shelf-edge buildups were detected here after seismic sequences were carefully defined to show (1) abrupt change in slope, (2) onlap onto shelf edges, and (3) thinning of sequences.

Upper Pennsylvanian (Cisco-Canyon) and Lower Permian bank edges are stacked in transgressive series through middle Wolfcamp, then in regressive fashion through late Wolfcamp and Wichita. Carbonate banks and bank edges are porous in this area, but lack of lateral closure prevented hydrocarbon accumulation. Banks of same age are prolific producers in Horseshoe atoll area, and locally in Central basin platform.

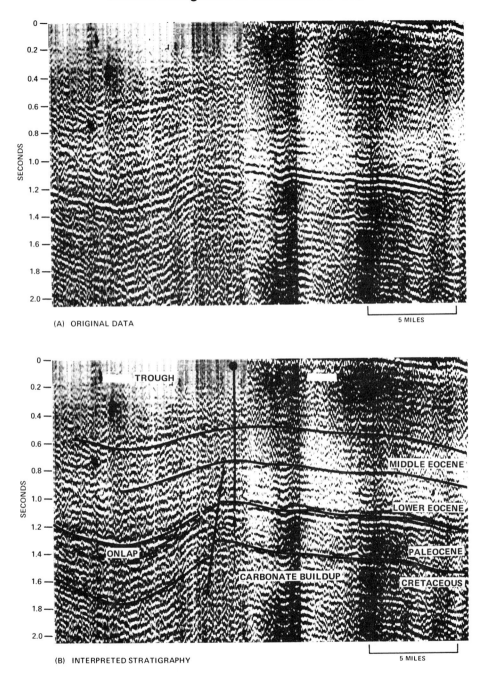

(A) ORIGINAL DATA

5 MILES

(B) INTERPRETED STRATIGRAPHY

5 MILES

FIG. 16—North Africa (3-fold CDP seismic data). Presence of carbonate buildup can only be inferred from basin architecture. Upthrown side of normal fault block would be logical position for shallow-water carbonates to accumulate under proper conditions. In seismic data, no parameter change was noted along fault block even though there is pronounced facies change from porous shallow-water carbonates on platform to shales and micritic limestones in trough.

Field produces from shallow-water Paleocene carbonate "pipe" or buildup in formations on upthrown side of major fault that separates platform from trough. Cumulative production of over 1 billion bbl of oil.

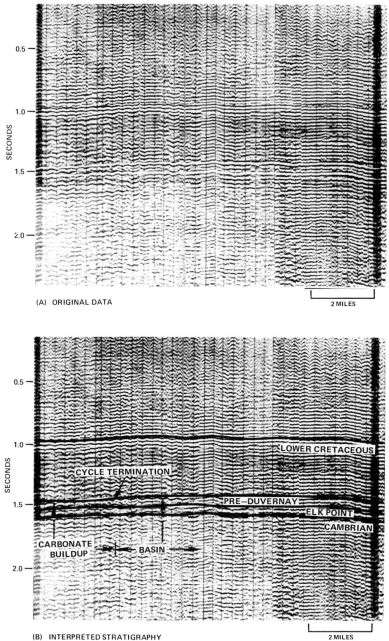

(A) ORIGINAL DATA

2 MILES

(B) INTERPRETED STRATIGRAPHY

2 MILES

FIG. 17—Judy Creek area, Alberta (singlefold conventional seismic data). Presence of shallow-water carbonate buildup indicated only by termination of single cycle. Termination occurs near point where, to southwest in structurally high position, shallow-water carbonates formed. To northeastward, or basinward within same time span, shale was deposited. After discovery well was drilled, fair success was achieved in outlining limits of this and similar nearby buildups using cycle termination as diagnostic parameter.

Judy Creek reef and bank complex is thin (75 m thick) carbonate mound within Beaverhill Lake Group (Upper Devonian) encased by argillaceous limestones and shales. Judy Creek field (discovered 1958) had cumulative production to June 1968 of 43 million bbl of oil; estimated recoverable reserves are about 131 million bbl of oil. Field covers about 115 sq km, and pay thickness averages 21 m.

Basin Architecture—In some instances, carbonate buildups can be inferred as likely to occur in a preferred location along a seismic profile, based on seismic and other geologic evidence of basin architecture, such as fault-block edges, position of hinge line, or contemporaneous structural highs. The interpretation is enhanced by a thorough knowledge of the geologic setting and history of the area, and of those time-stratigraphic units prone to the development of buildups or marked by extensive or local carbonate sedimentation.

INTERVAL VELOCITY IN CARBONATE SECTIONS

The development of high speed computers and the advent of widespread digital processing have facilitated the estimation of seismic interval velocity and its use for interpretation of lithology. Although applied mostly to estimating sandstone/shale ratios, some interval velocity work has been done with stratigraphic sections containing carbonate, sandstone, and shale. Carbonates do not have unique interval velocities, and the interval velocity of a particular buildup is dependent on a wide variety of factors including porosity and burial history. A search of the literature on carbonates in a variety of basins around the world shows a wide variation in the interval velocity of carbonate rocks (Fig. 4). Within each local area, therefore, all available well data should be analyzed carefully to aid in the interpretation of stratigraphy from interval velocity.

SOME LIMITATIONS AND QUALIFICATIONS

Many seismic reflection configurations interpreted as carbonate buildups are not unique, but may be exhibited by other geologic features similar in size or shape. Examples of such features include salt intrusions or pillows, igneous intrusions, volcanic cones, unconformity surfaces, and slump deposits. Interpretation of carbonate buildups from seismic data must be merged with all available geologic and geophysical data of the area. In particular, data on the occurrence and distribution of known buildups and carbonate units in adjacent parts of the basin are useful.

Resolution limitations must be recognized. This is particularly true in older carbonate sequences with high interval velocities. Interval velocities of 4,500 to 6,400 m/sec are common for many Paleozoic carbonate strata. Thus, a carbonate buildup 130 m thick with an interval velocity of 5,500 m/sec would be represented by only 47 msec, two-way time. There may not be sufficient response for recognition of this 130-m buildup, depending on such parameters as (1) the frequency content of the input pulse, (2) the amount of attenuation of frequency of the pulse with increasing travel time, and (3) the processing and filtering both in the field and before display of the final seismic section. A cycle breadth of 50 msec for 20-Hz data might not be uncommon and the buildup described would be represented by one cycle. Reflections from directly overlying beds might mask entirely any response of the seismic data to the buildups.

EXAMPLES OF CARBONATE BUILDUPS ON SEISMIC SECTIONS

Thirteen seismic profiles (Figs. 5-17), mostly documented by wells, illustrate the geophysical criteria for recognizing carbonate buildups shown diagrammatically on Figure 3. These criteria are summarized in Table 1 and are discussed in each figure description, along with pertinent geologic and hydrocarbon-production data. These exam-

Table 1. Summary Sheet of Seismic Examples.

Examples	Criteria for Recognizing Buildup									Type of Buildup		Production Associated with Buildup	
	Direct			Indirect									
			Amplitude Frequency, or Continuity Change		Effects		Basin Architecture						
Figure Number	Reflection Outline	Onlap		Drape	Velocity	Spurious Events	Change in Slope	Basin Position	Shelf Margin	Barrier	Pinnacle	Yes	No
North Africa	X	X		X	X						X	X	
Offshore W. Africa	X	X	X	X		X	X		X				X
Gulf of Papua	X	X	X				X		X				X
North Africa		X	X								X		X
Cen. Basin Plat.			X				X		X				X
Zama				X							X	X	
Golden Spike			X	X							X	X	
Innisfail			X			X					X	X	
Yekau Lake		X		X	X	X				X		X	
South Texas	X						X	X	X				X
Crosby Co., Texas	X						X	X	X				X
North Africa							X	X			X	X	
Judy Creek							X	X			X	X	

ples provide a spectrum of criteria for recognition of carbonate buildups.

REFERENCES CITED

Cumings, Edgar R., 1932, Reefs or bioherms?: Geol. Soc. America Bull., v. 43, p. 331-357.
Klement, K. W., 1967, Practical classification of reefs and banks, bioherms and biostromes (abs.): AAPG Bull., v. 51, no. 1, p. 167.
Lowenstam, H. A., 1950, Niagaran reefs of the Great Lakes area: Jour. Geology, v. 58, no. 4, p. 430-487.
Nelson, H. F., C. W. Brown, and J. H. Brineman, 1962, Skeletal limestone classification in W. E. Ham, ed., Classification of carbonate rocks—a symposium: AAPG Memoir 1, p. 224-252.

Seismic Stratigraphy and Global Changes of Sea Level, Part 11: Glossary of Terms used in Seismic Stratigraphy [1]

R. M. MITCHUM, JR.[2]

INTRODUCTION

Following are definitions of some terms most commonly used in seismic stratigraphy. Because of the lack of standardized language to describe seismic stratigraphic phenomena, this informal terminology has evolved by necessity over a period of years. The evolution is not considered complete by any means.

The glossary is presented here not only to aid the reader, but to solicit comments and criticisms for improvement. With each term is a reference to a numbered part of the Vail et al suite of papers, and (where possible) to figures in that part where the term is best discussed and illustrated. Standard geophysical terms are not included in this glossary, but are defined in Sheriff (1973). Geologic terms are defined in Gary, McAfee, and Wolf (1972).

THE GLOSSARY

Aggradation—*See* onlap.

Apparent onlap—*See* onlap.

Bank seismic facies unit—a seismic facies unit with a bank-like external form: essentially a sheet-like form with a clinoform or "bank-front" leading edge that commonly marks the shelf-slope break (part 6, Fig. 14; part 10).

Baselap—a term describing termination of strata along the lower boundary of a depositional sequence, used only where discrimination between onlap and downlap is difficult or impossible (part 2).

Base-discordance—*See* discordance.

Carbonate buildup seismic reflection configuration—a seismic reflection configuration, usually a mound, interpreted as carbonate reefs or banks. Carbonate buildups of sufficient size to be detected seismically include barrier buildups, pinnacle buildups, shelf-margin buildups, and patch buildups (part 10, Figs. 1, 3).

Chaotic seismic reflection configuration—discontinuous, discordant seismic reflection patterns suggesting a disordered arrangement of reflection

surfaces. They are interpreted either as strata deposited in a highly variable, relatively high-energy setting; or as initially continuous strata which have been deformed so that continuity has been disrupted (part 6, Figs. 11, 12a, b).

Chronostratigraphic correlation chart—a stratigraphic summary chart on which geologic time is plotted as the vertical scale, and distance across the area of interest as the horizontal scale, and on which a variety of stratigraphic information is brought together (part 7, Fig. 2).

Clinoform surface—a sloping depositional surface, commonly associated with strata prograding into deep water (part 6, Figs. 8, 9).

Coastal aggradation—*See* onlap.

Coastal encroachment—*See* onlap.

Coastal deposits—those sediments deposited at or near sea level (littoral, paralic, or coastal nonmarine deposits) (part 3, Fig. 2).

Coastal nonmarine deposits—those sediments laid down on the coastal plain above high tide (part 3).

Coastal onlap—*See* onlap.

Coastal toplap—*See* toplap.

Complex sigmoid-oblique reflection configuration—a seismic reflection configuration consisting of a combination of variably alternating sigmoid and oblique progradational reflection configurations within a single seismic facies unit (part 6, Figs. 8d, 9d). The upper (topset) segment of the facies unit is characterized by a complex alternation of horizontal sigmoid topset reflections and segments of oblique configuration with toplap terminations. In other respects this configuration is similar to the sigmoid configuration. *See* progradational reflection configurations, sig-

[1]Manuscript received, January 6, 1977; accepted, June 13, 1977.

[2]Exxon Production Research Co., Houston, Texas 77001.

moid reflection configurations, and oblique reflection configurations.

Concordance—parallelism of strata to sequence boundaries, with no consequent stratal terminations against the boundary surfaces (part 2, Fig. 2). *See* discordance.

Conformity—a surface that separates younger strata from older rocks, but along which there is no physical evidence of erosion or nondeposition, and no significant hiatus (part 2, Fig. 1). *See* unconformity.

Convergent seismic reflection configuration—a seismic reflection pattern in which a seismic unit thins laterally by nonsystematic termination of individual reflections, due to thinning of individual strata below seismic resolution. This internal convergence may occur anywhere within seismic sequences (part 6, Fig. 2), and should not be confused with terminations along sequence boundaries. Same as *divergent seismic reflection configuration,* with emphasis on thinning rather than thickening of strata.

Cycle of relative change of sea level—the interval of time during which a relative rise and fall of sea level takes place (part 3, Fig. 1). Regional cycles and global cycles are those recognized on a regional or global basis, respectively. Most regional cycles are eventually recognized as global events, but their use in regional stratigraphic studies does not depend on this recognition. First-, second- and third-order cycles are recognized. In strict usage, one cycle of rise and fall would be a third order cycle (part 4, Figs. 1-3). *See also* eustatic cycle.

Depositional sequence—a stratigraphic unit composed of a relatively conformable succession of genetically related strata and bounded at its top and base by unconformities or their correlative conformities (part 2, Fig. 1). *See* sequence.

Discordance—the lack of parallelism of strata to sequence boundaries, with consequent stratal terminations against the boundary surfaces (part 2, Fig. 2). *Top-discordance* (truncation and toplap) occurs at the upper sequence boundary, and *base-discordance* (onlap and downlap) occurs at the lower sequence boundary.

Distal downlap—*See* downlap.

Distal onlap—*See* onlap.

Divergent seismic reflection configuration—a seismic reflection configuration characterized by a laterally thickening wedge-shaped unit in which most of the lateral thickening is accomplished by splitting of individual reflection cycles within the unit, rather than predominantly by onlap, toplap, or erosion at base or top (part 6, Figs. 6, 7c, d).

Downlap—a base-discordant relation in which initially inclined strata terminate downdip against an initially horizontal or inclined surface (part 2, Fig. 2). *Distal downlap* is downlap in the direction away from the source of clastic supply. *Seismic downlap* is downlap interpreted from a seismic section. It is a relation in which seismic reflections interpreted as initially inclined strata terminate downdip against a reflection discontinuity interpreted as an initially inclined or horizontal unconformity. *Apparent downlap* on seismic sections may occur where reflections representing units of inclined or tangential strata terminate downdip, but the strata themselves actually flatten and continue as units which are so thin that they fall below the resolution of the seismic tool (part 6).

Downward shift of coastal onlap—a downslope seaward shift from the highest position of coastal onlap in a given stratigraphic unit to the lowest position of coastal onlap in the overlying unit (part 3, Fig. 8). It is used to recognize relative falls of sea level.

Encroachment—*See* onlap.

Erosional hiatus—*See* hiatus.

Erosional truncation—*See* truncation.

Eustatic change—pertaining to worldwide changes of sea level that affect all the oceans (Gary, McAfee, and Wolf, 1972, p. 241); or a relative change of sea level on a global scale, produced either by a change in the volume of sea water, or a change in the surface area of the ocean basins, or both (Fairbridge, 1961).

Eustatic cycle—the interval of time during which a eustatic rise and fall of sea level takes place (part 4, Fig. 7). *See* eustatic change.

Fill seismic reflection configuration—seismic reflection configuration interpreted as strata filling a negative-relief feature in the underlying strata (part 6, Figs. 14, 17). Underlying reflections may show either erosional truncation or concordance along the basal surface of the fill unit. Fill units may be classified by external form (channel fill, trough fill, basin fill, or slope front fill; part 6, Fig. 14), or by internal reflection configurations, such as onlap, mounded onlap, divergent, prograding, chaotic, or complex (part 6, Fig. 17; part 9, Fig. 9).

First-order cycle—a cycle of relative or eustatic change of sea level that has a duration in the order of 100 to 200 million years (part 4, Fig. 1).

Geochronologic chart—a chart of the geologic time scale.

Geotectonic subsidence—subsidence that affects an area greater than the area of study and can not be compensated for by determining coastal aggradation (part 3, Fig. 8).

Global cycle of relative change of sea level—an interval of geologic time during which one relative rise and fall of mean sea level takes place on a global scale. In our usage this would be a third order cycle (part 4, Figs. 2, 3). *See* cycle of relative change of sea level.

Global paracycle of relative change of sea level—an interval of geologic time during which a relative rise and stillstand of sea level takes place on a global scale, followed by another relative rise with no intervening relative fall.

Global relative change of sea level—a relative change of sea level that occurs on a global scale during a specific interval of geologic time (part 4, Figs. 2, 3).

Global relative fall of sea level—a relative fall of sea level that occurs on a global scale during a specific interval of geologic time (part 4, Figs. 2, 3). *See* relative fall of sea level.

Global relative rise of sea level—a relative rise of sea level that occurs on a global scale during a specified interval of geologic time (part 4, Figs. 2, 3). *See* relative rise of sea level.

Global relative stillstand of sea level—a relative stillstand of sea level that occurs on a global scale during a specified interval of geologic time (part 4, Figs. 2, 3). *See* relative stillstand of sea level.

Global supercycle of relative change of sea level—a group of global cycles of relative change of sea level in which a cumulative relative rise to a higher position of mean sea level is followed by a cumulative relative fall to a lower position. Rarely, a minor setback occurs (part 4, Figs. 1-3). A global supercycle is a second-order cycle.

Hiatus—the total interval of geologic time which is not represented by strata at a specified position along a given stratigraphic surface. If the hiatus encompasses a measurable interval of geologic time, the stratigraphic surface is an unconformity (part 2, Fig. 1). A *nondepositional hiatus* is attributed to termination of strata at their original deposition limits by onlap, downlap, or toplap. The nondepositional hiatus refers to the geologic-time interval during which no strata were deposited at the depositional surface. An *erosional hiatus* is attributed to the erosional truncation of strata. The erosional hiatus refers to the geologic-time range of the strata which were removed by erosion, and not the time at which the erosion occurred.

Highstand—the interval of time during a cycle or cycles of relative change of sea level when sea level is above the shelf edge in a given local area (part 3, Figs. 9a, 13a).

Hinterland sequence—a depositional sequence that consists entirely of nonmarine deposits laid down at a site interior to the coastal area, where depositional mechanisms are controlled only indirectly or not at all by the position of sea level (part 3). *See* sequence.

Hummocky clinoform reflection configuration—a seismic reflection configuration that consists of irregular discontinuous subparallel reflection segments forming a practically random hummocky pattern marked by nonsystematic reflection terminations and splits. Relief on the hummocks is low, approaching the limits of seismic resolution. This pattern commonly grades laterally into larger, better-defined clinoform patterns, and upward into parallel reflections (part 6, Figs. 8f, 10b, c).

Internal convergence—*See* convergent seismic reflection configuration.

Interregional unconformity—an unconformity resulting from erosion and nondeposition occurring during a global relative fall and lowstand of sea level, hence global in its occurrence, although in some areas of continuous deposition the hiatus may be too small to detect paleontologically or seismically and the surface may be defined as a conformity. Major interregional unconformities occur at second-order cycle (supercycle) boundaries (part 4, Fig. 7). Minor interregional unconformities occur at third-order cycle boundaries. *See* unconformity.

Lapout—lateral termination of strata at their depositional pinchout (part 2). Lapout may occur at the upper boundary of a sequence, where it is called toplap, or at the lower boundary, where it is called onlap or downlap.

Lens seismic facies unit—a seismic facies unit with a lens-like external form: a unit bounded by converging surfaces, at least one of which is curved, thick in the middle, and thinning out toward the edges (part 6, Fig. 14).

Lowstand—the interval of time during a cycle or cycles of relative change of sea level when sea level is below the shelf edge. A *comparative lowstand* occurs where sea level is at its lowest point on the shelf during the deposition of a series of sequences (part 3, Figs. 9b, 13a).

Marine onlap—*See* onlap.

Maritime sequence—a depositional sequence that consists of genetically related coastal nonmarine, littoral, and/or marine deposits. The major depositional mechanisms at the landward limit of a maritime sequence containing littoral or nonmarine deposits are controlled by the position of sea level as a base level (part 3). *See* sequence.

Migrating wave—a seismic reflection configuration consisting of a number of superposed wave-shaped reflections, each of which is progressively offset laterally from the preceding reflection, interpreted as a series of migrating sediment waves moving across a generally horizontal surface (part 6, Figs. 15, 16).

Mound seismic reflection configuration—a seismic reflection configuration interpreted as strata forming an elevation or prominence rising above the general level of the surrounding strata. Most mounds are topographic buildups resulting from either clastic or volcanic depositional processes or organic growth. Deepsea fans, lobes, slump masses, some deepsea current and contourite deposits, carbonate buildups and reefs (part 10), volcanic piles, and other features could have mounded two-dimensional configurations (part 6, Fig. 15).

Non-depositional hiatus—*See* hiatus.

Oblique seismic reflection configuration—a seismic reflection configuration interpreted as strata which form a prograding clinoform pattern consisting ideally of a number of relatively steeply dipping strata terminating updip by toplap at or near a nearly flat upper surface and downdip by downlap against the lower surface of the facies unit. Successively younger foreset segments of strata build almost entirely laterally from a relatively constant upper surface characterized by lack of topset strata and by pronounced toplap terminations of foreset strata (part 6, Figs. 8b, c; 9b, c). In the *tangential oblique* pattern, dips decrease gradually in the lower portion of the foreset strata, forming concave-upward strata which pass into gently dipping bottomset strata. In the *parallel oblique* pattern, the relatively steeply dipping parallel foreset strata terminate downdip at a relatively high angle by downlap against the lower surface.

Offlap—a term commonly used by seismic interpreters for reflection patterns generated from strata prograding into deepwater. The A.G.I. glossary defines offlap as "the progressive offshore degression of the updip termination of the sedimentary units within a conformable sequence of rocks (Swain, 1949, p. 635) in which each successively younger unit leaves exposed a portion of the older unit on which it lies" (Gary, McAfee, and Wolf, 1972, p. 490).

Onlap—a base-discordant relation in which initially horizontal strata terminate progressively against an initially inclined surface, or in which initially inclined strata terminate progressively updip against a surface of greater initial inclination (part 2, Fig. 2). *Proximal onlap* is onlap in the direction of the source of clastic supply. *Distal onlap* is onlap in the direction away from the source of clastic supply (part 2). *Marine onlap* is onlap of marine strata (part 3, Fig. 9b). *Coastal onlap* is the progressive landward onlap of the coastal (littoral or coastal nonmarine) deposits in a given stratigraphic unit (part 3, Fig. 2). *Coastal aggradation* and *coastal encroachment* are the vertical and horizontal components of coastal onlap, respectively (part 3, Fig. 2). *Seismic onlap* is the seismic expression of onlap (part 6, Figs. 2, 4). *Apparent onlap* is onlap observed in any randomly oriented vertical section which may or may not be oriented parallel to depositional dip. It is possible that apparent onlap on a given section could be a component of true downlap. If two sections intersecting at right angles both show apparent onlap, then *true onlap* is the likely relationship (part 2). A downward shift of coastal onlap (used to recognize relative falls of sea level) is a downslope seaward shift from the highest position of coastal onlap in a given stratigraphic unit to the lowest position of coastal onlap in the overlying unit (part 3, Fig. 8).

Onlap fill—*See* Fill seismic reflection configuration.

Paracycle of relative change of sea level—the interval of time occupied by one regional or global relative rise and stillstand of sea level, followed by another relative rise, with no intervening relative fall (part 3, Fig. 1).

Parallel oblique seismic reflection configuration—*See* oblique seismic reflection configuration.

Parallel seismic reflection configuration—seismic reflection configuration interpreted as strata deposited in a parallel geometric pattern (part 6, Figs. 6, 7a).

Progradational reflection configuration—a number of complex seismic reflection configurations interpreted as strata in which a significant portion of the deposition is due to lateral outbuilding or prograding. Sigmoid, oblique, complex sigmoid-oblique, shingled, and hummocky progradational reflection patterns are recognized (part 6, Fig. 8).

Proximal onlap—*See* onlap.

Reflection configuration—the geometric patterns and relations of seismic reflections that are interpreted to represent configuration of strata generating the reflections.

Reflection-free areas—homogeneous, nonstratified, highly contorted, or very steeply dipping geologic units may be expressed as essentially reflection-free areas on seismic data (part 6, Figs. 11c, 12c).

Region—in the sense used here, an entire oceanic domain, a continental margin, an inland sea, or a strip of its coast. The main requirement for regional analysis is that the depositional sequences of strata had some original physical continuity, so that regional correlations are based on strata that have a definite relation in both time and space (part 3).

Regional cycle—an interval of geologic time during which a relative rise and fall of sea level takes place on a regional scale (part 3, Fig. 1). A regional cycle is a third-order cycle.

Regional paracycle—an interval of geologic time during which a relative rise and stillstand of sea

level takes place on a regional scale, followed by another relative rise, with no intervening relative fall (part 3, Fig. 1).

Regional supercycle—a group of regional cycles of relative change of sea level in which a cumulative rise to a higher position of sea level is followed by a cumulative fall to a lower position. Rarely a minor setback occurs in the succession of rises within a supercycle. Commonly, one major fall, rather than a succession of them, is evident at the end of a supercycle (part 3, Fig. 1). A regional supercycle is a second-order cycle.

Regression—a seaward movement of the shoreline indicated by seaward migration of the littoral facies (part 3, Figs. 3, 4).

Relative change of sea level—an apparent rise or fall of sea level with respect to the land surface. Either sea level itself, or the land surface, or both in combination may rise or fall during a relative change. Generally speaking, a relative change may be operative on a local, regional, or global scale (part 3).

Relative fall of sea level—an apparent fall of sea level with respect to the underlying initial surface of deposition. It may result if sea level itself falls while the initial surface of deposition rises, remains stationary, or subsides at a slower rate, or if sea level remains stationary while the surface is rising, or if sea level rises while the surface is rising at a faster rate (part 3, Fig. 8). It is recognized by a downward shift of coastal onlap.

Relative rise of sea level—an apparent rise of sea level with respect to the underlying initial surface of deposition. It may result from: (1) sea level itself rising while the underlying initial surface of deposition subsides, remains stationary, or rises at a slower rate; (2) sea level remaining stationary while the initial surface of deposition subsides; or (3) sea level falling while the initial surface of deposition subsides at a faster rate (part 3, Fig. 2). It is recognized by coastal onlap.

Relative stillstand of sea level—an apparent constant stand of sea level with respect to the underlying initial surface of deposition. It may result if both sea level itself and the underlying initial surface of deposition remain stationary or if both rise or fall at the same rate (part 3, Fig. 6). It is recognized by the common presence of coastal toplap and a lack of coastal onlap.

Sand-prone seismic facies—a seismic facies unit generated by strata interpreted to have been deposited in a clastic depositional environment of sufficiently high energy to transport and deposit significant quantities of sand, assuming a supply of sand was available for transport to the area of deposition (parts 6; 9).

Sechron—the maximum interval of geologic time occupied by a given depositional sequence, defined at the points where the boundaries of the sequence change laterally from unconformities to conformities along which there is no significant hiatus (part 2, Fig. 1).

Second-order cycle—a cycle of relative or eustatic change of sea level that has a duration in the order of 10 to 80 million years. Supercycles are second-order cycles (part 4, Figs. 1, 2, 3, 7).

Seismic downlap—seismic expression of downlap (part 6, Figs. 4c, d).

Seismic facies analysis—the description and geologic interpretation of seismic reflection parameters, including configurations, continuity, amplitude, frequency, and interval velocity.

Seismic facies map—a map that shows the areal distribution, configuration, thickness, or other aspect of a given seismic facies unit or parameter (part 2, Figs. 4, 5).

Seismic facies unit—a mappable three-dimensional seismic unit composed of groups of reflections whose parameters, such as reflection configuration, continuity, amplitude, frequency, or interval velocity, differ from those of adjacent facies units. Once the interval reflection parameters, the external form, and the three-dimensional associations of the seismic facies unit are delineated, the unit can then be interpreted in terms of environmental setting, depositional processes, and estimates of lithology (part 6).

Seismic onlap—seismic expression of onlap (part 6, Figs. 4a, b).

Seismic sequence—a depositional sequence idenified on a seismic section. *See* sequence.

Seismic sequence analysis—the seismic identification and interpretation of depositional sequences by subdividing the seismic section into packages of concordant reflections separated by surfaces of discontinuity, and interpreting them as depositional sequences (part 6). *See* sequence.

Seismic stratigraphy—the study of stratigraphy and depositional facies as interpreted from seismic data (part 6).

Sequence—a term applied to a relatively conformable succession of genetically related strata bounded at its top and base by unconformities or their correlative conformities (part 2, Fig. 1). This is a modification of an earlier usage by Sloss (1963, p. 93). Several types of sequences are distinguished. A *depositional* sequence is the operational stratigraphic unit defined on seismic, well-log, or outcrop data (part 2, Fig. 1). A *seismic* sequence is a depositional sequence indentified on a seismic section (part 2, Fig. 4). A *well-log* sequence is a depositional sequence identified on a well-log cross section (part 2, Fig. 3). A *maritime* sequence is a sequence that consists of genetically related coastal nonmarine, littoral, and/or marine deposits. The major depositional mechanisms at the landward limit of a maritime sequence containing littoral or nonmarine deposits are controlled by the position of sea level as a base level (part 3). A *hinterland* sequence is one that consists entirely of nonmarine deposits laid down at a site interior to the coastal area, where depositional mechanisms are controlled only indirectly or not at all by the position of sea level (part 3).

Shale-prone seismic facies—a seismic facies unit generated by strata interpreted to have been deposited in a clastic depositional environment in which depositional energies are low and insufficient to develop significant sand accumulations (part 6).

Sheet seismic facies unit—a seismic facies unit with a sheet-like external form: a thin, widespread, tabular unit with a high width/thickness ratio, commonly large enough that the lateral edges of the unit are beyond the limits of much of the seismic control (part 6, Fig. 14).

Sheet drape seismic facies unit—a seismic facies unit whose external form is similar to the *sheet seismic facies unit* except that the blanket-like body is draped evenly over the pre-existing sedimentary surface without regard for pre-existing topography (part 6, Fig. 14).

Shingled reflection configuration—a prograding seismic pattern, within a thin unit, commonly with parallel upper and lower boundaries, and with very gently dipping parallel oblique internal reflectors, that terminate by apparent toplap and downlap. Successive oblique internal reflectors within the unit show very little overlap with each other. The overall pattern resembles that of the parallel oblique configuration, except that the thickness of the unit is just at the point of seismic resolution of the oblique beds (part 6, Figs. 8e; 10a, b).

Sigmoid configuration—a prograding clinoform pattern formed by a number of superposed sigmoid (s-shaped) reflections interpreted as strata with thin, gently dipping upper and lower segments, and thicker, more steeply dipping middle segments. The upper (topset) segments of the strata approach horizontality or have very low angles of dip, and are concordant with the upper surface of the facies unit (part 6, Figs. 8a, 9a).

Stratal configuration—the geometric patterns and relations of strata within a stratigraphic unit. These are commonly indicative of depositional setting and processes as well as later structural movement.

Stratal surfaces—surfaces that separate the principal sedimentary strata. They represent periods of nondeposition or a change in the depositional regime, and form practical time-lines through a depositional sequence.

Stratum—a tabular or sheet-like mass, or a single and distinct layer, or. . . sedimentary material . . . , visually separable from other layers above and below, by a discrete change in character. . . or by a sharp physical break in deposition, or by both. . . It has been defined. . . as a general term that includes both "bed" and "lamination" (McKee & Weir, 1953, p. 382). The term is more frequently used in its plural form, *strata* (Gary, McAfee, and Wolf, 1972).

Structural truncation—lateral termination of a stratum by structural disruption, produced by faulting, gravity sliding, salt flowage, or igneous intrusion (part 2).

Subparallel seismic reflection configuration—seismic reflection configuration interpreted as strata deposited in a subparallel geometric pattern (part 6, Figs. 6, 7B).

Supercycle—a group of regional or global cycles of relative change of sea level in which a cumulative rise to a higher position of sea level is followed by a cumulative fall to a lower position. *See* global supercycle and regional supercycle. A supercycle is a second-order cycle.

Supersequence—a group of sequences that successively reach higher positions of encroachment onto the underlying unconformity surface, followed by one or more sequences with lower positions of encroachment. As far as we know, most supersequences were deposited during second-order cycles (supercycles) of relative rise and fall of sea level (part 3, Fig. 13). *See* sequence.

Tangential oblique seismic reflection configuration—*See* oblique seismic reflection configuration.

Third-order cycle—a cycle of relative or eustatic change of sea level that has a duration in the order of 1 to 10 m.y. It is the fundamental cycle in that it represents only one rise and fall, whereas second- and third-order cycles are composites of more than one third-order cycle (part 4, Figs. 2, 3, 7).

Top-discordance—*See* discordance.

Toplap—termination of strata against an overlying surface mainly as a result of nondeposition (sedimentary bypassing) with perhaps only minor erosion (part 2, Fig. 2). Each unit of strata laps out in a landward direction at the top of the unit, but the successive terminations lie progressively seaward. *Coastal* toplap is toplap of the coastal deposits in a given depositional sequence (part 3, Fig. 6). Toplap occurs along the upper sequence boundary.

Transgression—a landward movement of the shoreline indicated by landward migration of the littoral facies in a given stratigraphic unit (part 3, Figs. 3, 4).

Truncation—termination of strata or seismic reflections interpreted as strata along an unconformity surface due to post-depositional erosional or structural effects (part 2, Fig. 2). It occurs along the upper sequence boundary. *Erosional* truncation implies the deposition of strata and their subsequent removal along an unconformity surface *See* Structural truncation.

Unconformity—in our usage, a surface of erosion or nondeposition that separates younger strata from older rocks and represents a significant hiatus (at least a correlatable part of a geochronologic unit is not represented by strata). Periods of erosion and nondeposition occur at each global fall of sea level, producing *interregional unconformities* (part 4, Fig. 7), although in some areas of continuous deposition, the hiatus may be too small to detect paleontologically or seismically, and the surface is defined as a conformity. Major interregional unconformities occur at supercycle boundaries and are listed in part 4 (Fig. 7). Minor interregional unconformities occur at all other cycle boundaries.

Wedge seismic facies unit—a seismic facies unit whose external form is wedge-shaped: a stratiform unit that thins out laterally (part 6, Fig. 14).

REFERENCES CITED

Fairbridge, R. W., 1961, Eustatic changes in sea level, *in* L. H. Ahrens et al, eds., Physics and chemistry of the earth: London, Pergamon Press, v. 4, p. 99-185.

Gary, M., R. McAfee, Jr., and C. L. Wolf, 1972, Glossary of geology: Washington, D.C., Am. Geol. Inst., 805 p.

McKee, E. D., and G. W. Weir, 1953, Terminology for stratification and cross-stratification in sedimentary rocks: Geol. Soc. America Bull., v. 64, p. 381-289.

Sheriff, R. E., 1973, Encyclopedic dictionary of exploration geophysics: Tulsa, Soc. Exploration Geophysicists, 266 p.

Sloss, L. L., 1963, Sequences in the cratonic interior of North America: Geol. Soc. America Bull., v. 74, p. 93-113.

Swain, F. M., 1949, Onlap offlap, overstep and overlap: AAPG Bull., v. 33, p. 634-636.

Seismic-Stratigraphic Interpretation of Depositional Systems: Examples from Brazilian Rift and Pull-Apart Basins[1]

L. F. BROWN, JR., and W. L. FISHER[2]

Abstract Seismic-stratigraphic interpretation has become an important element of exploration in basins with limited well control. This new direction in exploration imposes new responsibilities and qualifications on both the geologist and the geophysicist. Two general approaches are developing in response to exploration requirements—a physical approach involving processing and synthetic modeling, and a seismic-stratigraphic approach involving a new application of traditional facies geology.

Seismic-stratigraphic analysis of Brazilian offshore basins permits the development of approaches and concepts that can be applied to other basins. Analysis involved development of seismic-stratigraphic framework, interpretation of reflection patterns, chronostratigraphic correlation, mapping seismic-stratigraphic (depositional system) units, synthesis of depositional and facies interpretations, and, in many cases, strategic mapping of specific facies. Within Brazilian offshore basins, three principal depositional systems are recognized—delta and fan delta, carbonate platform and shelf, and slope. By integrating seismic and limited well data, it is possible to recognize on reflection seismic sections: (1) three deltaic facies—prodelta and distal delta, front or barrier; delta, front or barrier; and alluvial and delta plain; (2) two fan-delta facies—proximal and medial fan, and distal fan and prodelta; (3) three shelf and platform facies—neritic; reef, bank, shoal, and shelf edge; and submarine canyon fill; and (4) three principal arrangements of slope facies—offlap, onlap, and uplap.

Integration of conventional and seismic-stratigraphic analyses permits recognition of five fundamental types of rift and pull-apart basins in Brazilian offshore areas: early rift-fault basin, post-rift clastic basin with salt tectonism, post-rift basin with stable carbonate platform, pull-apart basin with passive clastic-carbonate offlap and onlap deposition, and pull-apart basin with deltaic sedimentation. Seismic-stratigraphic analysis permits extrapolation of limited well data to predict depositional systems tracts, tectonic elements, principal depositional modes, and source area and drainage characteristics. Similarly, the geologist can predict reservoir type and spatial distribution, stratigraphic and structural trap possibilities, and source bed and seal potential.

INTRODUCTION

General

Stratigraphic interpretation of seismic data has become an increasingly important element in exploration during the past 10 years, particularly in offshore areas. Exploration in frontier or poorly known continental shelf areas, where well control is unavailable or limited, has required a greater degree of stratigraphic interpretation of existing geophysical data. Integration of geologic and geophysical methods in response to exploration requirements has imposed some significant changes in approach and emphasis in basin analysis. Similarly, significant changes are occurring in the companion field of exploration geophysics. Probable reasons for the rapid emergence of seismic-stratigraphic capability at this time are (1) the dramatic advances in computer technology and seismic data acquisition, and (2) a similar breakthrough in basin analysis during the past 10 to 15 years resulting from the development of Holocene depositional models that advance the understanding of depositional processes, depositional environments, and facies interpretations of ancient deposits. The fortuitous and parallel advance in geophysics and basin-analysis concepts is providing some very useful exploration tools.

This new direction in exploration imposes new responsibilities on both the geologist and the geophysicist. Teams composed of both specialists are common today in many exploration groups, but it is increasingly obvious that a need exists for explorationists who possess expertise both in geophysics and facies analysis. Certainly, successful explorationists, geologists, and geophysicists will be involved in the emerging field called seismic stratigraphy.

Emerging from the seismic-stratigraphic "revolution" are at least two areas of specific interest and application: (1) a physical approach aimed at greater and more accurate discrimination and synthetic modeling of lithic composition, fluid content, and other similar properties, utilizing computer analysis of velocity, amplitude, and cy-

[1]Manuscript received, July 11, 1976; accepted, December 28, 1976.

[2]University of Texas at Austin, University Station, Box X, Austin, Texas 78712.

Many Petrobrás geologists and geophysicists were involved in the offshore basin analyses. These explorationists provided expertise in both geology and geophysics which contributed significantly to the program. The professional contribution and personal support of these associates are gratefully acknowledged: Hildeberto Ojeda y Ojeda, Ercilio Gama, Jr., C. S. Baumgarten, J. A. Estrela Braga, Roberto Morales, Henrique Della Piazza, G. Estrella, J. B. Gomes, M. V. Dauzacker, M. Saito, A. M. Fugita, and K. Tsubone. Special thanks are extended to Carlos Walter Campos, Chief of Exploration; Renato Pontes, Chief of Geology; Juarez Tessis, Chief of Geophysics; and Wagner Feriere, Subchief of the Risk Contract Program and formerly Chief of Geophysics. Permission to use illustrations and generalized information from the Brazilian basins for this report was granted by the Directorate of Petrobras.

cle parameters, etc., and (2) a stratigraphic-facies approach using reflection sections and density or sonic-log data to invoke facies interpretations and to integrate, spatially and chronologically, the depositional systems which fill basins.

Reflection seismic sections provide the experienced basin analyst with a vehicle for applying state-of-the-art facies geology to basins with limited well control. Although seismic sections have been used for many years for structural mapping and interpretation, the maximum stratigraphic significance and value of the section can be realized by appreciation of the facies fabric of basins of different tectonic settings. There are obvious pitfalls in inferring stratigraphic facies from seismic data, just as there are from making these inferences from other remotely sensed and indirect data such as well logs. Consequently, the successful application of seismic stratigraphy will involve the joint participation of geologists and geophysicists, or individuals adequately experienced in both areas.

Purpose

This report provides a general perspective of seismic-stratigraphic interpretations that have been made in offshore Brazilian basins. Interpretations have involved application of facies geology integrated with analysis of seismic-reflection data. Interpretations were made jointly by geologists and geophysicists.

The report is not a catalogue of seismic-facies patterns, but rather a general guide to an approach that organizes available seismic-stratigraphic data into an integrated view of a sedimentary basin. Neither is the report intended to convey the geology of any specific offshore Brazilian basin, but rather we are attempting (within proprietary constraints) to present some principles that were developed during detailed basin analysis using suites of well logs, paleontologic information, cores, and seismic data. Furthermore, we hope to illustrate the applicability of seismic-stratigraphic methods in frontier basins and further provide an independent example to compare with the studies in this volume by geologists and geophysicists of Exxon, U.S.A. (Vail et al, this volume).

Knowledge of facies geology is a fundamental requirement. Regional and local seismic reflections provide a seismic-stratigraphic framework. Distinctive seismic reflection patterns and available well data were the basis for facies interpretations and mapping.

Previous Studies

Although numerous abstracts and reports have been published on the physics of seismic-stratig-raphy, only a few abstracts are available that specifically address the stratigraphic and facies aspects of seismic interpretations (e.g., Vail and Sangree, 1971; Vail, Mitchum, and Thompson, 1974; and Sangree and Widmier, 1976). A recent report by Sangree et al (1976) summarizes most subjects previously covered by oral presentation. Reports by the above authors and others are included in this volume, and herein are referred to collectively as Vail et al. Reports by these geologists and geophysicists offer a comprehensive and critical analysis of seismic facies interpretation. In our Brazilian studies, presented as a series of unpublished company reports prepared for Petrobrás (Petroleo Brasileiro, S.A.) during the period from 1973-1976, we independently reached many similar conclusions based on data from different basins. Although differences in approach and interpretation exist between the studies by Exxon and Petrobrás, the observational differences are minor and, in part, may result from differences in the basins that were studied. Principal differences are interpretive and involve the use of relative sea-level control by Vail et al to explain formation of "submarine canyons" and onlap slope deposition. Variations in nomenclature occur, and will be equated where possible.

Extensive literature is available on facies interpretation, both Holocene and ancient: LeBlanc, 1972; Shelton, 1973; Reineck and Singh, 1973; Fisher and Brown, 1972; and others.

Source of Data

Since 1973, analyses of offshore Brazilian basins for Petrobrás provided an opportunity to apply depositional-systems (or facies) analysis to large basins with limited well control. Seismic reflection data provided the principal source of subsurface information which had to be integrated with information derived from well logs, samples, and rare cores.

Illustrations of seismic sections used in this report are generalized from actual sections, because no seismic sections are available for publication. *We are keenly mindful of the constraints and limitations this imposes on our presentation.* Although actual records would be ideal to illustrate the ideas presented in this report, it is hoped that readers can follow the presentation adequately using diagrammatic illustrations which show general attitudes and continuity of key reflections. The examples are a composite, selected from studies of most offshore Brazilain basins: Serigipe-Alagoas, Espirito-Santo, Mucuri-Cumuruxatiba-Jequitinhonha, Potiguar, Foz do Amazonas, Santos, and Barreirinhas basins. Because of proprietary constraints, specific aspects of the basins such as velocities, well control, paleontologic interpretations, and other factors are omitted in favor of a presentation of general reflection patterns and inferred depositional interpretations.

CONCEPT OF DEPOSITIONAL SYSTEMS

Stratigraphic interpretations utilizing seismic reflection data should be firmly based on an adequate understanding of the three-dimensional arrangement of lithofacies and their integration into depositional systems. It is essential, therefore, that the seismic stratigrapher understand depositional processes and facies models and conventional subsurface geology, as well as seismic geophysics, before attempting to infer stratigraphic and facies relations from a seismic section.

We believe one of the most useful concepts in seismic-stratigraphic analysis is that of "depositional systems." Fisher and McGowen (1967) defined depositional systems as three-dimensional assemblages of lithofacies, genetically linked by active (modern) or inferred (ancient) processes and environments. A depositional system is the stratigraphic record or analog of deposition within the myriad environments that constitute river, delta, barrier island, shelf, and slope systems, among others. The fundamental unit of the depositional system (as used in this report) is the lithofacies, a three-dimensional sediment or rock body bounded by depositional (or erosional) surfaces whose genesis is inferred from the interpretation of sedimentary structures, textural variations, bedding characteristics, internal and external stratigraphic relations, paleontology, and association with adjacent facies. A depositional-systems approach has been used by various workers using conventional subsurface data (Fisher and McGowen, 1967; Fisher, 1969; Brown, 1969; Guevara and Garcia, 1972; and Erxleben, 1975). A wide knowledge of various depositional systems permits prediction of component facies composition, geometry, and distribution using limited data. More importantly, an understanding of the spatial arrangement of depositional systems within various types of basins provides a fundamental tool in the stratigraphic interpretation of seismic reflection sections. When the spatial arrangement of facies or systems within various types of basins is understood, it is possible to infer with greater confidence stratigraphic relations and depositional patterns from subtle variations in reflection attitudes and continuity.

Contemporaneous depositional systems can be linked to produce what may be called a "systems tract." For example, fluvial, delta, shelf, and slope systems may be intergradational and, in part, contemporaneous. In addition, the tract defines paleoslope from basin margin to deep water. A basin is filled by deposition within a variety of systems tracts which evolve through time as tectonics and source areas change. Significant changes in the style or mode of deposition in the basin are commonly marked by regional, sometimes basinwide, seismic reflections (conformable or unconformable boundaries). The regional reflections constitute isochronous surfaces within a basin, except where they represent unconformities. The resulting reflection-bounded units composed of contemporaneous depositional systems (systems tracts) are herein called "seismic-stratigraphic units" (depositional and cyclic sequences, Sangree et al, 1976). The seismic-stratigraphic unit is the principal element of the seismic-stratigraphic framework of a basin. Recognition and delineation of principal and minor seismic-stratigraphic units will be discussed later.

EXAMPLES OF CONVENTIONAL BASINAL ANALYSIS

The general arrangements of facies and depositional systems within two types of basins, the Gulf basin of Louisiana and Texas and the Eastern Shelf of the Midland basin of Texas, provide examples of basin-fill style within a rapidly subsiding, oceanic-margin basin underlain by salt, and within an intracontinental basin that exhibited tectonic stability, respectively (Figs. 1, 2, 3). These examples (where the writers have had personal experience) are based principally on interpretations of conventional subsurface data in basins that are in mature stages of exploration. Many reports cover all aspects of these basins; references are not included, but the reader is referred to standard bibliographic sources. The two contrasting basins demonstrate similarities and differences in type and distribution of depositional systems within basins of significantly different tectonic style.

Basins such as these have been studied intensively, and constitute excellent "models" that may aid in the interpretation of other less-explored basins. Basins throughout the world can be grouped into a number of fundamental types (Klemme, 1971) which exhibit similar systems tracts and depositional history.

Gulf Basin of Louisiana and Texas

Although the early history of the Gulf basin is not well known, it was apparently marked by rift-basin tectonism characterized by fan, fan-delta, and salt deposition (Fig. 1A). Subsequently, a series of depositional systems tracts evolved, composed of Jurassic and Cretaceous shelf-platform carbonate, shelf-edge reef, and slope systems, and Cretaceous and Tertiary fluvial, delta, and slope systems.

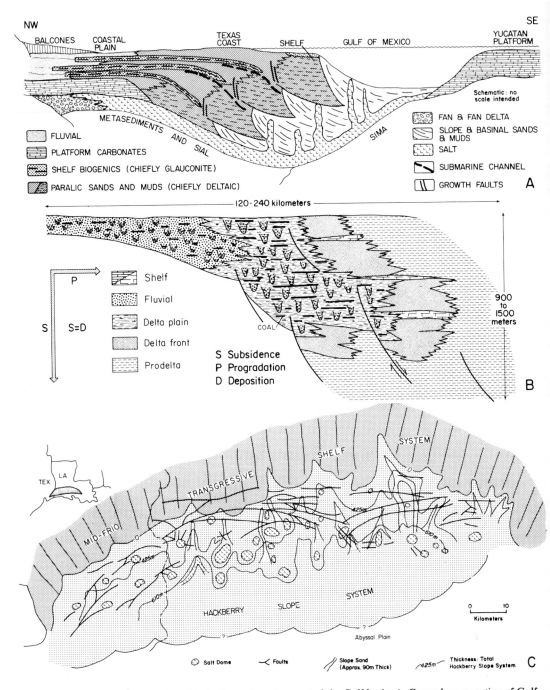

FIG. 1—General characteristics of facies in northwestern part of the Gulf basin. **A.** General cross section of Gulf basin showing principal depositional systems (Modified from Lehner, 1969). **B.** Dip cross section of Eocene delta system illustrating internal facies composition (after Fisher and McGowen, 1967). **C.** Regional map of inferred Hackberry (Oligocene) slope system (adapted from Paine, 1968).

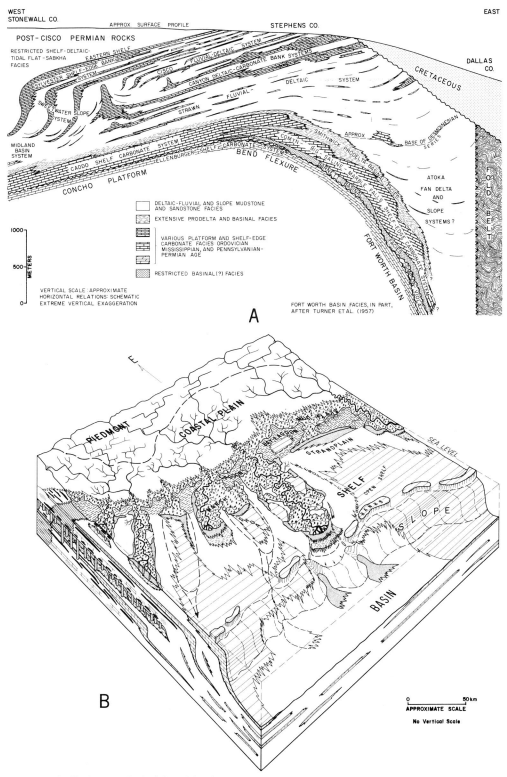

FIG. 2—Distribution of principal depositional systems in Fort Worth–Foreland basin and Midland basin during Late Paleozoic. **A.** General dip cross section illustrating change in depositional style during tectonic evolution of the basins (after Brown et al, 1973). **B.** Block diagram of Eastern shelf of Midland basin during Late Pennsylvanian time when deltaic, shelf, and slope systems were operative (after Brown, 1969).

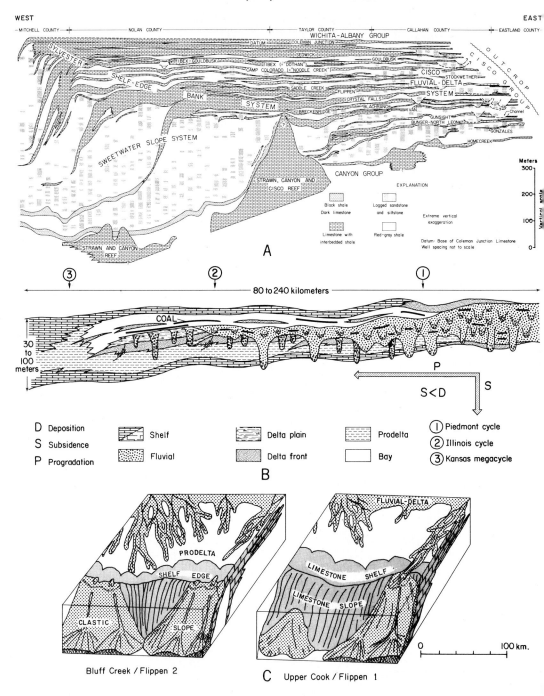

FIG. 3—General characteristics of Upper Pennsylvanian–Lower Permian facies in eastern part of Midland basin. **A.** Regional dip section showing deltaic, shelf, and slope facies (after Brown, 1969). **B.** Internal facies relations in a Pennsylvanian–Permian delta system that prograded across and intertongued with a carbonate shelf system (after Brown et al, 1973). **C.** Reconstruction of slope deposition by submarine fans during Late Pennsylvanian and Early Permian (after Galloway and Brown, 1972).

Tertiary deltaic deposition involved both river- and marine-dominated delta and associated slope systems, which constitute the largest volume of basin fill. Marine-dominated delta systems, such as the Oligocene Frio system, contrast with the Wilcox type by exhibiting extensive marine sandstone facies of barrier island origin. Delta systems in the Tertiary of the Gulf basin are characterized by thick, superposed facies, reflecting a balance between rates of deposition and subsidence (Fig. 1B). Growth faulting and salt or shale diapirism are also commonly associated with these thick delta and associated slope prisms.

Slope systems in the Gulf basin are generally beyond the drill, but several examples, such as the Oligocene Hackberry slope system of Louisiana and Texas (Fig. 1C) and the Yoakum system associated with the Upper Wilcox Group (Eocene) of Texas, provide some insight to the character of Tertiary deep-water facies. Slope systems in the Gulf basin were deposited in close association with contemporaneous salt diapirism, probably mobilized by the massive sedimentary load imposed by the larger prograding delta and associated slope systems. Plio-Pleistocene deltas have similarly mobilized salt tectonism in the modern slope of the Gulf of Mexico.

Eastern Shelf of the West Texas Basin

The eastern flank of the West Texas basin of Late Pennsylvanian and Early Permian age (Figs. 2, 3) provides a view of another type of basin where the fill is generally understood from conventional subsurface methods. The Fort Worth basin (Fig. 2A) was the site of westward prograding Late Mississippian and Early Pennsylvanian depositional tracts composed of fan, fan-delta, and slope systems. To the west (on the Concho Platform) were extensive carbonate environments with eastward-facing shelf edges. This fan, fan-delta, slope, basin, and carbonate platform systems tract evolved into Middle Pennsylvanian fluvial, delta, and carbonate bank tracts and ultimately into a Late Pennsylvanian–Early Permian fluvial, delta, carbonate-shelf, and slope tract (Fig. 2B). This final depositional style (Fig. 3) was responsible for filling much of the West Texas basin. Deposition was concluded by the Middle and Late Permian tidal and evaporite depositional episodes. Because the intracontinental basin subsided slowly, deltaic depositional rates far exceeded subsidence rates, resulting in widespread, thin cyclic sequences (Fig. 3B). Slope systems progressively filled the basin by deposition of a series of offlapping wedges of deeper water deposits (Fig. 3A, C).

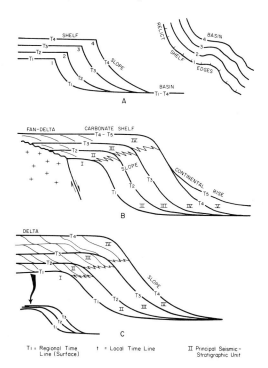

FIG. 4—Diagrammatic representation of isochronous lines (surfaces) in hypothetical basins, based upon regional and localized reflection surfaces. **A.** Progressive filling of a basin by shelf, slope, and basinal deposition. **B.** Fan-delta, carbonate-shelf, and slope progradation with periodic continental rise onlap. **C.** Delta-slope progradation (offlap).

Conclusions

The brief review of the Gulf and West Texas basins emphasizes a very important concept for either conventional basin analysis or seismic-stratigraphic analysis, which is that basins are filled principally by basinward accretion of the sedimentary prism (Fig. 4). The seismic stratigrapher must be aware that most basins "fill" laterally, rather than "fill up" through time. The degree of superposition of these wedges is controlled primarily by rates of basin subsidence, compaction, and salt or shale flowage. Although vertical aggradation occurs in many shelf environments and within some delta and slope environments, progradation of deltaic and slope systems occurs primarily along depositional surfaces that are inclined generally from 1 to 5°. These paleodepositional surfaces constitute isochronous surfaces at any instant in geologic time. Relief of the depositional surfaces is greatest in the slope environment and least in the delta environment (Fig. 4B, C). An appreciation of the configuration of various modern depositional surfaces or profiles is invaluable when inferring depositional topography from the configuration of reflection attitudes on seismic sections.

With this review of two types of basins for which extensive, conventional subsurface data exist, we will now focus on lesser-known basins where principal data consist of seismic records.

SEISMIC-STRATIGRAPHIC ANALYSIS

Following data acquisition and processing, several basic steps or factors are involved in a seismic-stratigraphic analysis: (1) recognition of regional (and certain minor) reflections or reflection discontinuities that subdivide the basin fill into seismic-stratigraphic units which represent distinctive episodes of deposition, constituting the basis for a stratigraphic framework of the basin; (2) integration of the seismic data (e.g., velocity data, reflection strength, continuity, attitude and evident anomalies) with any available well data (cores, logs, paleontology); (3) correlation of paleontologic data with regional and minor reflections to develop approximate chronostratigraphic correlation within the basin; (4) isopach mapping of principal and minor seismic-stratigraphic units using velocity-analysis data correlated with available well data; and (5) synthesis of a systems (or facies) tract and depositional models for each seismic-stratigraphic unit.

Seismic-Stratigraphic Framework

Development of a seismic-stratigraphic framework is similar to construction of conventional stratigraphic control using traditional subsurface data, except that seismic lines provide the opportunity for continuous correlation throughout the seismic grid. The writers agree with Sangree et al (1976) that reflections represent depositional surfaces except where they coincide with unconformities. Reflection-defined depositional surfaces constitute isochronous surfaces; the more widespread, continuous, reflections are fundamental stratigraphic markers within a basin (Fig. 5D). Regional reflections such as R_1-R_5 in Figure 5D bound principal seismic-stratigraphic units, commonly composed of two or more depositional systems; for example, fan-delta, carbonate-shelf, and slope systems constitute a tract of facies deposited within contemporaneous environments, extending from continental areas to deep water. The depositional history and facies composition of the total basin can thus be determined by sequential analysis of each individual seismic-stratigraphic unit in the basin.

The regional reflectors that separate seismic-stratigraphic units generally represent significant changes in reflection coefficient at the interface between: (1) thin, widespread marine transgressive facies and subjacent deltaic clastic facies, and (2) contemporaneous blankets of thin, widespread, hemipelagic muds and subjacent submarine fan complexes of the continental slope. These reflections commonly mark a hiatus in normal depositional rates within the basin, either because of eustatic rise in sea level or regional subsidence.

Inclined reflections commonly terminate downward and generally basinward along the upper surface of a regional reflection. This has been called "baselap" by Vail et al (this volume). This phenomenon marks the initial progradation or offlap of deltaic or slope facies over a transgressive shelf or onlapped shelf (respectively) of a subjacent seismic-stratigraphic unit. Consequently, each unit represents a cycle of progradation and transgression.

Within the principal seismic-stratigraphic units, lateral boundaries between contemporaneous depositional systems are generally distinctive, but gradational. Consequently, reflection boundaries such as those that separate contemporaneous fan-delta and shelf systems are normally gradational and intertonguing. However, systems boundaries such as those between horizontally stratified carbonate shelf systems and subjacent, inclined-slope systems are not so distinct; reflections may be slightly to prominently discordant—called "toplap" (Vail et al, this volume).

The general procedure for developing a basinal framework involves: (1) recognition of regional unconformities distinguished by onlap, baselap and/or truncated reflections; (2) extrapolation of these reflections throughout the basin into areas where the reflection coincides with conformable depositional surfaces; (3) recognition of intertonguing or discordant relations between minor, contemporaneous seismic-stratigraphic units such as shelf and slope systems or fan-delta and shelf systems, respectively; (4) recognition of minor discontinuities in reflection attitudes such as exhibited by onlapping reflections in minor continental rise units; and (5) recognition of minor erosional unconformities such as those which occur at the base of submarine canyon-fill deposits. These critical reflections are then correlated and traced through every dip and strike seismic line in the basin. As the study proceeds, other less prominent reflections may be recognized and traced, especially those that subdivide minor episodes of marine transgression over deltaic facies or slope reflections that mark a lengthy period of hemipelagic deposition.

The resulting seismic-stratigraphic framework may resemble the example shown on Figure 5D. On this framework, other more specific facies interpretations from well and reflection analysis will later be superimposed. Isopach maps can be constructed from the travel time and velocity values for each seismic-stratigraphic unit in the basin. The framework also provides the basis for chronostratigraphic correlation, calibrated with biostratigraphic data available from limited wells.

Interpretation of Reflection Signatures

Because reflections can not be directly interpreted in terms of lithofacies, it is possible, using limited well data, to make some surprisingly accurate estimates of facies type and composition, by analogy with other basins where extensive well control is available. Naturally the more experience the geologist has, the greater is his probability of an accurate interpretation. Similarly, the more well control that can be tied into the seismic framework, the better the odds become for valid interpretation.

Sangree et al (1976) described a wide variety of reflection patterns and respective sedimentary facies from the Pleistocene of the northern Gulf of Mexico. Papers by Vail et al (this volume) also provide a comprehensive classification of reflection patterns. The writers will apply similar terminology, but some variations will occur that should be obvious to the reader.

Reflection attitude, configuration, continuity, and amplitude, when combined with external geometry of the signature and spatial association with other signatures, as noted by Sangree et al (1976), provide a range of reflection pattern combinations that may permit interpretation of the facies from the seismic record. In some instances, zones of distinctive diffractions and sideswipe may prove to be indicative to certain relatively unique velocity variations related to facies geometry rather than of structural origin. Newer lines shot with higher energy sources (such as the air gun) may significantly improve the quality of subtle reflections.

Rather than interpret the Brazilian facies using a geometric classification scheme (like the one applied by Sangree et al, 1976; and by Vail et al, this volume), we interpreted the seismic facies signatures within the context of depositional systems. Most geologists and geophysicists are faced with evaluating large, sparsely drilled basins. This was true of our Brazilian experience, and we believe that under these kinds of exploration imperatives, the best analysis will involve the closest integration of traditional depositional systems analysis and seismic-stratigraphic information. We used

the seismic responses to extend and improve traditional basinal analysis interpretations, using available well data sometimes limited to 5 or 6 wells within 200,000 sq km. This combination of seismic data and depositional (process) analogues provides a very reliable tool for facies interpretation, as shown by subsequent drilling.

Stratigraphic Correlation

By assuming that seismic reflections are isochronous surfaces or unconformities, it is possible to establish the relative order or sequence of stratigraphic units in a basin (Fig. 4). The chronologic sequence of depositional and structural events thus can be determined accurately. Where these reflections are calibrated with biostratigraphic data from wells in the basin, the seismically based stratigraphy can be assigned to standard time-stratigraphic units (systems, series, stages). In this manner, limited paleontologic information can be applied throughout the basin.

Conformable, regional, seismic reflections which bound the principal seismic-stratigraphic units in a basin (Fig. 4) serve as approximate time-stratigraphic boundaries. However, within each seismic-stratigraphic unit, minor isochronous surfaces may be complexly arranged. For example, units bounded by prominent regional reflections may be composed of deltaic facies which internally exhibit complex offlapping reflections (Fig. 4C). Where nondeposition occurred during a given time interval, two or more reflections will merge (Fig. 4A, B). Reflections may also merge along unconformable surfaces, such as those developed during coastal onlap (Fig. 4B).

Very elaborate time-stratigraphic subdivisions can be delineated in basins using seismic-stratigraphic criteria. This aspect of analysis permits very accurate correlation, and provides the basis for precise determination of the chronologic order of geologic events within the entire basin.

Mapping Seismic-Stratigraphic Units

Where a basinwide seismic-stratigraphic framework has been established, and regional bounding reflection surfaces have been traced throughout the seismic grid (Fig. 5D), it is then possible to map the various seismic-stratigraphic units and their component depositional systems (Figs. 5, 6). Travel time can be converted to thickness at selected shot points by using well and velocity data; and contouring, with reference to the seismic section (e.g. faults, shelf-edge positions, submarine canyons), permits a basinwide picture of each seismic-stratigraphic unit and most component depositional systems.

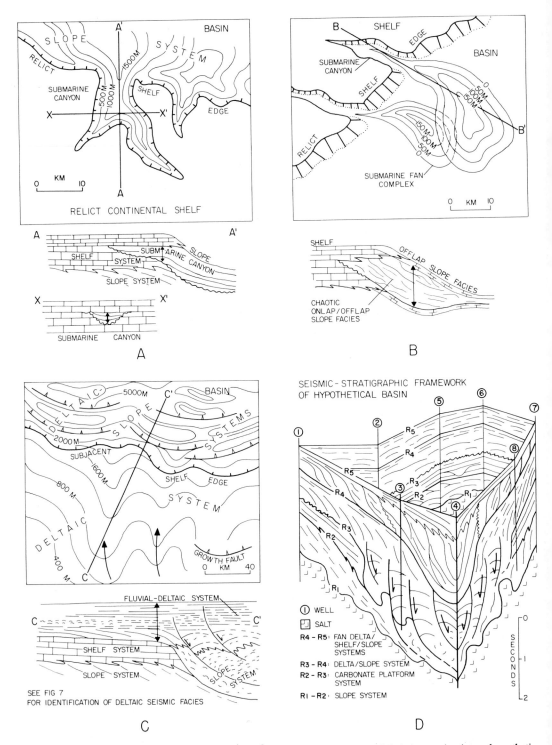

FIG. 5—General isopach patterns of seismic-defined depositional systems and their integration into a hypothetical, basinwide, seismic-stratigraphic framework. **A.** Seismic-stratigraphic isopach map of submarine canyon system. **B.** Seismic-stratigraphic isopach map of submarine fan complex (cone). **C.** Seismic-stratigraphic isopach map of deltaic and associated slope systems. **D.** Hypothetical seismic-stratigraphic framework integrating available wells and seismic data. On cross sections, terrigeneous clastic facies shown schematically by lines representing seismic reflection attitude and continuity.

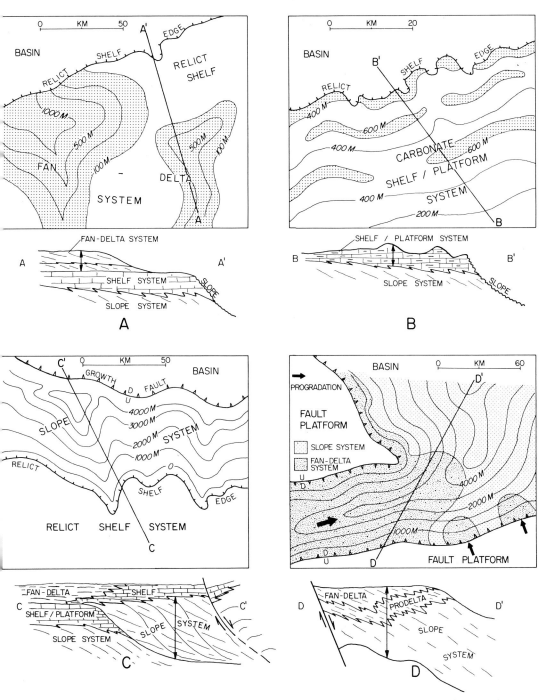

FIG. 6—General isopach patterns of seismic-defined depositional systems. **A.** Seismic-stratigraphic isopach map of a fan-delta system. **B.** Seismic-stratigraphic isopach pattern of a shelf carbonate system. **C.** Seismic-stratigraphic isopach map of a slope system. **D.** Seismic-stratigraphic isopach map of fan-delta and slope systems within rift basin. On cross sections, terrigenous clastic facies shown schematically by lines representing seismic reflection attitude and continuity.

Within the Brazilian basins, very distinctive (and predictable) isopach patterns characterize the various depositional systems that compose each unit. The isopach map of a delta system, that prograded across a continental shelf and over a contemporaneous slope system beyond the shelf edge (Fig. 5C), clearly delineates a subjacent shelf-edge position, principal dip-oriented depositional axes, and a distal growth-faulted complex. Seismic-stratigraphic isopach maps of shelf and platform carbonate systems (Fig. 6B), though, exhibit distinct isopach patterns that outline elongate trends parallel with the relict shelf edge, which probably represent reef or bank facies. Slope systems display isopach patterns that pinch out landward near the relict shelf edge (Fig. 6C) and exhibit exceedingly thick, closed contour values in areas of persistent submarine-fan deposition. Traced far into the basin, the slope system again thins into abyssal facies. Where slope systems were supplied with sediment via submarine canyons (Fig. 5A), the slope systems will extend landward into the canyon and pinch out in the relict shelf system. Fan-delta and associated slope systems that prograded into rift basins clearly outlined the basin geometry and may thicken toward contemporaneous faults. If the rift basin opened into the oceanic basin, the systems would thicken abruptly where they prograded into deeper water (Fig. 6D).

By mapping units defined by differences in reflection attitudes, continuity, or similar variations, it is possible to delineate a variety of specific facies (see later discussion). For example, by mapping the thickness of chaotic onlap reflection units within a slope system, fan-shaped isopach patterns emerge which define the limits of individual submarine fan complexes, or fan cones (Fig. 5B). Similarly, the mapping of units characterized by layered, but highly erratic, reflections that intertongue with layered reflections exhibiting extreme continuity, produces an isopach map of individual fan-delta lobes that prograded across and intertongued with a limestone shelf (Fig. 6A). Mapping of specific facies or individual depositional elements such as the submarine fan or fan-delta lobe, requires the recognition and delineation of relatively subtle reflection variations. With experience, nevertheless, the stratigrapher may gain insight to facies relations and distribution of economic significance. Criteria that help to identify relatively specific facies by seismic reflection patterns and relations may become subjective. The regional mapping of principal seismic-stratigraphic units is more objective but only provides a general guide to the potential reservoir facies.

Synthesis

A final phase of analysis involves a synthesis of conventional subsurface data and inferred seismic-stratigraphic information. The spatial arrangement and chronologic order of facies within each depositional system can be estimated, and the sequential events during deposition can be generally ascertained. Once a tract of contemporaneous depositional systems has been established and component facies identified, it is possible to predict potential traps and reservoirs within the seismic-stratigraphic unit.

The depositional mode interpreted for each successive seismic-stratigraphic unit provides the basis for inferring the overall tectonic and depositional evolution during basin fill and, in turn, points the stratigrapher toward potential prospects: types, stratigraphic position, geographic location and trend, structural situation, and reservoir character. At this point, some of the more sophisticated seismic analysis, involving velocity and amplitude variations, may be applied to prospective areas.

SEISMIC FACIES REFLECTIONS

A general description of the common seismic reflection patterns observed in Brazilian offshore basins will be described according to attitude, reflection configuration, continuity, amplitude, cycles, external geometry, associations, composition, and depositional mode. Seismic expression of a lithofacies has been called "seismic facies" by Sangree et al (1976); generally synonymous terms used here are "reflection unit" or "zone." An effort has been made to equate, where possible, differences in terminology for similar patterns.

Deltaic and Associated Systems

Probably the most important terrigenous clastic depositional system, in terms of volume of favorable reservoir facies, is the delta system (Figs. 1, 2, 3). A variety of Holocene and relict delta systems has been recognized and classified in several ways (LeBlanc, 1972; Fisher et al, 1969; Broussard, 1975). In general, a useful classification includes river-dominated, wave-dominated, and tidal-dominated deltas; naturally, many ancient and modern deltas developed in response to unique combinations of river and marine processes. Deltaic deposition was a principal mechanism for filling many sedimentary basins. Where combined with associated slope systems, the two systems account for most (by volume) of terrigenous clastic facies. Geologic studies have been focused on delta systems for the past two or three decades. Fan-delta systems that are characteristic of the Brazilian post-rift or pull-apart basins have received very limited attention, although these types of deltas are common throughout the world (Fisher and Brown, 1972).

Depositional Models

Two general types of delta systems have been recognized in the Brazilian offshore basins: delta systems tentatively inferred to be of the wave- and tide-dominated variety, and fan-delta systems.

A dip-oriented cross section and block diagrams illustrate the general arrangement of deltaic environments and associated lithofacies (Fig. 7A). The generalized cross section, which is uncomplicated by growth faults, schematically illustrates the common relative bedding attitudes within the delta system. The block diagrams demonstrate differences between deltas deposited primarily under the domination of river or marine processes. Specific differences involve composition and lateral continuity of river-dominated and marine-dominated delta facies, but the systems are similar in regional cross section and can be distinguished only by log patterns, cores, or large-scale, net-sandstone maps.

The seismic response to the two types of delta systems, along any dip line or profile, is similar, including growth faults, shale ridges, and inclined prodelta reflections. Thus, differentiation between an Eocene, river-dominated delta in the northern Gulf basin of Texas and a wave- and tide-dominated delta system like the Tertiary Niger system is not clear cut in seismic section, especially basinward of subjacent shelf edges where the delta progrades into deep ocean basins over associated slope facies. However, the Niger system is composed internally of barrier-bar, tidal-channel, and meanderbelt (point bar) sandstone facies (Weber, 1971), whereas the Gulf Coast Eocene deltas are composed of distributary-fill, channel-mouth bar, and delta-front sandstone facies (Fisher, 1969). Although reservoir characteristics will vary, both delta types commonly exhibit growth-fault and roll-over structural traps localized along delta depocenters (Fig. 7B). Updip pinchout of barrier sandstones is common in the wave-dominated system, and both delta types display downdip sandstone pinchout (See Le-Blanc, 1972; Fisher and Brown, 1972; Broussard, 1975; among others, for specific references on delta systems).

Fan-delta systems are alluvial fans that prograde into marine or lacustrine environments (McGowen, 1970). These systems are coarse-grained, principally braided-stream deltas deposited under the influence of higher gradients and higher bed load (Fig. 8). They commonly are associated with fault basins where short, high-gradient streams flow from nearby source areas (Fig. 8B). The fan delta may be associated with tidal-flat or strandplain environments, depending on the nature and intensity of marine or lacustrine processes (Fig. 8B). A diagnostic feature of the fan delta is the common association with carbonate shelf facies. Internally, the fan delta, like the alluvial fan, is composed of proximal, medial, and distal clastic facies that exhibit a basinward decrease in grain size. Distal fan facies are of transitional fan-marine origin (bars, barriers, tidal flats, or lagoons) and are developed by marine modification of fluvial facies (Fig. 8C). Prodelta facies are deposited from suspension beyond the distal sand deposits. Fan-delta facies may intertongue basinward and along strike with limestone and pelagic shale of the open-shelf environment.

Where faulting was contemporaneous with deposition, the fan-delta system thickens toward the fault and assumes a wedge-like geometry in longitudinal cross section (Fig. 8C). Brazilian fan deltas were probably wave and/or tidal-dominated (see Fisher and Brown, 1972, for references dealing with specific fan-delta processes and facies).

Seismic Facies Characteristics

Deltaic facies, recognized in offshore Brazil, can be grouped into several assemblages that exhibit reasonably diagnostic seismic reflection patterns: (1) prodelta and distal delta-front or barrier facies; (2) delta-front or barrier-bar facies; and (3) alluvial and delta-plain facies (Fig. 9A, B, C, F).

Prodelta and distal delta-front, barrier facies— Reflection patterns for these facies in dip sections are horizontal to steeply inclined, oblique, layered patterns within a zone that ranges from poorly layered to reflection-free or locally chaotic. Oblique reflections may converge (and baselap) downward (basinward). In strike sections, the facies commonly exhibit convex-upward, conformable drape-to-mounded-chaotic, or reflection-free, patterns with some evidence of channel or gully erosion. Beyond the relict shelf edge (Fig. 9C, F), reflections are strongly divergent and inclined toward growth faults; reflections define rollover structures in dip section, and mounded, chaotic-to-conformable patterns in strike sections. On the relict shelf, the prodelta reflections are discontinuous except for a few strong reflections, amplitudes are generally low except for reflections with moderate continuity, and spacing is

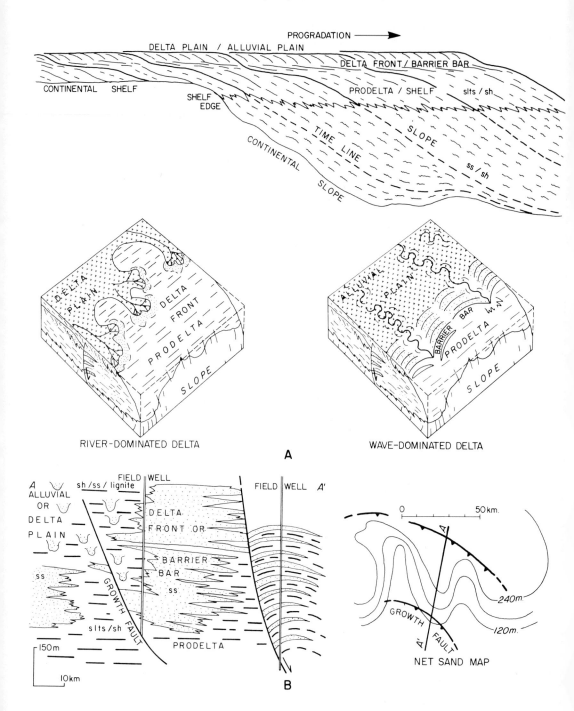

FIG. 7—Delta depositional model illustrating the distribution of reservoir facies. **A.** General cross section of prograding delta system. Plan views of wave-dominated and river-dominated deltas demonstrate variations in lateral distribution of reservoir sand bodies. **B.** Schematic example of reservoir/trap conditions possible in deltaic systems. These systems commonly are involved in extensive growth faulting and shale diapirism.

very erratic. Beyond the relict shelf edge (Fig. 9C, F), reflection continuity increases, amplitudes increase, and spacing becomes more uniform. On the shelf, the external geometry of the reflection-bounded patterns define individual wedge to mounded units arrayed in offlapping, imbricate arrangement; collectively, the reflections constitute a tabular zone that gradually thickens basinward. Beyond the relict shelf edge in dip sections (Fig. 9C, F), the reflections compose a wedge that thickens against growth faults; in strike sections, the unit is characterized by convex-upward, lobate mounds with chaotic-to-conformable, lenticular patterns. The reflection patterns are terminated abruptly upward (toplap) by relatively horizontal delta-front or barrier-bar reflections (Fig. 9A, F). Beyond the relict shelf edge, reflections are transitional downward into various types of slope reflections (Fig. 9C). The lithofacies that coincide with these reflections are massive units of laminated siltstone, mudstone, and some sandstone. The depositional mode of this reflection unit is inferred to represent, principally, suspension deposition on prodelta slopes with limited slumping and density flow—the unit is transitional between deposits of shallow-water deltaic, and/or barrier and deep-water slope environments. Depositional slope was approximately 1 to 5°.

Delta-front, barrier-bar facies—Reflection patterns in dip sections that coincide with these facies are horizontal to slightly inclined, parallel-layered near the base, grading upward irregularly into chaotic or reflection-free patterns with common convex-upward diffractions and poorly defined, mounded reflections (Fig. 9A, F). Subtle, inclined reflections within chaotic zones may represent delta-front or barrier-bar offlap and, hence, may constitute internal time lines. In a strike section, the basal reflections of the zone exhibit drape patterns and local chaotic, to reflection-free, zones display subtle, parallel-layered to draped, reflections and abundant diffractions (Fig. 9B). Basal reflections exhibit strong continuity, but continuity diminishes upward in the unit. The best continuity occurs in dip sections. Amplitudes are moderate to high in basal, high-continuity reflections, but low in chaotic intervals; spacing is moderately uniform in basal reflectors, but erratic in the upper part of the zone. The reflections collectively define a tabular zone with minor thickening on the downthrown side of growth faults (Fig. 9C). The reflection unit is transitional with subjacent prodelta and distal delta front or

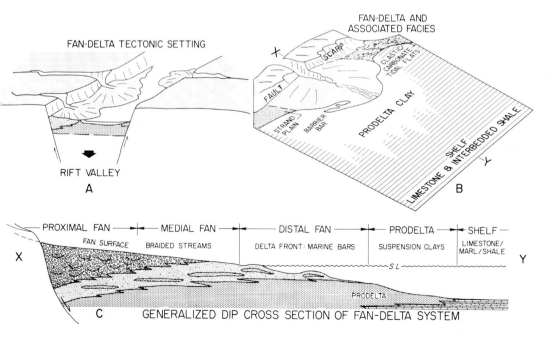

FIG. 8—General setting, facies, and facies associations that characterize fan-delta systems. **A.** Common setting in rift basins. **B.** Common association with carbonate facies and tidal flat-strandplain deposits. **C.** General cross section showing principal types of facies assemblages within the fan-delta system (after McGowen, unpublished).

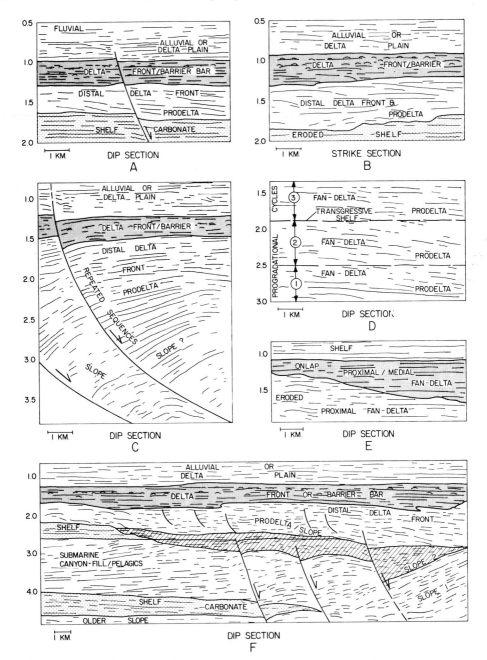

FIG. 9—Deltaic and associated facies patterns generalized from reflection seismic sections showing characteristic reflection attitudes and continuity. Vertical scale in seconds (two-way travel time): **A.** Dip section of delta system. **B.** Strike section of delta system. **C.** Dip section of deltaic and subjacent slope facies; growth faults are commonly associated with these systems. **D.** Dip section of fan-delta system showing periodic marine-transgressive reflections. **E.** Dip section of coastal onlap by fan-delta facies over unconformity. **F.** Regional dip section of delta system that prograded across subjacent shelf-edge and slope facies. Principal growth faults occur basinward (right) where delta prograded into abyssal depths.

barrier reflections and is abruptly terminated upward by alluvial and delta plain reflections (Fig. 9A, F). Lithofacies within the zone are inferred to be extensive sandstone and interbedded shale sequences at the base, and superposed, probably lenticular, sandstone bodies in the upper part of the zone.

The depositional mode inferred for the reflection unit is shallow-water marine and delta front, or barrier bar with superimposed fluvial distributary-channel deposition. Shifting distributaries and repeated cycles of progradation and abandonment resulted in superposition of several delta or barrier sequences within the zone, especially on the downthrown sides of growth faults. Wave and, possibly, tidal action redeposited fluvial sands laterally and into deeper water. High-continuity reflections are inferred to represent marine-reworked facies. Chaotic to poorly defined reflection zones may be in response to thick, massive delta-front or barrier sandstone facies. Marine transgressive facies may be responsible for extensive, strong reflections at the top of the sequence in many areas.

Alluvial, delta-plain facies—Reflection patterns in dip sections that characterize these facies (Fig. 9A, C, F) are principally horizontal, parallel, rarely divergent, layered to locally reflection-free; locally, erosional channels maybe inferred. In strike sections, the reflections are weak, parallel-layered to subtle-mounded, chaotic-to-drape patterns. Continuity of reflections ranges from excellent to fair in dip sections (Fig. 9A, F), but continuity is poor to fair in strike sections (Fig. 9B); amplitude is variable (high in continuous reflections and poor in chaotic zones); and spacing is very regular in zones of high-continuity reflections but irregular in the remainder of the unit. The reflections collectively define a tabular external geometry, the base of which rises in the section in a basinward direction. The reflections overlie, and are transitional with, the delta-front or barrier-bar patterns and are overlain by chaotic, locally parallel-layered reflections with variable amplitudes and continuity that, probably, characterize the fluvial system that supplied the delta system.

The lithofacies resulting in these reflection patterns are massive sandstones and shales with inferred local channel-fill deposits. Thin marine shale and marl facies intertongue with the massive facies, especially in the lower part. The depositional mode inferred for this reflection unit is delta-plain and alluvial-plain processes involving tidal, distributary-channel, and meanderbelt deposition, within floodbasin and perhaps tidal-basin environments. Marine and delta destructional

environments repeatedly transgressed the distal part of the delta plain.

Fan-delta facies in the Brazilian basins can be grouped into two assemblages that generally can be recognized from seismic data and verified by well information: (1) proximal- and medial-fan facies, and (2) distal-fan and prodelta facies. The intergradational facies can be separated only in a general, arbitrary manner by gradational changes basinward in seismic reflection configuration and continuity. These gradational changes are regional and can best be observed on regional dip sections. The following sections describe this basinal change in seismic characteristics (Figs. 9D, E; 10A, E, G).

Proximal, medial-fan facies—Reflection patterns that develop in response to these facies are poorly defined, parallel-layered to reflection-free in both dip and strike sections (Fig. 10A); they may exhibit coastal onlap of erosional surfaces (Fig. 9E). Reflection continuity is absent to very poor in dip and strike sections, amplitudes are generally low, and spacing is relatively uniform. External geometry of the reflection unit is wedge-shaped, thickening toward the source area or toward bounding basement faults. The reflection unit thins basinward by losing section at the base by intertonguing with reflections of the distal fan-delta facies. The reflection unit is composed of massive conglomerate and coarse-grained sandstone, and some thin shale. It is inferred that the depositional mode was that of braided stream and channel-fill deposition in response to high gradients on the proximal and medial fan surface (Fig. 8). A nearby elevated source area has been postulated.

Distal fan-delta, prodelta facies—This zone of reflections contains some poorly defined, inclined to horizontal, slightly divergent, layered reflectors alternating with reflection-free patterns. The number of reflections in the zone increases basinward and they may be gradational with well-developed shelf reflections (Fig. 10A, E, G). Some well-developed, inclined offlap reflections occur where the system progrades into deep water (Fig. 9D). The reflection continuity changes basinward from absent, to poor, to fair, and eventually grades into continuous shelf reflections (Fig. 10A). Amplitudes are variable and spacing is generally uniform. On stable shelf areas, the reflections collectively define a series of time-transgressive tabular units that overlap shelf carbonates basinward and are, in turn, overlapped by reflections of the proximal and medial fan-delta facies. In unstable basins (e.g., salt tectonic style) the reflection zone thickens basinward and may be subdivided into two or more units by thin, marine-

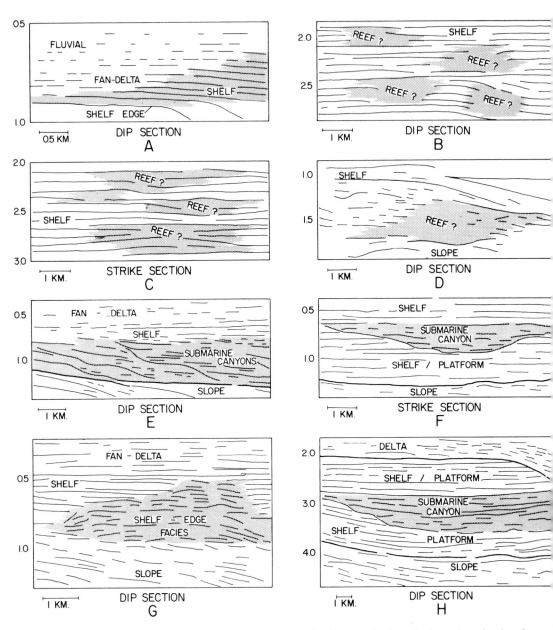

FIG. 10—Shelf and associated seismic facies patterns generalized from reflection seismic sections showing characteristic reflection attitudes and continuity. Vertical scale in seconds (two-way travel time): **A.** Dip section showing characteristic fan-delta/shelf transitional reflections. **B.** Dip section showing variations in reflection continuity and attitude that may represent reef or bank carbonate facies. **C.** Strike section of similar (**B.**) reflection variations. **D.** Inferred shelf-edge reef or bank outlined by termination of typical carbonate-shale reflections near shelf edge. **E.** Dip section showing periodic erosional surfaces in outer carbonate shelf; onlapping reflections terminate landward onto discontinuities. **F.** Strike section showing submarine canyon eroded into carbonate shelf and filled by subsequent channel-fill deposits. **G.** Dip section of complex shelf-edge facies showing distinctive change in carbonate/shale reflections near the shelf edge; inferred to represent a reef complex with probable diagenetic alteration. **H.** Dip section of onlapping submarine canyon-fill deposits within a carbonate shelf system.

transgressive shale and marl that coincide with strong, widespread reflections punctuating the progradational cycles (Fig. 9D).

On shelf areas, these reflections grade landward and upward into poorly defined proximal and medial fan reflections, and basinward and downward into well-defined shelf patterns. Where the fans prograde into deeper water (Fig. 9E), the reflections grade basinward into various slope reflections. The lithofacies that give rise to the reflections are composed of marine sandstone and laminated siltstone and shale; sandstone facies are locally glauconitic. These were deposited by braided streams during floods, and were subsequently reworked and redeposited by wave and/or tidal processes to produce shallow-marine and possibly tidal-flat facies. Suspended sediment was carried basinward and deposited as a thin blanket of prodelta; some suspended sediment probably was transported along strike. On stable shelf areas, these distal facies intertongue with shelf limestone and shale deposits. Where the fan delta prograded to the shelf edge, the distal facies were redeposited in deep water by density currents and submarine slump processes.

Shelf and Associated Systems

One of the most characteristic elements of pull-apart basins (Klemme, 1971) is the shelf or platform carbonate system that is contemporaneously associated with coarse-grained delta systems and with mixed carbonate and terrigenous slope systems. Because of unique tectonic style, these carbonate shelf and platform facies were deposited in association with terrigenous clastic environments. The seismic reflection patterns that occur in response to the widespread sequences of limestone, pelagic shale, and marl are very distinctive (Fig. 10). The evenly bedded sequences produce reflection patterns that exhibit great continuity, high amplitudes, and relatively uniform spacing over thousands of square kilometers. Where these diagnostic reflection patterns vary, it is generally in response to local variations such as shelf-edge shoal, reef or bank, and submarine canyon-fill facies. Recognizing and mapping progressive positions of the carbonate shelf edges in these basins enable the geologist to determine the gross distribution of depositional systems.

Depositional Models

Shelf systems in Brazilian offshore basins primarily are composed of: (1) widespread neritic limestone, marl, and pelagic shale, and localized reef, bank, or shelf-edge shoal facies that were deposited in slowly subsiding shelf environments that shifted basinward over prograding, principal-

ly terrigenous, clastic slope systems, supplied periodically by prograding fan-delta systems (Fig. 10); (2) shoal-water limestone, dolomite, and evaporite facies that were deposited in the absence of significant fan-delta deposition on relatively stable platforms, contemporaneous with limited deposition of mixed terrigenous and carbonate clastic and hemipelagic slope facies; and (3) thin, widespread, transgressive, biogenic, and terrigenous shelf clastic facies that were deposited during marine transgression (coastal onlap) of abandoned deltas and fan deltas. The first two carbonate shelf and platform types constitute a large volume of strata that compose independent depositional systems. The latter shelf type, which is a minor component of the delta or fan-delta system, is discussed elsewhere in this report.

Reflection characteristics in the carbonate shelf and platform systems are considerably less variable than those exhibited by delta or slope systems, because of widespread, relatively uniform depositional environments. Furthermore, many variations in carbonate facies can be distinguished only by textural, petrographic, and fossil composition. Brazilian geologists have recognized a wide variety of carbonate facies, but few of these variations can be recognized using seismic reflection data. It has been possible, nevertheless, to recognize three principal facies assemblages based on well and seismic data: (1) widespread, moderate to low-energy, neritic, shelf limestone and shale facies; (2) local to moderately widespread, high-energy, shoal-water limestone facies composing reef, bank, platform, and shelf-edge associations; and (3) submarine canyons filled with terrigenous and carbonate turbidites and pelagic facies.

Brazilian carbonate shelf systems resemble examples from the Upper Paleozoic of Texas in which deltaic clastic facies are restricted landward from carbonate-shelf and shelf-edge facies (Figs. 2, 3). However, Brazilian deltas associated with carbonate systems are principally of the fan-delta type. Brazilian platform carbonate systems, though, resemble Jurassic and Lower Cretaceous systems from the northern Gulf of Mexico (see Wilson, 1975, for references and analogous carbonate depositional systems).

Seismic Facies Characteristics

The shelf and platform facies of offshore Brazil are grouped into several assemblages that display distinctive seismic reflection patterns: (1) neritic shelf; (2) reef, bank, shoal-water, and shelf-edge facies; and (3) submarine canyon and channel fill (Figs. 9, 10, 15).

Neritic-shelf facies—Seismic reflections that occur in response to these facies are generally horizontal, parallel to slightly divergent or convergent, layered patterns. Reflection continuity is excellent, but pinchout occurs in convergent/divergent areas. Amplitudes of continuous reflections are high, and the reflections are uniform and closely spaced, except in local areas where reflectors diverge and converge. The reflections represent widespread, tabular units that are gradational updip with fan-delta patterns, and downdip with shelf-edge and various slope patterns. The patterns locally are terminated by submarine canyon erosion. The shelf sequence is composed of uniformly interbedded limestone, shale, and marl beds with great lateral continuity; thickness variations are slight and occur on a regional scale. These reflections are in response to facies that were deposited on broad, stable shelf areas away from the influence of fan-delta deposition. The biogenic and pelagic facies define widespread neritic paleoenvironments.

Reef, bank, shoal, and shelf-edge facies—Reflections in response to this group of facies are horizontal to steeply inclined, divergent to convergent, layered to reflection-free, or chaotic (Fig. 10B, C, D). Reflections exhibit poor continuity, variable amplitude, and irregular spacing. They compose a variety of lensoid to mounded units in dip sections; and in strike sections, the reflections generally compose an elongate unit that is parallel with the relict shelf edge. Reflections compose anomalous units within associated neritic shelf or platform reflections. Basinward, shelf-edge reflections may intertongue with subjacent slope reflections. Composition of the lithofacies responsible for these seismic reflections is massive limestone or dolomite, which may be highly altered by diagenetic processes. The massive carbonate facies are commonly composed of oolitic, algal, and calcarenitic limestone containing reef or bank fossils. The lithofacies are inferred to have been deposited in local, shallow-water reef, bank, and shoal environments subject to high wave and tidal energy. Anomalous reflections on middle and inner shelf areas probably represent lower energy facies than those on the relict shelf edges. Low energy lagoon facies also may be associated with the facies on the landward side of the seismic anomalies.

Submarine canyon-fill facies—Reflections that characterize these facies (Fig. 10E, F, H) are horizontal to gently inclined, parallel to divergent and convergent, and layered to chaotic patterns that generally exhibit onlap (onlapping fill, Sangree et al, 1976). Reflection continuity varies from excellent to poor, amplitude is variable, and spacing is not uniform. Collectively, the reflections compose lensoid, canyon-fill units in strike section, and el-

ongate, wedge-shaped units that thicken basinward in dip sections. Reflections occur within canyons cut into upper slope and shelf facies. The reflections may terminate by onlap against the walls of the canyon. Downdip, the reflections grade into onlap slope reflections.

Composition of the lithofacies varies from terrigenous to calcareous, clastic, and pelagic deposits. In map view, the reflection units may bifurcate updip and may be cut locally by growth faults. The reflections are inferred to represent submarine fan turbidites, slump deposits, hemipelagic, and, perhaps, neritic facies that were deposited in submarine canyons. Composition of canyon-fill deposits depends on the nature of the eroded shelf or fan-delta facies. Active fan deltas may have contributed sediment directly into the canyon during progradational episodes following canyon erosion.

Slope and Associated Systems

Beneath many continental shelf areas, and in many onshore basins, are slope systems that contain potential reservoir facies of variable quality and volume. Because these deep-water facies rarely crop out, except in highly complex orogenic areas, subsurface and seismic-stratigraphic methods are exceedingly important in their recognition and mapping. Considerable effort is being directed by marine geologists toward understanding modern slope processes and resulting sedimentary facies. By combining concepts of modern slope processes and facies with seismic geophysics, workers (such as Sangree et al, 1976) have begun to develop seismic-stratigraphic criteria for recognition of various deep-water facies. Seismic-stratigraphic methods provide an important tool for delineating, classifying, and predicting slope facies composition, spatial arrangement, and stratigraphic relations. Because sparse well control is a severe limitation in deep offshore basins, seismic methods must be utilized extensively in stratigraphic, as well as structural, exploration. Slope exploration will improve with continued drilling, but at this time it is important for the explorationist to develop conceptual models that may permit better prediction of reservoir quality, arrangement, and trap potential.

Depositional Models

Much has been published about turbidites and related facies, but effective exploration of slope systems will require much more information about slope processes, submarine fan composition, submarine canyon and channel characteristics, and the spatial arrangement of these facies. Integration of information obtained from modern

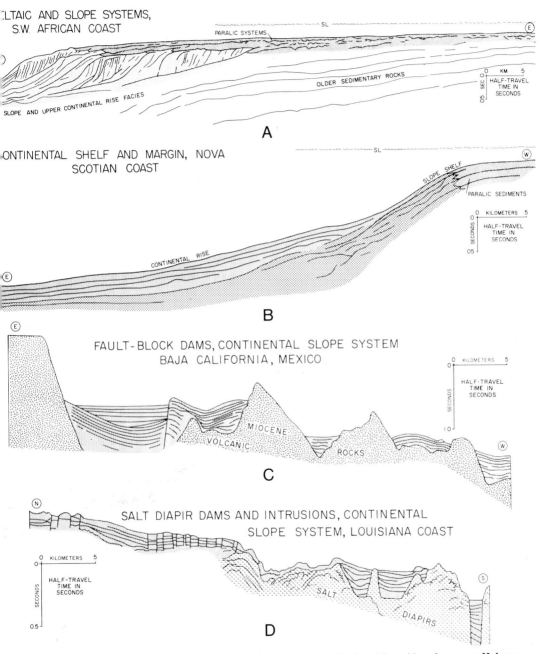

FIG. 11—General nature of seismic reflections that characterize several styles of deposition along some Holocene continental margins. **A.** Complex reflection patterns in deltaic and slope systems along southwestern African coast (after McMaster et al, 1970). **B.** Continental rise onlap along Nova Scotia coast (after Uchupi and Emery, 1967). **C.** Fault basins containing superposed (uplap) slope deposits dammed behind fault blocks, Baja California, Mexico (after Emery, 1970). **D.** Slope deposits trapped in salt basins and behind salt ridges along Louisiana coast, Gulf of Mexico (after Uchupi and Emery, 1968).

marine studies with results of conventional analysis of ancient slope systems is needed to support seismic-stratigraphic analysis of deep-water facies.

Conventional studies of deep-water reservoirs naturally have been focused on California basins for many years where the reservoirs have been so productive. Tertiary slope systems in Texas and Louisiana are being recognized (Fig. 1C), and Paleozoic intracontinental basins are being explored actively for deep-water reservoirs (Fig. 3A, C). It is natural that California basins were used as exploration models for deep-water reservoirs, but we suggest that the U.S. West Coast basins represent but one type of slope system, and that exploration of basins with different structural styles will require somewhat different concepts (see Middleton and Bouma, 1973; LeBlanc, 1972; Fisher and Brown, 1972; for references about slope systems).

Within the past 15 years, seismic cross sections of many continental margins have indicated the occurrence of various structural, stratigraphic, and sedimentary styles. Types of seismic reflections that we have observed in Cretaceous and Tertiary sequences in Brazilian offshore basins are herein called offlap/onlap, continental rise onlap, and fault-controlled and salt-controlled uplap; modern analogues are shown on Figure 11A, B, C, D, respectively. Recognition of these

and other types of slope systems in ancient basins provides insight to the structural style and potential slope reservoirs that may occur. The distribution of facies within complex sequences of submarine fan deposits is also significant in evaluating slope reservoirs. Figure 12 illustrates Upper Paleozoic submarine fan facies, interpreted by using well control and by analogy with modern fans. The manner in which submarine fans may shift in response to subsidence, *versus* sediment supply, is an important factor in predicting the stratigraphic arrangement of ancient slope reservoirs.

Two fundamental slope relations that can be observed on seismic section may be called "offlap" and "onlap" (Fig. 13A). Onlap is common along modern slopes (Fig. 11B) and is inferred to be in response to erosion and redeposition of shelf and slope facies in the absence of a sustained supply of either shelf-edge or paralic sediments. Dietz (1963) presented an interpretation of continental rise onlap (Fig. 13B). In Brazilian offshore basins, we observed that extensive onlap deposition was commonly accompanied by some degree of canyon erosion of the adjacent shelf edge (Fig. 13A). We inferred that where the sediment supply diminishes and the shelf-edge retreats under long-term submarine erosion, slope-depositional environments gradually shift landward in response to the retreating and diminish-

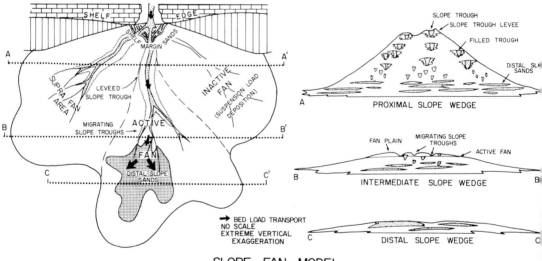

SLOPE FAN MODEL

FIG. 12—Slope submarine fan model illustrating general processes and resulting composition that typify these deep-water depositional systems. Successive fans may offlap or may onlap, depending upon a sustained or diminishing sediment supply, respectively. Submarine fan deposits may stack in vertical or superposed manner if subsidence rates exceed sediment supply, thus producing an uplap system (after Galloway and Brown, 1973; based on Shepard et al, 1969; Carlson and Nelson, 1969; and Normark, 1970).

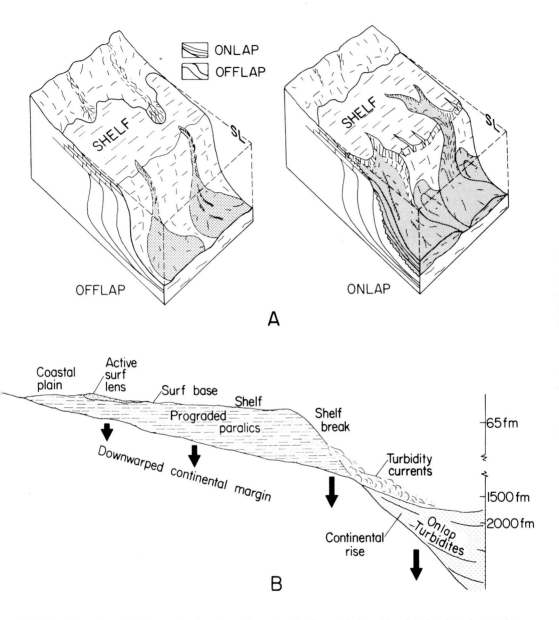

FIG. 13—The nature of offlap and onlap deposition—two fundamental depositional styles that characterize many slope systems. **A.** Block diagrams that illustrate general processes: offlap occurs during sustained sediment supply provided by deltas, fan deltas, and highly productive shelf-edge carbonate environments. Onlap conversely occurs when sediment supply diminishes and erosional processes rework shelf or paralic sediments, commonly via submarine canyons. Offlap reflections define the basinward progradation of slope deposits and onlap reflections mark periods of landward recession of slope depocenters. **B.** Schematic representation of onlap processes (after Dietz, 1963).

ing sediment source. Vail et al (this volume) referred to this type of reflection pattern as "marine onlap," and related canyon erosion and onlap to relative drop in sea level below the shelf edge (erosion), followed by a rise in relative sea level (onlap deposition). The progressive onlap of slope facies can be recognized in seismic sections.

Offlap seismic patterns in the Brazilian basins commonly coincide with periods of persistent deltaic progradation to the outer shelf, or to periods of excessive production of biogenic sediments along shelf-edge banks, reefs, and shoals. Therefore, offlap slope deposition is inferred to be in response to a sustained sediment supply that is greater than subsidence rates. During offlap, submarine fans and other slope facies shift progressively basinward as the basin is filled by slope deposition (Fig. 13A). Varieties of offlap (e.g., sigmoidal and oblique progradational, Sangree et al, 1976) probably result from variations in rates of progradation, versus subsidence. Onlap and off-

lap depositional models are illustrated on Figure 14 (I, II, III).

Another variation of slope sedimentation observed in Brazilian basins, as well as in seismic sections from elsewhere in the world, is herein called "uplap" (Fig. 14-I). Sangree et al (1976) called this slope relation "onlapping fill." It occurs in basins where subsidence rates are greater or equal to sediment supply, resulting in superposition of submarine fans and other slope deposits. Fault-controlled basins and salt grabens commonly exhibit this slope system, in which the deep-water facies onlap the flanks of the basin. Generalized examples of slope systems from Brazilian basins are shown in Figure 15; Figure 16 is a digrammatic representation of slope seismic reflections and inferred relations between slope type and the quality and distribution of potential reservoirs. We believe that seismic discrimination presently provides the best basis for predicting slope reservoirs and traps in frontier basins.

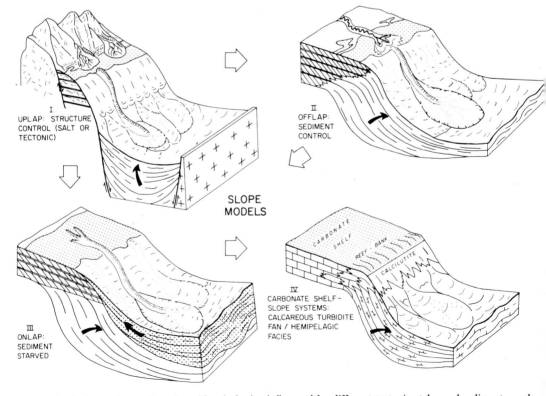

FIG. 14—Examples of slope deposition in basins influenced by different tectonic styles and sediment supply. During development of a basin, each depositional type may occur. The onlap type may develop where sediment supply diminishes periodically. In rift and post-rift basins, there is commonly a progression from Type I to Type II and eventually to Type IV, with periodic episodes during which Type III may develop.

SLOPE AND ASSOCIATED SEISMIC FACIES PATTERNS

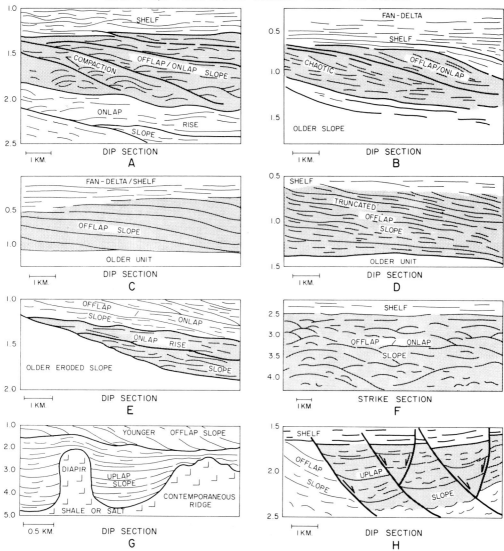

FIG. 15—Slope and associated seismic facies patterns generalized from reflection seismic sections. Heavier lines represent strong reflections and/or discontinuities. Vertical scale in seconds (two-way travel time): **A.** Dip section of a slope complex composed of older onlap rise facies and younger offlap/onlap slope facies. Onlap reflections terminate along strong local or regional reflections and/or discontinuities in reflection attitudes. **B.** Dip section through slope composed of chaotic reflections enveloped in sigmoid-shape offlap units. Complex offlap/onlap reflections shown in section A. may represent discrete, alternating episodes of offlap deposition followed by episodes of shelf-edge, slope erosion and extensive onlap. The sigmoid-shaped units composed of onlap and chaotic reflections may, on the other hand, reflect rapid offlap, oversteepening of slopes, extensive slumping, and local onlap of submarine fan and proximal slope deposits. **C.** Dip section of slope system showing relatively uniform continuity of offlap reflections (commonly calcareous/terrigeneous, hemipelagic, upper slope facies) representing relatively slow rates of deposition. **D.** Dip section of slope system showing relatively uniform, but truncated (sometimes called "oblique"), offlap reflections indicating relatively rapid progradation and minimum subsidence. **E.** Dip section through thick onlap slope units representing extensive continental rise deposition; internal reflections terminate along strong reflection discontinuities. An offlap/onlap complex overlies the rise deposits representing another depositional episode. **F.** Strike section of offlap/onlap slope system characterized by convex-upward reflectors representing hemipelagic blankets draping large slope fan complexes (or cones). **G.** Dip section through a slope system characterized by uplap reflections caused by rapid subsidence of salt basins. **H.** Dip section through a slope system composed of uplap reflections resulting from subsidence of minor fault block.

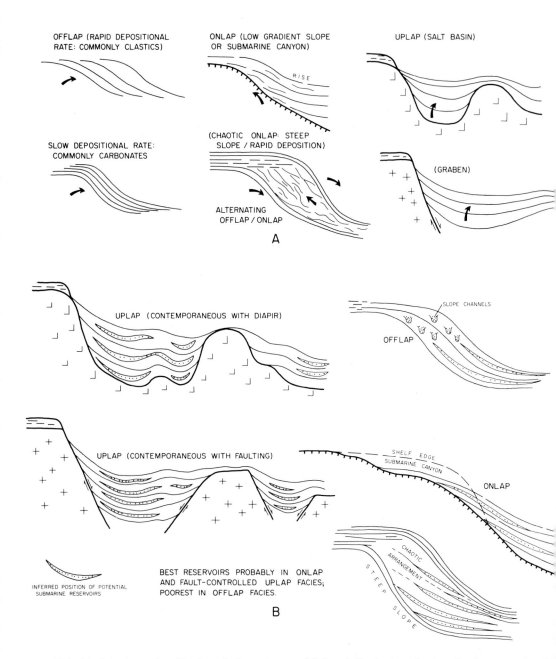

FIG. 16—Seismic-stratigraphic slope facies patterns and inferred distribution of submarine fan reservoirs. **A.** Schematic representation of reflection patterns that characterize offlap, onlap, and uplap slope facies. **B.** Inferred distribution of submarine fan sandstone facies in three principal types of slope systems.

Seismic Facies Characteristics

Seismic reflection patterns that characterize offlap, onlap, and uplap slope relations will be described and related to inferred lithofacies. Mixed offlap and onlap variations will be described separately (Fig. 15A, B).

Offlap slope facies—Both uniform offlap and truncated offlap reflections have been recognized (Figs. 15C, D; 16A). These patterns correspond, respectively, to sigmoid terrace and oblique progradational seismic facies (Sangree, et al, 1976). The reflections are inclined, parallel to divergent to convergent, layered with the dip of reflections decreasing near the base of the zone. In strike sections, the reflections commonly exhibit mounded and draped patterns. Continuity of reflections is high in uniform offlap and moderate in truncated offlap. Amplitude is high in uniform offlap reflections and high to moderate in truncated offlap reflections. Spacing is generally uniform in offlap reflections but more irregular in truncated offlap reflections. The reflections collectively compose a wedge to lensoid-shaped unit that pinches out updip and downdip (Fig. 16A); individual reflection units are inclined, generally recurved lenses or wedges that converge and pinch out basinward (baselap) and are terminated updip (toplap) by the basal shelf surface (Fig. 15C, D). In strike sections, the individual offlap reflection units are characterized by convex-upward, mounded reflections. The lithofacies that coincide with these reflections are hemipelagic shale and/or calcilutite, lenticular turbidite sandstone or calcarenite, and slump deposits. Offlap facies are inferred to have been deposited by turbidite deposition on submarine fans, by slumping, and by hemipelagic sedimentation over the entire slope surface. Sustained sediment supply was greater than subsidence, resulting in the offlap pattern of deposition. Calcareous slope deposits are common in the uniform offlap type of system; terrigenous sandstone and shale facies are common in truncated offlap reflection units.

Onlap slope facies—Onlap reflection patterns define thick, continental rise deposits (Figs. 15E; 16A), as well as elements within alternating offlap and onlap slope patterns (Fig. 15A, B). Within alternating offlap and onlap units, the onlap reflections may be well defined (Figs. 15A; 16A), or poorly defined to chaotic (Figs. 15B; 16A).

These reflections are slightly inclined to horizontal, onlapping, parallel-layered to chaotic patterns in dip sections. In strike sections, the reflections define horizontal to draped, parallel-layered to chaotic patterns. Reflection continuity is fair to excellent in thick, continental rise units, but fair

to poor in onlap units that alternate with offlap reflections (Figs. 15A, B; 16A). Amplitude is high to moderate, except in chaotic units, and spacing is relatively uniform in continental-rise facies, but less uniform in other onlap types. The reflections collectively compose a wedge to lensoid-shaped sequence of strata that pinches out or is truncated updip by shelf reflections (toplap), and pinches out downdip (baselap). In strike sections, the onlap unit exhibits a draped or chaotic mound geometry. Lithic composition is similar to offlap facies, although onlap lithofacies may be coarser grained if erosion has cut into fan-delta facies preserved on the relict shelf. Onlap facies are inferred to represent deposition under conditions of diminishing sediment supply. Vail et al (this volume) cited that this type of onlap, marine onlap, developed during sea-level rise when the continental shelf was subaerially exposed and rivers deposited sediment directly into deep water through eroded shelf-edge canyons. We inferred that sediment supply was derived primarily by submarine erosion of shelf, slope, or fan-delta deposits by submarine canyons. Some littoral sediments may have entered canyons if they extended into the longshore zones.

Continental rise units represent long-term episodes of onlap (Fig. 16A); alternating offlap and onlap reflections indicate shorter-lived periods of onlap, perhaps mixed with brief episodes of rapid, localized offlap. Chaotic onlap is common where slopes were steeper and submarine slumping was common; more regular onlap reflections occur on slopes or within canyons with lesser gradients. Submarine fan reservoirs in onlap facies are inferred to possess better potential than in offlap systems, since erosion may tap coarse, deltaic facies, and the onlap process provides for better updip pinchout possibilities (Fig. 16B).

Mixed offlap and onlap facies—These systems are common in Brazilian basins where the two slope-depositional modes commonly alternated during a prolonged progradational episode (Fig. 15A, B, F). Offlap deposition apparently terminated periodically, followed by extensive, perhaps local, submarine erosion of shelf edges and slope to supply onlapping slope environments. Renewed fan-delta deposition subsequently reactivated slope offlap or progradation. Although large volumes of slope facies are of the onlap variety, the onlap units are arranged in sigmoid-shaped offlap wedges (Fig. 15B). Widespread high-amplitude reflections can be traced from the shelf, through the shelf-edge facies, and downward through. the slope system. Onlapping slope reflections terminate updip along these discontinuities indicating that a long period of slope onlap coincided with shelf erosion (Fig. 15A, B). Seismic reflections within the offlap and onlap sequences are similar to those within independent systems previously described.

Uplap slope facies—To sustain slope deposition to produce a superposed arrangement of seismic reflections requires that subsidence and sediment supply are relatively balanced. Superposed slope-turbidite and hemipelagic facies of this type are common in fault basins and salt grabens that underwent subsidence during deposition (Figs. 15H; 16A). The reflections exhibit onlap against the steep flanks of the basins and may also display a variety of drape reflections over paleobathymetric highs.

The reflections are horizontal to slightly inclined, parallel-layered to rarely chaotic, or reflection-free patterns. Excellent continuity may be displayed, or continuity may be poor to missing in chaotic or reflector-free zones. Amplitude is high in zones with reflector continuity, and spacing is uniform within the layered, probably hemipelagic zones. The reflections compose irregular, concave-upward, lensoid to wedge-shaped units that may exhibit local drape structure (Fig. 16A). Composition of the facies is similar to offlap slope facies, but may be higher in coarse clastic sediment if the adjacent shelf is narrow and subjected to repeated fan-delta deposition (Fig. 14-I). Lithofacies exhibiting these types of reflections were deposited in subsiding, deep-water basins associated with faulting or salt tectonics. The basins were filled primarily during periods of sustained sediment supply, but deposition may evolve into an onlap phase. Because such basins commonly exhibit bathymetric relief, it is inferred that coarser, density-flow deposits will be concentrated within the paleobathymetric depressions, which may not coincide with structural highs (Fig. 16B).

SEISMIC-STRATIGRAPHIC MODELS OF OFFSHORE BRAZILIAN BASINS

Integration of the various delta, shelf, and slope depositional systems, characterized by facies that generally can be recognized by distinctive seismic reflection patterns, provided the basis for construction of a series of Brazilian basin models (Figs. 17-21). These models are generally typical of those that developed in early rift and subsequent pull-apart areas of the world, and are based principally on seismic criteria tied to available well control. They show the general variety and interrelation of depositional systems that developed in response to various tectonic styles since rifting was initiated. With variations, each basin developed through two or more of these

types during its post-rift history. The models are intended to illustrate an integration of well data and seismic-stratigraphic interpretation.

Early Rift-Basin Model

Basins that formed during early stages of rifting (Figs. 17; 18A), and which were isolated sufficiently from the proto-Atlantic, are commonly characterized by facies that indicate initial lacustrine environments that changed gradually to marine conditions. Subsequent marine environments were either normal or hypersaline. These basins were commonly sediment-starved initially, but an extensive sediment supply generally developed from nearby upthrown fault blocks; faulting was commonly contemporaneous with sedimentation.

Depositional systems tract—and Alluvial fan, fan delta, periodic transgressive shelf, and offlap-uplap slope systems compose the typical sequence or tract of depositional systems from source to deep basin.

Tectonic elements—The basins are bounded by faults and commonly tilted toward the continent. Faulting was contemporaneous with deposition, but slowly diminished in intensity. Shale diapirs developed in the basin in response to excessive loading of fan-delta sands on thick prodelta and slope facies; and shale movement continued until late in the history of the basin.

Principal depositional modes—Adjacent elevated source areas supplied large volumes of coarse-grained clastic sediment to the basin along short braided streams that crossed boundary faults to fan-delta depocenters. An integration of drainage supplied fan deltas that prograded along the axis of the basin from more distant sources. Rapid deposition and subsidence (including shale diapirism) precluded significant carbonate deposition, but thin biogenic shelf facies transgressed the fan-delta surface periodically in response to delta shifting, or minor eustatic or tectonic sea-level changes. Slope facies are principally of the offlap variety or (in cases of rapid basin subsidence) the uplap type. Submarine fan deposits were derived from slumping distal fan-delta, sand facies.

Source area and drainage basin—Sediment was derived principally from Paleozoic strata or basement rocks landward of the rift zone. The drainage system was poorly organized and originated in nearby elevated areas beyond the boundary faults.

Post-Rift Salt-Basin Model

Basins that formed during early stages of rifting and that were connected with the proto-Atlantic, via restricted openings, were commonly the site of extensive salt deposition (Figs. 17;

18B). As continental drift continued, normal marine circulation developed, and the basins became the sites of extensive clastic deposition when adjacent fault-bounded source areas were periodically elevated. Mobilization of the subjacent, bedded salt generated a relatively unique tectonic style and imposed a distinctive depositional mode on these basins. Rarely, in the absence of a clastic source, limestone was deposited directly on the bedded salt deposits, resulting in carbonate platform systems that exhibit down-to-the-basin growth faults.

Depositional systems tract—Alluvial fan, fan-delta, localized shelf-carbonate, and uplap slope systems typify the basinward sequence or tract of depositional systems in a post-rift salt basin. In areas where salt thickness diminishes, more normal offlap and onlap slope systems dominate.

Tectonic elements—These basins are bounded on the continental margin by extensive, regional fault zones that separate the basin from adjacent source areas. The reactivated basement faults originated during early rift stages of basin development. Faulting may have been contemporaneous with deposition, but more commonly, periodic fault activation occurred, accompanied by erosion of fan-delta facies along the basin margin, and followed by coastal onlap of fan-delta facies over the erosion surface. Salt diapirism was a dominant element in the tectonic history of these basins, when salt, mobilized by prograding fan-delta systems, moved basinward into salt ridges, and locally upward to form diapirs. Salt tectonic instability precluded development of well-defined carbonate shelf edges. Uplap slope facies were deposited in subsiding salt grabens and basins.

Principal depositional modes—Elevated source areas along the continental margin supplied fan-delta systems with coarse-clastic sediment. Periodic faulting rejuvenated the source areas. Fan-deltas prograded basinward, locally over extensive tidal-flat environments, but were trapped landward of large salt ridges. Periodically, bio-

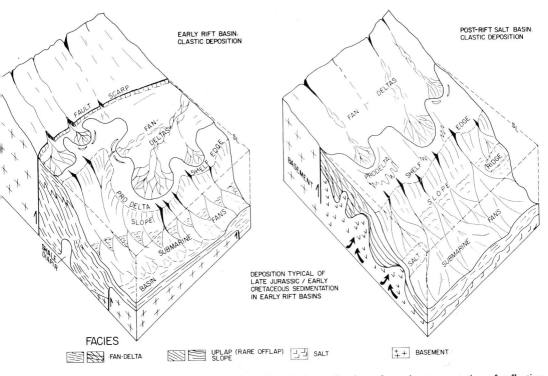

FACIES

FAN-DELTA   UPLAP (RARE OFFLAP) SLOPE   SALT   BASEMENT

FIG. 17—General depositional models of a rift basin and a salt basin showing schematic representation of reflection [ampli]tudes and continuity (see Figures 9, 10, 15 for additional detail of reflection attitudes and continuity). The rift-basin [exa]mple is characterized by rapid deposition of fan-delta and slope facies in contemporaneously faulted basins; slope systems [are] commonly represented by offlap and uplap reflections in response to rapid progradation and rapid subsidence. Similarly, [fan]-delta progradation across thick salt deposits produced salt mobilization resulting in salt dams and subsiding salt basins [con]taining uplap facies.

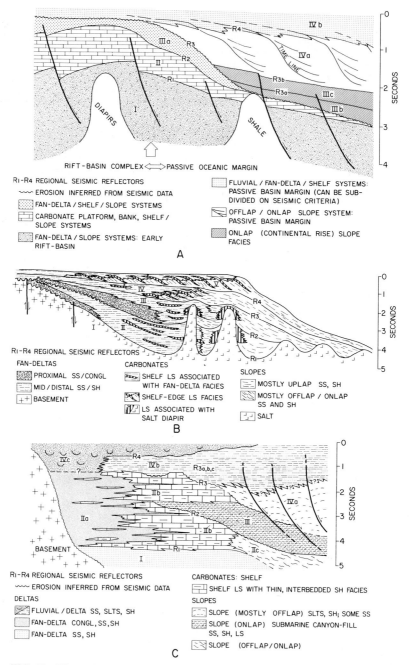

FIG. 18—Dip sections across margins of three pull-apart basins showing general distribution of depositional systems delineated and mapped, in part, using seismic-stratigraphic
criteria. **A.** Basin showing transition from rift-basin deposition to passive, ocean basin
deposition. **B.** Basin filled under the influence of salt tectonics and adjacent elevated source
area. **C.** Basin that evolved from rift type to passive offlap type to integrated deltaic/slope
type. Horizontal scale: schematic. Roman numerals signify principal seismic-stratigraphic
units delineated by regional reflections.

genic clastic and carbonate shelf facies transgressed the fan deltas; local platform carbonate (reef or bank) facies formed on positive structural elements. Distal fan-delta sediments provided the source of coarse-grained turbidite fan facies that were deposited in an uplap attitude within the subsiding salt basins and grabens. In salt basins without a clastic source, thick carbonate facies developed on top of the salt, generating local growth faults in the carbonate sequence.

Source area and drainage basin—Sediment was derived from elevated Paleozoic and basement rocks adjacent to the boundary faults. The drainage system was a short and poorly developed braided complex. Fan-delta systems exhibit extensive red-bed facies that developed on the fan surface during the course of basin development. The region may have evolved from an arid climate to a warm, humid climate with high rainfall, as a result of the growing size of the Atlantic Ocean.

Post-Rift Carbonate Platform Model

After initial rifting, basins with limited sediment supply were the sites of extensive aggradation or upbuilding by platform carbonate deposition (Figs. 19; 18A). Although limited terrigenous sediment reached the basin through fan deltas, it was generally trapped landward in rift basins that were undergoing final subsidence. The carbonate platform systems exhibit limited progradation, principally from deposition of hemipelagic carbonate deposits and carbonate turbidites. The platforms shifted basinward to produce a series of superposed, slightly offlapping shelf edges. Periodically, the platforms were subjected to inferred episodes of submarine canyon erosion. During platform carbonate deposition, the basin was essentially sediment-starved, except for in-situ biogenic sediment.

Depositional systems tract—Fan delta, carbonate platform, shelf edge, calcareous and offlap slope systems comprised the environmental tract at that time in basin history. Fan-delta systems were restricted near source areas, and the broad platform was the site of shallow-water carbonate deposition. Slopes were principally of the offlap variety, but some onlap sequences developed during periods of inferred submarine canyon erosion and deposition, especially terminal stages.

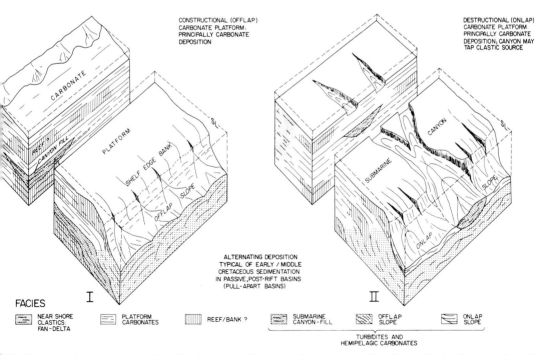

FIG. 19—General depositional model of carbonate platform complex in a post-rift pull-apart basin showing schematic representation of reflection attitudes and continuity (see Figures 9, 10, 15 for additional detail of reflection signatures). This early post-rift basin is characterized by offlaping shelf-slope systems of predominantly carbonate composition. **I.** Landward, minor fan-delta systems periodically supplied terrigenous sediment to the platform, but principal shelf/slope facies are limestone deposits. **II.** Shelf-edge reefs are common and the shelf edge prograded by alternating deposition of calcareous offlap slope facies and onlap slope facies commonly associated with submarine canyon systems.

Tectonic elements—Although bounded on the landward margin by extensive basement faults, the basin experienced slow, relatively uniform subsidence, resulting in deposition of thick limestone facies on the shallow platform. Where the carbonate platform sequence was underlain by salt or mobile, highly pressured shale, some tectonic instability was possible. Growth faulting associated with shale or salt diapirs (or ridges) resulted in deposition of thick sections adjacent to the contemporaneous faults. Conversely, movement on boundary basement faults resulted in the deposition of thin, commonly shoal-water facies on upthrown fault blocks; deeper water facies accumulated in downthrown fault blocks.

Principal depositional modes—In response to limited terrigenous clastic sediment supply and relatively uniform subsidence rates, the basin (Fig. 19) was the site of a variety of constructive carbonate facies tracts, including reef or bank, lagoon, open-shelf, and shelf-edge complexes. Algal facies denote commonly shallow-water environments, although neritic environments were common. Evaporite deposition also occurred in proximal areas. Carbonate facies intertongue updip with fan-delta or tidal clastic deposits. Downdip, the platform facies grade into steeply dipping calcareous or terrigenous clastic slope deposits. Slope facies commonly display relatively uniform offlap in which considerable reflection continuity exists from shelf to slope. Depositional rates were probably slow since aggradational platform depositional units commonly can be traced over the shelf edge and into the slope sequence. Calcarenites that are common on shelf edges and upper slopes grade basinward into calcilutites; however, calcarenite submarine fan deposition occurred locally on the lower slope. During periodic episodes (Fig. 19), inferred submarine canyon erosion developed along shelf edges and locally extended landward into the platform and shelf environment. The final canyon erosional episode commonly produced thick, onlapping continental rise deposits that lapped far up the eroded slope. Sediments eroded from carbonate shelf edges were transported by density flow to onlapping submarine fans. Hemipelagic carbonate sediment also settled on the slope, especially on upper and middle slope surfaces. Submarine fans onlapped the slope and, at least partially, filled submarine canyons. Pelagic carbonates also were deposited in the canyon systems.

Source area and drainage basin—Source areas were either low and unimportant or sediment was being effectively trapped within intermediate fault basins. Drainage was restricted or, perhaps, diverted into other adjacent basins.

Passive Offlap Model

When basement subsidence, rift faulting, and structural activity associated with salt and/or shale mobilization diminished in the pull-apart basins, the depositional mode slowly shifted from rift-related to passive-basin sedimentation (Figs. 20; 18A, B). When rift-related basement faulting terminated, depocenters shifted into marginal marine basins in the growing Atlantic Ocean. The depositional style or mode was dominated by extensive offlap of slope depositional systems onto deep oceanic crust. A relatively continuous sediment supply maintained these progradational episodes for long periods of time, interrupted only by periodic, perhaps lengthy, episodes of shelf-edge and slope erosion, and local submarine canyon development. Strata deposited in this style of basin comprise much of the sedimentary volume preserved along Brazilian coastlines, where an extensive, integrated drainage system was absent. Elevated continental marginal areas were the principal source areas.

Depositional systems tract—Fan-delta, carbonate-shelf, shelf-edge, and slope systems comprise the environmental tract that was responsible for depositing large volumes of carbonate and clastic facies. Slope systems exhibit alternating offlap and onlap varieties of deposition. Onlap slope deposition tended to dominate, especially in the older, landward slope systems, but younger slope facies commonly display better developed offlap (sigmoid type) of a more calacreous composition. Beneath many of these slope systems occur thick, onlapping continental rise sequences deposited in response to a long period of submarine erosion of older carbonate platform systems (see post-rift, carbonate platform model).

Tectonic elements—Depositional offlap into marginal ocean basins occurred with minimum structural control, other than regional subsidence. Compaction of thick clastic sequences also contributed significantly to regional subsidence. Source areas were maintained by relatively continuous uplift. In some basin, growth faulting developed progressively at the shelf edge–slope break, where thick sequences of undercompacted muds failed on steep slopes.

Principal depositional modes—Fan-delta systems repeatedly prograded across carbonate shelf environments, thus intertonguing the coarse deltaic clastics with open-shelf carbonate facies. Reworked deltaic sands and muds were transported by longshore currents along strike and, perhaps, basinward by tidal currents. Abandoned fan-delta lobes were periodically transgressed by open-marine carbonate environments. Carbonate facies

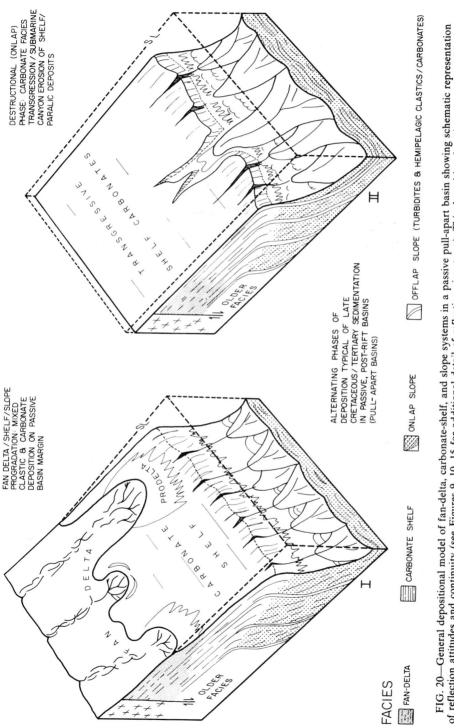

FACIES

| FAN-DELTA | CARBONATE SHELF | ONLAP SLOPE | OFFLAP SLOPE (TURBIDITES & HEMIPELAGIC CLASTICS/CARBONATES) |

FIG. 20—General depositional model of fan-delta, carbonate-shelf, and slope systems in a passive pull-apart basin showing schematic representation of reflection attitudes and continuity (see Figures 9, 10, 15 for additional detail of reflection signatures). This depositional setting was responsible for depositing a significant part of the fill in these basins. Active fan-deltas supplied terrigenous clastic sediment to the shelf area where limestone deposition dominated; this sustained episode of clastic/carbonate sediment supply produced a steady progradation of offlap slope facies **I**. Alternating with episodes of sustained progradation of shelf/slope environments were episodes of diminished sediment supply with corresponding erosion of shelf edges and onlap of calcareous/clastic slope facies **II**. The terrigenous clastic influence generally diminished through time, and slope systems became increasingly calcareous in composition.

were deposited in open-marine neritic and shelf-edge bank or reef environments. Slope systems exhibit complex offlap and onlap seismic patterns (generally oblique progradational) dominated by poorly defined onlap reflections (mounded chaotic) that commonly extend upslope into submarine canyons. The alternation of offlap and onlap facies indicates that the slope system was constructed by offlap of submarine fans, maintained by sustained terrigenous and carbonate sediment supply, and followed, perhaps, by long periods of erosion, slumping, and onlap of submarine fans (Fig. 20). Terrigenous clastic sediment supply slowly diminished during this depositional mode and an increasingly calcareous, dominantly offlapping slope (sigmoid progradational) characterizes the younger deposits.

Source area and drainage basin—Source areas were located along the basin margin as a series of elevated coastal ranges. Drainage systems were braided, and discharge was probably continuous

due to higher rainfall along the coastal ranges of the relatively broad south Altantic Ocean.

Late Tertiary Delta Model

Eventual integration of rivers into systems that drained large continental interior areas, focused immense volumes of sediment into a few large deltaic depocenters. In Brazil, this type of major oceanic delta differed from the fan-delta systems common in most post-rift, pull-apart basins. Discharge was higher, and bedload to suspended load ratios were much lower in the delta system than in fan-delta systems. The Brazilian delta prograded rapidly across a continental shelf and initiated shelf-margin deltaic sedimentation (Figs. 21, 18C). Redeposited deltaic sediments were transported into deep water by slumping and by turbidity flow, resulting in deposition of thick, generally offlapping slope facies over which the shelf-margin deltas prograded. Growth faulting and shale diapirism, which were common ad-

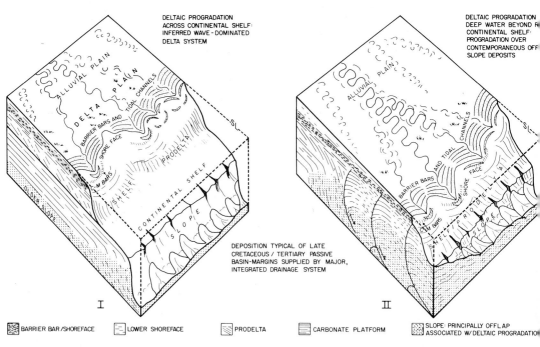

FIG. 21—General depositional model of delta/slope systems in Tertiary pull-apart ocean basins showing schematic representation of reflection attitudes and continuity (see Figures 9, 10, 15 for additional detail of reflector signatures). This style of deposition reflects development of an integrated drainage system in the pull-apart basin, resulting in deposition of thick deltaic and slope facies. Prograding deltas built across subjacent carbonate platforms **I.** to shelf-edge positions where they supplied sediment that was redeposited by slope processes into deepwater environments. Steep slopes generated by sustained deltaic deposition at the shelf edge resulted in development of growth faults and shale diapirs **II.** Delta systems may be either marine or river dominated; slope systems are principally of the offlap type.

juncts to deltaic and slope deposition, developed where water-saturated prodelta and slope muds failed under the sedimentary load. Wave and tidal processes are inferred to have dominated late Tertiary Brazilian delta facies.

Depositional systems tract—Fluvial, delta, and slope systems characterize the tract within this type of major oceanic depocenter. The systems prograded over submerged shelf systems until reaching the relict shelf edge, where deposition was directly into the deep oceanic basin. Minor and localized episodes of marine transgression occurred where deltaic lobes were temporarily abandoned by river shifting.

Tectonic elements—Distant source areas were probably affected by late Cenozoic tectonism in the Andes region. In the ocean basin, structural activity involved principally regional subsidence and sedimentary tectonics such as growth faulting and shale diapirism. The nature and spatial arrangement of deltaic and slope facies were significantly controlled by growth faulting. Slope and deltaic facies thicken dramatically toward contemporaneous faults. Slope shale, displaced by the fault blocks, migrated basinward into elongate shale ridges parallel with the basin margin.

Principal depositional modes—The delta system prograded basinward as a series of offlapping, imbricate wedges of sediment, deposited by shifting depocenters. Well-log patterns indicate that the delta systems were of wave- and tide-dominated variety, although evidence is not conclusive. The prodelta facies exhibits moderately well-developed offlap seismic reflections (oblique progradational) that are inclined toward the basin, beneath relatively horizontal reflections that represent delta-front or barrier and delta-plain facies (Fig. 21). Progradation rates exceeded subsidence rates on the continental shelf, but progradation probably diminished when the delta system reached deeper water beyond the shelf edge. Deltas remained relatively stationary near growth faults until displacement diminished, at which time they prograded basinward until another growth fault developed. Slope sedimentation was concentrated in front of the prograding delta where prodelta and delta front deposits were reworked and transported into deep water by density processes. Growth faulting and diapirism have significantly modified the primary attitude of deltaic and slope depositional surfaces. Marine environments locally transgressed inactive delta lobes, resulting in deposition of marl, shale, and glauconitic sandstone.

Source area and drainage basin—Source areas were within the continental interior, and the drainage system was an integrated complex that maintained relatively high discharge most of the year. The river transported dominantly suspended sediment.

CONCLUSIONS

The combined use of seismic reflection data and current basin-analysis concepts provides a potent exploration tool in basins where well control is limited. Even where well control is dense, the use of seismic-stratigraphic interpretations will permit maximum use of well data. Conventional facies interpretations, based on well data, can be tested and extrapolated throughout a basin using a time-stratigraphic framework constructed using regional seismic reflections that conform to depositional surfaces. The external geometry and distribution of zones that reflect distinctive configuration and continuity (seismic reflection units) can be calibrated with available well data to provide insight to the nature of the strata that fill the basin. Even where seismic reflections cannot be tied to well data, the character of the reflection unit and its spatial association with other groups of reflections may permit uncommonly accurate prediction of lithofacies composition and, hence, facies patterns within the basin. This prediction capability, permitted by use of seismic data in stratigraphic interpretation, can lead to logical interpretations of petroleum potential long before well control provides sufficient data for conventional stratigraphic interpretation.

REFERENCES CITED

Broussard, M. L., ed., 1975, Delta models for exploration: Houston Geol. Soc., 555 p.

Brown, L. F., Jr., 1969, Geometry and distribution of fluvial and deltaic sandstones (Pennsylvanian and Permian), North-Central Texas: Gulf Coast Assoc. Geol. Socs. Trans., v. 19, p. 23-47. Reprinted as Texas Univ. Bur. Econ. Geology Geol. Circ. 69-4.

———, A. W. Cleaves, II, and A. W. Erxleben, 1973, Pennsylvanian depositional systems in North-Central Texas, a guide for interpreting terrigenous clastic facies in a cratonic basin: Texas Univ. Bur. Econ. Geology Guidebook 14, 132 p.

Carlson, P. R., and C. H. Nelson, 1969, Sediments and sedimentary structures of the Astoria submarine canyon-fan systems, northeast Pacific: Jour. Sed. Petrology, v. 39, p. 1236-1282.

Dietz, R. S., 1963, Collapsing continental rises: an actualistic concept of geosynclines and mountain building: Jour. Geology, v. 71, p. 314-333.

Emery, K. O., 1970, Continental margins of the world, *in* F. M. Delaney, ed., The geology of the East Atlantic continental margin; Pt. 1, general and economic papers: Great Britain, Inst. Geol. Sci. Rept. 70/13, p. 3-29.

Erxleben, A. W., 1975, Depositional systems in Canyon Group (Pennsylvanian System), North-Central Texas: Texas Univ. Bur. Econ. Geology Rept. Inv. 82, 76 p.

Fisher, W. L., 1969, Facies characterization of Gulf Coast basin delta systems, with Holocene analogues: Gulf Coast Assoc. Geol. Socs. Trans., v. 19, p. 239-261.

——, and J. H. McGowen, 1967, Depositional systems in the Wilcox Group of Texas and their relationship to occurrence of oil and gas: Gulf Coast Assoc. Geol. Socs. Trans., v. 17, p. 105-125.

—— et al, 1969, Delta systems in the exploration for oil and gas: Texas Univ. Bur. Econ. Geology Spec. Pub., 212 p.

—— et al, 1970, Depositional systems in the Jackson Group of Texas: Gulf Coast Assoc. Geol. Socs. Trans., v. 20, p. 234-261.

——, and L. F. Brown, Jr., 1972, Clastic depositional systems—a genetic approach to facies analysis; annotated outline and bibliography: Texas Univ. Bur. Econ. Geology Spec. Rept., 230 p.

Galloway, W. E., and L. F. Brown, Jr., 1972, Depositional systems and shelf-slope relationships in Upper Pennsylvanian rocks, North-Central Texas: Texas Univ. Bur. Econ. Geology Rept. Inv. 75, 62 p.

Guevara, E. H., and R. Garcia, 1972, Depositional systems and oil-gas reservoirs in the Queen City Formation of Texas: Gulf Coast Assoc. Geol. Socs. Trans., v. 22, p. 1-22. Reprinted as Texas Univ. Bur. Econ. Geology Geol. Cir. 72-4.

Klemme, H. Douglas, 1971, To find a giant, find the right basin: Oil and Gas Jour., v. 69, no. 10, p. 103-110.

LeBlanc, R. J., 1972, Geometry of sandstone reservoir bodies, in Underground waste management and environmental implications: AAPG Mem. 18, p. 133-190.

Lehner, P., 1969, Salt tectonics and Pleistocene stratigraphy on continental slope of northern Gulf of Mexico: AAPG Bull., v. 53, p. 2431-2497.

McGowen, J. H., 1970, Gum Hollow fan delta, Nueces Bay, Texas: Texas Univ. Bur. Econ. Geology Rept. Inv. 69, 91 p.

McMaster, R. L., J. De Boer, and A. Ashraf, 1970, Magnetic and seismic reflection studies on continental shelf off Portuguese Guinea, Guinea, and Sierra Leone, West Africa: AAPG Bull., v. 54, p. 158-167.

Middleton, G. V., and A. H. Bouma, eds., 1973, Turbidites and deep water sedimentation: SEPM Pacific Sect., Short Course, Anaheim, 157 p.

Normark, W. R., 1970, Growth patterns of deep-sea fans: AAPG Bull., v. 54, p. 2170-2195.

Paine, W. R., 1968, Stratigraphy and sedimentation of subsurface Hackberry wedge and associated beds of southwestern Louisiana: AAPG Bull., v. 52, p. 322-342.

Reineck, H. E., and I. B. Singh, 1973, Depositional sedimentary environments: New York, Springer-Verlag, 439 p.

Sangree, J. B., et al, 1976, Recognition of continental-slope seismic facies, offshore Texas-Louisiana, in A. H. Bouma, G. T. Moore, and J. M. Coleman, eds., Beyond the shelf break: AAPG Marine Geology Comm. Short Course, v. 2, p. F1-F54.

——, and J. M. Widmier, 1975, Interpretation of depositional facies from seismic data (abs.): Geophysics, v. 40, p. 142.

Shelton, J. W., 1973, Models of sand and sandstone deposits—A methodology for determining sand genesis and trend: Oklahoma Geol. Survey Bull. 118, 122 p.

Shepard, F. P., R. F. Dill, and U. Von Rad, 1969, Physiography and sedimentary processes of LaJolla submarine fan and fan-valley, California: AAPG Bull., v. 53, p. 390-420.

Turner, G. L., 1957, Paleozoic stratigraphy of the Fort Worth basin: Abilene and Fort Worth Geol. Socs. Joint Guidebook, p. 57-77.

Uchupi, E., and K. O. Emery, 1967, Structure of continental margins off Atlantic coast of United States: AAPG Bull., v. 51, p. 223-234.

——, 1968, Structure of the continental margin off Gulf Coast of the United States: AAPG Bull., v. 52, p. 1162-1193.

Vail, P. R., and J. B. Sangree, 1971, Time stratigraphy from seismic data (abs.): AAPG Bull., v. 55, p. 367-368.

——, R. M. Mitchum, Jr., and S. Thompson, III, 1974, Eustatic cycles based on sequences with coastal onlap (abs.): Geol. Soc. America Abs. with Programs, v. 6, no. 7, p. 993.

Weber, K. J., 1971, Sedimentological aspects of oil fields in the Niger Delta: Geol. en Mijinbouw, v. 50, p. 559-576.

Wilson, J. L., 1975, Carbonate facies in geologic history: New York, Springer-Verlag, 471 p.

Seismic Facies and Sedimentology of Terrigenous Pleistocene Deposits in Northwest and Central Gulf of Mexico[1]

CHARLES J. STUART[2]and CHARLES A. CAUGHEY[3]

Abstract Clastic sedimentary facies in the northern and central Gulf of Mexico were delineated using borehole and seismic data. The present continental shelf is underlain by deltaic deposits which appear as intervals of discontinuous, strong reflectors on seismic profiles. Fluvial, prodelta, and transgressive marine deposits are characterized by weak reflectivity. Near the present shelf edge, rapid progradation of mud and sand formed wedges of inclined layers (inclined-reflector seismic facies).

The continental slope of the northwest Gulf of Mexico is strongly affected by salt tectonics. Thick sedimentary sequences preserved between massifs, stocks, and spines of salt are characterized by interbedded, continuous, strong reflectors and irregular chaotic units. Turbidity-current and hemipelagic settling mechanisms initially formed these deposits, followed by downslope mass movements and bed disruption. The structure and thickness of sedimentary layers on the slope vary according to sedimentation rates, upward diapiric movement, and erosion. At least one deep erosional valley, the Mississippi Trough, formed on the slope during a low stand of sea level. The trough probably formed by slumping, bottom-current, and turbidity-current mechanisms, and is thought to have channeled sediment from the shelf to the Mississippi fan.

Slope deposits grade downslope into the continental-rise deposits. The Mississippi fan, the major rise feature in the Gulf of Mexico, consists of widespread chaotic layers interbedded with strong-reflector intervals. Some of the interbedded strong reflectors are folded or are terminated by slump faults, suggesting that rapidly deposited, low-strength mud was disrupted repeatedly by gravity processes.

Continental-rise deposits grade into abyssal-plain sequences in the deepest part of the gulf. Undeformed deposits represented by intervals of continuous, strong reflectors underlying the abyssal plain onlap or interfinger with the adjacent slope and rise deposits. The undeformed layering was not affected by salt diapirism or faults; these deposits were formed by turbidity-current and pelagic processes.

This study of Pleistocene deposits suggests that seismic profiles can provide data of stratigraphic and sedimentologic importance. Similar interpretive techniques may also provide geologically useful data in other basins.

INTRODUCTION

The stratigraphic and sedimentologic framework of the northwest Gulf of Mexico continental shelf is well defined because of the data available from many oil and gas exploration tests. However, near the shelf edge and in slope and at greater water depths, wells and boreholes are sparse, and the geologic framework is not as well known. The deeper water part of the northwest Gulf of Mexico has been surveyed extensively by the petroleum industry, universities, and government agencies; thus a wealth of geophysical and near-surface geological data (Figs. 1, 2) is available. As a part of a regional study of the Pleistocene Series, we utilized these data to extend Quaternary stratigraphic datums across the continental slope into abyssal-depth parts of the gulf, and to interpret Pleistocene sedimentology. Seismic facies—zones of different seismic character—were delineated on profiles and interpreted geologically on the basis of facies geometry, waveform pattern, and borehole data.

INTERPRETATION METHODS AND CONTROL

This study consisted of several parts: (1) identification and mapping of Quaternary stratigraphic units in the deep-water gulf, (2) sedimentologic interpretation of these intervals, and (3) interpretation of Pleistocene geologic history. Log and paleontologic data from exploration boreholes on the continental shelf of the northern gulf provided geologic control for identifying Pleistocene datums. The base-of-Pleistocene horizon corresponds to the top of the Terrebonne shale, a widespread subsurface marker on the northwestern shelf (Caughey et al, 1976). A mid-Pleistocene datum was picked also, but it was correlated only in the shelf and upper-slope regions. Borehole data (Fig. 1) from the Deep Sea Drilling Project (DSDP) provided points of control for identifying the base-of-Pleistocene horizon in the abyssal gulf. Holocene sediments are thin to absent in most parts of the gulf and, in this study, were not differentiated from Pleistocene deposits. The selected datums were identified on seismic profiles and carried over the network illustrated in Figure 2.

Methods of unit correlation and interpretation using seismic profiles are discussed by Vail et al in this volume. Seismic sequences are defined by widespread correlatable reflections or reflection

[1]Manuscript received, October 4, 1976; accepted, March 17, 1977.

[2]Exploration Research Division, Continental Oil Company, Ponca City, Oklahoma 74601.

[3]Continental Oil Company, Lafayette, Louisiana 70501.
Permission to publish this paper was granted by Continental Oil Company. Robert Brenner, Arthur Whipple, and Edward Parma offered suggestions that improved the manuscript.

249

groups and unconformities. This method required modification for this study because of the structural complexity of the continental slope of the northwest Gulf of Mexico and the scarcity of widespread unconformities. Reflections on seismic profiles were followed until they were truncated by a fault or a diapir. Correlations across the disrupting feature were made by comparing profile segments on each side of a diapir or a fault, and then by aligning reflections. In most cases this method worked satisfactorily and it was possible to check correlations by examining crossing lines that were unaffected by diapirism or faulting. In the abyssal gulf, reflectors extend over long distances without break, thus providing good correlation (see Watkins et al, 1976).

Seismic facies were discriminated on the basis of (1) reflector continuity, (2) reflector amplitude, (3) geometric relations between reflectors, and (4) overall reflector form—i.e., flat, folded, faulted, etc. Geologic significance of seismic facies at some localities was established directly by use of borehole data, which included proprietary logs and paleontologic interpretations from the northwest gulf and published information elsewhere. Indirect verification was provided by borehole data from either similar seismic facies in other areas or stratigraphically higher, analogous sequences. Boreholes drilled on the northwest gulf slope by Exxon, Shell, and Superior (data summarized by Lehner, 1969; Caughey and Stuart, 1976; Woodbury et al, 1976; Sangree et al, 1976;

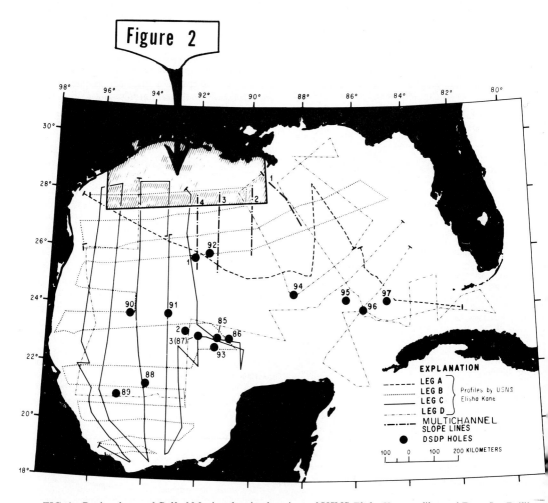

FIG. 1—Regional map of Gulf of Mexico showing locations of USNS *Elisha Kane* profiles and Deep Sea Drilling Project boreholes (modified from Garrison and Martin, 1973). Legs A through D are profiles completed by USNS *Elisha Kane*.

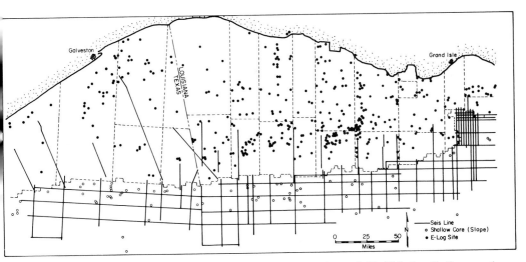

FIG. 2—Map of seismic-profile grid and borehole data located in northwest Gulf of Mexico. Shallow core (open circles) descriptions are summarized by Caughey and Stuart (1976).

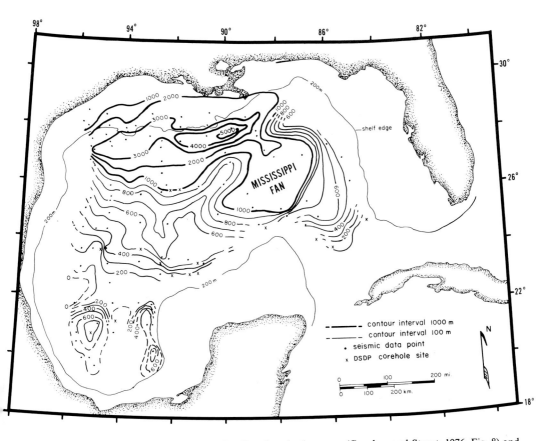

FIG. 3—Pleistocene isopach map, Gulf of Mexico. Based on isotime map (Caughey and Stuart, 1976, Fig. 8) and conversion factor of 2,300 m/sec.

Skolnick, 1976) and DSDP boreholes (Worzel et al, 1973) provided the most useful geologic data. Many of these boreholes, however, do not penetrate the entire Quaternary section; thus control is incomplete.

The seismic data used in this study include a gulf-wide grid of single-channel sparker profiles, four six-fold regional slope lines in the northern gulf, and a grid of twenty-four-fold profiles in the middle and upper slope (Figs. 1, 2). The sparker data were obtained by the USNS *Elisha Kane* in 1969. The multichannel profiles are proprietary lines obtained between 1967 and 1973. Interval-velocity data were calculated from stacking velocities at several localities in the northwest gulf; and the average velocity of the Pleistocene Series was assumed to be about 2.30 km/sec throughout the Gulf of Mexico on the basis of these calculations. This value was used to determine interval thickness and depth from seismic travel time.

The time interval between the sea bottom and the base-of-Pleistocene horizon was determined at selected points over the seismic grid, then converted to depth based on the 2.30 km/sec Pleistocene velocity. The resulting Pleistocene isopach map (Fig. 3) is supplemented with stratigraphic control from DSDP and exploration boreholes. For simplicity, the local effects of diapirs were not considered in mapping thickness trends. Salt diapirs and growth faults, however, both affect sediment thickness significantly and require consideration in detailed mapping.

GEOLOGIC SETTING

Geological and geophysical studies indicate that the Gulf of Mexico is a relatively old ocean basin which has remained tectonically undisturbed, except for regional subsidence, since the late Paleozoic or early Mesozoic (Antoine and Bryant, 1969; Ladd et al, 1976). Ocean crust underlies the gulf and is buried beneath a mass of Cenozoic deposits ranging in thickness from about 5,000 m in the center of the basin to more than 18,000 m along portions of the northwestern basin margin (Antoine and Ewing, 1963, Fig. 4; Antoine and Pyle, 1970, Fig. 1 and p. 485).

Sedimentation caused partial filling of the original gulf basin and molded its present form. Major geomorphic elements of the western gulf (Fig. 4) resulted from the interaction between sedimentary processes and salt tectonics. Shallow-water carbonate rocks predominate in stable platform areas of Florida and Yucatan, forming the eastern and southeastern basin margins. The Florida and Yucatan peninsulas and adjacent continental-shelf areas are underlain by more than 3,000 m of Cretaceous and Paleogene limestone, dolo-

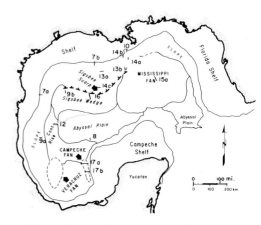

FIG. 4—Geomorphic provinces of Gulf of Mexico. Line segments and numbers locate following seismic-profile figures. Modified from Caughey and Stuart (1976, Fig. 1); published courtesy of Gulf Publishing Co.

mite, and anhydrite (Rainwater, 1971; Wilhelm and Ewing, 1972, p. 588-594). Neogene reefal carbonate deposits and nearshore clastic deposits less than 300 m thick veneer the older rocks. Lower Cretaceous shelf carbonate rocks separate filled epicontinental basins of the coastal plain of Mexico and the United States from the oceanic gulf.

The western Gulf of Mexico is a terrigenous clastic province with surrounding landmasses which serve as prolific sediment sources. Late Cretaceous (Laramide) disturbances initiated a flood of detritus that was funneled basinward through the ancestral Mississippi and Rio Grande drainages and through smaller rivers fringing the northwestern gulf. Shifting deltaic depocenters spilled over the Cretaceous reef tracts and prograded into the basin throughout the Tertiary and Pleistocene, the Florida and Yucatan platforms remained stationary. Vertical accretion of carbonates occurred there while clastic accumulations extended the northwestern shelf progressively basinward.

Pleistocene deposits form the final increment of significant shelf-edge extension in the northwest gulf. There are two offlapping sediment lenses preserved in the Pleistocene section (Caughey, 1975), each consisting of a thin interval of sandy deposits along the inner shelf and expanding basinward into a thick sequence of mud-rich deposits underlying the outer shelf. Relations among these two Pleistocene units, the subjacent Pliocene deposits, and major salt structures are illustrated on Figure 5, a schematic cross section ex-

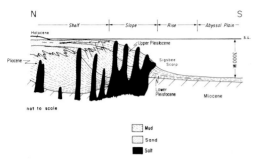

FIG. 5—Schematic profile of offlapping Pliocene-Pleistocene sediment wedges showing sand-mud transitions. Based on borehole and seismic data.

tending from the Louisiana coastline southward to the abyssal Gulf of Mexico. The distribution of sand and the depositional facies of the Pleistocene section are shown by Figure 6.

Broad areas of coastal Texas and Louisiana were exposed to erosion during low stands of sea level associated with Pleistocene glaciation. Fluvial channels funneled detritus across the emergent inner continental shelf to a series of delta lobes located in the present outer-shelf area (Figs. 5, 6). Thick sequences of paralic deposits accumulated rapidly during deltaic progradation, causing basinward accretion of the shelf-platform.

Growth-fault systems developed along the outer margin of the platform in response to continued loading on a mobile substrate of uncompacted mud and salt (Caughey, 1975). The deltaic overburden displaced salt into hummocky swells on the adjacent continental slope. Basins between the salt swells trapped pelagic sediments, and mass-movement debris slumped from deltaic deposits.

Filling of the upper-slope basins caused additional salt displacement into nearby salt massifs. These rising structures shed detritus back into the slope basins. Salt-massif development culminated basinward in the Sigsbee scarp, a salt wall.

Rising sea levels accompanying the onset of interglacial periods caused the abandonment of outer-shelf delta lobes. During high-level stillstands, sedimentation was concentrated along the outer part of the present coastal plain. Stoping of shelf-edge deposits and resedimentation in continental-slope basins probably continued at a slower rate.

SEISMIC FACIES

Classification

Pleistocene seismic facies in the Gulf of Mexico are characterized broadly by parallel strong reflector units, weak or reflectorless intervals, and

irregular chaotic intervals. These reflector types are further distinguished by lateral variations in reflector strength and continuity, geometric relations between overlying and underlying layers, and gross form (i.e., flat, irregular, channelized, folded, or faulted). The facies classification is basically descriptive in that it utilizes reflector variations, and partly genetic in that it groups facies on the basis of their inferred origins as primary sedimentation facies and secondarily modified facies (e.g., slump, slide, fold, fault types). For the most part, the secondary modifying influences are readily shown by features on seismic profiles. The descriptive part of the classification is similar to that of Sangree et al (1976). Table 1 summarizes the characteristics of Pleistocene seismic facies, their distribution, and their origin.

Reflector strength is interpreted qualitatively by its visual impact. Strong reflectors form heavy lines or bands on a seismic profile indicating strongly contrasting acoustic impedances between layers. Weak reflectors do not stand out and indicate low acoustic-impedance contrasts. Reflector strength (amplitude) is also a measurable parameter, and it can be analyzed quantitatively. Harms and Tackenberg (1972), Hilterman (1975), Rice and Waters (1976), and Sheriff (1976) have discussed analytical methods and the geologic significance of detailed waveform analysis.

Primary Sedimentation Units

Strong-Reflector Facies

Parallel-bedded reflectors—Strong-reflector facies occur throughout the Gulf of Mexico from delta-marine to abyssal-plain environments. Reflectors in shelf-platform sediments are mostly parallel-bedded and vary laterally in intensity and continuity (Fig. 7), in contrast to the more uniform and continuous reflectors underlying the slope and abyssal plain (Figs. 8, 9).

Core descriptions from boreholes in the upper slope (e.g., see summary in Caughey and Stuart, 1976) show that upper Pleistocene sediments consist of clay, silt, and scattered sand. The sand and silt apparently were derived from shelf environments via turbidity-current mechanisms; the clay mud is hemipelagic. Exploration boreholes on the shelf (see Caughey, 1975) indicate that shelf sediments consist of intercalated sand, silt, and clay of deltaic origin. The sands, especially, are discontinuous, although numerous.

Therefore, the visual contrast in seismic response between shelf and slope deposits is probably caused by discontinuous and complex bedding of intercalated sand, silt, and clay layers in

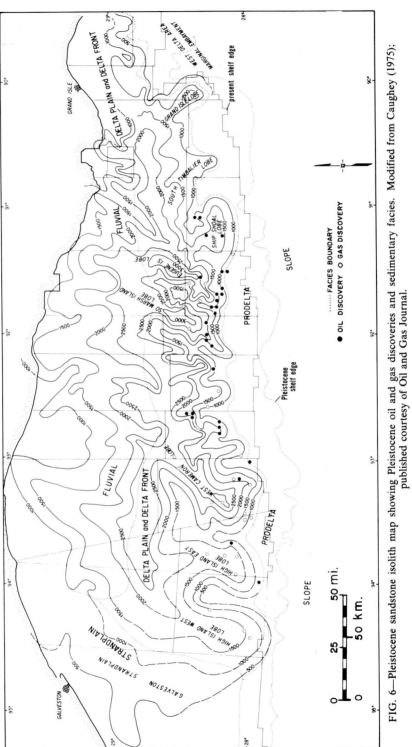

FIG. 6—Pleistocene sandstone isolith map showing Pleistocene oil and gas discoveries and sedimentary facies. Modified from Caughey (1975); published courtesy of Oil and Gas Journal.

Seismic Facies Classification

Seismic Character	Regional Distribution	Special Character	Primary Transport and Depositional Mechanisms	Secondary Modifying Mechanisms
Strong-Reflector Intervals				
A. Primary Features				
1. Concordant (parallel bedding to gradual convergence)	shelf	discontinuous, variable amplitude	traction currents: distributary channel, tidal channel, bottom currents; wave agitation	gravity-fault dislocation; small-scale mass movements; plastic flowage
	slope-rise (fan)	continuous reflectors, locally truncated by salt structures; strong secondary modification; reflectors may converge toward salt diapirs	turbidity current, pelagic settling, traction-current reworking	salt intrusion; creep, slump, slide, debris flow, plastic flowage
	abyssal plain	undeformed continuous reflectors; reflectors onlap plain margins	as above	none
2. Discordant				
a. Inclined stratification	shelf edge	merges updip into flat-lying shelf layers, and downdip into slightly inclined upper-slope deposits	prograded marine shelf wedge; prodelta	creep, slump, slide
	slope	encased by undeformed continuous reflectors	turbidity current, traction current; submarine-fan progradation	none
b. Drape (angular discordance)	slope, rise	drapes older topographic-structural features; localized over diapirs	pelagic settling or onlap by turbidites	may be deformed if deformation of underlying strata continues
B. Secondary Features				
1. Folded reflectors (creep or incipient slump)	shelf edge, slope	no visible discrete glide (fault) planes	various mechanisms	downslope creep of plastic sediment, commonly clay-rich mud
2. Faulted reflectors (slump)	shelf edge, slope	occurrence of curved glide planes	various mechanisms	slumping or rotational block faulting of coherent strata
Weak or Reflectorless Intervals				
A. Primary Features (parallel bedding)	shelf, slope, rise, abyssal plain	discontinuous, weak internal reflectors; flat boundaries; locally with irregular basal boundary (channel); may be weakly chaotic in channel fill	continuous deposition of sand or mud without sharp changes in seismic impedance, by traction-current or pelagic mechanisms	gas formation after deposition, e.g., gassy mud; gravity-fault dislocation; mass-movement mechanisms; plastic flow
B. Secondary Features (chaotic bedding)	slope basins, continental rise (fan)	irregular to lenticular chaotic layer; sharp basal boundary, hummocky upper surface; diffractions abundant, commonly define upper boundary; diversely oriented internal reflectors; commonly composite with laterally continuous strong reflectors separating chaotic zones; locally associated with upward salt movements and shelf-edge oversteepening.	various mechanisms	slide origin: mobilization of bedded sediment by gravity processes; possibly a continuous sequence of creep-slump-slide-debris flow - turbidity-current mechanisms.

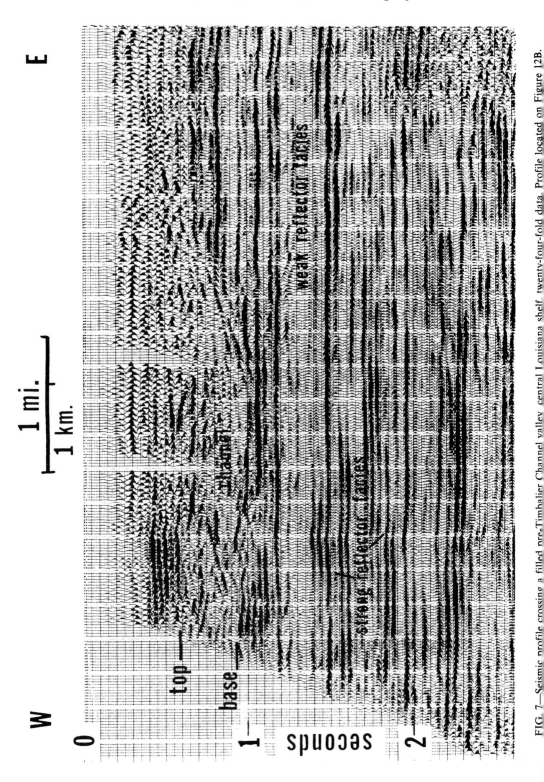

FIG. 7—Seismic profile crossing a filled pre-Timbalier Channel valley, central Louisiana shelf, twenty-four-fold data. Profile located on Figure 12B.

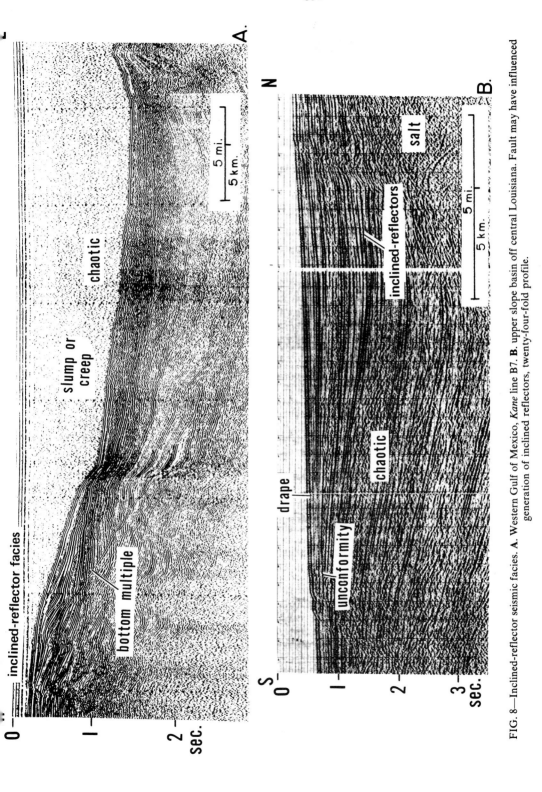

FIG. 8—Inclined-reflector seismic facies. **A.** Western Gulf of Mexico, *Kane* line B7. **B.** upper slope basin off central Louisiana. Fault may have influenced generation of inclined reflectors, twenty-four-fold profile.

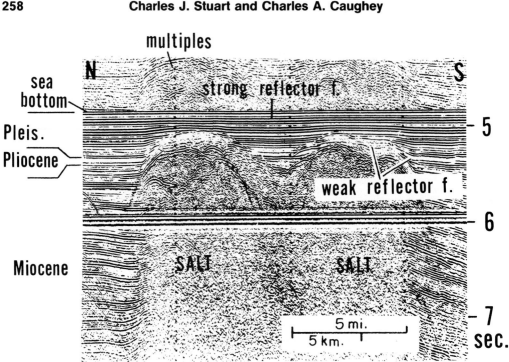

FIG. 9—Seismic profile crossing Sigsbee Knolls. Strong band at 6.0 sec is drum splice and not acoustic reflector. Multiples at top also reflect mechanical recording procedure and should be ignored. Correlations are based on nearby DSDP boreholes. *Kane* line D6.

shelf sequences, and laterally continuous layers of mud and scarce fine sands in slope environments.

Inclined-reflector patterns—Layering of strong-reflector units in some areas is oriented at an angle to normal bedding planes and is referred to as "sigmoid-terrace" and "oblique-progradational seismic facies" by Sangree et al (1976). Both types are characterized by an "S" shape, but the updip end of the oblique progradational facies is terminated by a flat reflector. In the present study, these differences were not noted, and the term "inclined-reflector" was used to identify this facies. Inclined-reflector facies occur in Pleistocene sediments near the shelf edge and in the upper slope off western Louisiana, Texas, and northeastern Mexico. Good examples of this seismic character occur off South Texas (Fig. 8A) and are probably of late Wisconsin age (Lehner, 1969, Fig. 39). Shelf-edge inclined-reflector intervals merge updip into nearly horizontal layers, and merge downdip tangentially into beds that dip gently basinward. The distal parts are also affected to a variable extent by salt diapirs and gravity-induced mass movements. Shelf-edge inclined reflectors dip a maximum of 2 to 4° and occur within a zone about 8 km wide. They consist mostly of mud, but contain thin intercalations of fine sand (Lehner, 1969, p. 2469-2470; Woodbury et al, 1976, p. C-13).

Inclined reflectors also occur at the margins of some slope basins. For example, Figure 8B shows an offlapping wedge south of a diapir-associated fault. The distal part of the interval onlaps the downslope side of the basin. True dips of these layers are on the order of 2 to 3°.

The significance of inclined-reflector layering appears to be that it is evidence of rapid sedimentation at a relatively sharp change in bottom gradient. The lateral extent of inclined-reflector zones varies with the nature of the break in slope—e.g., a regional hinge line or the rim of a local slope basin. Inclined-reflector layering alone does not indicate shelf-edge sedimentation, even though it may be best developed in that setting.

Drape unconformity—Angular discordance of layers overlying diapiric structures, chaotic seismic units, and other topographic irregularities are common in the northwest continental slope and rise portions of the gulf. In these occurrences a zone of small diffractions commonly separates a parallel-bedded sequence above from a deformed or tilted unit below. The simplest drape relation is where the overlying layers are continuous and uniformly thick over irregular topography (e.g., see Sangree et al, 1976, Fig. 7). In other cases, basin-fill sequences (where layers onlap the flanks of a basin) may underlie the more continuous drape layers.

Drape results from the widespread accumulation of hemipelagic or pelagic sediments over preexisting topography. Hemipelagic sediments consist of terrigenous particles that settle through the water column from near-surface turbid layers like those off river mouths. Other pelagic layers (oozes, red clays) consist primarily of plankton tests and atmospheric dust. Pelagic deposits generally accumulate slowly, although more rapid deposition may occur below plumes of turbid water off delta distributaries.

Weak-Reflector Facies

Seismic facies characterized by weak internal reflectors (or by no reflectors) and by parallel, flat boundaries, occur in some shelf-platform, upper-slope, and abyssal-plain sequences. Weak-reflector facies are thickest in shelf-platform sequences (Fig. 7) and thinnest, but most continuous, in abyssal-plain environments (Fig. 9). They are associated with, or compose the major part of, shelf-edge inclined-reflector facies (Fig. 8A), drape layers and interdiapir slope-basin fill (Fig. 8B), the Mexican Ridge province (Fig. 10A), and Mississippi Trough fill (Fig. 11).

Weak-reflector facies on the shelf-platform are not sharply separated from strong-reflector sequences (Fig. 7); the distribution is patchy with areas of lateral interfingering. This contrasts with the distinct boundary between weak-reflector facies and overlying strong-reflector facies in the abyssal gulf (Fig. 9). The differences reflect the lateral continuity of layers and the sharpness of lithologic or compaction changes in deep-water environments.

Boreholes (e.g., Fig. 12) suggest that at least some weak-reflector intervals on the shelf-platform are massive shales devoid of significant thicknesses of sandstone or other acoustically different layers. Wells in Figure 12 are close to the seismic profile in Figure 7, and the filled erosional channel noted in each figure is the same. Based on this correlative interval, it appears that the channel is cut into a sandstone-shale sequence (strong-reflector facies). Outside the channel, the sand-rich interval is overlain by a shaly sequence that correlates with a weak-reflector interval. The sandstone also overlies a thick shale layer below the log segments shown on Figure 12, and corre-

sponds to a weak-reflector interval on the profile. The sandstones and shales are Pleistocene deltaic sediments, so the overall seismic character reflects the abrupt lithologic changes characteristic of the Mississippi system.

Weak-reflector seismic facies may also result from massive sandstone deposits such as those in fluvial sequences. Internal impedance contrasts (e.g., those from sandstone-shale layering in other sedimentary facies) are weak; thus strong reflectors are not present.

Weak-reflector facies in the abyssal plain were cored at several DSDP sites. At Site 3 (Fig. 1), for example, pelagic, plankton-rich sediments of Pliocene age were recovered. The Pleistocene sequence overlying these pelagic sediments consists mostly of turbidites and is a strong-reflector facies. Thus, the overall seismic character illustrates changes in sedimentation rate and type.

Secondary Modification of Sediments

Seismic facies resulting from postdepositional movements of sediment include a spectrum of types and geometric relations. These features are discussed below in terms of their interpreted origins and in order of increasing effect on the original sediment; their relations to seismic characteristics are summarized in Table 1.

Creep

Some strong-reflector facies of the upper slope are folded to varying degrees (Fig. 8A). Downslope gliding along bedding planes or nondiscernible cross-cutting glide planes likely produced the folding. Some creep occurrences may represent incipient slump of low shear-strength sediments. These sediments deform plastically rather than remaining as a coherent slump block. Engineering studies of shelf sediments (e.g., Roberts et al, 1976; Watkins and Kraft, 1976) have shown that creep and other forms of plastic deformation are strain effects primarily of gravity-induced stress. Gravitational stress may be supplemented by stress originating from storm waves, hydraulic (pore fluid) gradients, and salt diapirism.

Slump

Blocks of sediment are rotated along curved glide planes (Figs. 10B) in continental-slope and rise-fan areas. These structures differ from creep deformation in that discernible glide planes are formed and sediments usually are not folded except at slump "toes." Growth faults are similar to slumps in seismic sections and may represent slumps of enormous scale contemporaneous with rapid deltaic deposition. In some areas, closely spaced slump blocks with small vertical offsets

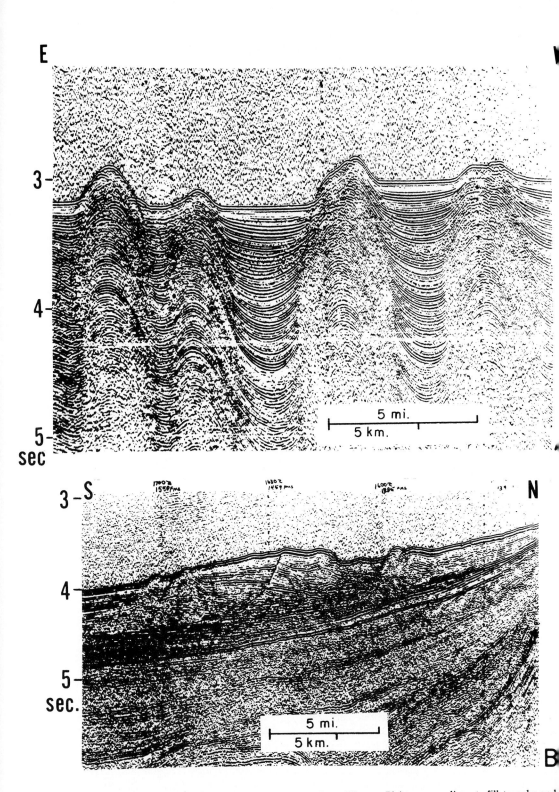

FIG. 10—**A.** seismic profile crossing Mexican Ridge province. Pliocene-Pleistocene sediments fill troughs and onlap ridges. Folded sediments are Miocene in age. No-record zones may result from steep dip or sediment disruption. Folding is example of very large-scale creep. *Kane* line B13. **B.** slumped sediments on lower slope off South Texas. *Kane* line D1.

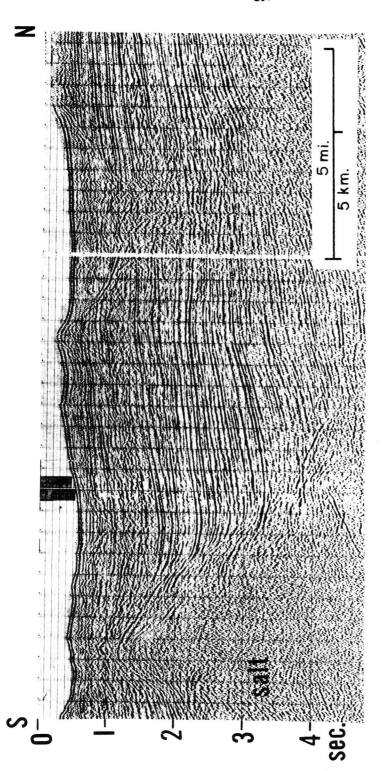

FIG. 11—Seismic profile crossing Mississippi Trough. Smaller, mud-filled scour channels occur in central part of trough, twenty-four-fold profile.

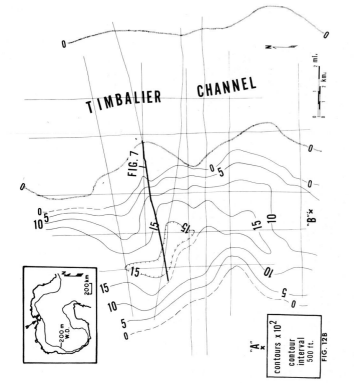

FIG. 12—**A.** electric logs in area of pre-Timbalier Channel valley located on **B;** log "A" is outside channel. **B.** Isopach map (in feet) of sediment fill in pre-Timbalier Channel valley shown in Figure 7 based on seismic-profile data. Time-depth conversion of 2,010 m/sec was used in constructing map. Boundaries of Timbalier Channel are also shown. "A" and "B" locate boreholes shown in **A.**

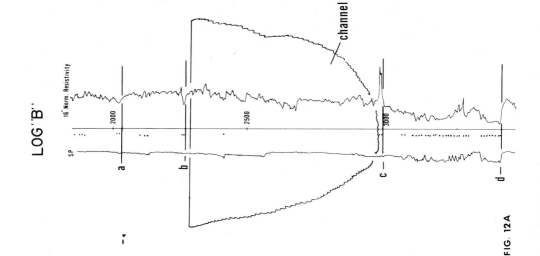

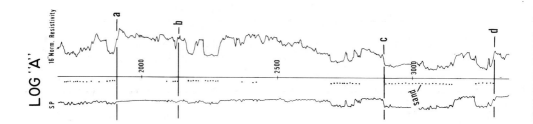

produce wave or megaripple patterns on seismic profiles (Fig. 13). These may be difficult to differentiate from large primary sedimentary structures or from secondary folding and creep.

Slide

Slide deposits occur abundantly on slopes underlain by muddy terrigenous sediments (Fig. 14; also Fig. 8) and as talus piles at the base of the Campeche carbonate slope. These units form irregular zones characterized by broken chaotic reflectors or else no coherent reflectors. The upper boundary of these units is hummocky, and the lower contact commonly is flat and sharp. Small diffractions are concentrated along the hummocky upper surface but are common throughout the unit. Chaotic units form irregular layers interbedded with strong-reflector units in interdiapir basins on the slope (Figs. 8, 14-17) and extensive units in rise-fan sequences (especially those of the Mississippi fan; Fig. 18). Chaotic layers in the Sigsbee wedge (Fig. 19) are widespread and, in contrast to most slope examples, have flat, smooth boundaries. Chaotic seismic units have been penetrated by DSDP and petroleum industry coreholes in the Sigsbee wedge (DSDP Site 1), and in the upper part of the northwestern continental slope (e.g., see Woodbury et al, 1976, p. C6-C13; Sangree et al, 1976, p. F41-F52). They commonly consist of sand in upper-slope occurrences but probably are mud in downslope areas.

Lenticular, chaotic (slide) deposits in many northwestern slope basins probably formed as a result of upward salt movements and shelf-edge slumping. As sediments were elevated, or in other ways reached the point where gravitational stress exceeded their shear strength, downslope movement by a combination of slump and slide mechanisms occurred. The hummocky character of these deposits apparently resulted from rapid deposition of mobilized debris. Extensive chaotic deposits in the middle part of the Mississippi fan (Fig. 18B; Stuart and Caughey, 1976) may have resulted from sediment oversteepening caused solely by high sedimentation rates on axial parts of the fan, since salt structures are not present in this part of the fan. The more uniform boundaries of chaotic units making up the Sigsbee wedge, and distal portions of Mississippi fan units, may be related to water entrainment. A more fluid debris slide (or debris flow), being less viscous, probably would remain in motion longer than more viscous slides, and would form a widespread layer with smooth boundaries. Irregular chaotic units, therefore, probably formed from less fluid masses of coherent sediment blocks and mud.

Structural Modifications and Erosion

Structural and erosional events affect sediment continuity and seismic response in addition to the gravity-driven mass-movement mechanisms already discussed. Diapirs, diapir-related tension faults, growth faults, and wide erosional channels affect sediments in many parts of the gulf. Salt diapirism is the single most effective cause of

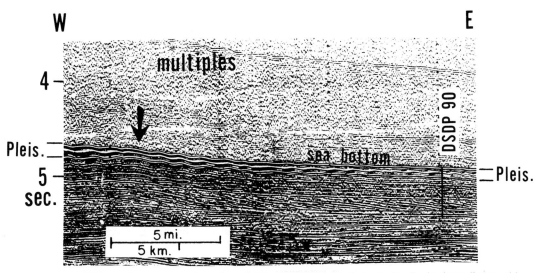

FIG. 13—Pseudomegaripple pattern (arrow) caused by closely spaced slump faults. Fault planes flatten with depth and penetrate Pleistocene to Miocene strata (correlated with DSDP 90). These features suggest slump fault rather than sedimentary megaripple origin. Multiples at top result from recording method. *Kane* profile B11.

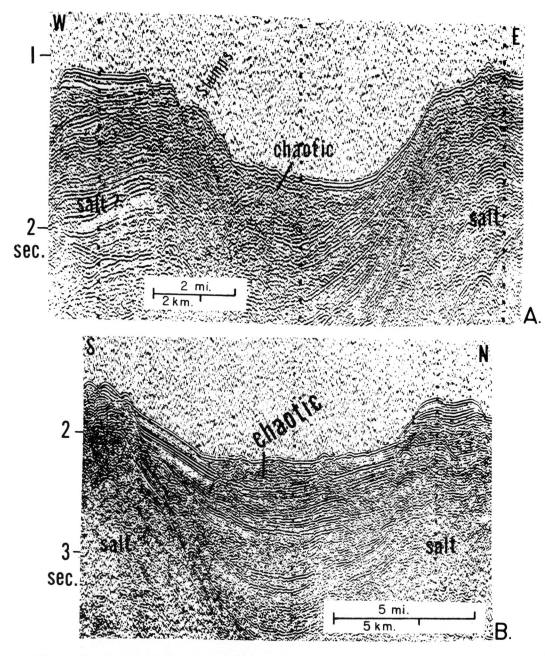

FIG. 14—Chaotic seismic facies, western Louisiana slope. **A.** chaotic facies caused by deformed sediments (slump and slide deposits) associated with diapiric uplift to left. *Kane* line B3. **B.** chaotic facies in center of interdiapir slope basin; probably was transported from a point outside plane of profile. Geologic control points not located on profile segments shown. *Kane* line B2.

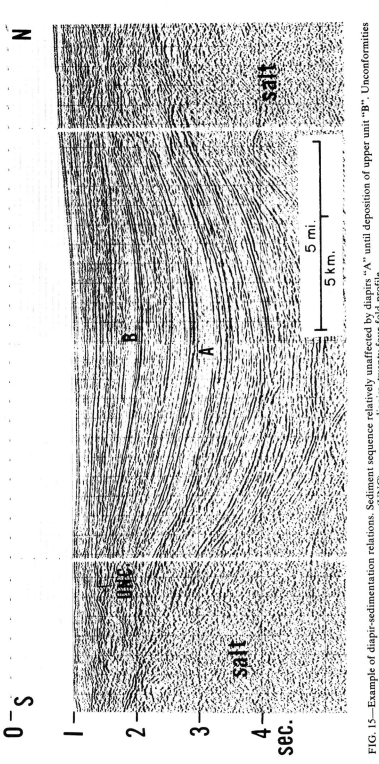

FIG. 15—Example of diapir-sedimentation relations. Sediment sequence relatively unaffected by diapirs "A" until deposition of upper unit "B". Unconformities (UNC) occur over diapirs, twenty-four-fold profile.

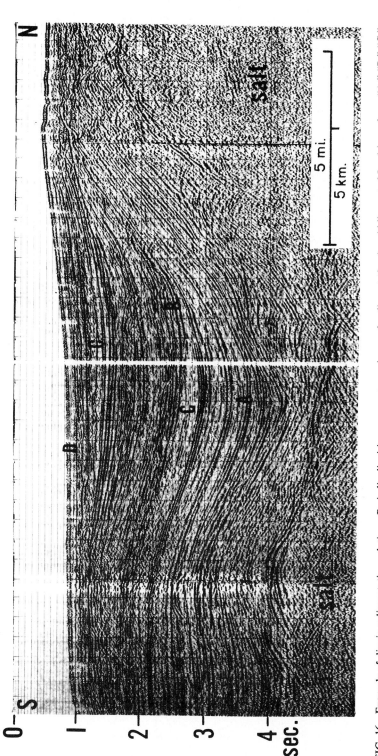
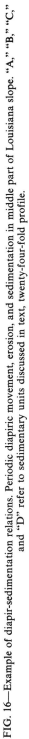

FIG. 16—Example of diapir-sedimentation relations. Periodic diapiric movement, erosion, and sedimentation in middle part of Louisiana slope. "A," "B," "C," and "D" refer to sedimentary units discussed in text, twenty-four-fold profile.

postdepositional deformation in the northwest gulf—bending, faulting, dislocating, probably compacting overlying sediments, and generating new sedimentary units (slumps and slides). Continuous reflectors commonly turn upward close to diapirs or are sharply truncated by them. Some sediment intervals thin toward diapirs (convergent layering) because of onlap and erosion related to diapir growth during sedimentation, or because of mud compaction. Compaction is feasible only if excess pore water can be drained from the sediment. Faults or thin, relatively permeable layers (silts and sands) may provide drainage paths.

Where sediments have been elevated and then eroded by bottom currents, waves, and mass-movement mechanisms, unconformities occur. On seismic profiles, unconformities are recognized by reflector discordance and diffractions. Diffracted energy at the unconformity is caused by surface roughness, or by edges of layers in the underlying sequence.

Some reflecting horizons are offset by well-defined tension faults above salt diapirs; other reflective intervals thicken noticeably where large growth faults are crossed. Most growth faults, however, are limited to the outer shelf and uppermost slope.

Deeply incised channels occur on the shelf-platform and on the continental slope. The Timbalier Channel, which crosses the South Timbali-er and Grand Isle platted areas of offshore Louisiana, and the contiguous Mississippi Trough south of it, are examples of erosional channels. The major effects of channeling are to truncate previously deposited beds and to confine the distribution of channel-fill deposits. The channel-fill deposits illustrated in Figure 7 are of weak-reflector to chaotic type, and they contrast sharply with the more continuous layering of the prechannel deposits. Secondary processes such as in-situ biogenic gas formation, faulting, slumping, and sliding may affect channel-fill sediments differently than the host sediments, further distorting the appearance of channel sediments on seismic profiles.

DEPOSITIONAL SYSTEMS

Seismic profiles are two-dimensional displays that can provide almost continuous data between points of geologic control. By utilizing the seismic characterization of lithologic facies, the interrelations and character of facies changes can be studied, and the geologic history of a stratigraphic sequence can be inferred with a minimum amount of geologic control. Economic implications are important and varied. For example, the lateral changes of known oil and gas reservoirs can be studied, providing better data for development drilling. In exploration, potential reservoir and source-rock facies may be delineated prior to wildcat drilling; this could significantly improve

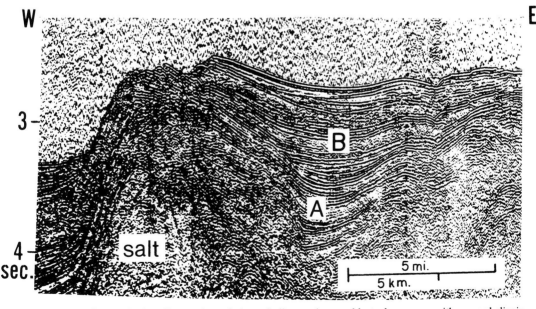

FIG. 17—Examples of diapir-sedimentation relations. Sedimentation unable to keep pace with upward diapir growth. "A" and "B" refer to sedimentary units discussed in text. *Kane* line B5.

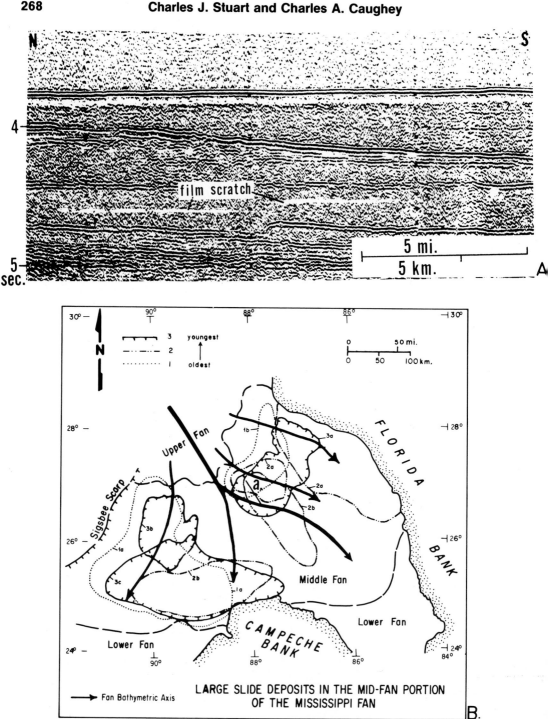

FIG. 18—Chaotic seismic units in mid-Mississippi fan. **A.** seismic profile shows stacked chaotic units and inter-bedded strong-reflector facies (inferred to be turbidites). Profile located in center of eastern group of chaotic units shown in part B. *Kane* line D22. **B.** distribution of composite chaotic units in mid-Mississippi fan. From Stuart and Caughey (1976); published courtesy of Gulf Coast Association of Geological Societies.

chances for discovery with minimum drilling. In addition, studies emphasizing the stratigraphic and sedimentologic history of widespread facies tracts can be made.

Shelf-Platform Fluvial-Deltaic Systems

Pleistocene sediments underlying the continental shelf consist mainly of fluvial deposits in the inner and middle shelf areas and a suite of deltaic and delta-margin deposits along the outer part of the shelf (Figs. 5, 6). Deep channels were cut into fluvial-deltaic sediments during low stands of sea level (Fig. 12) when deltas occupied the shelf edge.

Fluvial-deltaic sediments are characterized by irregular seismic response because of the lack of uniformity in depositional facies (migrating delta lobes produce discontinuous and overlapping facies sequences). Discontinuous, variable-amplitude strong-reflector facies identify delta-plain and delta-front sedimentary facies (Fig. 7), which consist of interbedded sandstones and shales. Sandstone lenticularity and small-scale channeling may result in small diffractions from layer edges, giving delta-plain facies a more erratic appearance than delta-front facies. Massive sandstones and shales of fluvial and prodelta facies, respectively, both appear as weak-reflector units because of the lack of reflective interfaces. Ma-

rine shale sequences, however, may be distinguished from fluvial sequences by their relatively continuous, moderate-strength internal reflectors. The relative position of each facies on a facies tract may also be used to distinguish them.

The change from shelf to slope environments is marked by the change of strong reflectors from discontinuous on the shelf-platform, to continuous in the upper slope, and by inclined-reflector seismic facies at the shelf edge. The upper-slope boundary shown in Figure 6 is based on this seismic change.

Deep erosional valleys formed on the shelf-platform during low stands of sea level. The Timbalier Channel (Fig. 12B) was probably cut during the Wisconsin glacial period and filled during the subsequent rise in sea level. The older, parallel channel presumably formed and filled in a similar way.

The older channel shown in Figures 7 and 12 is about 8 km wide and is filled with up to 450 m of clay and silt, in contrast to typical sand-filled distributary channels in deltaic environments. Numerous diffractions or postdepositional deformation of the fill gives it a chaotic appearance. The thickness of fill (channel depth) was mapped from seismic profiles for the older channel using a time-to-depth conversion of 2,010 m/sec. Figure 7 illustrates that there was no structural influence

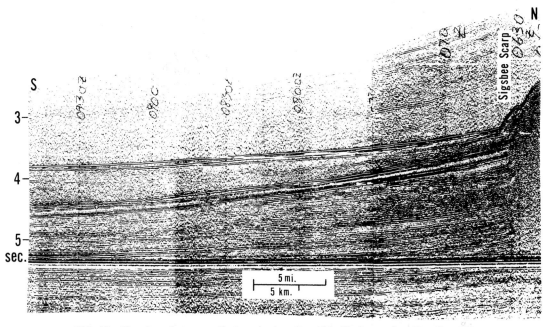

FIG. 19—Chaotic and strong-reflector seismic units within Sigsbee wedge. *Kane* line C22.

on valley depth and suggests a totally erosional origin. The depth of about 450 m is far greater than what would be expected from a eustatic fall in sea level, which should be on the order of 150 to 180 m.

Another possibility is that channel cutting during a high stand of sea level formed a submarine valley or canyon. We do not know of paleontologic data which might provide independent evidence of water depth. The younger Timbalier Channel (Fig. 12B) is also about 8 km wide and is mud-filled; the fill is unconsolidated and probably gas-saturated, making it difficult to obtain seismic data from below the channel. Time delay caused by the low-velocity fill results in depression of subchannel reflections (shown in Fig. 7).

Continental Slope

The continental slope in the northern Gulf of Mexico is dominated by salt ridges, massifs, stocks, and spines. Sediments deposited on the slope were intruded and uplifted by these salt structures and were subsequently eroded and remobilized (slumps and slides). In addition, large-scale erosional features, notably the Mississippi Trough, locally affect the continuity and type of sedimentation.

Relative activities of upward salt movement, rate of sedimentation, bottom-current erosion, and mass-flow or turbidity-current generation all affect the type of sedimentation, thickness, and geometry of Pleistocene deposits on the continental slope. Figures 8 and 15 through 17 show examples of slope sediments that were affected in various ways by mass movements (slumps and chaotic seismic facies), turbidite and hemipelagic sedimentation, and salt movements.

The shelf-edge/upper-slope profile (Fig. 8A) is only peripherally affected by diapirs, but oversteepening caused by rapid progradation of the shelf edge resulted in sediment instability—small slumps followed by sliding. Sediment deformation apparently became more pervasive downslope.

A seismic profile midway down the Louisiana slope (Fig. 15) illustrates a slope basin that was unaffected by diapir movement until late stages of sedimentation. The fact that layers in the lower interval (A) are of relatively uniform thickness suggests that diapirs were inactive. An unconformity at the top of the lower sequence indicates that later sediments were not deposited fast enough to keep pace with upward diapir movement; uplift and erosion ensued. A topographic basin formed and was filled during a later stage of sedimentation, as indicated by onlapping re-

flections (B). Tension faults associated with the diapirs offset the seafloor and indicate that another unconformity may be forming.

Alternating strong-reflector facies and weak-reflector to chaotic facies characterize this part of the slope. Strong-reflector facies probably consist of turbidites, and weak-reflector facies of pelagic to hemipelagic muds. The near-chaotic character of some intervals suggests downslope sliding or slumping.

Figure 16 illustrates an upper-slope basin that has undergone periodic diapir growth in an area of relatively rapid slope sedimentation. Layering is uniform in the lower interval (A), indicating that the diapir had not yet become active. Reflector convergence near the middle of the sediment column (B) indicates onlap of a topographic high, erosion, or variable compaction caused by upward diapiric movement. Basin-fill sequences (C) and convergent strong-reflector sequences followed by a draping layer (D) near the present sea bottom indicate the periodic occurrence of basin formation and filling, sedimentation in balance with diapir movement, and erosion. The basin axis appears to have moved northward with time, suggesting northward movement of the salt mass. The diapir is still active as indicated by shoaling above it and by down-to-the-north tension faults that offset the seafloor.

The profile in Figure 17 shows the effects of a strong imbalance between sedimentation and upward diapiric movement. Diapiric activity did not greatly influence sedimentation of most strong reflector facies (A; turbidites and hemipelagic deposits), but became increasingly important during deposition of the upper third of the sequence (B). Chaotic facies (slide deposits) and tilted reflectors indicate more intense diapiric activity. Sedimentation slowed after deposition of the uppermost part of the sequence but diapiric movement continued. The result is high topographic relief across the diapir and structurally dominated topography of the clastic sequence.

Mississippi Trough

The Mississippi Trough is located in the upper slope west of the Mississippi delta. The trough is the slope continuation of the Timbalier Channel (Fig. 12), which crosses the central Louisiana shelf. The trough deeply notches the shelf edge and continues downslope to a water depth of about 2,100 m (Woodbury et al, 1976, Fig. 19), merging with the Mississippi fan system. The erosional nature of the upper trough is documented by irregular topography and small erosional channels (Fig. 11). Its location may have been influenced by damming effects of the shelf-edge/upper-slope salt structures west of it. The trough is partially filled with clay, silt, and scarce sand, and the muds may (in part) be gas-saturated (see Woodbury et al, 1976; Stuart and Caughey, 1976, Table 1). Trough-fill sediments are weak-reflector and chaotic seismic facies. Chaotic response may be caused by noise associated with inhomogeneous gas saturation or by lenticular silts and sands. The trough is cut into upper Pleistocene parallel-bedded, strong-reflector sequences. Pleistocene sediments on the profile are probably turbidites and hemipelagic deposits, but may be deltaic toward the north margin of the trough.

The Timbalier Channel and Mississippi Trough formed as the ancestral Mississippi River became entrenched during the Wisconsin low stand of sea level. A narrow delta-marine shelf zone located near the present shelf edge accumulated large volumes of sediment. Slumping, bottom-current erosion, and turbidity-current erosion were probably the mechanisms responsible for submarine-trough development. During and after its formation, the trough funneled large volumes of mud to the upper Mississippi fan, where it was remobilized and redeposited on lower parts of the fan. The trough was partially filled during the post-Wisconsin rise in sea level. The Timbalier Channel and Mississippi Trough are records only of the Wisconsin sea-level lowering and subsequent sea-level rise. However, the great thickness of Mississippi fan sediments and occurrence of pre–Timbalier Channel valleys on the shelf (Fig. 12; also Woodbury et al, 1976, Fig. 19) suggest that similar events occurred during other periods of lowered sea level. Evidence of these other events on the slope has been masked by intensive disruption of slope deposits by salt structures.

Continental Rise-Fan Systems

The continental rise is a convex-upward sediment prism that occurs as a transition zone between the slope and abyssal plain. The rise is well developed in the east-central part of the gulf where high rates of sedimentation associated with the Mississippi River have occurred, and also at the base of the western and southwestern slopes of the gulf basin. Continental rises form as coalesced submarine fans at the base of the slope, or they can form as single fan complexes. The Mississippi fan is the largest rise-fan deposit in the gulf; it extends southward and westward for a distance of about 600 km before merging with abyssal plains. The Sigsbee wedge, Campeche "fan," and Veracruz "fan" are other smaller, but significant rise-fan deposits of Pliocene-Pleistocene age.

Mississippi Fan

The Mississippi fan is a regional topographic and stratigraphic feature in the east-central part of the Gulf of Mexico (Fig. 4). The fan is shown by sediment thickening on the Pleistocene isopach map (Fig. 3), and grades downdip into abyssal-plain environments to the southwest and west. The West Florida and Campeche carbonate banks abut distal parts of the fan.

Mississippi fan deposits of Pleistocene age are as thick as 4,600 m in the uppermost slope (a complex of shelf, slope, and fan systems), and they thin to about 230 m where the lower fan merges with the Sigsbee plain. The middle part of the fan is 600 to 1,200 m thick. Areas of thickest sediment accumulation in the middle and lower parts of the fan correspond to its present bathymetric axes; thus these axes represent the average depositional loci of the fan. This thick wedge locally downwarps the basin floor, but mostly fills a preexisting basin.

The upper fan is poorly defined because of similar structures and sediment types in the adjacent slope. It is partly bounded on the west by the Sigsbee scarp, and grades eastward into relatively coarse, terrigenous slope sediments derived from rivers flowing into the northeast gulf. The Mississippi Trough is present in the upper-fan area. Remobilized deltaic sediments apparently crossed the continental slope through the submarine trough and were deposited on the fan. However, the fan itself may have begun forming in Pliocene time. The chaotic seismic facies in both the Pleistocene and upper Pliocene deposits underlying the fan are evidence of high sedimentation rates.

The sea bottom becomes smoother and of lower gradient in the middle fan. Slide deposits (chaotic seismic facies), slightly contorted turbidite and pelagic layers (parallel-bedded, strong-reflector facies; Fig 18), and discontinuous channels or open slump scars characterize this part of the fan. Composite slide sequences extend over large areas in mid-fan zones flanking the primary depositional loci. Deposits of this type occur throughout the Pleistocene and upper Pliocene parts of the fan, and at least two slide deposits are exposed on the sea bottom (Walker and Massingill, 1970). The relative ages and distribution of slide deposits in three local Pleistocene stratigraphic subunits (Fig. 18B) suggest that slide activity occurred sporadically in response to gravitational instability caused by rapid deposition. Strong, continuous reflectors (probably turbidites) within chaotic se-

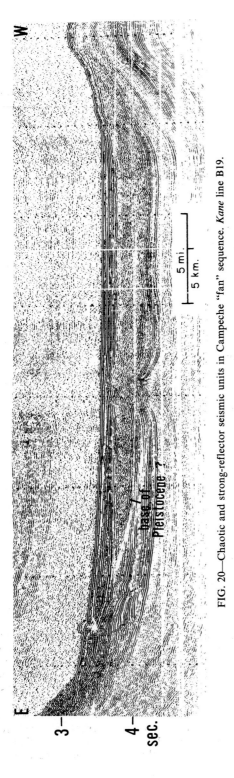

FIG. 20—Chaotic and strong-reflector seismic units in Campeche "fan" sequence. *Kane* line B19.

quences (Fig. 18A) and local contortions of overlying strong reflectors suggest that sliding and slumping occurred periodically, even within previously deposited slides, alternating with quiescent periods of widespread turbidite and pelagic deposition.

The lower fan is smooth and of very low gradient; it grades downslope into the Florida plain and Sigsbee plain (Fig. 4). Reflector continuity, lack of postdepositional sediment movement, and low dip angles characterize this part of the fan. Seismic character and core-hole descriptions suggest that Pleistocene sediments in this area are primarily turbidity-current and pelagic silts and clays.

Sigsbee Wedge

The Sigsbee wedge is a continental-rise segment that extends southward from the Sigsbee scarp (Figs. 4, 19). It grades eastward and southward into the lower unit of the Mississippi fan and Sigsbee plain, respectively, and westward into sediments derived from the Rio Grande system. A shallow bathymetric trough separates the Sigsbee wedge from the Mississippi fan to the east, although Wilhelm and Ewing (1972) suggested that the two features intertongue. The Sigsbee wedge appears as a regional thinning trend on the Pleistocene isopach map (Fig. 3) and is gradational with the slope to the north, although it is now terminated updip by the Sigsbee scarp.

The Sigsbee wedge consists primarily of strong-reflector facies and widespread chaotic facies that offlap basinward (Fig. 19). DSDP hole 1 (Fig. 1) encountered deformed bedding at the top of a chaotic seismic interval. Overlying deposits consist of turbidite silt and clay of strong-reflector character. Upward movement of salt may have caused resedimentation of turbidites or, alternatively, the slide deposits may be related to depositional oversteepening caused by rapid sedimentation.

Campeche "Fan" and Veracruz "Fan"

Sediments similar to Mississippi fan and Sigsbee wedge deposits occur in the southwest Gulf of Mexico, but are much less extensive. Both the Campeche "fan" (Figs. 20, 21) and Veracruz "fan" complexes are localized within submarine troughs that deepen basinward to the north. Their lateral boundaries are poorly defined because of the effects of salt intrusion and downslope sediment deformation. The troughs are thought to be filled with Pleistocene(?) sediments, which are characterized by alternating strong-reflector and chaotic intervals. Chaotic (slide) intervals pinch out to the east and west (Fig. 20), suggesting northward movement within the confining troughs. Enclosing strong-reflector intervals presumably are turbidite and pelagic deposits.

Abyssal Plain

Pleistocene sediments thin markedly in the flat parts of the Sigsbee plain and the Florida plain (Figs. 3, 4). Thickness variations are not uniform, however, because of lobate extensions of the Mississippi fan and Sigsbee wedge.

The seismic character of Pleistocene sediments (Sigsbee Unit; Watkins et al, 1976) in most parts of the abyssal plain is typically of parallel-bedded, strong-reflector type. The underlying Pliocene interval (Cinco de Mayo Unit) is a parallel-bedded, weak-reflector facies. Somewhat chaotic weak-reflector facies overlie Sigsbee plain salt structures (Fig. 9).

Most Pleistocene sediments are graded silts and clays interpreted as turbidites; the Pliocene interval consists of pelagic muds (e.g., DSDP Site 3). Sediments cored at the top of one salt structure (DSDP Site 2) consist of late Miocene to Pleistocene pelagic deposits, in contrast to the abundance of turbidites surrounding the diapirs (Burk

et al, 1969). Thus the diapirs have been positive bathymetric features since late Miocene time.

Sedimentologic studies of abyssal-plain deposits (e.g., Davies, 1972; Devine et al, 1973; Beall et al, 1973) indicate that the major source of Pleistocene and Holocene sediments in the abyssal Gulf of Mexico was the Mississippi River, via the Mississippi delta, slope, and submarine-fan distribution systems. Sediments were transported primarily from east to west via the Mississippi fan. Periodic influx of Mississippi sediment from the north crossed the Sigsbee wedge and added to abyssal-plain deposits. Other circum-gulf sources and pelagic settling provided additional terrigenous and carbonate sediments.

SUMMARY OF PLEISTOCENE DEPOSITION

Pleistocene delta systems are generally restricted to the present continental-shelf area of the northwest Gulf of Mexico (Figs. 5, 6). Prodelta mud lobes grade vertically and laterally into thick sequences of slope mud interpreted as turbidites and hemipelagic deposits. Continuous layering of sediments in intervals within continental-slope basins results in laterally continuous seismic reflectors and contrasts with discontinuous reflec-

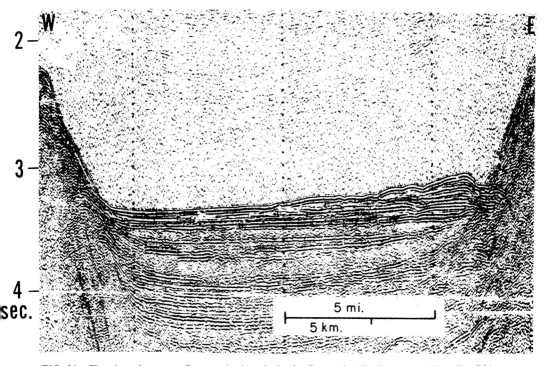

FIG. 21—Chaotic and strong-reflector seismic units in the Campeche "fan" sequence. *Kane* line B21.

tors characteristic of deltaic sequences. Extensive salt structures deformed the slope sediments, in part causing slumping and accumulation of irregular to lenticular, reflector-poor chaotic masses in basins between diapirs. Most diapirism occurred during deposition of upper Pleistocene sediments and continues today. Some sediment layers are faulted and folded. Slope sediments grade downdip into rise-fan systems at the base of the slope. The Mississippi fan and Sigsbee wedge are the most significant rise-fan deposits. They consist of alternating continuous reflector units and reflector-poor chaotic units. Turbidity currents, hemipelagic settling, slumping, and sliding apparently were the predominant transport mechanisms. Rise-fan systems grade downdip into flat-lying abyssal-plain deposits. Laterally continuous reflectors and thin reflectorless layers characterize this region; turbidity currents and pelagic settling are thought to be the dominant sedimentary processes.

CONCLUSIONS

The study of Pleistocene deposits of the northwest and central Gulf of Mexico has suggested the following conclusions, which also have significant application to exploration problems in other areas.

1. Seismic facies are zones of distinctive seismic character that can be identified from single-channel and stacked multichannel seismic profiles.

2. Seismic waveform patterns result from differences in lithology, layer thickness and continuity, pore-fluid type (gas, oil, or water), internal geometric features, and overall morphology. These geologic features result from various primary and secondary (diagenetic and deformational) sedimentary processes.

3. Geologic control at key locations, as well as the use of inferred facies-tract sequences based on depositional models, is necessary to resolve ambiguous interpretations of sedimentary facies versus seismic facies.

4. Seismic profiles provide almost continuous two-dimensional data that can help resolve local and regional problems in correlation and sedimentary-facies distribution and character.

REFERENCES CITED

Antoine, J. and W. R. Bryant, 1969, Distribution of salt and salt structures in Gulf of Mexico: AAPG Bull., v. 53, p. 2543-2550.

———— and M. Ewing, 1963, Seismic refraction measurements on the margins of the Gulf of Mexico: Jour. Geophys. Research, v. 68, p. 1975-1984.

———— and T. E. Pyle, 1970, Crustal studies in the Gulf of Mexico: Tectonophysics, v. 10, p. 477-495.

Beall, A. O., Jr., et al, 1973, Sedimentology in Worzel, J. L., et al, Initial reports of DSDP: Washington, D.C., U.S. Govt. Printing Office, v. 10, p. 699-729.

Burk, C. A., et al, 1969, Deep-sea drilling into the Challenger knoll, central Gulf of Mexico: AAPG Bull., v. 53, p. 1338-1347.

Caughey, C. A., 1975, Pleistocene depositional trends host valuable Gulf oil reserves: Oil and Gas Jour., v. 73, no. 36, (Sept. 8, 1975), p. 90-94 (part I), and v. 73, no. 37, (Sept. 15, 1975), p. 240-242 (part II).

———— and C. J. Stuart, 1976, Where the potential is in the deep Gulf of Mexico: World Oil, v. 183, no. 1, p. 67-72.

———— and C. A. Ames, Jr., 1976, Pleistocene stratigraphy and depositional history, northwest Gulf of Mexico: Geol. Soc. America Abs. with Programs, v. 8, no. 1, p. 11-12.

Davies, D. K., 1972, Deep sea sediments and their sedimentation, Gulf of Mexico: AAPG Bull., v. 56, p. 2212-2239.

Devine, S. B., R. E. Ferrell, Jr., and G. K. Billings, 1973, Mineral distribution patterns, deep Gulf of Mexico: AAPG Bull., v. 57, p. 28-41.

Garrison, L. E., and R. G. Martin, Jr., 1973, Geologic structures in the Gulf of Mexico basin: U.S. Geol. Survey Prof. Paper 773, 85 p.

Hampton, M. A., 1972, The role of subaqueous debris flow in generating turbidity currents: Jour. Sed. Petrology, v. 42, p. 775-793.

Harms, J. C., and P. Tackenberg, 1972, Seismic signatures of sedimentation models: Geophysics, v. 37, p. 45-58.

Hilterman, F. J., 1975, Amplitudes of seismic waves—a quick look: Geophysics, v. 40, p. 745-762.

Ladd, J. W., et al, 1976, Deep seismic reflection results from the Gulf of Mexico: Geology, v. 4, p. 365-368.

Lehner, P., 1969, Salt tectonics and Pleistocene stratigraphy on continental slope of northern Gulf of Mexico: AAPG Bull., v. 53, p. 2431-2479.

Rainwater, E. H., 1971, Possible future petroleum potential of peninsular Florida and adjacent continental shelves, in I. H. Cram, ed., Future petroleum provinces of the United States: AAPG Mem. 15, v. 2, p. 1311-1345.

Rice, G. W., and K. H. Waters, 1976, Some statistical techniques to optimize the search for stratigraphic traps on seismic data: 8th Offshore Technology Conf., Preprints, p. 445-456.

Roberts, H. H., D. W. Cratsley and T. Whelan, III, 1976, Stability of Mississippi delta sediments as evaluated by analysis of structural features in sediment borings: 8th Offshore Technology Conf., Preprints, p. 1-28.

Sangree, J. B., et al, 1976, Recognition of continental-slope seismic facies offshore Texas-Louisiana, in Beyond the shelf break: AAPG Marine Geology Comm. Short Course, v. 2, p. F-1 to F-54.

Sheriff, R. E., 1976, Inferring stratigraphy from seismic data: AAPG Bull., v. 60, p. 528-542.

Skolnick, H., 1976, Late Pleistocene—impact of unique continental climatic condition on sediment accumulation on slope of northern Gulf of Mexico, in Beyond the shelf break: AAPG Marine Geol. Comm. Short Course, v. 2, p. J-1 to J-32.

Stuart, C. J., and C. A. Caughey, 1976, Form and composition of the Mississippi fan: Gulf Coast Assoc. Geol. Socs. Trans., v. 26, p. 333-343.

Walker, J. R., and J. V. Massingill, 1970, Slump features on the Mississippi fan, northeastern Gulf of Mexico: Geol. Soc. America Bull., v. 81, p. 3101-3108.

Watkins, D. J., and L. M. Kraft, Jr., 1976, Stability of continental shelf and slope off Louisiana and Texas; geotechnical aspects, *in* Beyond the shelf break: AAPG Marine Geology Comm. Short Course, v. 2, p. B-1 to B-33.

Watkins, J. W., J. L. Worzel, and J. W. Ladd, 1976, Deep seismic reflection investigation of occurrence of salt in Gulf of Mexico, *in* Beyond the shelf break: AAPG Marine Geology Comm. Short Course, v. 2, p. G-1 to G-34.

Wilhelm, O., and M. Ewing, 1972, Geology and history of the Gulf of Mexico: Geol. Soc. America Bull., v. 83, p. 575-600.

Woodbury, H. O., J. H. Spotts, and W. H. Akers, 1976, Gulf of Mexico continental slope sediments and sedimentation, *in* Beyond the shelf break: AAPG Marine Geology Comm. Short Course v. 2, p. C-1 to C-28.

Worzel, J. L., et al, 1973, Initial reports of the Deep Sea Drilling Project: Washington, D. C., U.S. Govt. Printing Office, v. 10, 748 p.

Stratigraphic and Seismic Evidence for Late Cretaceous Growth Faulting, Denver Basin, Colorado[1]

ROBERT J. WEIMER[2] and T. L. DAVIS[3]

Abstract Interpretation of 250 mi (400 km) of reflection seismic data in conjunction with surface maps and well data along the east flank of the Denver basin, reveals two distinct types of Late Cretaceous faulting. An early Laramide, basement-controlled fault system is the dominant structural style in the zone of flank deformation. Basinward from the fault system is an associated new tectonic style which has now been recognized in the Cretaceous foreland basin. Deltaic sedimentation and overpressured shale masses initiated a shallow-depth growth fault system similar to the tectonic style of many Cenozoic sequences along continental margins.

The shallow growth fault system is approximately 10 mi (16 km) wide and 30 mi (48 km) long, and affects the uppermost Cretaceous strata. Seismic data indicate three or four major trends of listric normal faults that do not appear to extend below a depth of 5,000 ft (1,524 m) in the Pierre Shale. Antithetic horst-graben fault blocks are found on the basinward side of each major fault. Near-surface growth fault movement is indicated by a five-fold thickening of the Fox Hills Sandstone from a normal 75 ft (22.9 m) to 400 ft (121.8 m), and the presence of thicker mineable coal beds in the Laramie Formation in downthrown blocks.

Recognition of growth fault systems will play an important role in future exploration for petroleum and coal in the Rocky Mountain region.

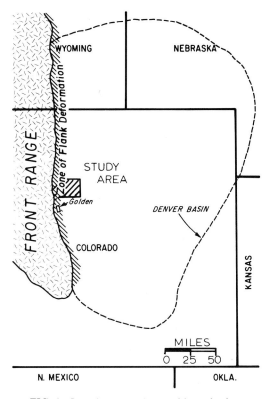

FIG. 1—Location map of area of investigation.

INTRODUCTION

In mineral exploration, geologists and geophysicists search for anomalies (i.e., areas where geologic conditions differ from normal). This paper describes an area on the west flank of the Denver basin, Colorado, just east of the Front Range uplift in parts of Boulder, Weld, and Adams Counties (Figs. 1, 2), that contains several geologic and geophysical anomalies.

Traditionally, all Late Cretaceous faulting in the Rocky Mountain region has been considered as basement-controlled. A fault zone, mapped at the surface, was commonly extended to the basement and related to deformation during the Laramide orogeny after the sediments were deposited. The geologic and geophysical anomalies in *this* study cannot be explained by the traditional concepts. The observational data document a new tectonic style in the Cretaceous foreland basin: shallow growth faulting associated with deltaic sedimentation—its recognition is relevant to exploration for petroleum and coal in the Rocky Mountain region.

The west flank of the Denver basin has a long and sustained history of exploration and production of petroleum and coal. Petroleum production, from both structural and stratigraphic traps, has had pay intervals in the upper Dakota Group and in the Terry and Hygiene Sandstones of the middle Pierre Shale (Fig. 3). The oldest field in the Denver basin, the Boulder oil field (Fig. 2), was discovered in 1902 (Fenneman, 1905) and has produced oil from fractured Pierre Shale (Hygiene Sandstone).

Recent discovery and development of the Wattenberg field (Fig. 2), part of which extends into the eastern part of the study area, has renewed exploration interest in the area. The Wattenberg

[1]Manuscript received, September 24, 1976; accepted, February 11, 1977.

[2]Colorado School of Mines, Golden, Colorado 80401.

[3]Department of Geophysics, University of Calgary, Calgary, Alberta, T2N 1N4, Canada.

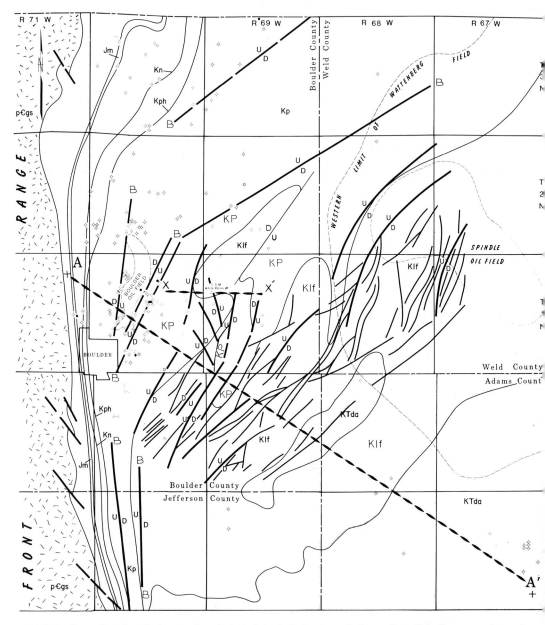

FIG. 2—Tectonic and geologic map of study area, Seismic faults shown by heavy lines; light lines are major surface or subsurface faults found in mines. *B* indicates basement faults. Symbols shown in Figure 3.

field produces gas from the "J" sandstone of the Dakota Group at depths of 8,000 to 9,000 ft (2,438 to 2,742 m). During development of the Wattenberg field, oil production was established from the Terry and Hygiene Sandstones of the Pierre Shale (Fig. 3) in Spindle and nearby fields (Moredock and Williams, 1976); producing depths are 4,500 to 5,000 ft (1,371 to 1,524 m).

Coal production from the Boulder-Weld coalfield began before 1900. Three main coal beds were mined from the lower 125 ft (34 m) of the Laramie Formation, whereas Spencer (1961) reported coal bed thickness ranging from a wedge edge up to 40 ft (12 m).

Exploration and production of energy resources in the study area continues, and the geologic and geophysical concepts presented herein should aid in future programs.

GEOLOGY OF WEST FLANK, DENVER BASIN

Structure

The area of investigation is on the west flank of the Denver basin (Fig. 1). The regional geologic setting of the area can be described in terms of two crustal blocks: the upthrown Front Range block, exposing a Precambrian metamorphic and igneous complex flanking the western part of the area, and the downthrown Denver basin block, an asymmetric structural basin that contains up to 13,000 ft (4,262 m) of sedimentary section ranging in age from Paleozoic to the present. The Upper Cretaceous, up to 9,000 ft (2,743 m) in thickness, dominates the sedimentary section (Fig. 3). In a 3 to 6 mi (5 to 10 m) wide zone between the two blocks, the sedimentary sequence is deformed into an east-dipping monocline cut by high angle basement faults (Figs. 2, 4). Dip of the strata ranges from 15° to overturned. Part of the stratigraphic section may be cut out by faulting in this zone of deformation.

One of the largest structural anomalies in the Denver basin is the horst-graben faulting in T1S, Rs 69-70W, and Ts 1-2N, Rs 67-70W (Fig. 2). Much of the shallow depth information about individual faults is from coal mines within the fault zone. Colton and Lowrie (1973) compiled the mine data in an area they referred to as the Boulder-Weld coalfield. In following this designation, the overall area of faulting is herein referred to as the Boulder-Weld fault zone.

The regional structural dip in the area is 2 to 4° southeast; the structural strike is northeast. This gentle flank of the Denver basin is broken by the fault zone which is 10 mi (16 km) wide, 30 mi (48 km) long, and consists of numerous branching en-echelon faults with near-surface displacements

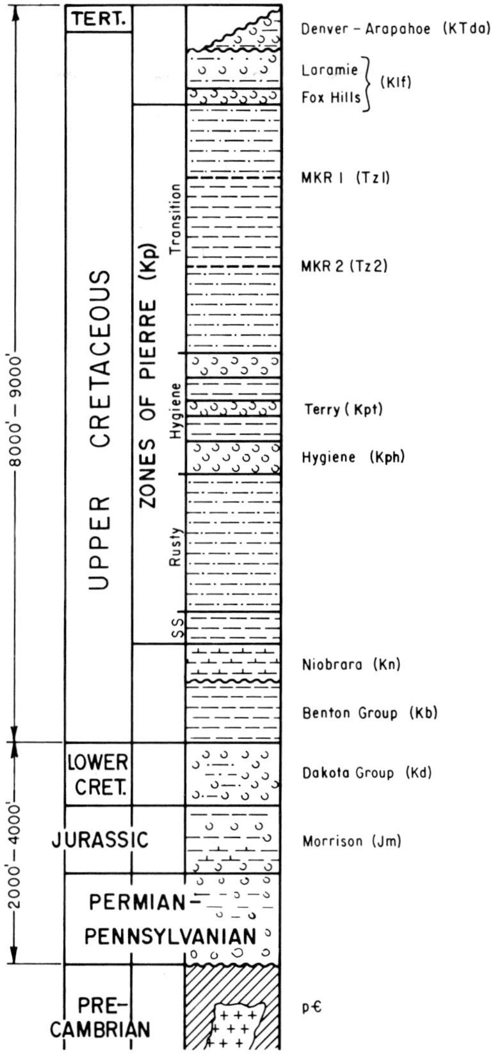

FIG. 3—Generalized diagram showing formations and seismic marker horizons, west flank Denver basin (not to scale).

from a few feet up to 500 ft (152 m). Although the overall fault zone is northeast-trending, individual faults vary from dominant northeast trends to minor north-northwest trends (Fig. 2). The structural style is one of horst-graben fault blocks that range from ¼ to 2 mi (0.4 to 3.2 km) in width and are several miles in length. The fault planes are not well exposed, but where observed they are high angle and either reverse or normal. One of the best descriptions of the faulting is by Spencer

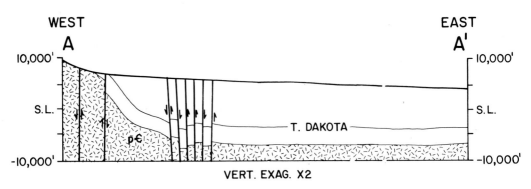

FIG. 4—Geologic cross section showing surface faults in Boulder-Weld fault zone extending to Precambrian (modified after Haun, 1968). Location approximately along A-A' (Fig. 2).

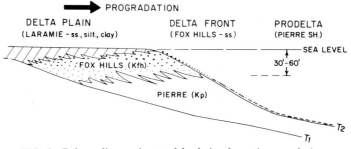

FIG. 5—Delta sedimentation model relating formations to facies to environments of deposition (after Weimer, 1973).

FIG. 6—Aerial view of Fox Hills Sandstone at White Rocks (Sec. 18, T1N, R69W). Arrow points to location of CSM core hole shown on Figure 7. *F* denotes fault; *D* is downthrown block. View to north.

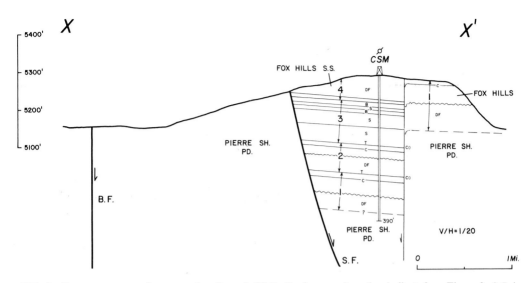

FIG. 7—East-west structural cross section through White Rocks area. Location indicated on Figure 2. *S.F.* is nearsurface expression of a seismic fault. *B.F.* is seismic basement fault. Four sedimentation cycles in Fox Hills Formation are shown by CSM core hole. *DF*, delta front; *C*, channel fill; *CO*, coal; *S*, shoreface; *B*, beach; *PD*, prodelta; *T*, marks transgression.

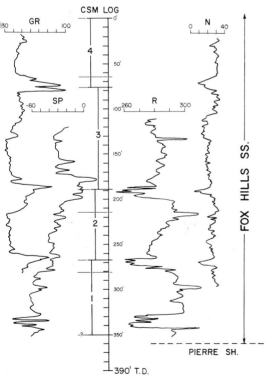

FIG. 8—Mechanical logs from CSM core hole for Fox Hills Sandstone. *GR*, gamma ray; *SP*, spontaneous potential; *R*, resistivity; *N*, neutron (compiled from Bedwell, 1974). Four sedimentation cycles determined from cores are indicated.

(1961) in the Louisville Quadrangle which covers the southwest part of the fault zone.

Strata within the fault blocks are warped into anticlines and synclines. The larger anticlines have been drilled as prospects for petroleum, but none have been productive.

Stratigraphy

The Boulder-Weld fault zone is identified by offsets mainly in the upper Pierre, Fox Hills, and Laramie Formations (Figs. 2, 3). The zone contains three stratigraphic anomalies: (1) unusual thicknesses of the Fox Hills Sandstone; (2) mineable thickness of coal in graben areas compared to thinner coal beds over the horst areas, and (3) high rates of sedimentation in the uppermost Cretaceous sequence.

For the western Denver basin, the depositional model for the Pierre, Fox Hills, and Laramie Formations has been reconstructed by Weimer (1973) as a regressive deltaic sequence as illustrated on Figure 5. The Pierre Shale is a shelf-prodelta deposit of thick shales with lesser amounts of siltstone and minor sandstone occurrences. The Fox Hills Sandstone is fine to medium-grained sandstone interpreted as shallow-marine delta-front or beach and shoreface deposits. The coal bearing Laramie Formation consists of sandstone, siltstone, and clay of delta plain origin. The environments of deposition are largely fresh water although oysters and thin burrowed beds suggest minor incursions of brackish to marine water.

In this study the first stratigraphic anomaly was observed in the White Rocks area (Secs. 7, 18, T1N, R69W). Excellent surface exposures

A

1   2

ROCKY MTN. STD. INC.   WOODWARD OIL INC.
\#1 Gallager   \#1 Gorce
Sec.9, T1N-R70W   Sec. 10, T1N-R70W
1 Mi. NE offset   2.8 Mi. NE offset

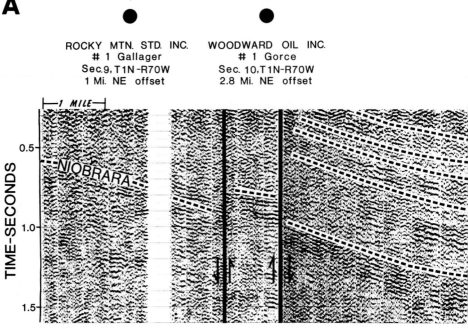

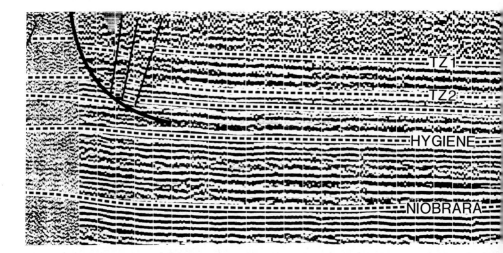

FIG. 9—Northwest-southeast composite regional seismic line. Location of line indicated on Figure 2. Dista

3

DAVIS OIL CO.
1 Iannacito
Sec. 6, T1N-R69W
4.7 Mi. NE offset

4

CONTINENTAL OIL CO.
1 Borra
Sec. 5, T1S-R69W
0.5 Mi. SW offset

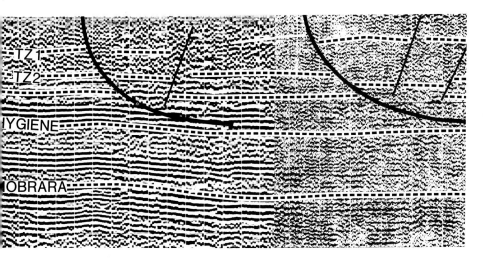

5

T OIL CO.
1 Koch
4, T1S-R68W
. NE offset

6

U.S. CORPS OF ENGINEERS
I Rocky Mtn. Arsenal
Sec. 26, T2S-R67W
0.5 Mi. NE offset

A'

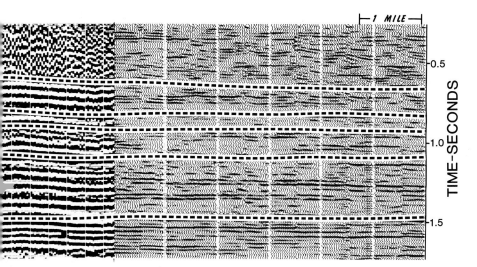

ach segment of seismic line from A-A' is noted beneath drill hole location. Travel time is two-way.

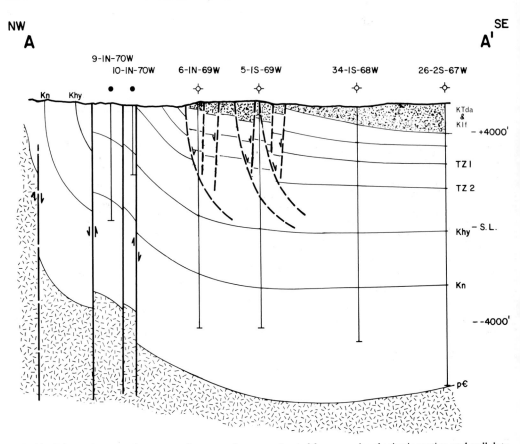

FIG. 10—Schematic composite structural cross section reconstructed from a regional seismic section and well data. Line of section is approximately A-A' on Figure 1. Vertical exaggeration is approximately 5 to 1.

(Fig. 6) indicate two regressive cycles of Fox Hills Sandstone (Weimer, 1973, p. 93-95) with a total thickness of approximately 160 ft (50 m). The section was measured on both sides of a north-trending high-angle normal fault (Fig. 6). A core hole was drilled by the Colorado School of Mines Geophysics Department as a part of a research project in interpreting depositional environments from borehole measurements (Bedwell, 1974). The hole was cored continuously to 343 ft (104.5 m) and then drilled to total depth of 390 ft (118.8 m). Core descriptions were reported by Weimer (1973) and the environmental interpretations from his paper are summarized on Figure 7. Mechanical logs for the core hole are shown on Figure 8. Four regressive cycles of dominantly shallow-water marine sandstone are clearly delineated in the cores and logs. Cycles range in thickness from 80 to 120 ft (24.4 to 36.6 m). According to surface mapping by Trimble (1975),

the top of cycle 4 is thought to be at least 70 ft (21 m) above the ground elevation of the core hole. Whereas, a normal Fox Hills section in the western Denver basin has one or two sedimentation cycles ranging from 60 to 150 ft (18.3 to 45.7 m), the anomalous Fox Hills section at White Rocks contains four stacked cycles with a total thickness estimated to be 430 ft (131 m). Weimer (1973) suggested that local growth faulting could best explain the Fox Hills section at White Rocks.

Elsewhere in the Boulder-Weld fault zone, exceptional thicknesses for the Fox Hills in the Louisville Quadrangle were reported by Spencer (1961) and Rahmanian (1975). From surface measurements and drill hole data, the Fox Hills ranges in thickness from 140 to 400 ft (42.7 to 122 m). In an 8 sq mi (20.7 sq km) area southeast of Boulder (T1S, R70W), Rahmanian (1975) described the Fox Hills as being thicker in major graben areas than in the area of horst blocks.

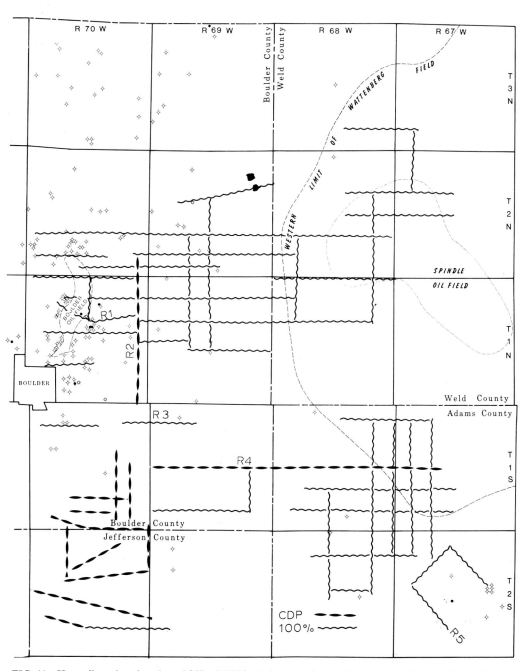

FIG. 11—Heavy lines show location of 250 mi (400 km) of seismic lines utilized in study. R_1, R_2, R_3, and R_4 mark location of composite regional seismic line (Fig. 9).

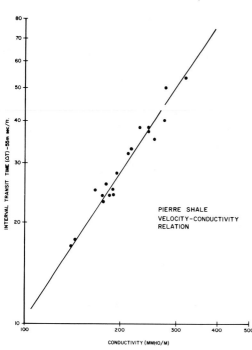

FIG. 12—Velocity-conductivity cross-plot.

A third stratigraphic anomaly relates to high rates of sedimentation for the uppermost Cretaceous. The 7,500 to 8,000 ft (2,286 to 2,438 m) section of Pierre Shale in the Golden-Boulder area has long been recognized as the thickest Pierre section in the Denver basin. Scott and Cobban (1965) described the Pierre Shale in the Front Range area and mapped the distribution of faunal zones. Recently, Obradovich and Cobban (1975, p. 36) have published potassium-argon radiometric dates for bentonite beds associated with faunal zones in the Upper Cretaceous. Together with ammonite correlations in the Pierre Shale, these data suggest that the upper 4,000 ft (1,219 m) of the Pierre Shale (transition zone, Fig. 3) was deposited in a 1 to 2 million year interval from approximately 69 to 67 m.y. Even though these dates may be subject to revision, the rates of sedimentation are high when compared to other

These data are similar to White Rocks and also suggest fault movement in the Boulder-Weld fault zone at the time of Fox Hills deposition.

A second stratigraphic anomaly in the area is thickness variation of coals in the lower 125 ft (38 m) of the Laramie Formation. In the Louisville Quadrangle, Spencer (1961) reported mining of sub-bituminous "B" rank coal from three main coal beds. The map compiled by Colton and Lowrie (1973) indicates that the mines in the Boulder-Weld coalfield are in an area 6 mi (10 km) wide and 25 mi (40 km) long, generally corresponding to the southeast margin of the overall fault zone (Fig. 2). The majority of the old abandoned underground coal mines are clearly located in graben blocks of the fault system. Mineable thickness of coal is generally thought to have been 4 or 5 ft (1.2 to 1.5 m). Although present on the horst blocks, the coal is much thinner and, therefore, was not mined. Like the underlying Fox Hills, thicker coals in the graben fault blocks compared to the horst blocks indicate penecontemporaneous (growth) fault movement. The faults were active during deposition of peat in freshwater swamps. A greater thickness of peat accumulated on the downthrown side of faults compared to the upthrown side.

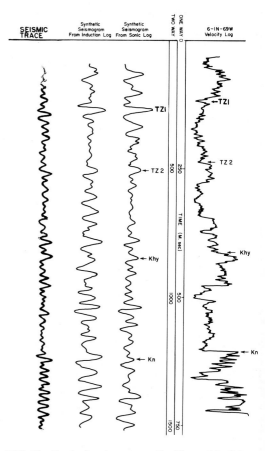

FIG. 13—Synthetic seismograms. See Figure 3 for lithologic correlation.

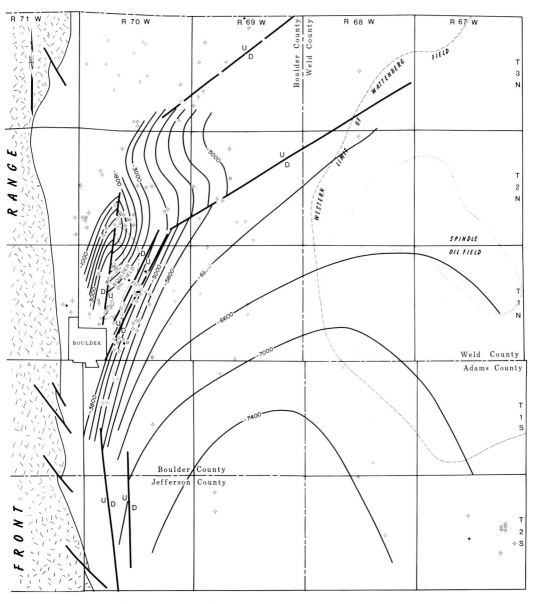

FIG. 14—Top of Precambrian structure contour map from seismic and well data. Contour interval is 400 ft (122 m).

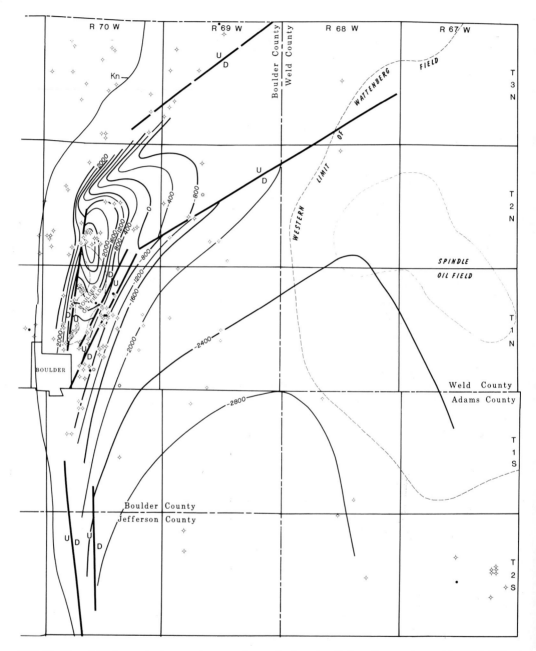

FIG. 15—Top of Niobrara structure contour map from seismic and well data. Contour interval is 400 ft (122 m).

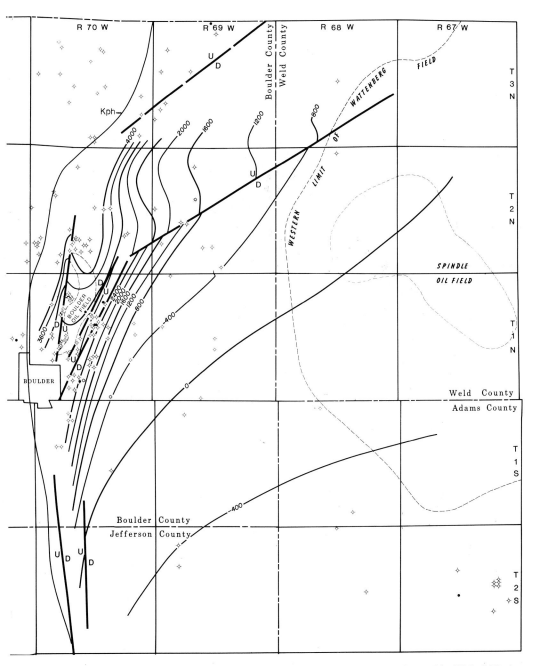

FIG. 16—Top of Hygiene structure contour map from seismic and well data. Contour interval is 400 ft (122 m).

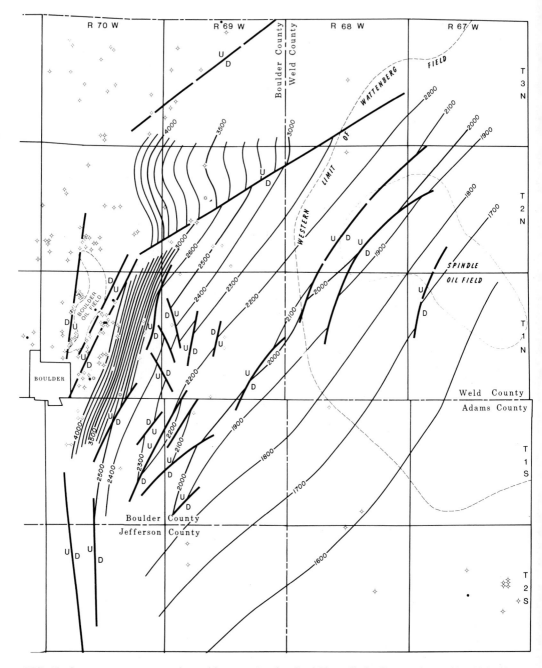

FIG. 17—Structure contour map of transition zone (marker 2) of Pierre Shale. Contour interval is 100 ft (30 m).

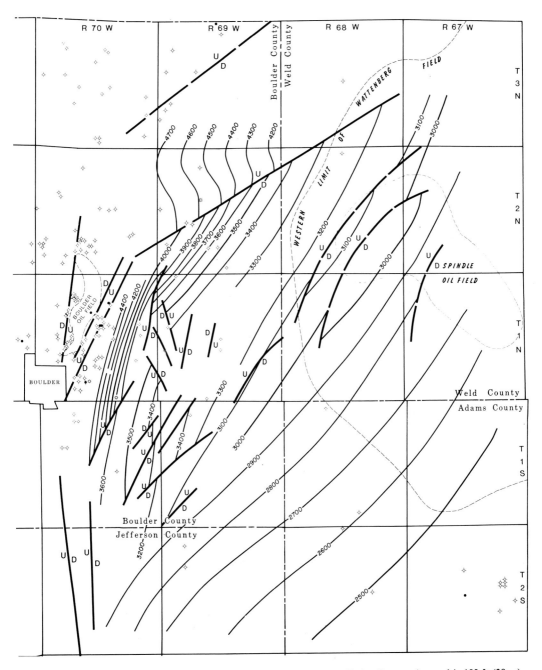

FIG. 18—Structure contour map of transition zone (marker 1) of Pierre Shale. Contour interval is 100 ft (30 m).

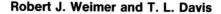

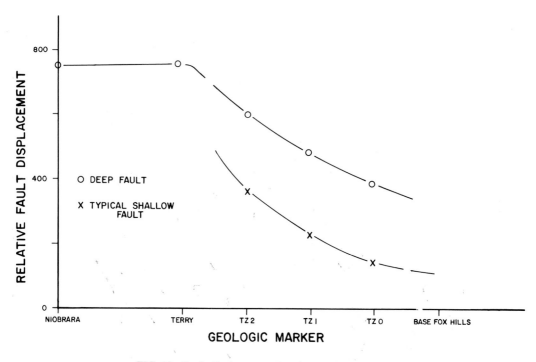

FIG. 19—Fault displacement of various geologic markers.

parts of the Upper Cretaceous. The overlying Fox Hills and lower Laramie, having a combined thickness of 1,000 to 1,500 ft (305 to 457 m), probably had similar high rates of sedimentation.

In the Tertiary of the Gulf Coast, Curtis (1970) and Bruce (1973) reported growth faults associated with deltaic depocenters. Growth faulting occurs in high constructional Cenozoic deltas throughout the world.

The three stratigraphic anomalies described above suggest a genetic relation between high rates of deltaic sedimentation and growth faulting in the Boulder-Weld fault zone.

SEISMIC INVESTIGATION

Structure

An important aspect of the Boulder-Weld fault zone anomaly is whether or not the faults extend to the basement as suggested by earlier workers (e.g., Spencer, 1961; Haun, 1968). To resolve the question a seismic survey of the area was undertaken by Davis (1974). Solely on the basis of geologic data, a geologist may draw a typical cross section across the faulted area as shown on Figure 4. All faulting would be assumed to be associated with the Laramide orogeny comprising basement-controlled block faulting and associated

horst-graben systems. Interpretation of a regional seismic line (Fig. 9) shows that basement-controlled fault systems exist only in the northern and western extremities of the area. The seismic data, however, do not appear to support a basement-controlled tectonic style for the fault system in the central part of the area. Persistent reflections occur from both the Niobrara and Hygiene horizons. Faulting at these levels cannot be detected underlying the prominent surface fault zone, but subtle anticlinal features appear at the Hygiene level and below. Thus the seismic data reveal that two distinct fault systems are present in the area. The deep, basement-controlled fault system occurs mainly within the zone of flank deformation with near-vertical dip on the fault planes. A shallow, listric (curved fault plane, flattening at depth), normal fault system extends basinward from the edge of the deep fault system.

Three or four major trends of listric normal faults are mapped by seismic methods within the Boulder-Weld fault zone (shown by heavy lines on Fig. 2). Fault planes have high dips near the surface which diminish with depth passing into bedding planes. The fault planes dip east and blocks are downthrown in that direction. These faults extend over a vertical distance of approximately 5,000 ft (1,524 m) dying out somewhere in or above the Pierre-Hygiene zone. Seismic evidence suggests that antithetic faults occur on the basinward side of many of the major shallow faults. However, as suggested by the surface fault pattern there are numerous curved-surface antithetic faults with displacements too small and curvature too slight to be interpreted from seismic data. On the basis of seismic information and well control, a generalized regional geologic cross section depicts the structure as shown on Figure 10.

Interpretation of 250 mi (400 km) of seismic data from throughout the area (Fig. 11) was facilitated by reprocessing the data with the intent of enhancing the shallow data and integrating seismic and well-log information to establish adequate subsurface velocity control. These data were provided through the auspices of various industry groups and consisted of 25 mi (40 km) of multifold dynamite and Vibroseis® seismic coverage and 200 mi (320 km) of 100% tape data, some of which had to be transcoded from original nontape data. In addition, 25 mi (40 km) of multifold Vibroseis® data collected by the Geophysics Department of the Colorado School of Mines during 1975 and 1976 has been used in this investigation.

A standard processing sequence consisted of the application of static and normal moveout corrections through careful control of the seismic velocities, filtering to remove groundroll and highline (60 cycle) noise, deconvolution for pulse compression, detailed velocity analysis for optimum stacking, automatic trim statics, and digital gain control with scaling for trace amplitude equalization.

Because of the scarcity of velocity information from lack of sonic logs or check-shot velocity surveys in the area, a "boot trap" process of obtaining velocity information from induction logs proved useful. This scheme is amenable to areas of uniform lithology in which velocity information is limited relative to conductivity (or resistivity) data. Using only those wells which had both sonic and induction logs, an empirical logarithmic relation between conductivity and velocity (reciprocal of transit time) was derived by crossplotting techniques as shown on Figure 12. Interval velocities were derived in wells which had only induction logs and were used to supplement the existing velocity control. Synthetic seismograms were generated from both sonic and induction log data (Fig. 13), thus enabling the determination of the nature of reflected seismic events and their spatial location in the subsurface.

This velocity information in conjunction with some seismic velocity analysis data enabled the preparation of structure contour maps of several horizons (Figs. 14 to 18).

The structure contour maps illustrate the areal extent of the basement-controlled fault system within the zone of flank deformation in the west and northwest part of the area. The greatest displacement on this fault system occurs in the extreme western part of the area. The axis of the Denver basin is reflected by the reversal in structural trend in the south-central map area. An arcuate fault system is observed within the transition zone at TZ1 and TZ2 levels. Displacements on this shallow fault system are greatest in the western part of the area as well.

Growth Faulting

A possible clue to the nature of the fault systems present along the west flank of the Denver basin is the concept of fault displacement, a measure of which is the relative offset of a marker horizon across the fault. If the displacement is the same for every marker horizon then the fault can be considered as occurring in a "single event." However, if the displacement differs across the fault from marker horizon to marker horizon, then the fault can be interpreted as having undergone recurrent movement and growth. Relative displacement thus provides a means of analyzing time and rate of fault movement. A plot of relative displacement of several marker horizons (based on both seismic and well information; Fig. 19) reveals:

1. Displacement (thickness) changes of several hundred feet occur basinward across the deep, basement-controlled fault system (T2N, R69W) in the uppermost Cretaceous. The initial fault movement during time of deposition of the upper Hygiene zone may mark the beginning of the Laramide orogeny along the east flank of the central Front Range. Recorded displacement changes and recurrent fault movement extends from the upper Hygiene zone throughout the uppermost Cretaceous.

2. Displacement on the major shallow listric faults increases with depth. Thickness changes of several hundred feet occur across these faults at least throughout post-TZ2 deposition. These faults die out into bedding planes or zones of shale flowage in the lower transition zone or upper Hygiene zone, thus recording the initial development of these faults as occurring during a corresponding time interval. Growth of these features continued with subsequent Late Cretaceous sedimentation.

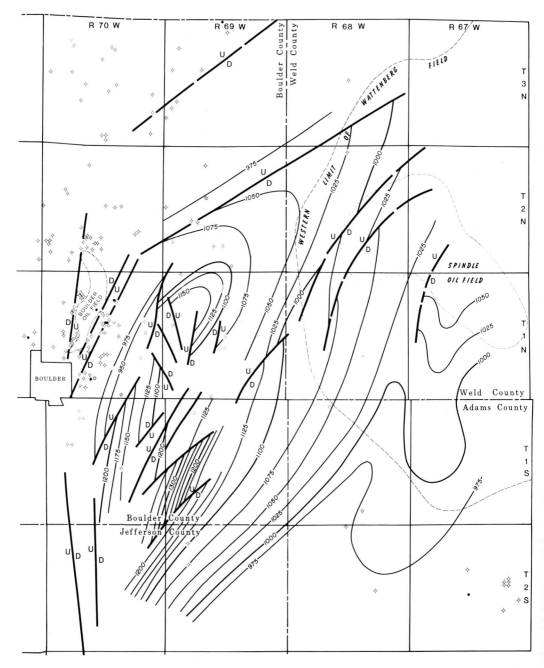

FIG. 20—Isopach map of interval between transition zone marker 1 (TZ1) and marker 2 (TZ2) of Pierre Shale (from seismic and well data). Contour interval is 25 ft (7.6 m).

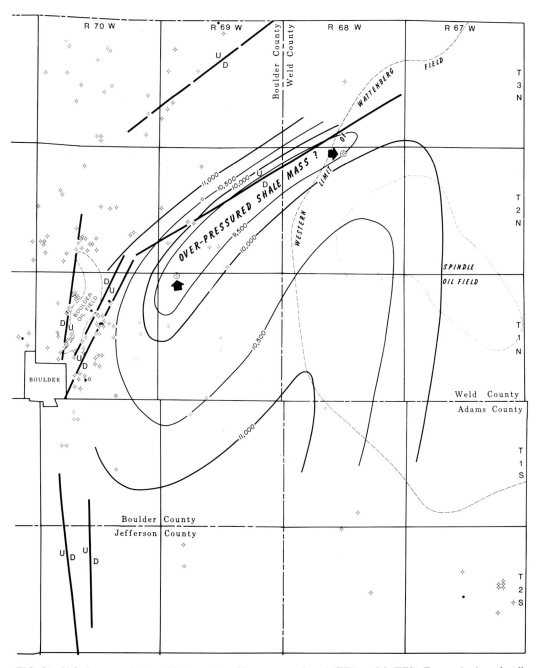

FIG. 21—Velocity map of interval between transition zone markers 1 (TZ1) and 2 (TZ2). From seismic and well data. Contour interval is 500 ft/sec (152 m/sec).

3. The similarity of the displacement curves for the deep and shallow fault systems reveals that both systems recorded recurrent movement and growth during the Late Cretaceous, even though the nature of the faulting differs.

Overpressured Shale Masses

An isopach map between the TZ1 and TZ2 marker horizons is shown on Figure 20. This interval encompasses the most uniform shale interval of the transition zone. The isopach map documents substantial thickening of this interval on the downthrown sides of the major shallow growth fault systems and also depicts a local TZ1-TZ2 isopach thick-zone in the central map area. Note also the TZ1-TZ2 isochron variation on the regional seismic line (Fig. 9). This localized thick-zone coincides with a zone of low TZ1-TZ2 interval velocity (Fig. 21). In Tertiary delta systems throughout the world, abnormal fluid pressures are often associated with thick shale masses and low interval velocities. Mechanisms thought to account for these anomalous thickness and velocity features in this area include the formation of a local depocenter during transition zone deposition and generation of a low-density, overpressured shale swell.

Examination of sonic and induction logs from two wells, Sec. 6, T1N, R69W, and Sec. 3, T2N, R68W (Fig. 21), within the anomalous low-velocity region reveal abrupt transition zone velocity and conductivity trend reversals. The velocity reversal within the TZ1-TZ2 uniform shale interval is documented on the velocity log of Figure 13. The magnitude of the velocity anomaly is $-1,400$ ft/sec (-426.7 m/sec) and corresponds to a $+60$ mmho/m conductivity anomaly.

A technique analogous to Reynolds' (1974) enables the computation of fluid pressure gradients from interval velocities at depths within the anomalous interval. Under the linear assumption inherent in this technique, fluid pressure gradients are the maximum predicted to occur within the overpressured zone. The technique used involves the following procedures:

1. Interval velocities versus depths are plotted on semi-log paper as shown on Figure 22.

2. A line is drawn through the interval velocities of the normally compacted intervals. Additional well information is shown on Figure 22 to support the normal compaction trend.

3. Undercompacted, low-density, overpressured zones exhibit departures from the normal trend. Interval velocity in the undercompacted section (at depth D_2) is then compared vertically with a point on the normal compaction trend (at

depth D_1) having the same interval velocity and assumed same effective stress.

4. The depths D_1 and D_2 are used in conjunction with the normal fluid pressure gradient of 0.465 psi/ft and an overburden pressure of 1.0 psi/ft to obtain the fluid pressure gradient (P_2) for the undercompacted section:

$$P_2 = \frac{D_1 \cdot 0.465 + (D_2 - D_1) \cdot 1.0}{D_2}$$

Fluid pressure gradients of 0.7 psi/ft were computed for the TZ1-TZ2 interval for wells in Sec. 6, T1N, R69W and Sec. 3, T2N, R68W. This gradient is approximately 0.2 psi/ft greater than normal and is thought to be indicative of overpressuring within the transition zone section of these two wells, but no drill-stem tests are available to test the accuracy of these calculations. Overpressuring appears to exist within the transition zone and does not extend to depths greater than the base of this unit. Pressure gradients and the areal extent of overpressuring as indicated may have been much greater in the past than at present. Differential loading and the progressive development of abnormal fluid pressures probably controlled the areal extent and limits of growth faulting.

NEW ROCKY MOUNTAIN TECTONIC STYLE

Basement-controlled fault systems have generally been regarded as the dominant central Rocky Mountain structural style. An associated new tectonic style must be invoked in the Denver basin to explain the nature of the shallow fault system as outlined from stratigraphic and seismic data. A

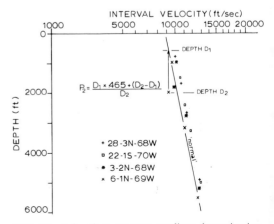

FIG. 22—Subsurface pressure gradient determination for Pierre Shale.

model, following concepts summarized by Curtis (1970) and Bruce (1973) and designed to explain the nature of the shallow fault system, involves formation of a regional growth fault system through the processes of tectonism and deltaic sedimentation. Successive phases of the model as depicted on Figure 23 include:

1. A deltaic depocenter was formed by tectonism related to uplift of the Front Range. The depocenter formed on the basinward side of the deep, basement-controlled fault system during deposition of the lower transition zone. Once established by fault control, the depocenter localized sedimentation throughout the remainder of the Late Cretaceous.

2. A thick-zone of sediment was recycled into the Pierre seaway along the east flank of the uplift area, and extremely high rates of sedimentation occurred in depocenter systems along the flank of the uplift. It is estimated that the entire transition

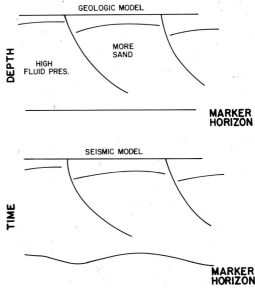

FIG. 24—Seismic anomaly pitfalls associated with shallow growth faulting in the Denver basin.

zone of the Pierre Shale was deposited in 1 to 2 million years. Prograding sequences of deltaic sediments derived from the Front Range uplift continued to fill the depocenter. Sediment loading, subsidence, and recurrent movement on the deep fault system resulted in the generation of high fluid pressures and the initiation of growth faulting.

3. Continued progradation of deltaic sediments basinward throughout the remainder of Cretaceous time resulted in differential loading of the underlying sediments, continued growth faulting, and subsequent development of smaller scale horst-graben structures as antithetic faults.

APPLICATION TO PETROLEUM EXPLORATION

Apart from possibly creating fracture permeability and porosity in the developing Spindle field (northeast end of the study area; Moredock and Williams, 1976), the growth of faulting described in this paper does not involve potential petroleum reservoir rock. This style of faulting may exist and involve potential reservoir rock at different stratigraphic levels and in different areas of the Rocky Mountain region, however. For example, it is anticipated that a similar style of growth faulting may exist in the Frontier Formation of western Wyoming, the Mesaverde Formation of Colorado and Utah, and the Eagle Sand-

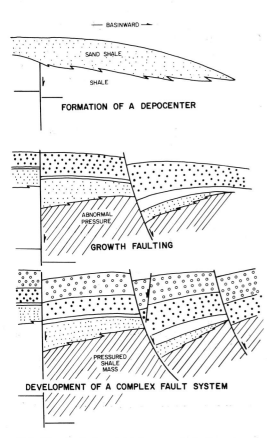

FIG. 23—Geologic model for growth fault system in Upper Cretaceous of Rocky Mountains.

stone and Judith River Formation in Montana. In each, growth faulting may play a primary role in the localization of traps. To explore for these traps and to adequately evaluate their potential, the basic geologic and geophysical model is essential and has been developed herein.

The fact that growth faulting can occur at different stratigraphic levels and/or be in superposition with other fault systems would make proper identification of growth faulting difficult in the subsurface. Growth faulting at a stratigraphic level above the main exploration interval of interest can lead to pitfalls in the interpretation of seismic data. For example, Figure 24 illustrates the effects that lithologic interval thickness and abnormal pressure variations associated with growth faulting have on the seismic definition of a marker bed beneath the level of growth faulting. In this model, structure can be artificially created on the seismic time-section representation of a pre-growth faulting horizon, by the overlying lateral and vertical velocity variations within the overlying growth section. Figure 9 illustrates subtle anticlinal features at the Hygiene and Dakota levels on the regional seismic section from the Denver basin study area. These apparent anticlinal features with 30 to 40 msec of time-structural relief are entirely created by the overlying combination of overpressured shale masses and growth faults in the upper Pierre transition zone.

An integrated approach is the only way by which growth faulting can be recognized and sought in the subsurface, particularly in areas of tectonic overprinting. Such an integrated approach must involve explorationists who are willing to combine geophysical and geologic information to arrive at an accurate representation of the subsurface geology.

CONCLUSIONS

From this study the following conclusions are made:

1. Based on seismic and geologic data, Late Cretaceous faulting along the east flank of the central Front Range illustrates a new Rocky Mountain tectonic style. The new tectonic style involves the interrelation of shallow depth penecontemporaneous growth faulting with basement-controlled fault systems.

2. Within the Denver basin a Late Cretaceous depocenter was formed by fault control along the east flank of the central Front Range. Rapid deposition of sediment and recurrent movement on the basement-controlled fault systems resulted in the generation of overpressured shale masses and associated shallow growth fault systems within the uppermost Cretaceous.

3. Similar tectonic features probably occur elsewhere in depocenters of the Cretaceous foreland basin. For example, growth faults may be associated with the Frontier Formation in western Wyoming, the Mesaverde Formation in western Colorado and Utah, and the Eagle Sandstone and Judith River Formation in Montana. In areas of extensive younger Laramide tectonic overprinting, the recognition of early growth fault systems will be more difficult. Integrated geologic and seismic exploration investigations are the principal way to identify growth fault systems.

4. Recognition of growth fault systems will play an important role in future petroleum exploration and development in the Rocky Mountain region. Growth faults provide early traps for migrating gas and may control the thickness and quality of reservoirs. Because of tensional antithetic faults and fractures associated with the listric normal faults, production may be enhanced by natural fracturing.

5. Coal exploration and development in the lower Laramie has been greatly affected by influence of growth faulting on coal bed thickness. Growth-fault concepts may play an important role in developing coal production associated with delta-plain environments of deposition in many areas of the Rocky Mountain region.

REFERENCES CITED

Bedwell, J. L., 1974, Textural parameters of clastic rocks from borehole measurements and their application in determining depositional environments: PhD thesis, Colorado School of Mines, 215 p.

Bruce, C. H., 1973, Pressured shale and related sediment deformation—mechanism for development of regional contemporaneous faults: AAPG Bull., v. 57, p. 878-886.

Colton, R. B., and R. L. Lowrie, 1973, Map showing mined areas of the Boulder-Weld coal field, Colorado: U.S. Geol. Survey Misc. field studies map MF-513.

Curtis, D. M., 1970, Miocene deltaic sedimentation, Louisiana Gulf Coast: SEPM Spec. Pub. 15, p. 293-308.

Davis, T. L., 1974, Seismic investigation of Late Cretaceous faulting along the east flank of the central Front Range, Colorado: PhD thesis, Colorado School of Mines, 65 p.

Fenneman, N. M., 1905, Geology of the Boulder District, Colorado: U.S. Geol. Survey Bull. 265, 98 p.

Haun, J. D., 1968, Structural geology of the Denver basin—regional setting of the Denver earthquakes: Colorado School Mines Quart., v. 63, no. 1, p. 101-112.

Moredock, D. E., and S. J. Williams, 1976, Upper Cretaceous Terry and Hygiene sandstones—Singletree, Spindle, and Surrey fields—Weld County, Colorado, *in* R. E. Epis and R. J. Weimer, eds., Studies in Colorado field geology: Colorado School of Mines Prof.

Contr., no. 8, p. 264-274.

Obradovich, J. D., and W. A. Cobban, 1975, A time-scale for the Late Cretaceous of the western interior of North America, *in* W. G. E. Caldwell, ed., The Cretaceous System in the western interior of North America: Geol. Assoc. of Canada Spec. Paper No. 13, p. 31-54.

Rahmanian, V. D., 1975, Deltaic sedimentation and structure of the Fox Hills and Laramie Formations, Upper Cretaceous, southwest of Boulder, Colorado: Master's thesis, Colorado School of Mines, 83 p.

Reynolds, E. B., 1974, Seismic interpretation for drilling: Oil and Gas Jour., v. 72, no. 10, p. 112-124.

Scott, G. R., and W. A. Cobban, 1965, Geology and biostratigraphic map of the Pierre Shale between Jarre Creek and Loveland, Colorado: U.S. Geol. Survey Misc. Geol. Inv. Map, I-439.

Spencer, F. D., 1961, Bedrock geology of the Louisville Quadrangle: U.S. Geol. Survey Geol. Quad. Map, GQ-151.

Trimble, D. E., 1975, Geologic map of the Niwot Quadrangle, Boulder County, Colorado: U.S. Geol. Survey Geol. Quad. Map, GQ-1229.

Weimer, R. J., 1973, A guide to uppermost Cretaceous stratigraphy, central Front Range, Colorado: deltaic sedimentation, growth faulting and early Laramide crustal movement: Mtn. Geologist, v. 10, no. 3, p. 53-97.

Application of Amplitude, Frequency, and Other Attributes to Stratigraphic and Hydrocarbon Determination[1]

M. T. TANER and R. E. SHERIFF[2]

Abstract Improvements in seismic data acquisition and processing techniques make it possible to observe geologically significant information in seismic records which has not been evident in the past. New types of measurements help in locating and analyzing geologic features, including some hydrocarbon accumulations. Analysis of a seismic trace as a component of an analytic signal permits the transformation to polar coordinates and the measurement of quantities called "reflection strength" and "instantaneous phase." These, plus several other quantities derived from them, are called *attribute measurements* and can be coded by color on seismic sections. Such color displays permit an interpreter to associate measurements and changes in measurements with structural and other features in the seismic data. They thus facilitate identification of interrelations among measurements. A series of examples shows how such analysis and display helps in locating and understanding faults, unconformities, pinchouts, prograding deposition, seismic sequence boundaries, hydrocarbon accumulations, and stratigraphic and other variations which might be misinterpreted as hydrocarbon accumulations.

INTRODUCTION

The concept of correlation based on seismic "character" has been used by geophysicists since the beginning of reflection exploration. Interpreters have learned that certain reflecting sequences or depositional situations are characterized by a distinctive waveform and have used this for jump correlation, correlation across faults, and verification that reflections have been picked correctly. They have sometimes noted changes in the waveform and have correlated these with changes in the thickness of intervals, in facies, in the number and thickness of beds, etc. Character interpretation has been an art—a geophysicist with long experience in one particular area has learned the significance of certain pattern changes, whereas an equally good geophysicist without local experience would not (a) recognize the changes or, (b) be able to explain their significance even if recognized.

Today geophysicists more easily recognize changes attributed to subsurface conditions due to three factors:

1. Better recording and processing techniques, so that data are less distorted and less obscured by noise. Data must be recorded and processed so as to minimize nongeologic variations. Amplitude information and a wide band of frequencies must be preserved, the seismic waveform shortened and stabilized, multiple energy and other noise removed, and the data repositioned (migrated) without amplitude or waveform distortion. Variations which remain on a seismic record must be associated with variations in the subsurface to derive stratigraphic meaning.

2. Availability of more types of measurements made on seismic data, providing more ways of looking at the data. The arrival time of events is the most common measurement made on seismic data, followed by measurements of the amount of normal moveout (variation of arrival time with source-to-geophone distance) of amplitudes and, to a lesser extent, of the dominant frequency (or dominant period). An objective of this paper is to show how additional meaningful measurements can be made.

3. Better displays for communicating measurements to an interpreter so that he can see significant interrelations. Commonly we have too many measurements (position along the seismic line, arrival time, normal moveout, amplitude, dominant frequency, etc.) to permit an intelligible display on two-dimensional graphs (record-sections). Multiple displays are then used, such as record sections at different amplitudes, or sections using different stacking velocities, or those using different filters. Separate displays are also made of velocity analyses, spectral plots to show frequency content, and other quantities. The multiple display technique often makes it very difficult to relate measurements from one display to another. A second objective of this paper is to show how color displays help in this communication problem.

SEISMIC WAVES AS ANALYTIC SIGNALS

Seismic waves which we ordinarily detect and record can be thought of as an analytic signal with both real and imaginary parts, of which only the real part is detected and displayed. Another way of expressing this point of view is to call it a "time-dependent phasor" (Bracewell, 1965). This

[1]Manuscript received, October 1, 1976; accepted, January 19, 1977.

[2]Seiscom Delta Inc., Houston, Texas 77036.
Many contributed to this paper and to the work on which it is based. Special acknowledgment must go to N. A. Anstey, F. Koehler, and Seiscom Delta management.

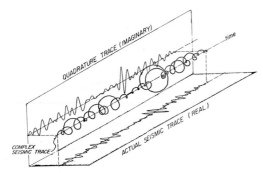

FIG. 1—Complex seismic trace as generated by a vector whose length varies with time, rotating as a function of time. The actual seismic trace is the projection of this vector onto the real plane and the quadrature trace is the projection onto the imaginery plane..

point of view looks on the observed seismic trace g(t) as expressed by:

$$g(t) = R(t) \cos \theta(t).$$

The quantity R(t) is the envelope of the seismic trace and $\theta(t)$ is the phase.

One can imagine a vector perpendicular to the time axis (Fig. 1) whose length varies as a function of time; this vector also rotates about the time axis as a function of time. The projection of the head of this rotating vector on the real plane gives the conventional seismic trace, g(t). The head of the vector can be projected onto the imaginary plane to give the *quadrature trace*. The quadrature trace, h(t), is expressed by:

$$h(t) = R(t) \sin \theta(t).$$

Hilbert transform techniques (Bracewell, 1965) permit us to generate the quadrature trace from the observed real trace so that both portions are available for analysis.

The seismic trace is usually a measure of the velocity of motion (with geophones) or of pressure variations (with hydrophones) which result from the passage of seismic waves. A seismic wave involves the moving of particles of matter from their equilibrium positions and thus involves kinetic energy. Hence the conventional seismic trace, g(t), may be thought of as a measure of kinetic energy. The particle motion is resisted by an elastic restoring force so that energy becomes stored as potential energy. As a particle moves in response to the passage of a seismic wave, the energy transfers back and forth between kinetic and potential forms. The quadrature trace, h(t),

may be thought of as a measure of potential energy. Ordinarily, geophone output is proportional to the velocity of particle motion, which means that the kinetic energy is proportional to the square of the amplitude of the quadrature trace. Observed in this way, the quantity R(t) in the above equations can be thought of as being proportional to the square root of the total energy of the seismic wave at any given moment.

REFLECTION STRENGTH AND INSTANTANEOUS PHASE

We can solve the foregoing equations separately for R(t) and $\theta(t)$. We call R(t) the "reflection strength" and $\theta(t)$ the "instantaneous phase:"

$$R(t) = [g(t)^2 + h(t)^2]^{1/2}$$
$$\theta(t) = \tan^{-1} [h(t)/g(t)]$$

These equations are solved for every sample point so that R(t) and $\theta(t)$ have independent values at each point rather than being averages over a number of samples.

On arrival of the seismic reflection, the reflection strength first increases then decreases, thus a reflection is evidenced by a local maximum in the reflection strength. Polarity at the time of the maximum may be either positive or negative, depending on whether the reflection coefficient is positive or negative, how the interference of successive reflections affect the waveform, what conventions have been assumed in the recording-processing-display procedures, and what phase shifts have been introduced by recording and processing. If the magnitude of the reflection strength maximum is displayed in relation to the phase at the time of the maximum, polarity as well as the amplitude of the envelope will be illustrated. In subsequent figures (8, 19, 20), reflection strength is color-coded and superimposed on the real trace to accomplish this purpose.

Reflection strength may have its maximum at phase points other than peaks or troughs, especially where the reflection is the interference composite of several subreflections. Thus maximum reflection strength associated with an event is more meaningful than merely the amplitude of the largest peak or trough and reflection strength measurement differs from conventional amplitude measurement in a fundamental way.

Observing where (within an event) the maximum reflection strength occurs gives additional information. The color-coded reflection strength display provides a measure of reflection character. It is sometimes an aid, for example, in distinguishing reflections from massive reflectors and those which are interference composites. Reflections from massive interfaces tend to remain constant over a large region. Such reflections provide the best reference for smoothing or flattening data or for measuring time-thickness variations which might indicate differential compaction, local or regional thinning, facies changes, velocity variations, etc. Reflections which result from the interference of several separate reflections tend to vary along a seismic line as the thickness or contrast of the individual component reflector changes. Variations which are systematic with structure may indicate growth during deposition. Unconformities often show changes in reflection strength character as the subcropping beds change. This may be the indicator for unconformities which are otherwise difficult to detect. The quantitative aspect of reflection strength measurement may aid in the lithologic identification of subcropping beds if it can be assumed that deposition is constant above the unconformity (so all the reflection coefficient changes can be attributed to the subcropping bed). Seismic sequence boundaries tend to have fairly large reflection strength.

The instantaneous phase is a quantity independent of reflection strength. Phase emphasizes the continuity of events; in phase displays (shown later in Figs. 9, 13, 16, 18), every peak, every trough, every zero-crossing has been picked and assigned the same color so that any phase angle can be followed from trace to trace. Weak coherent events thus are brought out. Phase color displays use the color wheel such that $+180°$ and $-180°$ are the same color because they are the same phase angle. Such phase displays are especially effective in showing pinchouts, angularities, and the interference of events with different dip attitudes.

FREQUENCY

The time derivative of the instantaneous phase is called *instantaneous frequency*. Like instantaneous phase, it is a value appropriate to a point rather than being an average over some interval. The instantaneous frequency can vary abruptly, which is sometimes an advantage because abrupt changes do not get lost in an averaging process. It is also sometimes a disadvantage because there may be so many changes that the interpreter cannot comprehend them. (Instantaneous frequency is displayed in Figs. 10, 15).

It is useful to smooth frequency measurements; smoothing can be done in many ways, such as using "time windows" of varying shape and length in time. One particularly useful scheme is

to use a weighting according to reflection strength, which produces "averaged weighted frequency." (This quantity is displayed in Figs. 11, 17, 21).

The above methods of determining frequency—both instantaneous frequency and weighted frequency—are different from the more familiar Fourier-transform methods in which the data over an appreciable length of trace are fitted with sine wave curves of different frequencies, amplitudes, and phase shifts. The amplitudes of the different frequency components are thus averages over the entire part of the trace being fitted, rather than values appropriate to a single instant in time.

In some areas frequency has been a good indicator of condensate reservoirs; these are associated with a characteristic low-frequency anomaly directly underneath them. Use of weighted frequency as an indicator of a condensate reservoir is an empirical relation based on a number of observations. The mechanism by which such zones attenuate high frequencies is not known.

POLARITY

The sign of the seismic trace (whether positive or negative) when the reflection strength has its maximum value is determined and called *polarity*. A magenta or blue color is assigned to suggest that the reflection coefficient (if the event indicates a single dominant interface) is positive or negative. The intensity of the color is varied according to the magnitude of the reflection strength (polarity displays shown, Figs. 12, 14).

USE OF COLOR SECTIONS FOR DISPLAY

An interpreter usually faces a major problem in assimilating large masses of data. The ability to extract more from the data compounds this problem by requiring more data for optimum comprehension and interpretation.

Color displays help show the significance of measurements and interrelations. Color effectively adds another dimension to comprehension; color-coded quantities can be superimposed on a conventional record-section plot so that both the conventional data and the color-coded quantity can be seen simultaneously, thus making interrelations easier to see.

Color has become increasingly important, though the literature on the subject is limited. Balch (1971) discussed the use of color seismic sections as an interpretation aid, and occasional advertisements in *Geophysics®* have illustrated limited use of color. Most uses of color on seismic sections have been either simple, using few colors, or difficult to reproduce. Seis-Chrome® displays of seismic attribute measurements have been used for several years but little has been published. The Seis-Chrome® process color-codes data retaining the fidelity of digital processing; a number is associated with each location, according to the attribute being displayed, and a color is assigned to each number or range of numbers. Usually a one-to-one correspondence is established so that each number is represented by a different color. A color key is commonly provided to show the numerical value which a color represents, and thus to permit quantitative interpretation of color changes. (The set of color keys usually used is shown in Fig. 7, and a diagramatic representation of the encoding is shown in Fig. 22.) Different or additional colors could be used; whatever color code is defined produces exactly the same color whenever the same number occurs.

The seismic measurements most commonly color-coded are: (1) reflection strength, (2) phase, (3) frequency, either instantaneous or weighted, (4) polarity, and (5) velocity. However, quantities displayed in color are not restricted to this particular assortment; other quantities such as dip, rate of change of dip, or cross-dip have also been displayed.

EXAMPLES AND GEOLOGIC MEANING

Geologic interpretation of color-coded displays of attribute measurements is illustrated with examples of five seismic lines. Interpretation of some features is shown on the conventional black and white seismic sections in Figures 2 through 6, where letters locate features referred to subsequently (color displays of these lines are shown in Figs. 8-22). All of the color sections were made by the Seis-Chrome® process and are copyrighted by Seiscom Delta Inc.

Ideally, displays of various attributes are available for each line because different displays bring out different features; there is useful information where the same feature is indicated, and also where different features are indicated. However, only selected displays are shown here because the intent is to illustrate their use rather than to make a complete interpretation.

Figure 7A illustrates the color code often used for reflection strength. The color steps indicate "dB" less than the maximum reflection strength on the section (or in the area, if a number of sections are being processed to allow meaningful line to line comparisons); thus "O dB" indicates the maximum amplitude. Since attention usually is focused on the strongest reflections, these are assigned the red end of the spectrum, although the

color assignment is arbitrary and a different color assignment could be made.

Figure 7B shows the code commonly used for phase. The colors represent a color wheel, that is, $-180°$ is the same color as $+180°$, so that colors are continuous.

Figure 7C shows frequency in 2-Hz steps. Frequencies lower than 6 Hz, including occasional negative frequencies (when the phasor temporarily reverses its sense of direction), are left uncolored.

The bar graph of Figure 7D is used for polarity displays; it shows whether the phase is positive or negative at the points where reflection strength has a local maximum. The color hues are divided into five ranges according to the reflection strength. Where the sense of polarity is known, magenta is used to indicate a positive reflection coefficient and blue shows a negative one.

OFFSHORE LOUISIANA EXAMPLE

Figures 8 and 12 show data for the offshore Louisiana line shown in Figure 2; this line is 10 mi (16 km) long. These (and all of the following sections) are "squash plots," that is the data have been compressed horizontally so as to include greater length of line. The vertical exaggeration which results is often helpful in stratigraphic interpretation because many miles of data can be seen at a glance. The squash plot does, however, distort structure. For example, faults appear to be steeper than they are. A black and white gain-equalized plot (Fig. 2) forms the background of the color sections except for phase; a phase plot in black and white is sometimes used as a background.

Figure 8 shows reflection strength; several gas and oil accumulations are known to occur along this line. Gas accumulations often show as "bright spots," high-amplitude reflections, indicated by red and yellow colors (events A, B, C in Fig. 2). The shallow gas accumulations (A, B) are noncommercial and the bright spot at 1.35 sec, (C), corresponds to a commercial gas field. Below this gas field there is condensate production at 2.1 sec, (D). The amplitude anomaly associated with this condensate zone is not especially obvious although a flat spot attributed to the water level surface can be seen at about 2.2 sec.

The phase display (Fig. 9) emphasizes the continuity of events; it is especially useful where the signal to noise ratio is poor, although this benefit is not well-illustrated in this example. Discontinuities, faults, angular unconformities, pinchouts, zones of thickening and thinning offlap, onlap, interfering events, and diffractions stand out

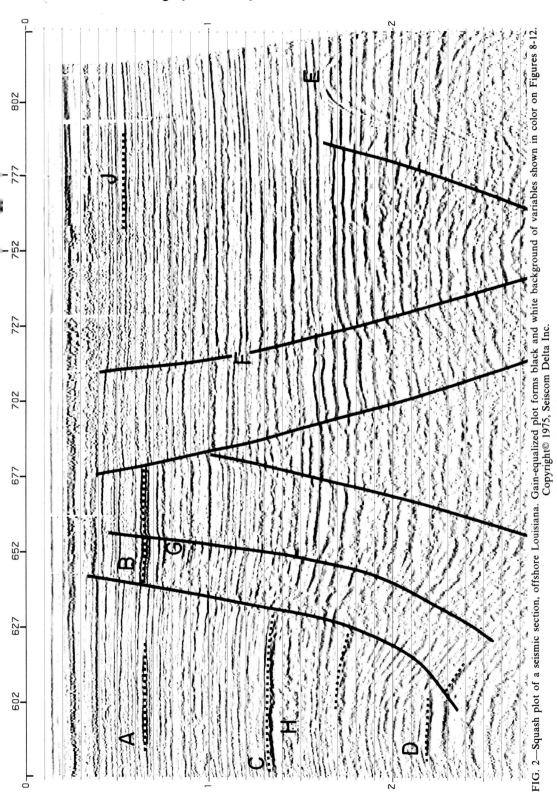

FIG. 2—Squash plot of a seismic section, offshore Louisiana. Gain-equalized plot forms black and white background of variables shown in color on Figures 8-12. Copyright© 1975, Seiscom Delta Inc.

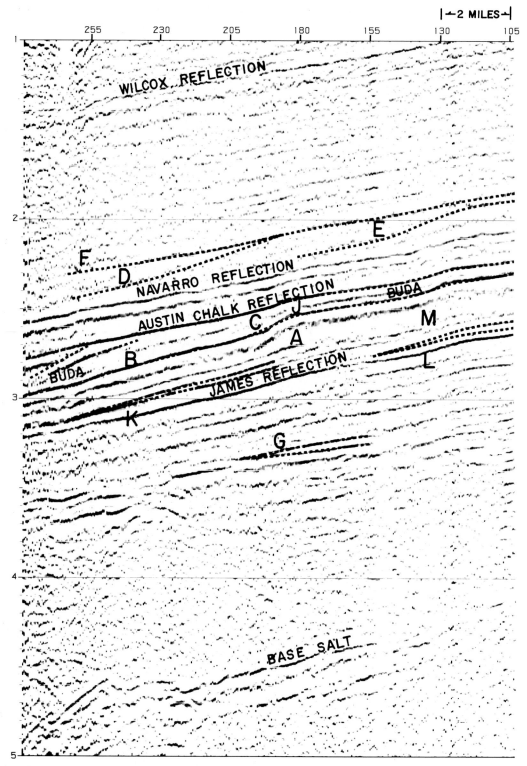

FIG. 3—Seismic section of north-south dip line, East Texas. Black and white section matches color Figures 13-15.
Copyright© 1975, Seiscom Delta Inc.

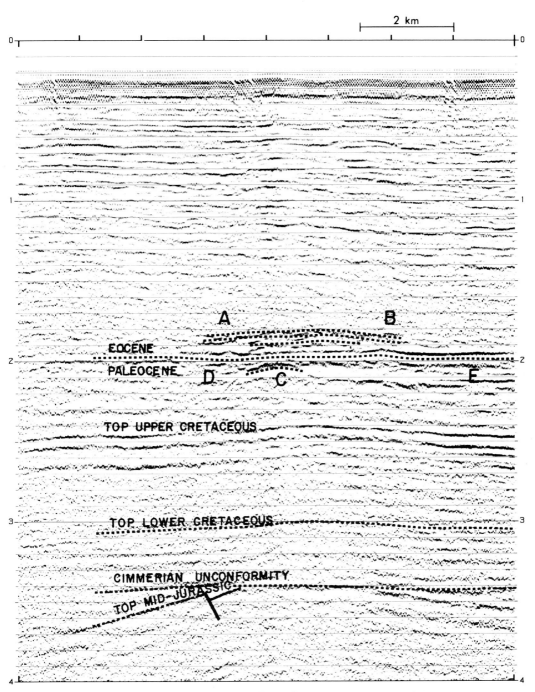

FIG. 4—Seismic section of line in North Sea. Corresponds to color Figures 16-17. Copyright© 1976, Seiscom Delta Inc.

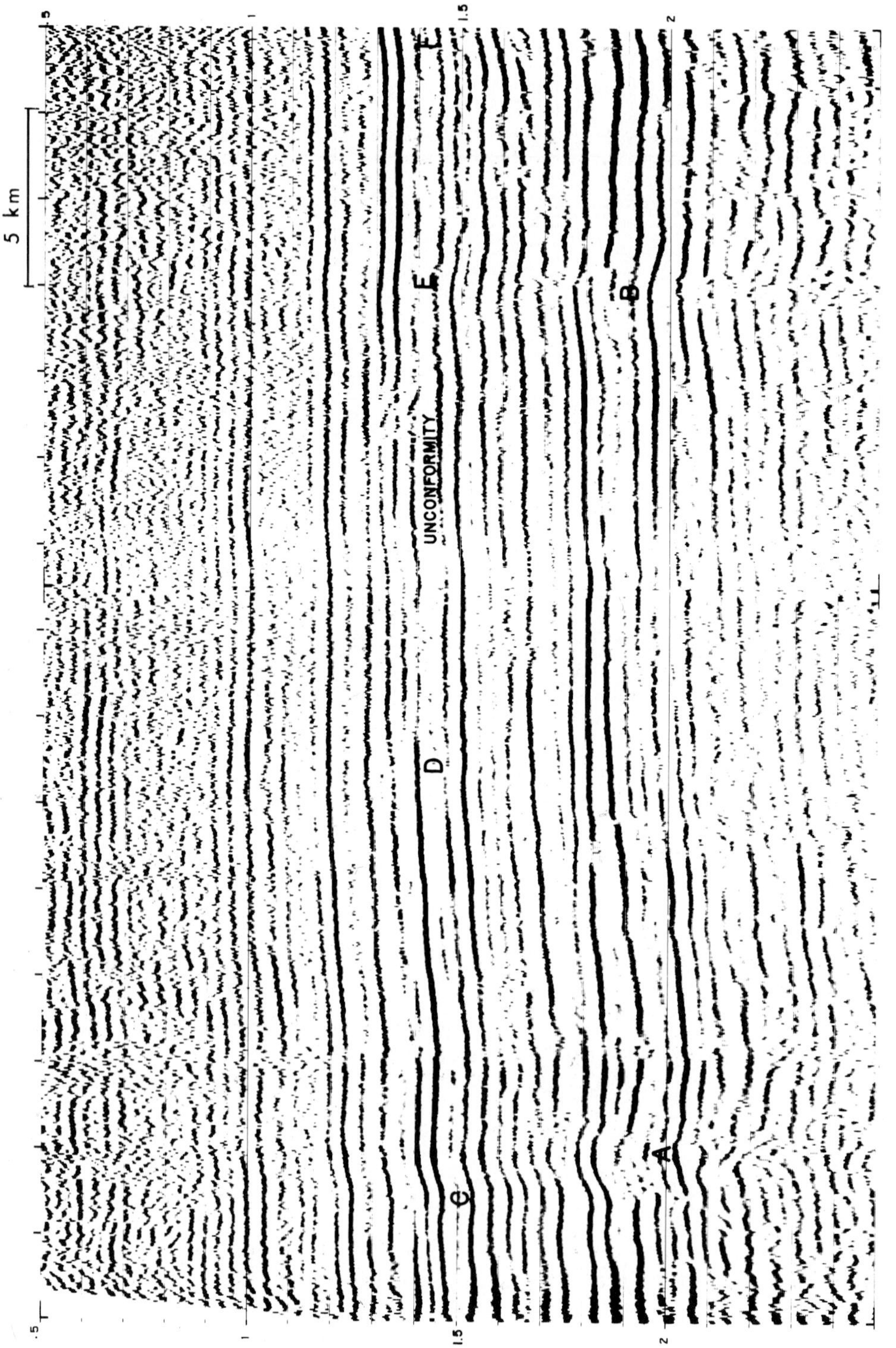

FIG. 5—Seismic section from line in Western Canada. Black and white corresponds to color Figures 18-19. Copyright© 1975, Seiscom Delta Inc.

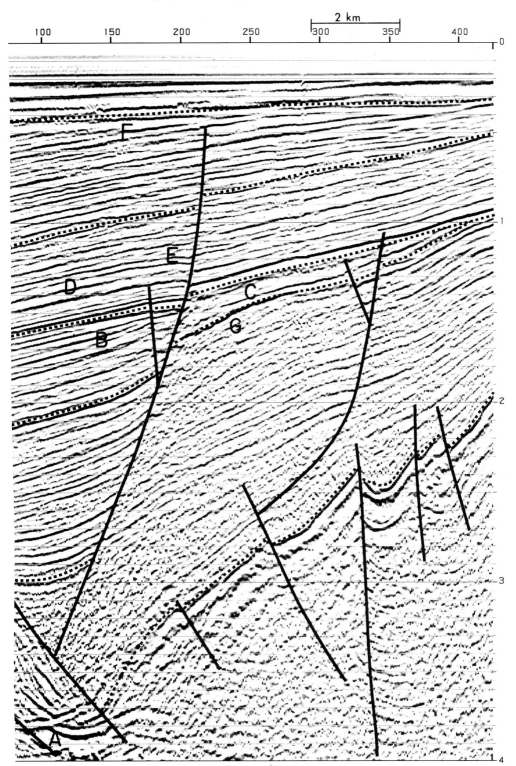

FIG. 6—Seismic section of line offshore Alaska. Interpretation is speculative due to lack of well control. Figure corresponds to color Figures 20-21. Copyright© 1975, Seiscom Delta Inc.

clearly in the phase display. The diffractions cresting at 1.65 sec at the right of the section (E) come from off (to the side of) the line and so do not indicate anything about the geology along this line.

Figure 10 shows instantaneous frequency. The color patterns help in correlating across faults (e. g., F). Lateral changes in the color pattern indicate that something has changed about the reflection. Such changes focus the interpreter's attention on the places where a change occurs even though the nature of the change may not be indicated.

Figure 11 shows the frequency weighted by reflection strength and then smoothed. The rapid variation of frequency which is sometimes distracting on instantaneous frequency displays is thus averaged out. Sharp shifts toward lower frequencies underneath the gas sandstones (G, H) indicate loss of high frequency in the gas zones. The condensate accumulation (D) shows as a distinctive low-frequency anomaly; such anomalies often characterize accumulations in this area.

Polarity for this line is shown in Figure 12. This display makes clear that the magenta bright spot at about 0.55 sec (J) is of a different kind than the blue bright spots associated with gas accumulations (A, B, C). The magenta bright spot indicates a positive reflection coefficient, an increase in acoustic impedance (perhaps an increase in calcium carbonate content), whereas gas accumulations usually have lowered acoustic impedance because of the lower velocity and density of the gas sandstones. The flat spot at 2.2 sec at the left of the section (D) appears to be associated with a positive reflection coefficient.

EAST TEXAS EXAMPLE

Figures 13, 14, and 15 show a north-south dip line in East Texas; this section is shown in black and white in Figure 3. Figure 13 shows phase. The outstanding feature on this line is the nonporous Edwards reef, the left edge of which is just to the right of the center at 2.6 sec (A in Fig. 3). The event which extends from 2.75 sec at the left edge to 2.3 sec at the right is the Austin Chalk reflection. Below the Austin Chalk to the left of the Edwards reef, the Woodbine Sandstones pinch out (B, C). The fore-reef Woodbine Sandstone produces gas in this area. The phase display clearly shows the prograding depositional pattern of the Woodbine.

The prograding pattern of the Midway section (D, E) is seen for about 0.2 sec above the Navarro reflection. Regions of onlap and offlap show so nicely in the phase display that it is helpful in picking seismic sequence boundaries. Such sequence boundaries include F and the Navarro, the top and base of the prograding Woodbine; the James, evidenced by the onlapping pattern above it as seen at K and L; and the onlap pattern at G.

Figure 14 is a polarity display. Note the change in polarity (change from blue to magenta) of the Austin Chalk reflection above the Edwards reef (J). To the right, the reflection coefficient is positive, associated with limy, high-velocity rocks, whereas to the left it is negative because of the lower velocity sandstone-shale Woodbine sequence. The changes in polarity as determined from the phase at peak reflection strength generally agree with the results from interval velocity calculations based on normal moveout along this line. Strong reflectors at the bottom of the section are interpreted as the base of the Louann Salt.

Mapping the time between reflectors (isotime maps) is an important interpretive tool to delineate geologic features. Its reliability deteriorates if the waveshape of the reflections being timed also varies. The lateral constancy of the color pattern of the polarity, reflection strength, or other measurements is helpful in selecting good reference reflections (where there is a choice) for such isotimes. For example, the lateral change of pattern of the James reflection makes it clear that isotime maps referenced to it will include the effect of variations which involve the James as well as variations elsewhere.

Figure 15 shows instantaneous frequency along this line. Note the distinctive pattern in the Edwards reef zone (A to M).

NORTH SEA EXAMPLE

Figure 16 shows phase for a line in the North Sea. A black and white copy of this section is shown on Figure 4. Production is derived from turbidite sandstones of Paleocene and Eocene age which piled up as mounds. The strong reflector at 1.95 sec, approximately the Eocene–Paleocene boundary, shows little structural relief. Turbidite sandstones (A, B on Fig. 4) appear to have been deposited on top of this interface although irregularities just under it (C) may indicate Paleocene turbidites. The turbidites have distinctive character in the phase display.

Figure 17 is the weighted frequency display of the same line. The zones of low frequency just beneath the turbidites are attributed to loss of higher frequencies in the gas accumulations. Thus the boundaries of this low frequency zone (D, E) may indicate the limits of the production. The low frequency anomaly is not uniform across the field, probably indicating variabilities in the accu-

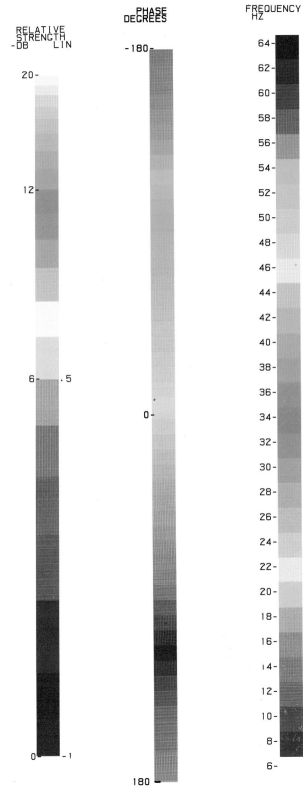

RELATIVE
STRENGTH
-DB   LIN

20-

12-

6-   .5

0-   -1

PHASE
DEGREES

-180-

0-

180-

FREQUENCY
HZ

64-
62-
60-
58-
56-
54-
52-
50-
48-
46-
44-
42-
40-
38-
36-
34-
32-
30-
28-
26-
24-
22-
20-
18-
16-
14-
12-
10-
8-
6-

APPARENT
POLARITY

+

—

Figure 7 - Color Keys

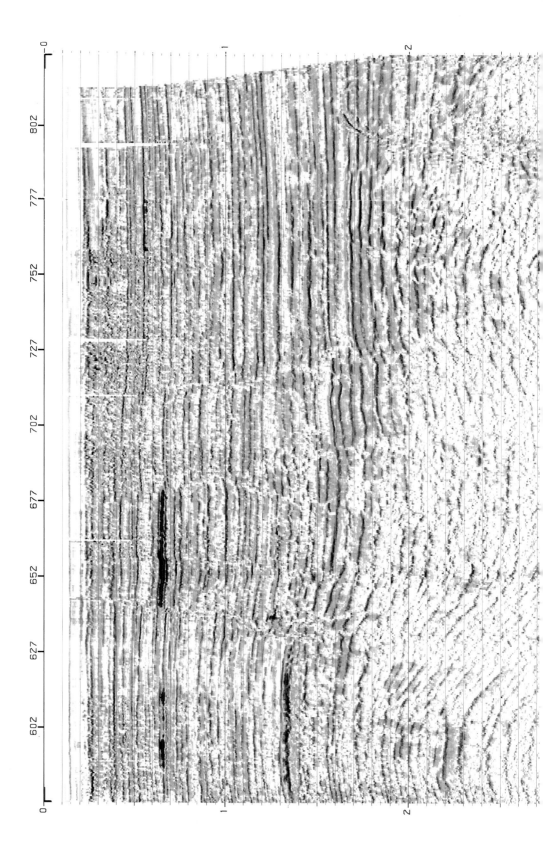

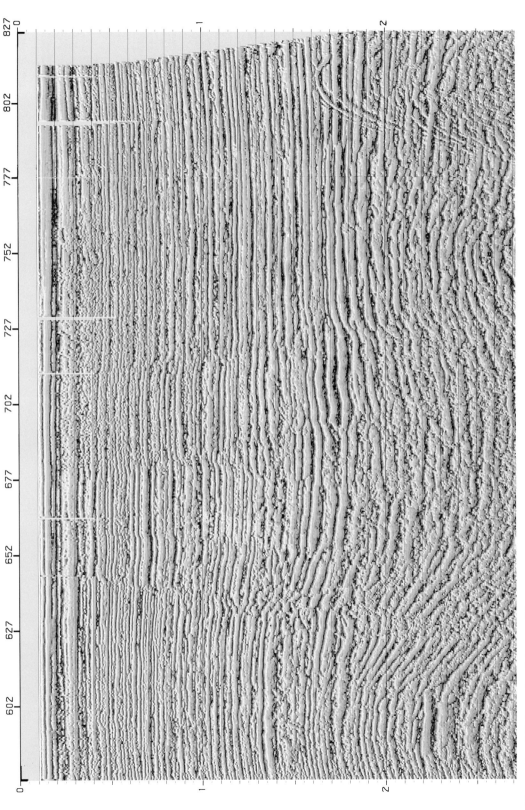

Figure 9 - Offshore Louisiana - Phase

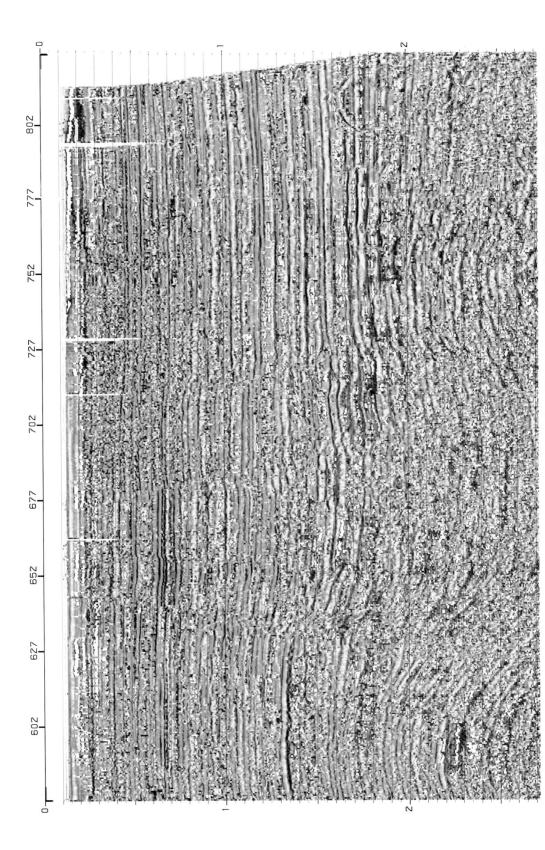

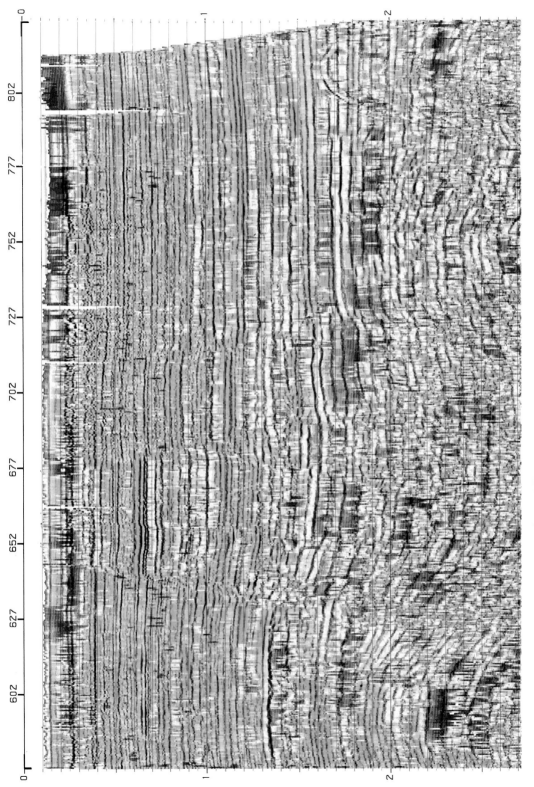

Figure 11 - Offshore Louisiana - Average Weighted Frequency

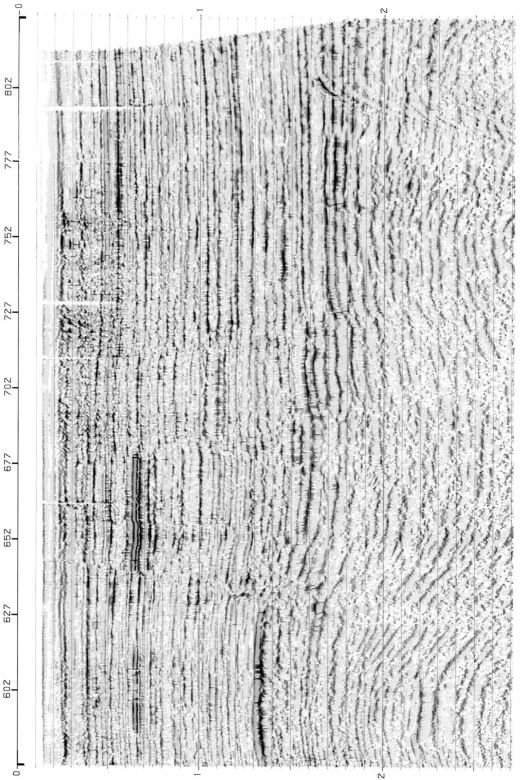

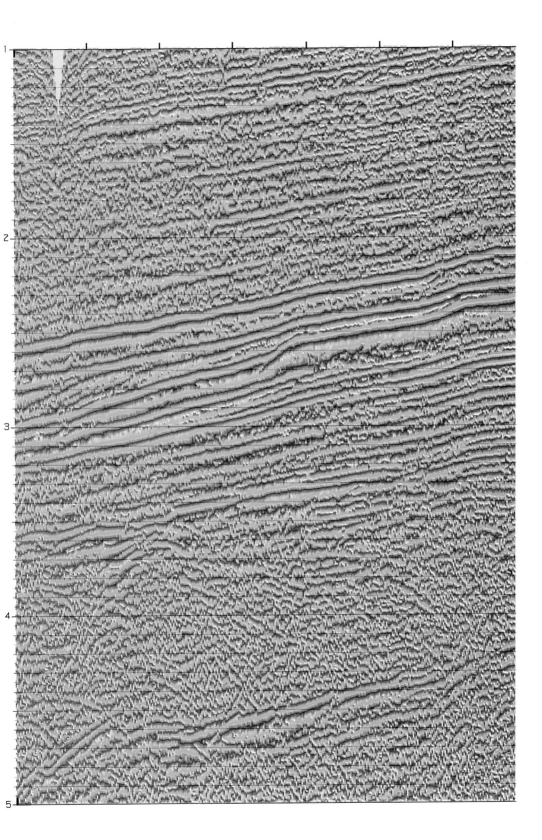

Figure 13 - East Texas - Phase

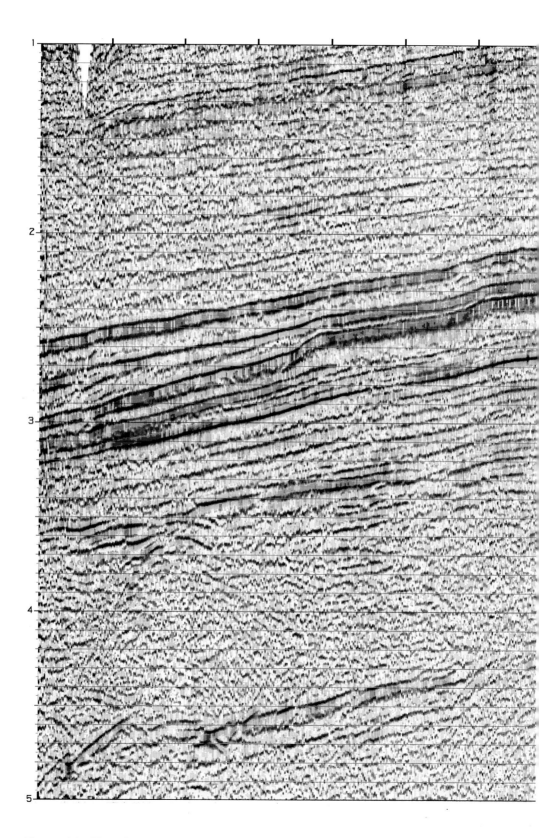

Figure 14 - East Texas - Apparent Polarity

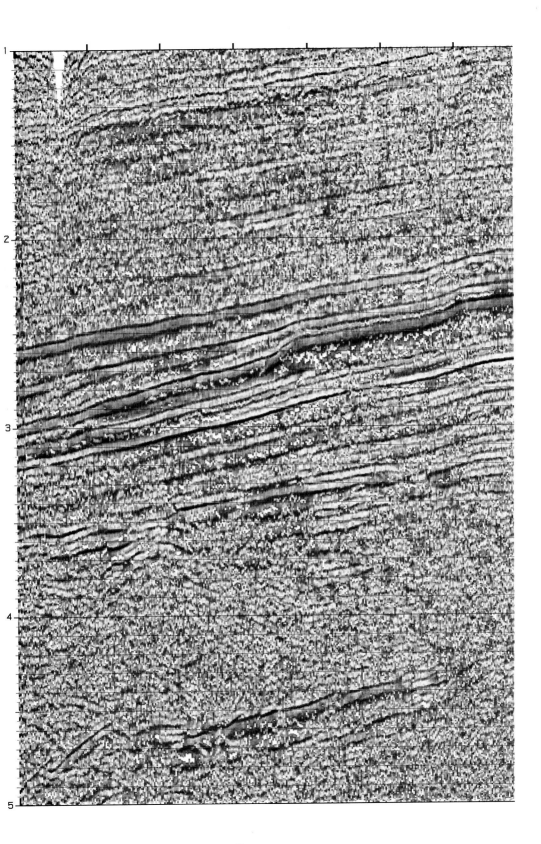

Figure 15 - East Texas - Instantaneous Frequency

Figure 16 - North Sea - Phase

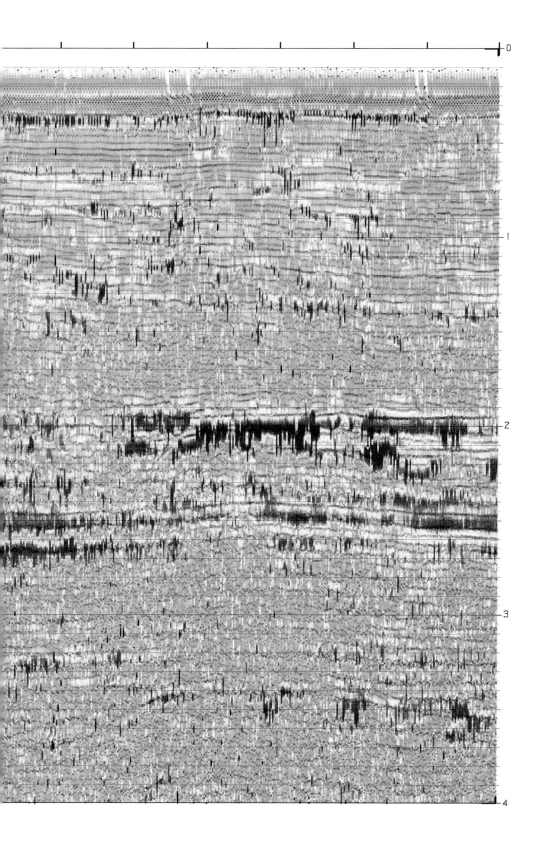

igure 17 - North Sea - Average Weighted Frequency

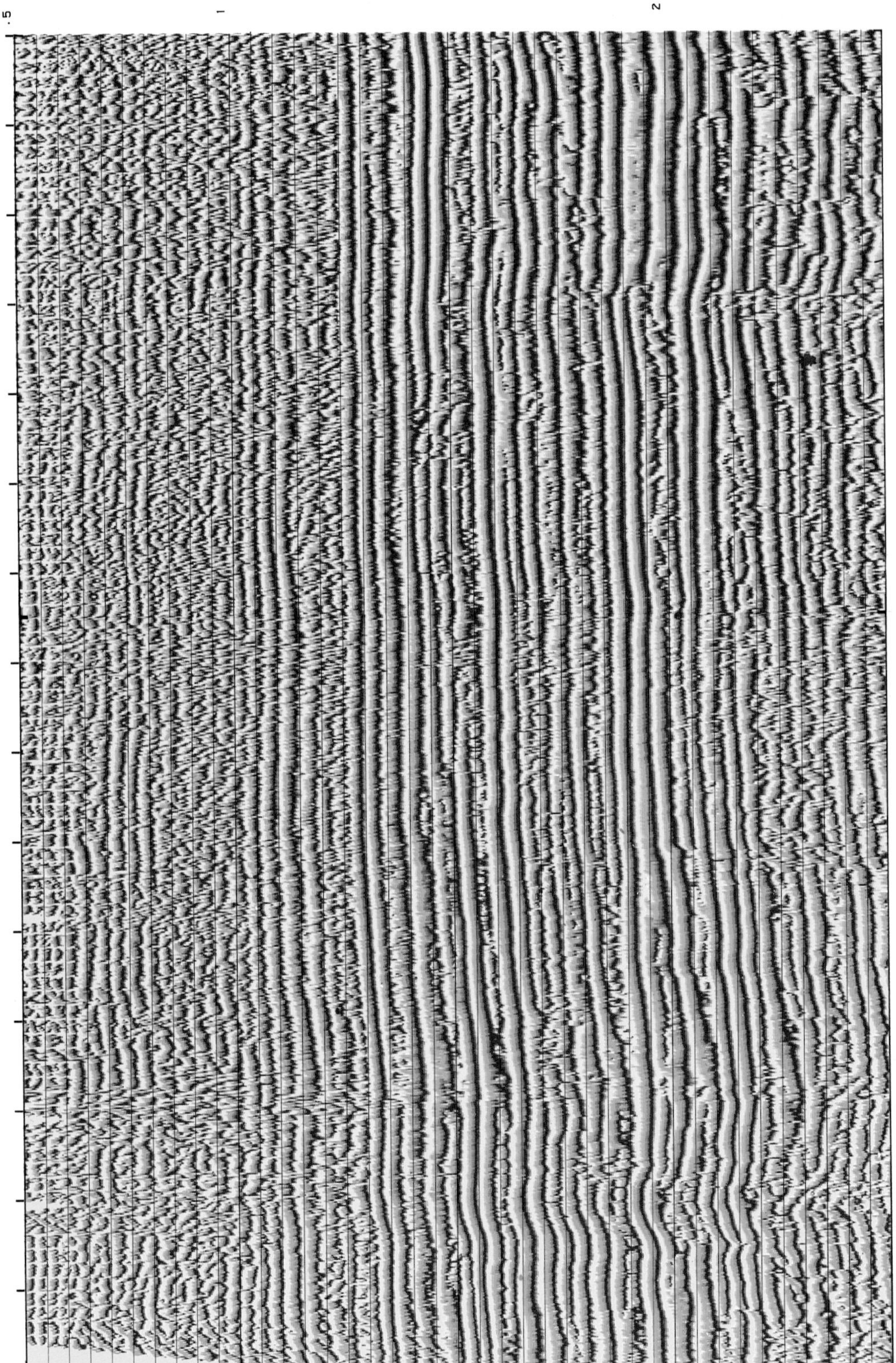

.5
1
2

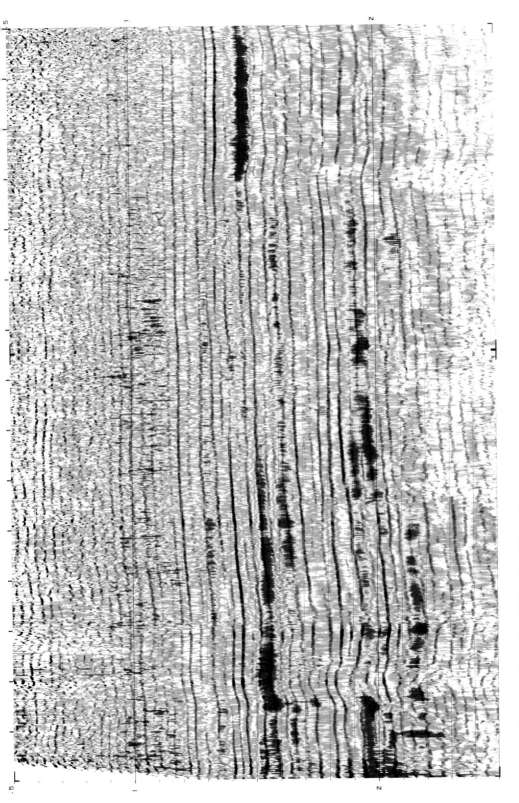

Figure 19 - Western Canada - Reflection Strength

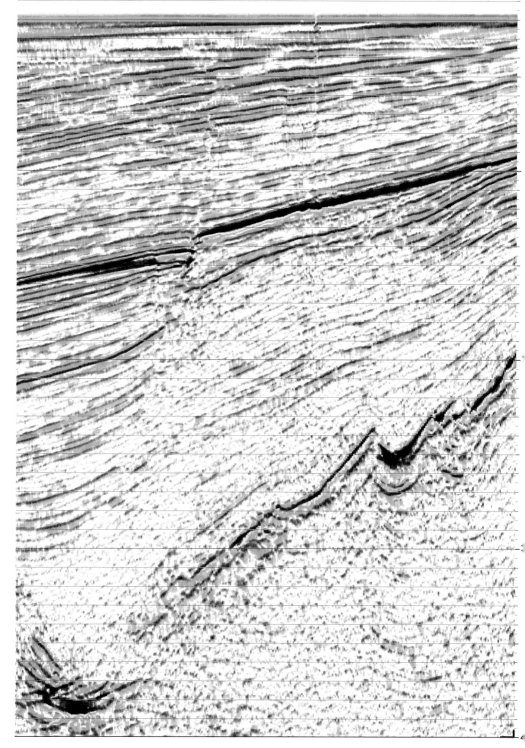

Figure 20 - Offshore Alaska - Reflection Strength

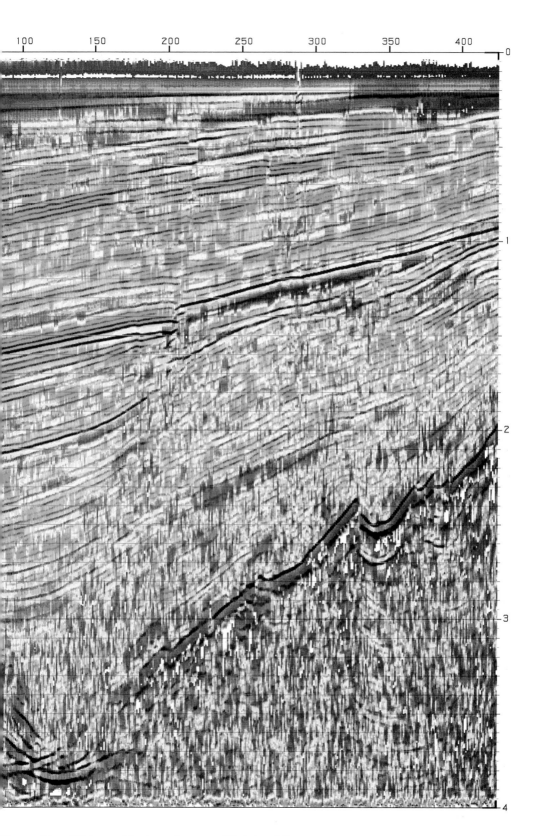

Figure 21 - Offshore Alaska - Average Weighted Frequency

REFLECTION STRENGTH   PHASE   INSTANTANEOUS FREQUENCY   AVERAGE FREQUENCY   APPARENT POLARI

Figure 22 - Attributes for Same Portion of One Trace

mulation. This line shows over 50 msec of structure at the top of the Cretaceous at about 2.4 sec.

WESTERN CANADA EXAMPLE

Figures 18 and 19 show phase and reflection strength for a seismic line across reefs in western Canada; this section is also shown on Figure 5. There is a reef just below 2 sec toward the left side of the section (A). Differential compaction over and off the reef has produced drape in the sediments for an appreciable distance above the reef. The reef is a weak reflector; it shows mainly as an interruption in the lateral continuity. Another smaller, less obvious reef is interpreted toward the right side of the section (B), similarly indicated by changes in lateral continuity of reflection strength. Reflection strength appears to be greater at the reef level on the seaward (left) side of the reefs.

An unconformity lies at about 1.55 sec at the left side of the section and 1.45 sec at the right side. A basal sandstone lying on this unconformity is absent in the regions where the reflection strength is particularly large (C to D and E to F). The bright spots indicating the absence of the basal sandstone appear to terminate abruptly above reefs (C, E), suggesting that differential compaction over the reef produced some relief on the unconformity surface at the time of deposition of these basal sandstones so that they are absent for some distance in the landward direction. The abrupt seaward termination of these bright spots thus provides additional evidence for the underlying reefs.

This bright spot interpretation could not be made without considerable well control in the area. Much more meaning can be obtained from seismic sections when well information can be combined with the seismic interpretation. Color-attribute sections extend well information to the surrounding region, such as that of mapping the distribution of this basal sandstone. (Likewise, noting the lateral extent of the color pattern measures changes from what is shown in the well.)

OFFSHORE ALASKA EXAMPLE

Figures 20 and 21 show a line offshore Alaska. A black and white representation of this line is shown on Figure 6, but this interpretation is speculative because there is no well control. These data have been migrated by wave equation migration which preserves amplitudes and reflection character. Migration has sharpened features and resolved buried foci and conflicting dips; however, there is appreciable cross-dip which has not been allowed for in the migration (which assumes

that the data are two-dimensional). The interfering events at (A) differ in cross-dip.

Figure 20 shows reflection strength. The strong reflecting event extending from 1.6 sec at the left to 0.9 sec at the right is an angular unconformity. Change in polarity of the reflection across the fault is indicated by the red color being superimposed on a trough between the black lines of the background gain-equalized section (B) to the left of the fault, and on a peak (C) to the right of the fault.

Figure 21 shows averaged weighted frequency. Commonly entirely different features stand out in one display compared to another. Thus the strong unconformity downthrown (B) does not stand out whereas a shallower reflection (D) does. The lack of an upthrown counterpart to (D) in the frequency-character is attributed to rapid growth of the fault at the time of deposition. Confirming this interpretation, note the considerable downthrown rollover into the fault here and just above this event (E), which is nearly gone in the shallower beds, although the fault clearly extends much higher in the section. Several seismic sequence boundaries have visible effects on the frequency patterns (as at F, G), as well as the unconformity cited above (B, C). Sequence boundaries are most evident in phase displays (the phase display for this line is not included) but they also show effects on reflection strength, frequency, and other displays. More information is obtainable by using a set of displays of different attributes synergetically than by interpreting them individually.

APPENDIX: INTERRELATION OF ATTRIBUTES

An enlarged portion of a seismic trace (actually part of one of the traces shown on Figs. 3, 13, 14, 15) is shown in Figure 22 along with the color representations of the reflection strength, phase, instantaneous and weighted frequency, and polarity.

REFERENCES CITED

Bracewell, R. N., 1965, The Fourier transform and its applications: New York, McGraw-Hill, p. 268-271.
Balch, A. H., 1971, Color sonagrams—a new dimension in seismic data interpretation: Geophysics, v. 36, no. 6, p. 1074-1098.
Reilly, M. D., and P. L. Greene, 1976, Wave equation migration: 46th Ann. Mtg., Soc. Exploration Geophys., Houston.
Sheriff, R. E., 1976, Inferring stratigraphy from seismic data: AAPG Bull., v. 60, p. 528-542.
Taner, M. T., et al, 1976, Extraction and interpretation of the complex seismic trace: 46th Ann. Mtg., Soc. Exploration Geophys., Houston.

Seismic Exploration for Stratigraphic Traps[1]

MILTON B. DOBRIN[2]

Abstract The capability, as of the late 1960s, of the seismic reflection method for the location of stratigraphically entrapped hydrocarbons was evaluated by Lyons and Dobrin. Limitations of the data resolution that could be expected from seismic reflection made it appear unlikely that significant improvement could be expected in its historically poor performance as a tool for finding stratigraphic oil and gas.

Since that time, significant developments in seismic data acquisition and processing have resulted in better definition and hence resolution of the basic seismic signal. Among these developments are the use of seismic amplitudes in defining stratigraphic features. Model studies and tests in areas where the geology is known give encouraging indications that these developments should improve the effectiveness of seismic reflection techniques. But we still have no way of knowing whether these capabilities overcome limitations of the method to the extent that discovering stratigraphic accumulations of hydrocarbons has actually been improved. It is hard to assess the value of geophysical data in stratigraphic discoveries that have involved extensive coordination of geophysics and geology. Moreover, few case histories are available on discoveries where the most modern seismic techniques have been used.

Recent interpretive techniques developed by Vail et al enable us to recreate depositional history and deduce depositional environments by analysis of reflection patterns on record sections. Such analysis can isolate areas that are environmentally most prospective for hydrocarbon accumulation, making it possible to locate stratigraphic entrapments with a minimum of additional seismic and geologic investigations.

Where technological improvements have increased the potential of the reflection method for finding stratigraphic oil and gas directly, there still exists a need for sophisticated integration of seismic and geologic data, particularly from wells that correlate with seismic lines. Most important discoveries of stratigraphic oil and gas (attributable to seismic reflection) have used such integration—many with seismic data that are considered primitive by today's standards.

Case histories present the performance of seismic reflection in exploring for various types of stratigraphic entrapment features, including carbonate bodies, truncations of clastic layers, sandstone bodies, and facies transitions.

INTRODUCTION

The status of the seismic method as a tool for discovering stratigraphically trapped oil and gas was reviewed by Lyons and Dobrin (1972) and covers the state of the art as of the late 1960s. Since that time, there were dramatic improvements in seismic techniques and in data processing. Many of these improvements are discussed in other papers of this volume.

These developments enhanced the resolution of seismic mapping and are expected, in principle at least, to increase the capability of the seismic method for locating stratigraphic traps. The principal limitation of the method as it affects this type of exploration has been in its resolution of detail, and any improvement in seismic resolving power should lead to better success in the seismic search for stratigraphically trapped hydrocarbons.

Yet, the difficulties that must be resolved are so great that a careful evaluation of the actual performance is needed before we can take for granted that our improved capabilities have led to an increase in the percentage of successful stratigraphic wildcats since the 1960s.

Ideally, we could determine our success by comparing the percentage of seismically located wildcat drilling (that results in stratigraphic discoveries) with the percentage that was successful in the late 1960s, but such a comparison is not feasible. Statistics of this type have not been published and there is doubt whether they even exist on a scale widespread enough to be meaningful.

Case histories provide another means to evaluate the performance of the seismic method in stratigraphic exploration, but few case histories have been released in which modern seismic techniques have had a part. Those that were published in AAPG *Memoir* 16 were mostly on fields that were previously discovered, so that most geophysical data collected in exploration used analog recording (much of it predating magnetic tape).

As a preliminary reconnaissance, the study of depositional history and depositional environments using patterns of seismic reflections offers considerable promise as an exploration tool. Several other papers in this volume are concerned with the mechanics of such analysis; here I show how this approach can be implemented in typical situations.

The effectiveness of the seismic method varies for different kinds of stratigraphic features. For example, many types of reefs can be located with consistent success on seismic record sections. Potentially productive features such as truncations, sandstone bodies, or facies changes are more difficult to observe on seismic records directly, but

[1]Manuscript received December 3, 1976; accepted June 6, 1977.

[2]Geology Department, University of Houston, Houston, Texas 77004.

can often be located by proper coordination of the seismic data with pertinent geologic information. A judicious combination of seismic-stratigraphic modeling and analysis of available case histories should make it possible to establish criteria for predicting the success of the reflection method in finding various types of stratigraphic targets and to establish the critical parameters for each.

USE OF REFLECTION PATTERNS IN EXPLORATION FOR STRATIGRAPHIC FIELDS

Methods

A knowledge of the depositional history and depositional environment in a prospective area is invaluable in searching for stratigraphically entrapped hydrocarbons. More detailed exploration approaches such as waveform amplitude analysis (Neidell and Poggiagliolmi, this volume) velocity-log synthesis (Lindseth, 1976), or even stratigraphic drilling, can be applied in subsequent exploration of areas thus isolated.

Many depositional and erosional processes such as onlap, progradation, and truncation give rise to characteristic patterns on seismic records. Several papers in this volume demonstrate such relations. These studies show that an important key to depositional history is the nature of cycle terminations with respect to adjacent reflection events. The terminations make it possible to locate the boundaries between what Vail (this volume) has termed *sequences,* which are zones where specific types of deposition, each associated with characteristic reflection patterns, have taken place. The degree of convergence or divergence of reflections is also diagnostic of environmental conditions during deposition within a sequence. Such mechanisms as delta formation, transgression and regression of sea level, and tilting of strata are associated with patterns on record sections that make it possible for the interpreter to recreate the geologic history of sedimentary areas.

A somewhat different indicator of depositional environment is interval velocity (Domenico, 1976) which can be measured with a precision never before obtainable from digital analysis techniques (Hilterman, 1976). Such information makes it possible under favorable conditions to identify lithology and to determine such quantities as sandstone-shale ratios.

The use of reflection patterns on record sections for deducing depositional history is not really new, but little information on the subject has been available in the literature. Only recently, have the techniques involved been written up in teachable form so that each geophysicist does not have to "re-invent" them himself.

This type of stratigraphic interpretation was recently facilitated by the introduction of acquisition and processing techniques that have suppressed much of the noise which would otherwise obscure subtle features (such as cycle terminations that provide a key to depositional mechanisms). By the elimination of scattered energy, multiple reflections, and reverberations such as those in water layers, each reflection cycle can be identified with a single lithologic boundary or with a closely spaced succession of such boundaries. Older data can be especially useful when the entrapment is due to a combination of structural and lithologic factors. The structural pattern can commonly be ascertained from less-than-ideal seismic data, thus the lithology can be determined from geologic data such as are found in outcrops or well logs.

In the early stages of seismic exploration of an area, depositional features such as deltas or pinchouts can be identified from reflection patterns on record sections, but actual entrapments are rarely detailed well enough to permit selection of drilling locations from these studies alone. Ordinarily the study of the patterns can only narrow the search to areas in which the depositional environment appears favorable for stratigraphic accumulation of hydrocarbons.

Typical Applications

To illustrate how patterns of reflections on seismic sections might be used for stratigraphic evaluations, seismic lines were selected from a number of typical offshore areas where no drilling information is available. Any subsurface information that may exist in these areas is not accessible to the author, so only seismic data are used in the illustrations that follow.

The first example (Fig. 1) is a simple one. The section (from Kendall, 1967) was shot on the continental shelf of the Gulf of Mexico. The most conspicuous feature on it is a major unconformity on which onlap deposition has taken place. If any of the onlapping beds are favorable reservoir rocks, and if they are overlain by impermeable material, the updip pinchouts of these beds would be promising targets for drilling. The first stage in further exploration could be a stratigraphic test that correlates with the seismic line. This would identify any potential reservoir beds and locate them with respect to the reflections on the section. At such a well, all acoustic properties of the penetrated formations can be evaluated in terms of their relation to lithology.

Figure 2 is a record section made over the shelf edge of a delta of a major river system. The river source is in high mountains elevated during the

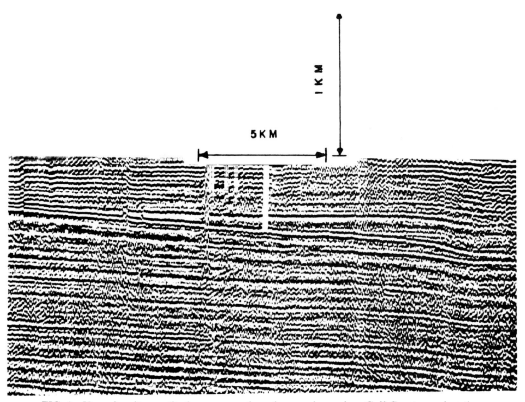

5 K M

1 K M

FIG. 1—Unconformity constituting sequence boundary as observed on Gulf Coast record section
(from Kendall, 1967).

Cenozoic. The line extends seaward across the present deltaic platform and terminates a short distance down the continental slope. The delta has built upward more than 3,000 ft (914 m) and outward about 20 mi (32 km) along the line represented by the seismic section. The shelf edge advanced outward horizontally from the right edge of the section to a point about 5 mi (8 km) shoreward of the present shelf edge. Then deposition began on the shelf and the edge moved upward and outward with time. The most likely explanation for the greater rate of deposition on the shelf is a relative rise in sea level. Although if this took place, finer-grained sediments associated with deep-water deposition would be expected to overlie the deltaic formations deposited when the water was shallow. The supply of sediments carried by the river system also would affect the lithologies of deposits on the shelf and slope. The area where the prograding sediments are buried below the horizontal shelf formation should be favorable for hydrocarbon entrapment. The truncation of the deltaic sandstones exhibiting the oblique progradational patterns against overlying layers (that are likely to be fine-grained silts or shales)

should make the area behind the present shelf edge favorable for further exploration.

A third example is from the Atlantic shelf off the eastern United States. The Baltimore Canyon line crosses an unusual structure feature in about 200 ft (67 m) of water (Fig. 3). It is an anticline which was truncated by an unconformity in the upper Eocene; the truncational surface and the formations directly above it show a more gentle reversal about the same axis than is observed in the beds below the unconformity. A magnetic high coincident with the structural axis makes it very likely that the structure resulted from an elongated igneous intrusive. But the mechanism by which the unconformable surface and the beds overlying it acquired their reversal in dip is not clear. The question has an important bearing on the origin of any hydrocarbons that might be trapped in the structure.

One possible explanation for the relief above the unconformity is incomplete beveling of the original anticline at the time of truncation, leaving a ridge on which subsequent deposition took place. A second explanation is a renewal of deformation after the sediments were deposited hori-

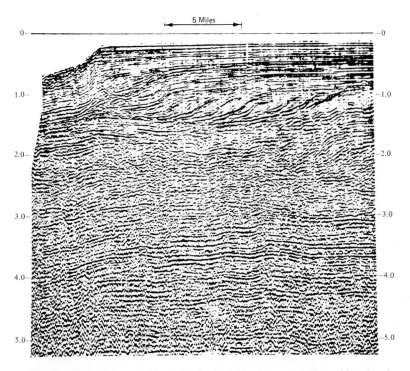

FIG. 2—Forward progression of deltaic deposition patterns. Note sudden rise of shelf edge to present position after long period of horizontal movements (Proprietary section of United Geophysical Corp.).

zontally over a completely beveled erosional surface. A third is differential compaction over the igneous ridge, indicated by magnetic data, to lie several thousand feet below the unconformity.

If the first hypothesis is correct, we should observe terminations of reflection events against the unconformity, which indicate onlap against this surface. But no such pattern is seen, as all cycles directly above the surface are continuous. The second explanation would require an onlap pattern with cycle terminations that begin farther up in the section than the unconformable surface. Although there appears to be a small amount of poorly-defined onlap starting well above the unconformity, there is not enough to account for the present relief. The third possible mechanism, differential compaction, would not call for cycle terminations, but rather thinning of intervals between reflection cycles in the direction of the structural axis. The reflections show thinning, although their poor quality makes it hard to say how much. Such an analysis should have considerable practical significance because of the relation between the mechanics and timing of the de-

formation and the migration of any hydrocarbons that might be trapped within the structure.

NEW DEVELOPMENTS THAT BEAR ON STRATIGRAPHIC TRAP LOCATION

Recognition of Limitations

Lyons and Dobrin (1972) showed how the primary limitation of the seismic method (for locating stratigraphic entrapment features) is its resolution. The higher the frequency of the seismic pulse, the greater is the resolution that can be obtained. Because of the selective attenuation of high frequencies as the pulse travels through earth materials, it is not generally possible to improve the resolution simply by generating higher frequency pulses, or by filtering out lower frequency components of the source signal. Except for carbonate reservoirs, most stratigraphic entrapments are in sandstone layers which are much thinner than a seismic wavelength. It is difficult to detect such features on seismic reflection records; moreover, the velocity and density contrasts between the oil-bearing sandstones and the shales that provide stratigraphic seals for oil are often

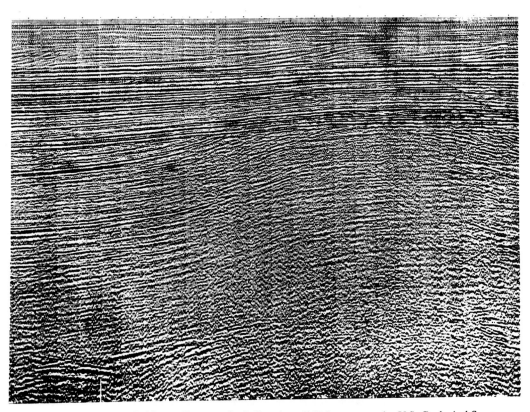

FIG. 3—Structure along Baltimore Canyon seismic line shot off U.S. east coast by U.S. Geological Survey.

very small, so that the reflectivities, and hence amplitudes of reflections, will be so low that the events may not be observable above noise.

A more serious limitation is the difficulty of isolating reflections from boundaries of reservoirs where there are other boundaries less than a wavelength away. The reflections from the top and the bottom of a converging sandstone, for example, are so subject to interference from similar reflection events that originate short distances above and below it, that the convergence may not be identifiable. Consider the model section in Figure 4, composed of materials having six discrete velocities separated by five horizontal boundaries having spacings considerably less than a wavelength. The reflection observed from the complex of boundaries is generated by superposition of the reflections from the individual interfaces. The reflected wave (bottom of figure) is longer and more complicated than the source pulse and it is unlikely that changes in the thickness or velocity of any of the individual layers (such as will occur in a pinchout) will be deduced

from changes in the waveform observable at the surface.

New Techniques

Certain complex layering patterns may be associated with characteristic waveforms that can be identified by statistical techniques. Mathieu and Rice (1969) and Waters and Rice (1975) have reported on experiments in which patterns have been identified in reflections from a stratigraphic interval of varying lithology. The zone, which is within the Pennsylvanian Morrow Formation in Oklahoma, has been penetrated by a series of wells, across which seismic lines have been laid out. Synthetic seismograms were made from the velocity logs obtained in the wells. Figure 5 compares a set of velocity logs, synthetic traces, and field traces for this interval at five of the wells, each showing a different kind of lithology within the interval. Pattern recognition techniques involving the use of factor analysis were applied to records along lines between the wells and the various kinds of lithology at each shot point were

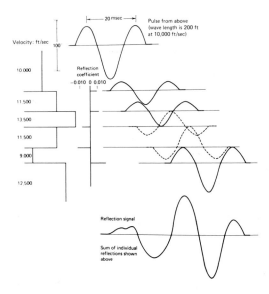

FIG. 4—Synthesis of reflection from five interfaces all less than a seismic wavelength apart (from Dobrin, 1976).

identified and mapped (Fig. 6). The productive sandstone labeled S-1 appears to be a channel deposit (sinuous configuration). Mathieu and Rice (1969) applied discriminatory analysis to differentiate between sandstone and shale within a specified stratigraphic interval in another area. Although techniques of this type were successful in the two studies that were reported, the authors note that there were other instances where the statistical approach did not predict lithology reliably.

Zero-phase wavelet extraction techniques (Neidell and Poggiagliolmi, this volume) simplify the complex waveforms generated by marine sources so that the reflection signals will have the greatest possible resolution. The time shifting in the zero-phase manipulations of the signals and the symmetry in the resultant waveform make it possible to associate the wavelet peak with the actual reflecting surface. The basic limitation in resolution is still the attenuation of the amplitude, which falls off exponentially both with frequency and depth to the reflecting surface. The deeper this surface, the lower will be the dominant frequency of the signal that is returned from it, and hence the poorer the resolution.

Some types of stratigraphic traps allow isolation of the reflections that come from boundaries of producing zones. Extraction techniques are most likely to be effective in identifying reservoirs such as pinchouts or sandstone bodies where the

formations immediately surrounding them are homogeneous over vertical distances that are large compared with seismic wavelengths. Oil- or gas-bearing sandstones on the Gulf of Mexico and the North Sea are commonly so isolated that reflections from their tops and bottoms are not subject to interference from adjacent reflections. However, the stratigraphic setting of surrounding formations results in closely spaced reflecting boundaries such that the productive sandstone can not be isolated on seismic records.

SEISMIC DELINEATION OF HYDROCARBON TRAPS

Carbonate Traps

The easiest type of stratigraphic trap to locate by seismic means is one in which the reservoir rock consists of carbonates rather than clastics. Reefs are the most common of the hydrocarbon-bearing carbonate features, but other limestone buildups such as the progradational deposits in the Abo reef trend in New Mexico are productive.

Reefs and other prospective carbonate bodies are more likely to be observable on seismic record sections than most sandstone entrapments because of the high contrast between their seismic velocity and that of surrounding formations, particularly those of clastic composition. Where the reef contrast generally will be much smaller. Car-

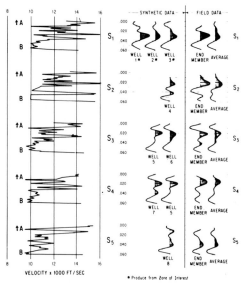

FIG. 5—Velocity logs and synthetic and recorded waveforms for reflections from lithology types indicated by numbers S1 to S5 for stratigraphic interval A-B (from Waters and Rice, 1975).

bonate bodies are generally thicker than most sandstone reservoirs so that the resolution of the seismic pulse does not have to be so great. Finally, limestone is not as compactible as most clastic rocks so draping structures which can be identified seismically are more likely to be observed over limestone bodies than over comparable clastic entrapments. Thus, stratigraphic traps in carbonates have been easier to locate by the seismic reflection method than traps in clastics.

For the first 15 years after oil was discovered at Leduc in 1947, two primary indicators of reefing were sought on seismic records. One was structural draping over the reef mass; the other was an apparent structural high below the level of the reef resulting from a higher velocity in the reef limestones than in the shales they replaced. It was rare at that time to see a reef directly on seismic records, usually because systems of that time could not adequately suppress noise. Now this noise can be removed effectively, using field techniques such as common-depth point recording and digital filtering. It should be possible to recognize reefs directly on relatively noise-free record sections by noting the absence of reflections from their interiors, and in many cases by observing reflections from their top surfaces. One good example of this is the Horseshoe atoll in West Texas, which shows up on seismic sections (as illustrated in Fig. 7).

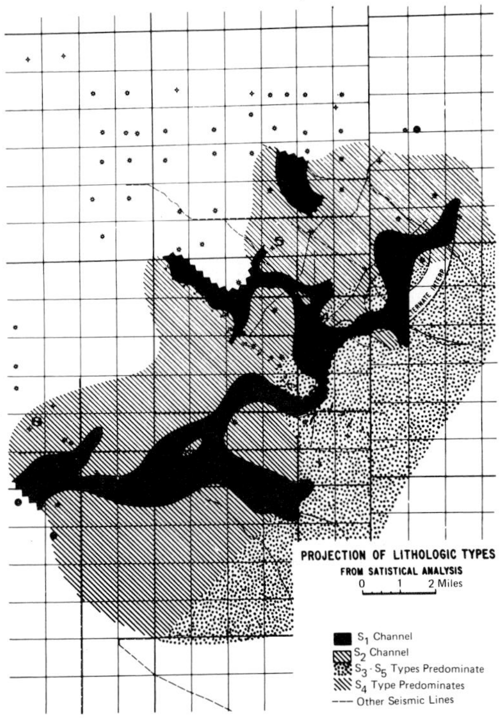

Niagaran reefs from Michigan also are characterized by an absence of reflections (Fig. 8). Over the reef itself there is a characteristic weakening of reflection strength which seems to be diagnostic of this kind of reefing. Keg River reefs in the Lower Devonian of northwestern Alberta and northeastern British Columbia also can be identified on record sections because of the interruptions they cause in the regular reflection patterns.

In the search for both the Michigan and Keg River reefs, various digital processing procedures such as deconvolution, automatic statics, interval velocity analysis, and seismic modeling have been useful in showing up diagnostic distortions of beds above and alongside the reefs, and the zones in their interiors from which no reflections originate (Evans, 1972; Caughlin et al, 1976).

Truncational Entrapments

Stratigraphic traps commonly are associated with erosional truncations such as pinchouts below unconformities or subcrops against erosional surfaces. Limitations in resolution discussed earlier make it particularly difficult to detect accumulations in reflection data where the truncational surface makes a small angle with the unconformity. The erosional surface against which the entrapment takes place will rarely yield a reflection that is directly discernible on a record section, commonly because of interference with other interfaces near the unconformity. Where the angle is large, it is more likely that divergences will be observed that can be used to map the unconformity indirectly.

The Dogger productive zone in the Hohne field of West Germany is a truncational entrapment that was located using seismic data. Production is at the subcrop of Dogger sandstone against an unconformable contact between the Cretaceous and the Jurassic (Hedemann and Lorenz, 1972). The unconformity overlies a previously established structural oil entrapment in the Lias. Prospects for Dogger production at the subcrop were recognized on seismic cross sections plotted from old-fashioned paper records of the type that were shot in the 1950s. Such a section passing through the Dogger discovery well (Hohne 108) is shown on Figure 9. The Dogger reflection was identified from well ties, and its termination against the angular unconformity indicated by the reflections was the basis for selecting the location for Hohne

PROJECTION OF LITHOLOGIC TYPES
FROM SATISTICAL ANALYSIS
0   1   2 Miles

■ S_1 Channel
▨ S_2 Channel
▧ S_3 - S_5 Types Predominate
▩ S_4 Type Predominates
— — Other Seismic Lines

FIG. 6—Map showing lithology variation over stratigraphic interval A-B of Figure 5, based on statistical analysis of waveforms recorded along seismic lines (Waters and Rice, 1975).

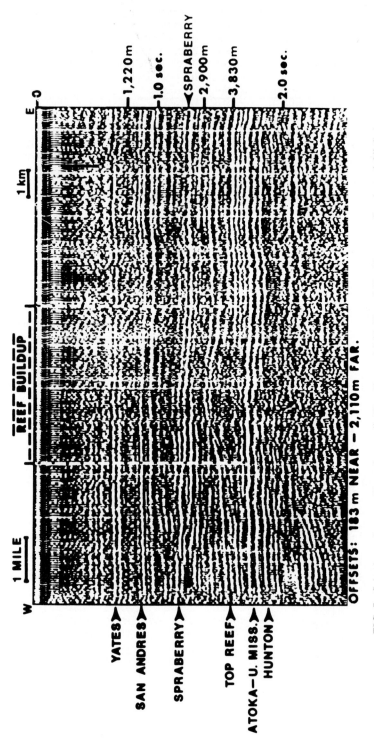

FIG. 7—Seismic section over Horseshoe atoll, West Texas, 6-fold, 38 to 8 Hz (Courtesy, Continental Oil Co.).

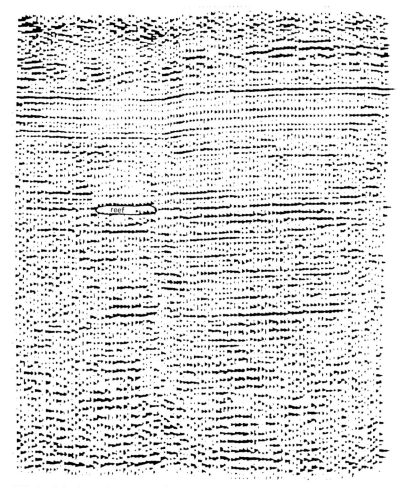

FIG. 8—Seismic section over productive reef in Michigan (Courtesy John W. Mack).

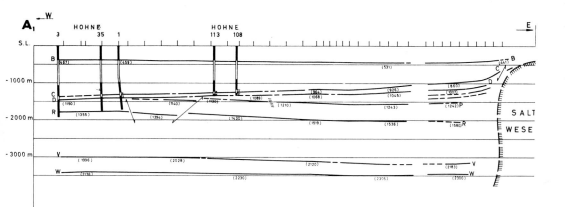

FIG. 9—Section obtained by plotting seismic reflections over Hohne oil field, West Germany. Well 108 was discovery well for Dogger production, trapped by updip truncation of Dogger sandstone at Cretaceous–Jurassic unconformity (Hedemann and Lorenz, 1972).

Conventional stack

Geoglogical section

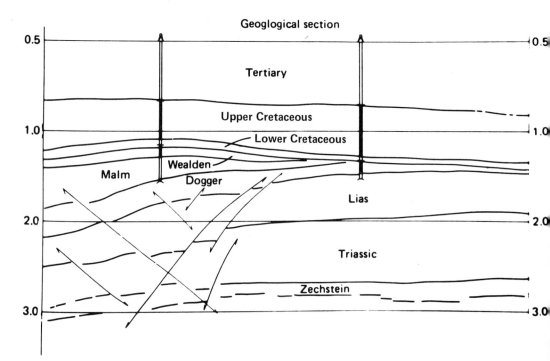

FIG. 10—Modern seismic section over Hohne field and geologic cross-section showing identification of reflections (Prakla-Seismos).

108. Figure 10 shows a digitally recorded and processed seismic section over the field which was shot long after drilling was complete. The geologic cross section identifies the reflectors.

Oil accumulations at Prudhoe Bay are trapped on one side by a truncational surface between the impermeable Cretaceous and the productive Jurassic–Mississippian below (Morgridge and Smith, 1972). The surface shows up clearly on the NE-SW seismic section of Figure 11; the original analog data were digitized and reprocessed to give the sections shown. The size of the reservoir and the large angle of divergence made by the truncational surface cause the erosional entrapment feature to be conspicuous on the section.

Not all truncational entrapments show up on seismic records. The Boscán field, near Lake Maracaibo in Venezuela, was discovered by a well (7F-1, Fig. 12) located on the basis of a seis-mically determined structural closure that turned out to be unrelated to the actual entrapment (Sutherland, 1972). Development drilling revealed a much larger productive area than was expected from the seismic map (Fig. 13). It was realized that the accumulation was not trapped structurally at all but rather by an updip pinchout of the Boscán sandstone against an erosional unconformity at its north edge. The production is bounded on the east by a fault that was not observed on the seismic records. If the well had been located 200 m east of its actual position, it would have penetrated the sandstone on the wrong side of the fault and would have been dry. The entire area probably would have been abandoned.

Figure 14 is a cross section of the field. It is questionable whether the unconformity could be located by modern shooting and processing pro-

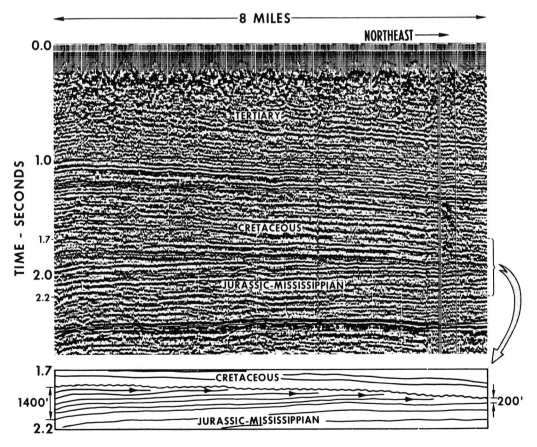

FIG. 11—Northeast-southwest trending record section across Prudhoe Bay oil field made by digitization and computer processing of data originally recorded on analog tape. Note truncational surface between Cretaceous and Jurassic–Mississippian (Morgridge and Smith, 1972).

cedures because of the small dip of the beds and
the thinness of the sandstone near the actual
pinchout. However, it is probable that the fault
on the east could be detected by modern seismic
techniques, but it is, of course, a less critical element of the entrapment.

Two truncational accumulations are illustrated
at sites where the seismic method showed conditions favorable for stratigraphic hydrocarbon ac-

cumulation, and a third where seismic shooting
led to a discovery on the basis of pure serendipity.
The angle between the truncational surface and
the surfaces of bedding is probably the most important factor that determines whether a potentially productive pinchout can be observed on a
seismic section. Even where the angle is large, as
in Figure 1, the proper geologic identification of
the reflection corresponding to the productive

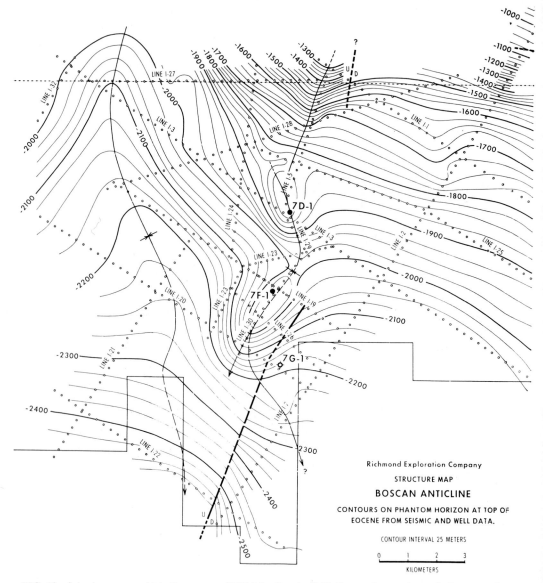

FIG. 12—Seismic map on which discovery well 7F-1 for Boscán field, Venezuela, was located. Contours of top of
Eocene. (Sutherland, 1972).

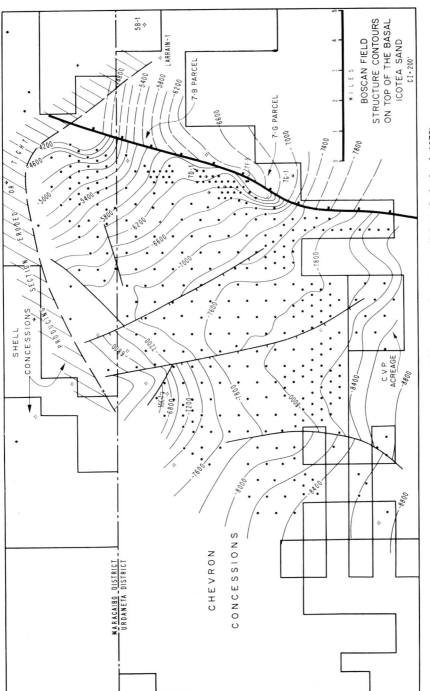

FIG. 13—Present structure map of Boscán field contours based on well data (Sutherland, 1972).

Milton B. Dobrin

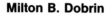

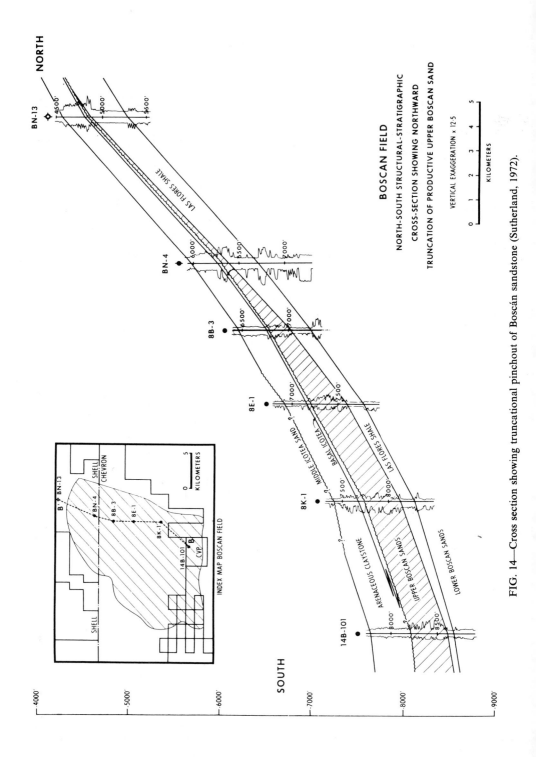

FIG. 14—Cross section showing truncational pinchout of Boscán sandstone (Sutherland, 1972).

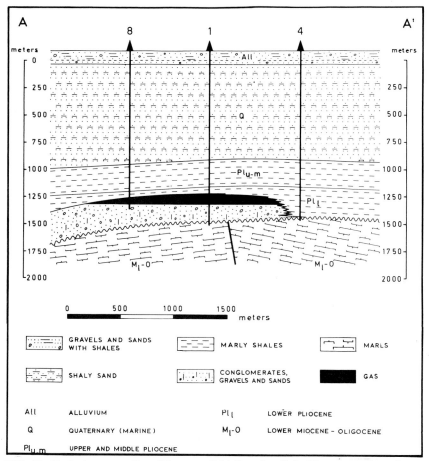

FIG. 15—Geologic cross section of Sergnano field, Po valley, Italy. Discovery well, no. 1, was located from seismic data. (Rocco and d'Agostino, 1972).

body is an obvious requisite for locating a successful drilling site; such an identification requires ties with wells or with surface data.

Entrapment in Sandstone Bodies

Stratigraphic traps of oil or gas are in sandstone bodies of various types. These may be buried stream channels, sandstone lenses, or sealed updip terminations of sandstone onlapping against unconformities. Sandstone bodies in general are the most difficult of all stratigraphic features to map by the seismic reflection method. One reason is that the surfaces bounding a sandstone body are not commonly conformable with other nearby interfaces. Thus a reflection from such surfaces can not be built up by summation or reinforcing events from parallel surfaces in a way that will allow the body itself to be outlined by reflection on record sections. Another difficul-

ty arises from the poor velocity contrast commonly observed between productive sandstone bodies and surrounding shale formations that give rise to the entrapment.

The seismic data associated with three types of sand-body entrapments: a turbidite deposit, a series of sandstone lenses, and a buried stream channel will all be considered here.

The Sergnano field (Rocco and d'Agostino, 1972), in the Po Valley of Italy, is a large gas field discovered using a seismic reflection survey performed in 1953. Interpretation of the seismic and subsurface data from the field, that was published by the authors, indicates the gas production to be from a conglomerate mass which changes to shale along time lines shown on Figure 15. However, a record section from a seismic line (Fig. 16) shot after the field was developed shows a reflection dipping northeastward between wells 1 and 4

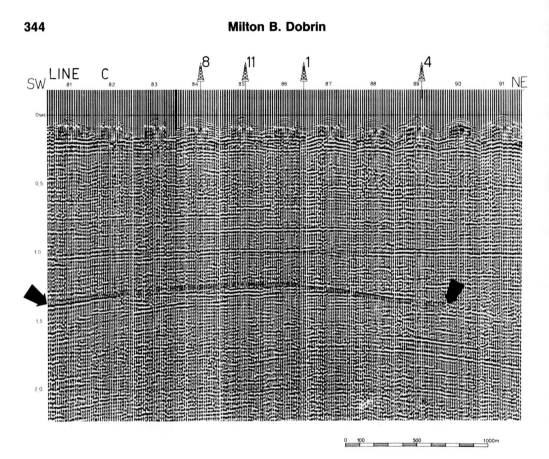

FIG. 16—Analog record section over Sergnano field with reflection indicating top of gas-bearing sandstone body indicated by shading. (Rocco and d'Agostino, 1972).

across almost horizontal events that appear to be multiple reflections. The dipping reflection, according to modern concepts, must follow a time line and hence must represent the top of a northward-thinning conglomeratic body (shown by the shaded line) which a recent reinterpretation of well data indicates is a turbidite deposit. Common-depth point recording and digital processing should show the reflection from this surface more clearly.

Geophysics performed ambivalently in the discovery of the Candeias field (Vieira, 1972) in the Reconcavo basin of Brazil, even though the discovery well was located on the basis of seismic data. Figure 17 shows the seismic map, indicating a well-defined anticlinal axis. The location of the initial well, 1C-1-BA, was based on this map. After the well turned out to be productive, further drilling revealed no relation between the oil entrapment and the seismic "structure." The production turned out to come from four overlapping sandstone lenses (Fig. 18) which were not indicated on the seismic records. The seismic "an-

ticline" of Figure 17 could be accounted for by velocity pullup resulting from an anomalously high speed in the massive sandstone at the surface. The discovery well penetrated the lower edge of the uppermost sandstone, which by good fortune yielded oil. If the well had been located a few hundred meters farther to the northwest, it could have missed the sandstone body altogether.

The original seismic work at Candeias was done about 1940. In the late 1960s, experimental shooting was carried out to ascertain whether modern seismic methods of recording and processing could highlight the productive sandstone lenses. No indication of the sandstones was found in the record sections obtained from the digital records. The negative results from the more recent work demonstrate the difficulty that is commonly encountered in delineating lenticular sandstone bodies encased in other clastics, even where modern seismic techniques are applied.

In the South Ceres field of Noble County, Oklahoma, the production occurs in a buried channel sandstone. Although this field was not initial-

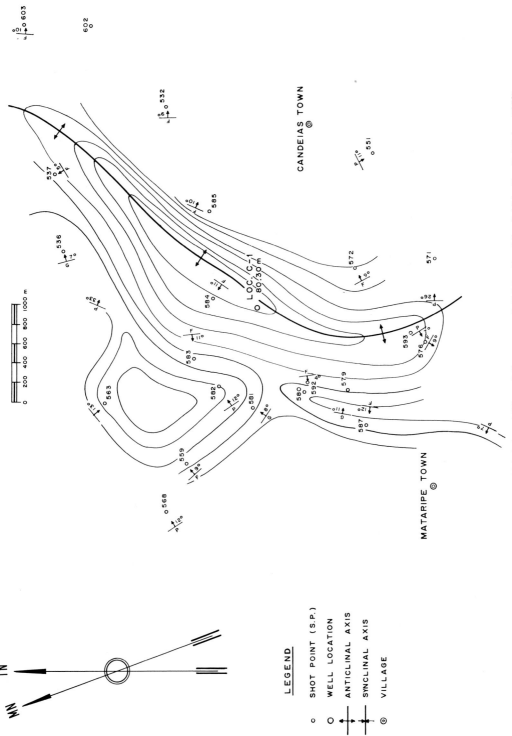

FIG. 17—Seismic structure map which led to drilling of discovery well C-1 for Candeias field in Brazil. (Vieira, 1972).

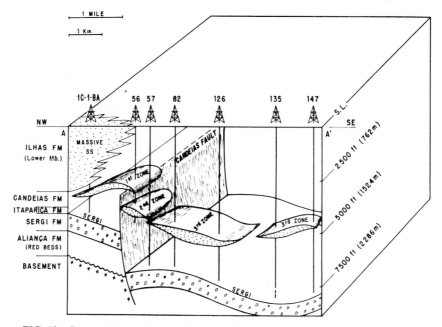

FIG. 18—Cross section of Candeias field after development. Production is in sandstone bodies. Seismic high in Figure 17 was caused by velocity lensing of "massive sandstone." (Vieira, 1972).

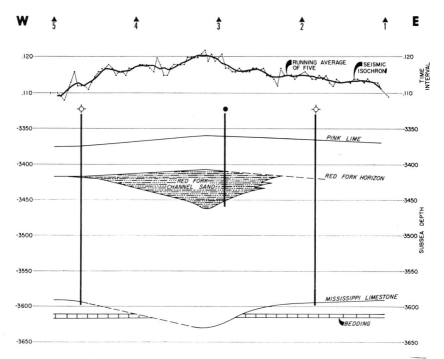

FIG. 19—Cross section across South Ceres field, Noble County, Oklahoma. Production is in Red Fork channel sandstone (Lyons and Dobrin, 1972).

ly discovered from seismic shooting, it was developed with a minimum number of dry holes by using seismic data. The geophysical results are described in detail by Lyons and Dobrin (1972) and are summarized briefly here.

The sandstone channel from which production is obtained in this field is hairpin-shaped and very narrow. Any well drilled more than a few hundred feet from the axis of the channel generally is dry. The problem of locating the channel seismically was complicated by the fact that no reflections were observable from the top or bottom of the productive Red Fork sandstone. A reflecting bed above the sandstone, the Pink Lime, drapes over it as shown in Figure 19 and the reflecting surface of the Mississippian limestone, below the sandstone, has a channel cut into it with an axis underlying that of the Red Fork formation. Thus the two reflectors have their greatest separation along the axis of the channel, and this shows up as a maximum time interval between the respective reflections. The time anomaly is marginal because of the small relief involved and lack of significant velocity contrasts, but careful trace-by-trace picking of reflection times and smoothing of statistical irregularities yielded a characteristic pattern on the time-interval profile which located the channel axis precisely on the seismic line.

It is evident from these examples that sandstone bodies are difficult to observe directly on seismic sections unless overlying strata are conformable with the top surface of the sandstone. In marine areas, gas at the top of a sandstone body might result in a strong reflection which should allow detection of the accumulation by the "bright spot" technique.

In some Montana and Wyoming fields, oil is trapped in sandstone lenses with upper surfaces that are not conformable with overlying strata, as was observed at Candeias. The areas over some of these fields were surveyed by reflection seismic methods before their discovery, but no indications of the sandstone bodies were found—the discovery wells having been located on the basis of nongeophysical considerations. The Bell Creek and Recluse fields are in this category.

Entrapment by Facies Changes

Where the accumulation of hydrocarbons is governed by lateral changes (along time lines) of lithology from permeable to impermeable facies, the seismic method has had varying degrees of success, depending on the nature of the facies changes involved. It is difficult to see a permeability barrier as a discrete reflection on a record section because reflections will follow time lines rather than lithology boundaries and termina-

tions of characteristic wave forms can sometimes be used to trace the facies boundary on a seismic map. Where facies boundaries cross anticlinal axes at right angles, so that entrapment is structural in one direction and stratigraphic in the opposite, seismic data can be used to delineate the structural feature although it is rare for the seismic records to reveal, in addition, the facies changes responsible for the entrapment perpendicular to the anticlinal axis. To locate such accumulations in this way requires careful coordination of seismic data with geologic information from surface or subsurface sources.

The Red Wash field in the Uinta basin of Utah (Chatfield, 1972) is the largest oil field in the world producing from lacustrine rocks. Surface geology in the area of the field (Fig. 20) indicated a northwest to southeast succession of depositional environments in the Green River Formation where marginal lacustrine facies change to offshore fine-grained lacustrine facies. A weak anticlinal trend perpendicular to the subsurface projection of this facies boundary was suggested by the surface observations, and a seismic survey was performed to find out if the trend continued into the subsurface. As shown in Figure 21, the anticline has accentuated relief at the depth of the Green River, and a productive well was drilled along its axis somewhat downdip from the projected boundary between the sandstone and the shale facies. Subsequence drilling showed that reservoir conditions within the sandstone were variable, resulting in the segregation of the oil into a number of discrete sandstone bodies separated by zones of silt- and clay-sized detritus. It is unlikely that such subtle variation of lithology could have been detected seismically.

The San Emidio Nose field in California (Bazeley, 1972), with production in the Miocene Stevens sandstone, has a similar type of combined structural-stratigraphic entrapment, but numerous deep holes were drilled before the facies boundary could be defined with enough precision to pinpoint a successful well location. A period of 24 years lapsed between the first seismic shooting and the discovery of the field.

Commonly a change in facies causes the loss of a reflection. This effect was observed in the Bramberge field in West Germany (Roll, 1972). Here, in the Bentheim Sandstone, entrapment of oil resulted from the grading of the sandstone into shale. The initial seismic work over the field showed a structural closure against a fault, but two wells were drilled outside the area of closure to satisfy drilling commitments. The first of these was dry, but the second encountered an oil-bearing sandstone, 17 m thick, that would not have been expected from the seismic structure map.

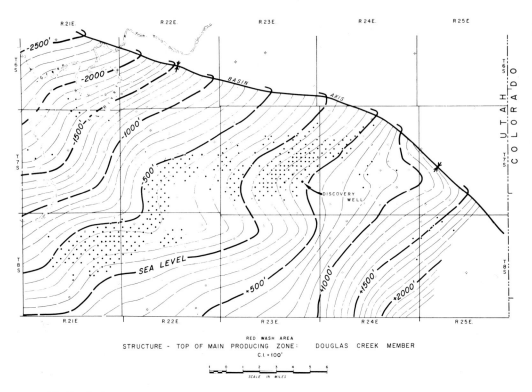

FIG. 20—Surface structure map over area of Red Wash field, Uintah County, Utah (Chatfield, 1972).

The next well (also outside the zone of closure) penetrated a thicker oil column, so it became evident that the entrapment was not structural but stratigraphic. Later wells proved that an updip transition from sandstone to shale was responsible for the accumulation.

An experimental seismic survey was performed to determine whether the facies change could be detected on the reflection records. A definite correlation was observed between the character of the Bentheim reflection and its lithology (Fig. 22). Three lines were shot—two (B and C) crossed the boundary as established by well control, and the third (A) was shot entirely over the sandstone facies. The disappearance of the Bentheim reflection as the line crosses the transition zone into shale is well defined on lines B and C, although not on A. Such a clear correspondence between lithology and the presence and absence of a reflection is rather rare, and it should not be surprising that those developing the field did not trust the results of this seismic survey and drilled some dry holes by disregarding its implications.

DISCUSSION

These case histories illustrate the performance of the seismic reflection method in exploring for different types of stratigraphic traps, and show that the results of seismic surveys over a stratigraphic entrapment may fall into any of the following five categories.

1. Seismic data does not show any anomaly, and the subsequent discovery is credited either to wildcat drilling or to geology alone. Bell Creek and Recluse fields are examples.

2. Discovery is made by drilling on a seismically determined structural anomaly which turns out to be either spurious or unrelated to the actual accumulation. The Boscán and Candeias fields are typical of this category.

3. A structural axis such as an anticline is mapped seismically which is expected (using geologic considerations) to cross a facies change or truncation that would give rise to an entrapment that is both structural and stratigraphic. A combination of the seismically derived structural information and the lithologic pattern based on geologic data makes it possible to select the proper location for the discovery well. Prudhoe Bay, Red Wash, and San Emidio were discovered using this type of coordination.

4. Configuration of the entrapment can be deduced from reflections which do not necessarily originate from boundaries of the oil accumulation

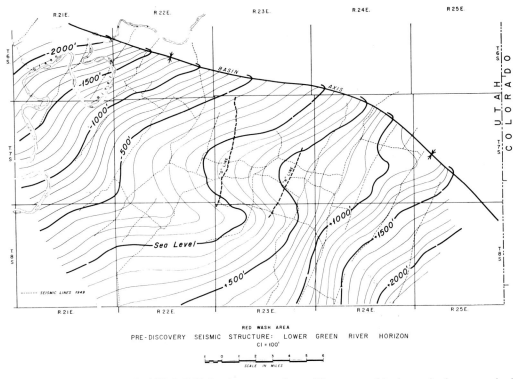

FIG. 21—Seismic map over Red Wash field showing structural nose. Discovery well is along axis of nose on seismic line "N." (Chatfield, 1972).

itself, but from boundaries (or groups of boundaries) conformable with surfaces of entrapment. The Hohne field, for example, was discovered because a seismically observable angular truncation of Upper Jurassic beds by the Cretaceous-Jurassic unconformity led to the recognition that there might be entrapment in the Dogger sandstone at its updip termination against the unconformity.

5. A lens-shaped productive zone is directly indicated on the seismic section. This case is best represented by limestone bodies which show up directly (even if only by an absence of reflections) on the second section. Examples are the Horseshoe atoll in West Texas and various productive reefs in Michigan. It is uncommon for productive sandstone bodies to show up on seismic records in this way.

Two questions are raised by these observations: (1) Why did the seismic method fail to show oil accumulations in fields such as Bell Creek? and (2) Are there recent improvements in seismic technology, or are there any in development, that would have given positive indications if they were used before the discoveries were made? In other words, to what extent will recent developments, such as zero-phase wavelet pro-

cessing, multivariate analysis of reflection waveforms, true-amplitude presentation, stratigraphic modeling techniques, or synthetic sonic logs constructed from seismograms, make it possible to identify productive stratigraphic features such as the Bell Creek or Candeias sandstone lenses or the Boscán pinchout, that could not have been distinguished on records obtained before the availability of such developments? It should be noted that many successful efforts to locate productive stratigraphic entrapment features seismically (either before or after the discovery of hydrocarbons) made use of seismic techniques that would be considered primitive by today's standards. This success was partly attributable to favorable geologic characteristics of the target and partly to imaginative interpretation of the seismic data that made full use of geologic information.

The most promising type of stratigraphic deposit for location by the reflection method is stratigraphically tapped gas in offshore accumulations. It is now possible to detect such deposits by newly established indicators such as bright spots, which should be diagnostic either of structural and stratigraphic gas where the records are suitably processed.

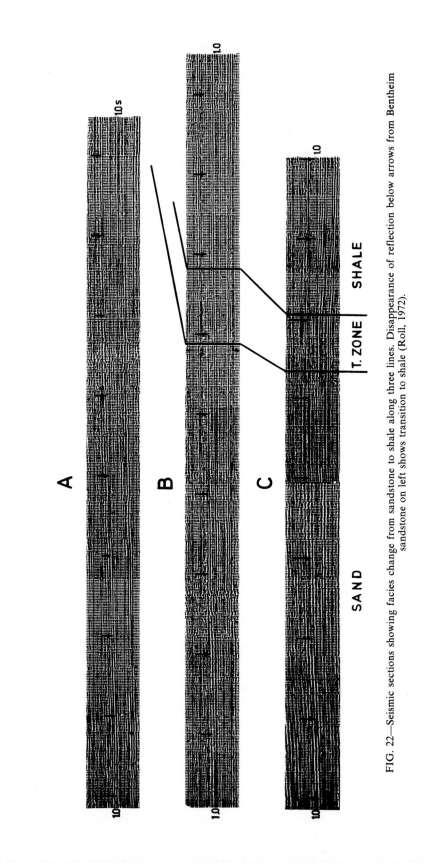

FIG. 22—Seismic sections showing facies change from sandstone to shale along three lines. Disappearance of reflection below arrows from Bentheim sandstone on left shows transition to shale (Roll, 1972).

The improvements in resolution made possible by such processing techniques as zero-phase wavelet extraction and analysis based on amplitudes (such as seismically derived velocity logs) should make it easier than ever before to locate productive entrapment features where the geologic setting is favorable. Resolution should be particularly promising where the reservoir consists of a sandstone body embedded in rock materials that are homogeneous above and below it for vertical distances appreciably greater than a wavelength. The use of wavelets extracted from the data should make it feasible to follow such beds closer to their termination than was previously possible. Further improvement in noise reduction techniques should make it easier to observe weak reflections from critical interfaces where the velocity contrasts are so small that they can not be seen above noise at the present state of the art.

CONCLUSIONS

Many limitations, which historically have made the seismic method less effective for locating stratigraphically trapped oil and gas than for finding structural entrapments, are intrinsic in the physics of the reflection process and in the nature of reflecting boundaries within the earth. However, improvements have been made in the past few years that promise to increase the capabilities of the seismic method. New developments that decrease noise and increase seismic resolution should make it possible to detect many, although not all stratigraphic entrapment features that were not previously detectable on seismic records.

The application of newly introduced principles of seismic stratigraphy should enhance the effectiveness of seismic data in narrowing the search for stratigraphically trapped hydrocarbons by isolating favorable target areas. And most importantly, increasing coordination of geophysical and geologic information should improve the likelihood that the new technological improvements in seismic prospecting will be applied as effectively as the art allows, to the resolution of the geologic problems encountered in seismic exploration.

REFERENCES CITED

Bazeley, W., 1972, San Emidio Nose oil field, California, *in* R. E. King, ed., Stratigraphic oil and gas fields: AAPG Memoir 16, p. 297-312.

Chatfield, J., 1972, Case history of Red Wash field, Uintah County, Utah, *in* R. E. King, ed., Stratigraphic oil

and gas fields: AAPG Memoir 16, p. 342-353.

Caughlin, W. G., F. J. Lucia, and N. L. McIver, 1976, The detection and development of Silurian reefs in northern Michigan, Geophysics, v. 41, p. 646-658.

Dobrin, M. B., 1976, Introduction to geophysical prospecting, 3d ed.: New York, McGraw-Hill, 630 p.

Domenico, S. N., 1977, Lithology and velocity *in* stratigraphic interpretation of seismic data: AAPG-SEG School Manual (unpublished).

Evans, H., 1972, Zama—A geophysical case history, *in* R. E. King, ed., Stratigraphic oil and gas fields: AAPG Memoir 16, p. 440-452.

Hedemann, H. A., and Lorenz, H., 1972, Truncation traps on northwest border of Gifhorn trough, East Hanover, Germany, *in* R. E. King, ed., Stratigraphic oil and gas fields: AAPG Memoir 16, p. 532-547.

Hilterman, F. J., 1976, Lithologic determination from seismic velocity data *in* Stratigraphic interpretation of seismic data: AAPG-SEG School Manual (unpublished).

Kendall, R., 1967, The role of the geophysicist in expanding man's domain: Geophysics, v. 32, p. 1-17.

Lindseth, R. O., 1976, Seislogs *in* Stratigraphic interpretation of seismic data: AAPG-SEG School Manual (unpublished).

Lyons, P. L., and M. B. Dobrin, 1972, Seismic exploration for stratigraphic traps, *in* R. E. King, ed., Stratigraphic oil and gas fields: AAPG Memoir 16, p. 225-243.

Mathieu, P. G., and G. W. Rice, 1969, Multivariate analysis used in the detection of stratigraphic anomalies from seismic data: Geophysics, v. 34, no. 4, p. 507-515.

Morgridge, D. L., and W. B. Smith, Jr., 1972, Geology and discovery of Prudhoe Bay field, eastern Arctic slope, Alaska, *in* R. E. King, ed., Stratigraphic oil and gas fields: AAPG Memoir 16, p. 489-501.

Neidell, N. S., and E. Poggiagliolmi, 1977, Stratigraphic modeling and interpretation—geophysical principles and techniques: this volume.

Rocco, T., and O. D'Agostino, 1972, Sergnano gas field, Po basin, Italy—a typical stratigraphic trap, *in* R. E. King, ed., Stratigraphic oil and gas fields: AAPG Memoir 16, p. 271-285.

Roll, A., 1972, Bramberge field, Federal Republic of Germany, *in* R. E. King, ed., Stratigraphic oil and gas fields, AAPG Memoir 16, p. 286-296.

Sutherland, J. A. F., 1972, Boscán field, western Venezuela *in* R. E. King, ed., Stratigraphic oil and gas fields: AAPG Memoir 16, p. 559-567.

Vail, P. R., et al, 1977, Seismic stratigraphy and global changes of sea level: this volume.

Vieira, L. P., 1972, Candeias field—typical stratigraphic traps, *in* R. E. King, ed., Stratigraphic oil and gas fields: AAPG Memoir 16, p. 354-366.

Waters, K. H., and G. W. Rice, 1975, Some statistical and probabilistic techniques to optimize the search for stratigraphic traps using seismic data: 9th World Petroleum Cong., Proc. (Tokyo), panel discussion 9, paper 1.

Analysis of High Resolution Seismic Data[1]

HERMAN C. SIECK and GEORGE W. SELF[2]

Abstract High resolution data analysis can provide the detail commonly absent in deep seismic shooting. Greater detail provides a means for seafloor mapping, subcrop mapping, and mapping of shallow structures, contours, and hydrocarbon accumulations and seeps. Accurate shallow mapping also offers a means to select a drilling site by providing data for a potential drilling and construction hazard map, and helps the engineer plan for the task by providing geotechnical information for foundation design.

High resolution geophysics should more properly be termed, *high resolution acoustics* or else *continuous acoustic profiling.* Acoustic systems provide analysis by using sound-generating devices, sound receivers, and graphic recorders which define the water depth and provide a cross-sectional display of the sea bottom and subsurface lithology. Commonly, more than one acoustic device is used simultaneously to achieve different data for different studies. Such a system is typically referred to as a *multisensor acoustic system.*

Different acoustic systems provide different data. Water depth systems define water depth for bathymetric mapping; tuned transducers define bubble clusters, marine flora, and fish accumulation; side-scan sonar defines bottom irregularities, outcrops, and bubble clusters; while acoustic subbottom profilers penetrate the seafloor to provide subsurface data. Of course, filtering and stacking improve resolution on some acoustic systems just as they improve resolution in deep seismic shooting.

INTRODUCTION

Recent emphasis in resource exploration and development has shifted toward the world's offshore geologic basins. On land, systematic exploration programs traditionally begin with reconnaissance surface geologic mapping and sampling. This is done to unravel the geology of the basin, evaluate the petroleum potential of the basin, and qualify the more prospective areas for more detailed investigations.

Surface geologic conditions on land generally can be observed easily and directly. However, in the oceans, where personal geologic observations are usually impractical, indirect geophysical and seafloor-sampling methods are the most practical way to collect the data that are easily obtainable onshore by geologic field parties. Thus, a suite of marine *high-resolution* geophysical devices has been developed to examine not only the surface conditions of the seafloor, but also the overlying water column and the shallow subbottom geology.

Where seafloor sampling, water-column sampling, and shallow core hole information are combined with the high-resolution data, a wealth of knowledge concerning the stratigraphy, structure, and geochemistry of the area of interest can be gained. In some places hydrocarbon accumulations are revealed, and valuable information concerning foundation conditions and potential hazards to drilling and production facilities is provided.

This study demonstrates the wide range of exploration and engineering applications of high-resolution geophysical data. Obviously, a high-resolution geophysical study is only a part of the costly exploration and engineering program leading to the location and production of hydrocarbons, but it must be emphasized that such studies are just as important as surface geologic mapping and engineering site investigations on land.

DESCRIPTION AND USE OF HIGH-RESOLUTION SYSTEMS

The term *high-resolution geophysics* refers to the use of sound-generating devices, sound receivers, and graphic recorders which define the water depth and provide a cross-sectional display of the sediment and rock layers below the sea bottom. A more appropriate term is *high-resolution acoustics* or *continuous acoustic profiling.* Usually more than one acoustic device is used simultaneously in a particular program using different frequency responses to obtain a more complete picture from the water surface to a depth of several hundred meters below the seafloor. Typically a system with a combination of acoustic devices is referred to as a *multisensor acoustic system.*

Table 1 summarizes the acoustic systems which are generally high-frequency, low-energy devices whose primary purpose is resolution rather than penetration. All of the acoustic (seismic) systems listed in Table 1 operate on the principle whereby transmitted seismic energy incident on an acoustic interface is partly reflected from this interface. An acoustic interface is any interface across which there is a contrast in acoustic properties. Contrast is dependent on the acoustic impedance of the materials (a function of density and elastic properties) on each side of the interface. These acoustic interfaces then are displayed graphically by each of the acoustic systems. Generally the

[1]Manuscript received, August 25, 1976; accepted, May 23, 1977.

[2]McClelland Engineers, Inc., Houston, Texas 77081.

acoustic interfaces displayed on the subbottom profiles correspond to physical interfaces such as bedding planes, unconformities, faults, the top of hard rock, boundaries of gas zones, the surface of gas bubbles in the water column, and other similar features. Direct identification of sediment or rock materials usually can not be based on the subbottom reflection profiles alone, but bright spots (high-amplitude pulse phase reversals) are displayed on the profiles and may indicate the presence of zones containing gas or oil.

With high-frequency depth sounder and side-scan systems, virtually all of the transmitted energy is reflected from a single acoustic interface—the sea bottom. However, for the lower-frequency systems, the transmitted energy of which partly penetrates the subbottom, several interfaces commonly are detected. Thus, that part of the energy which penetrates the seafloor travels downward

to the next interface where the reflection process is repeated. At each interface, the amount of energy that travels downward to the next interface is reduced by the amount that is reflected. The reflectivity of an interface depends on the contrast in acoustic impedance between the two materials. Thus, the boundary between a very dense sandstone with a large acoustic impedance and a soft clay with a small impedance will act as a strong reflector. The limit of penetration is reached when no detectable energy remains. The actual limit, of course, varies and depends partly on the number and reflectivity of the acoustic interfaces involved.

Interfaces are displayed graphically, based on the time it takes the transmitted energy to travel from the source to each interface and back to the receiver. Once the speed of sound in the materials (interval velocity) is known, the depth to each in-

Table 1. Summary of Acoustic Systems.

Acoustic System	Frequency	Purpose
Water Depth System (Fathometer)*	12-80 kHz	Water depths, bathymetric maps
Water Column Bubble Detector (Tuned Transducer)	3-12 kHz	Bubble clusters, marine flora, fish, debris in water column
Dual Channel Side-Scan Sonar	38-250 kHz	Bottom irregularities, sea-bottom debris, outcrop, bubble clusters
Acoustic Subbottom Profilers:		
1. Tuned Transducers	3.5-7.0 kHz	Bubble detection, with 10 KW booster amp subbottom penetration to 30 meters
2. Electromechanical (Acoustipulse)*	0.8-5.0 kHz	Subbottom penetration to 120 meters, best resolution of shallow active and nonactive gas-charged zones
3. Sparker		
a. Standard	0.04-0.150 kHz	Subbottom penetration to 1,000 meters
b. Optically Stacked	0.04-0.150 kHz	Subbottom penetration to 1,000 meters, better horizontal resolution, and information preserved on magnetic tape
c. Fast Firing 4-KJ and 10-KJ	0.04-0.150 kHz	Penetration to 300 and 1,000 meters respectively, out-standing horizontal and vertical resolution of the sediment and geologic environment, recorded on magnetic tape
d. De-bubbled De-reverberated	0.04-0.150 kHz	4-KJ and 10-KJ energy levels yield resolution superior to standard and Optically Stacked Sparker, direct gas-charged sediment detection
e. Multichannel Digital	0.04-0.150 kHz	Subbottom penetration to 1,000 meters, data preserved on magnetic tape for multifold digital processing and velocity analysis

*Trade names.

terface can be calculated. However, in practice, the speed of the seismic wave is complex and depends on the acoustic properties of the materials, but in general it increases as the density increases. For this reason, the depth scale on the profiles can not be considered truly linear, and hence velocity or borehole data are required for accurate calibration. A close approximation of shallow subbottom depths can be made by assuming a velocity previously determined for other similar materials.

All acoustic systems, except the side-scan sonar system, are designed to provide data from directly beneath each transducer or transducer array; no data are received from either side of the line of profile, except rare, spurious side reflections ("side swipes"). In contrast, the side-scan sonar system does provide data from several tens or hundred of meters on both sides of the ship's trackline.

Water-Depth Systems

A commonly used depth-sounder system consists of a power supply, a transducer (a transceiver that alternately transmits and receives

sound), and a graphic recorder. The depth-sounder transducer usually is installed in the hull of the vessel, midship, and 2 to 3 m below the water line. The transducer converts electric energy to sound energy, and the sound energy is transmitted downward to the seafloor. When this energy strikes the seafloor (or any other object having acoustic properties different from water), a part of it is reflected back to the transducer (transceiver) as an echo. This sound energy then is reconverted to electric energy, then recorded on a graphic recorder. Assuming that the speed of sound in water is nearly constant, the amount of time which lapses between the pulse transmission and the echo reception is a measure of the distance traveled. Dividing this value in half converts the distance traveled, to depth. Figure 1 is an example of a depth-sounder record from Yakutat Bay, Alaska.

Because the expanded vertical scale is provided by the record, bottom microtopography (as little as 0.3 m relief) is clearly displayed. The high-frequency pulse has a wavelength much less than 0.3 m, which provides very high-resolution bottom profiles. Because of this high frequency, virtually

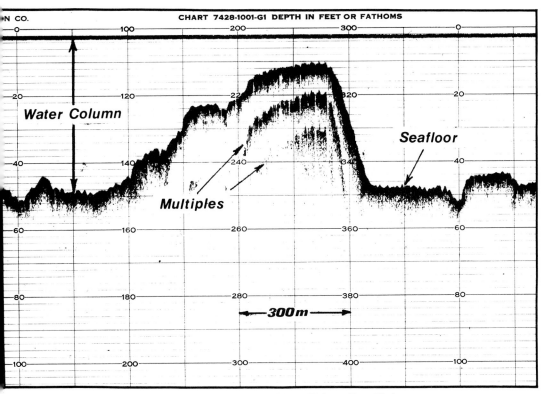

FIG. 1—Depth sounder record from Yakutat Bay, Alaska.

no subbottom penetration is obtained. This sytem is suitable for operating in water depths of a few meters to several hundred meters, and also can be used to detect gas bubbles.

The accuracy of a depth-sounder system is largely dependent on two variables: (1) the velocity of sound in water, which can vary with changes in temperature and salinity, and (2) the precision of the electronic timing circuits used. Both of these variables necessitate frequent calibration of the system for accurate bathymetric determinations.

Side-Scan Sonar Systems

Side-scan sonar systems provide graphic records that show a two-dimensional (map) view of seafloor topography (Fig. 2). Gas bubbles within the water column scanned are detected and displayed. These records are closely analogous to a low-oblique aerial photo. Features shown on the side-scan sonar commonly appear similar to their natural perspective.

A typical side-scan system consists of a towfish containing two arrays of transducers and a shipboard graphic recorder. The towfish typically is deployed so that it is towed above the bottom at a distance equal to about 10 to 20% of the range scale being used. The transducers emit 105-kHz pulses as fan-shaped beams on each side of the towfish. The beams are perpendicular to the direction of vessel travel and are broad enough in the vertical plan to extend from beneath the towfish to the full system range of 500 m (to each side of the towfish). The beam width is only 1.2° in plan view. Each pulse lasts 0.1 msec and the repetition rate varies depending on the range scale used. When using the 500-m range, the repetition rate is 667 msec.

Signals reflected from the bottom and from objects on it are displayed on a continuous graphic record produced on a two-channel, wet-paper recorder. The intensity and distribution of reflections received depend on the composition and surface texture of the reflecting object, its size,

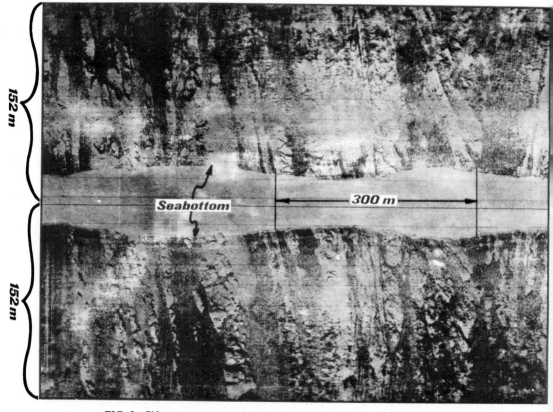

FIG. 2—Side-scan sonar record of rock outcrops from Cape Yakataga, Alaska.

and its orientation with respect to the transducers in the towfish. Due to the beam shape and the short length of the transmitted pulse, the side-scan sonar can resolve small objects on the sea-floor and details of topographic irregularities (Fig. 3).

Tuned Transducer Systems

One of the most commonly used transducer systems is a 7-kHz transducer and graphic recorder. The transducer can either be towed behind the survey vessel in a streamlined housing or mounted alongside the hull of the vessel. The tuned transducer system uses a piezoelectric transducer to provide an extremely short, high-powered pulse. This pulse produces one to two cycles of signal frequency with some transducer ringing and with a noticeable delay time for the signal. The returning signal is recorded on a graphic recorder with an option for constant gain or time variable gain.

The greatest value of the tuned transducer system is its ability to detect gas bubbles in the water column. Figure 4 is an example of a tuned trans-ducer record from offshore Texas. The system can be used for shallow investigation into the sea-floor, but there are other devices far superior to the tuned transducer for subbottom profiling. The tuned transducer is most useful when operated simultaneously with the sparker system to provide accurate water depths and gas detection in the water column.

Electromechanical Subbottom Profiling Systems

An electromechanical acoustic profiling system consists of power supplies, trigger banks, ship-board electronics package, filter, graphic rec-order, and a towed transducer array with an ac-companying towed hydrophone. One such system is Acoustipulse®, which is a single channel, high resolution, marine reflection seismic system de-veloped to provide both high resolution and pene-tration up to 100 m in areas underlain by sedi-ment and soft-rock strata. Typical subbottom penetration is of the order of 70 to 100 m. Pene-tration of as much as 150 m can be obtained un-der the most ideal acoustic conditions (commonly where the material being profiled is soft), but it is

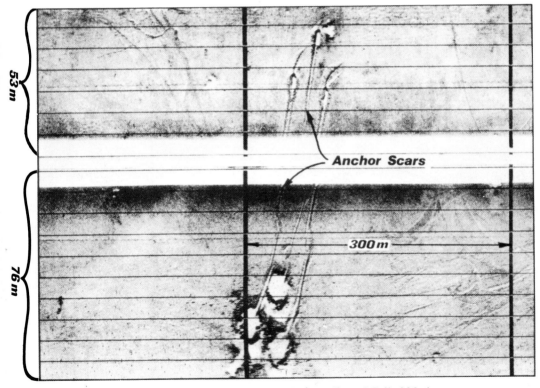

FIG. 3—Side-scan sonar record of rig marks on seafloor of Gulf of Mexico.

very limited in hard rock and in very dense sediment. Vertical resolution ranges from 0.3 to 1.0 m but typically is of the order of 0.5 m (Fig. 5). Because of the inherent high resolution of the Acoustipulse® system, the profiles also can be used to determine water depth in areas with water depths from 10 to 500 m.

The Acoustipulse® source produces a minimum phase, high frequency waveform with a minimal amount of source reverberation and no bubble pulse. The frequency of the output pulse is broadband, ranging from 200 Hz to 10 kHz. The Acoustipulse® waveform is produced by discharging stored electric energy into a potted wire coil. Eddy currents set up in the coil repel an adjacent aluminum plate. A partial vacuum created by the flexed plate returns the plate to its original position. The return motion is highly dampened by a rubber diaphragm, thereby minimizing source reverberations. The source is driven by a capacitive-discharge power supply capable of producing up to 1,000 joules per transducer.

Normally a source consists of three electromechanical transducers, all of which are fired simultaneously every [1/2] sec. Energy output of each transducer is variable, all of which are mounted on a 2.5-m long catamaran and towed at or near the water surface 30 m behind the survey vessel. Reflected signals are received by a 10-element linear hydrophone array towed at or near the water surface and abeam of the source transducers. Signals outside the desired frequency range are removed by frequency filtering. The seismic data are displayed graphically on a [1/4]-sec record. Vertical exaggeration of the subbottom profiles is typically 10 to 20 times. Signals also can be recorded on an analog tape recorder for laboratory playback.

As shown on Figure 6, both structural and stratigraphic features, as well as shallow hydrocarbon accumulations, can be identified. The nature of materials also can be inferred from their acoustic properties.

Sparker Systems

A sparker system is used primarily to obtain data from intermediate depths (100 to 1,000 m) and thus fill the gap between the shallow penetra-

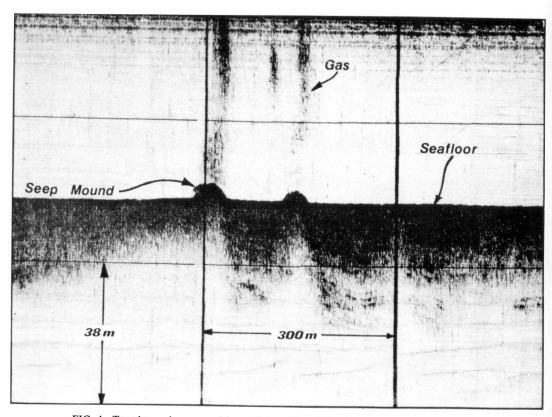

FIG. 4—Tuned transducer record from offshore Texas, showing gas seeps and gas mounds.

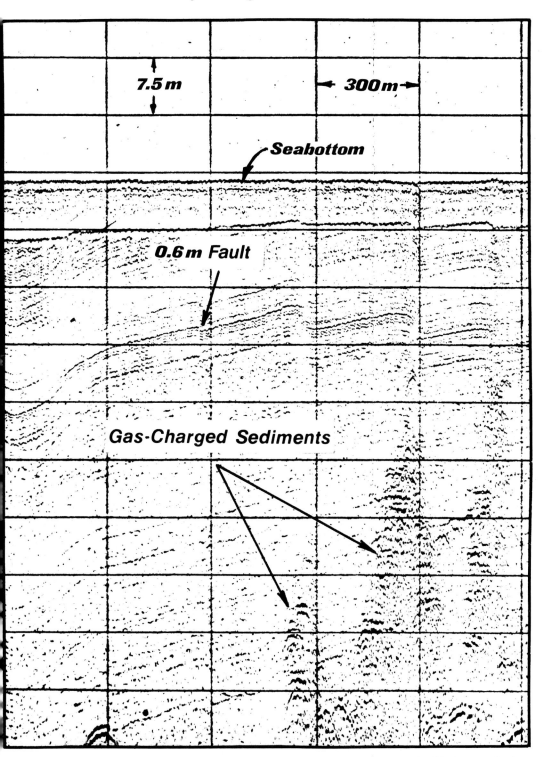

FIG. 5—Acoustipulse® record from Gulf of Mexico, showing degree of resolution attainable with electromechanical subbottom profiling systems.

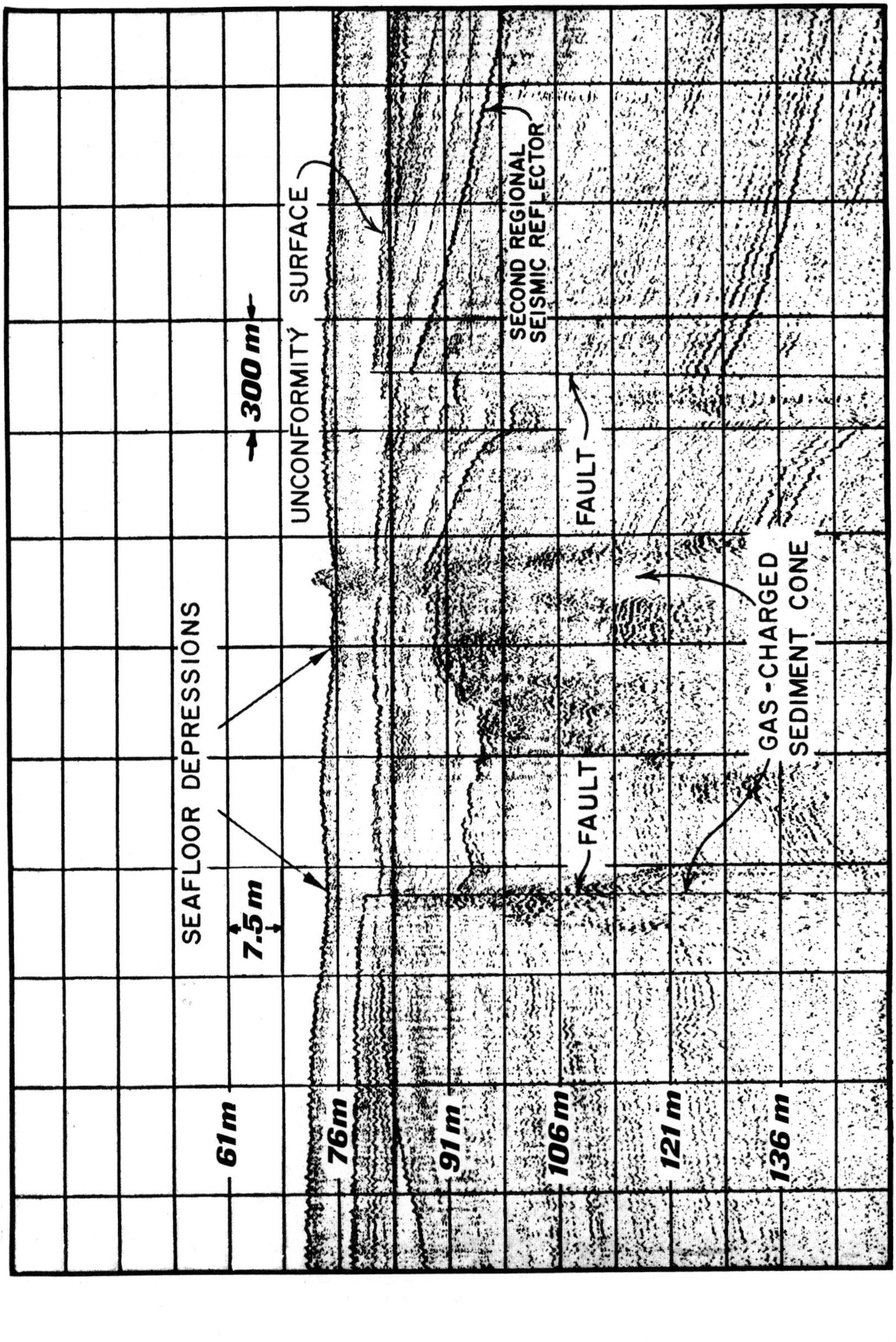
SEAFLOOR DEPRESSIONS
7.5 m
300 m
UNCONFORMITY SURFACE
SECOND REGIONAL SEISMIC REFLECTOR
FAULT
FAULT
GAS-CHARGED SEDIMENT CONE
61m
76m
91m
106m
121m
136m

tion Acoustipulse® system and multichannel, common depth point (CDP) seismic systems designed for deeper penetration. The vertical resolution attainable is about 7 m.

Sparker systems transmit signals by discharging stored electric energy through a spark gap. The resulting spark produces a steam bubble by vaporizing the surrounding water, which expands until the outward pressure of the bubble is overcome by the hydrostatic pressure of the water. The bubble then collapses, creating a "bubble pulse," which can be greater in amplitude than the initial impulse. This bubble pulse is repetitive for two or three cycles until its amplitude becomes negligible or the bubble reaches the water surface. There is an interval of 30 msec between the time of the initial bubble formation and the final bubble collapse. The "noise" produced during this interval generally masks reflections for 15 to 25 m at the seafloor and at each event (Fig. 7).

The capacitor-stored electrical power is discharged into a nine-electrode sparker cage. Various combinations of capacitors and power supplies can be used to yield the desired energy output. A system uses different numbers of power supplies and capacitors to obtain the desired subbottom penetration. Because of the frequency of the sparker system and the natural frequency of the earth, no matter how much energy is discharged, there is a maximum penetration due to attenuation of the signal. Thus, it is better to concentrate on resolution in the shallow subbottom with a sparker than to continue to increase the energy output. In reference to the firing rate equation (see appendix A), the best possible horizontal and vertical resolution for any geologic setting can be obtained by varying the number of power supplies and the number of pulses generated.

For the 10-KJ (kilojoules) system, the maximum rate of full-energy discharge is once every 2 sec. This firing rate typically provides one output pulse for every 6 m of horizontal travel at 6 knots. In shallow-water areas, output power commonly is reduced to 4-KJ to minimize the number and intensity of bottom multiples and the firing rate is

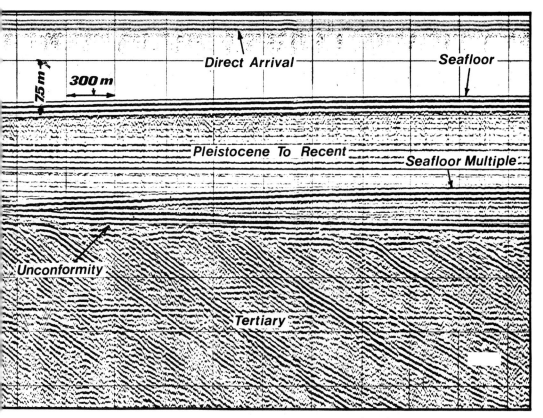

FIG. 7—Optically stacked sparker record from Gulf of Alaska.

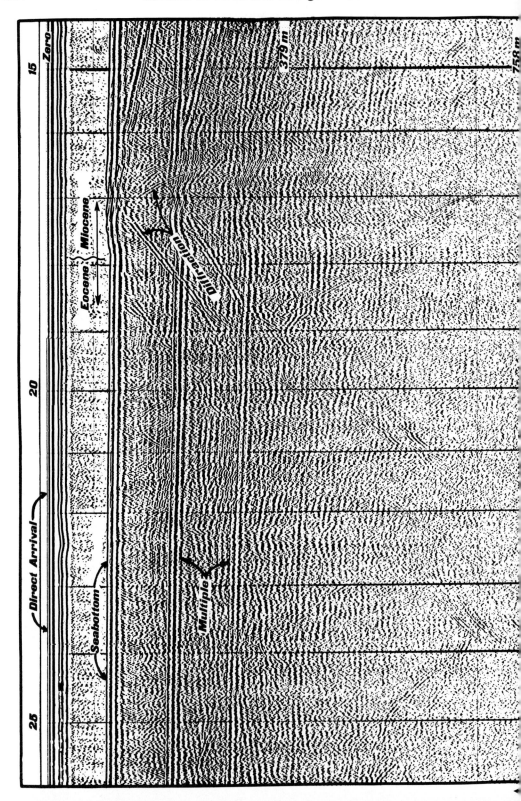

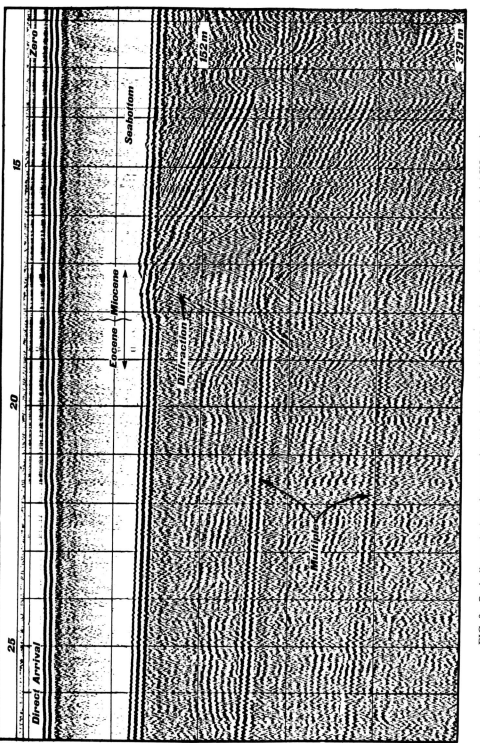

FIG. 8—Optically stacked sparker records: (A) using stacked 10-KJ sparker, and (B) using stacked 4-KJ sparker.

once per second, or about once every 3 m of horizontal distance. In either case, the horizontal resolution is typically several times greater than it is with standard 10-KJ sparker systems that are fired once every 4 to 6 sec.

Figure 8 shows the philosophy of firing rate versus energy output, using two sparker records (10-KJ and 4-KJ) from the same line of profile from offshore southern California. Better definition of the individual events within the Eocene material was obtained on the 4-KJ record because of the greater number of pulses, and in this case, without sacrifice in depth of penetration.

Thus, using the basic firing rate equations (see appendix A), the sparker system can be modified and arranged to allow for the best or most optimum data quality for a given area. In other words, an electrically stored energy system can be tailored relative to energy and firing rate for any geologic setting.

The horizontal resolution of a subbottom profile taken with a sparker system can be increased further by using an optical stacking technique. The reason for using the term *optical stacking* is that the common depth points are displayed side by side, rather then by being electrically summed into one trace. Thus, the optical stacking technique provides maximum output energy without sacrificing the effective horizontal and vertical resolution for any number of pulses discharged, while maximum penetration and horizontal definition are obtained. One practical advantage in using optically stacked data rather than data that has been electrically expanded is that true subsurface information is preserved in areas of steep dip, whereas by using data generated with an expander circuit, steeply dipping beds will have a fuzzy, stepped appearance. Another advantage inherent in optically stacked data is its low vertical exaggeration, typically about four times (see Fig. 7).

To properly accomplish optical stacking, the sparker cage and the hydrophone receivers are towed 30 to 60 m astern and on opposite sides of the survey vessel. They are deployed so that each is 4 m below the sea surface to take advantage of the constructive interference created by signals reflected from the sea surface (surface ghosts). Four meters is the optimum depth because it equals approximately ¼ wavelength of the system's peak frequency of 130 Hz. It is essential that the source and receiver are towed at the same depth for the signal to be in focus. Figure 9 is a sparker record from the Gulf of Alaska, which illustrates the effect of a source and receiver not towed properly. Note that the station intervals are approximately 2.5 cm apart and that the signal consists of a coherent sparker bubble pulse on the left half of the record. There is a noticeable deterioration of data quality from the left to the right side of the record. At first one can suspect that this difference in data quality may be due to a change in geology from left to right, just as occurred on Figure 8 in going from Miocene to Eocene sediments. But an examination of the distance between the station intervals, and of the breakup of the signal (bubble pulse) on the right side of Figure 9, reveals the true nature of the problem. The station intervals in all cases are to be 300 m. Where the station interval on the record increases, it indicates that the vessel is traversing the 300 m at a slower speed. The vessel has either encountered a stronger headwind, or a stronger current, or the engine RPMs have been decreased.

The slower speed allows the sound source and the hydrophone to sink, and because the sound source is heavier, it sinks more rapidly. Thus, the source and receiver are no longer at equal depths, and the sparker signal is no longer in focus.

Digital Processing (Debubbling and Dereverberation)

Further signal enhancement of the recorded sparker data is accomplished by digitally processing the analog data. The two most significant improvements that digital processing of the data provide are: reducing the 30-msec "bubble pulse" (debubbling technique) so that vertical resolution is significantly improved, and minimizing multiples of the seafloor (dereverberation technique). Either debubbling or dereverberation, or both, may provide a record of optimum quality. This digital processing can be applied to optically stacked data and permits conventional vertical stacking of the two traces. Figures 10 and 11 illustrate the difference between unprocessed and processed data from the western Gulf of Alaska.

In the debubbling process, the bubble pulses and surface ghosts are removed from the reflection data by digital filtering. This filtering is done by predictive deconvolution (Wiener filtering) using the Levinson algorithm. The resulting seismic record displays each reflector as a single, rather than multiple, pulse. Structural features and thin beds are no longer obscured. The debubbled data can recover true amplitude, which means that the amplitude of reflected signals is not destroyed in the signal enhancement process. This permits clear differentiation of low-impedance horizons (phase reversals), such as horizons containing gas, or clay layers ¼ wavelength (or greater) thick (illustrated on Fig. 12, a digitally processed sparker record from the Bering Sea, Alaska).

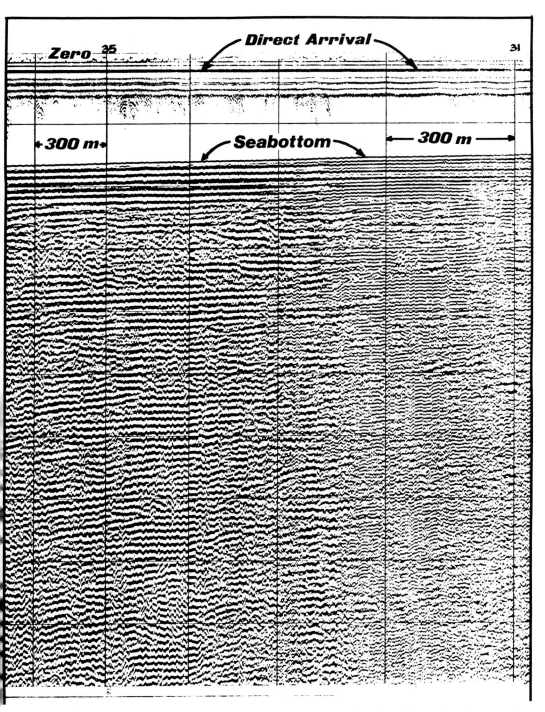

FIG. 9—Optically stacked sparker record from Gulf of Alaska, showing results of improper towing of source and receiver.

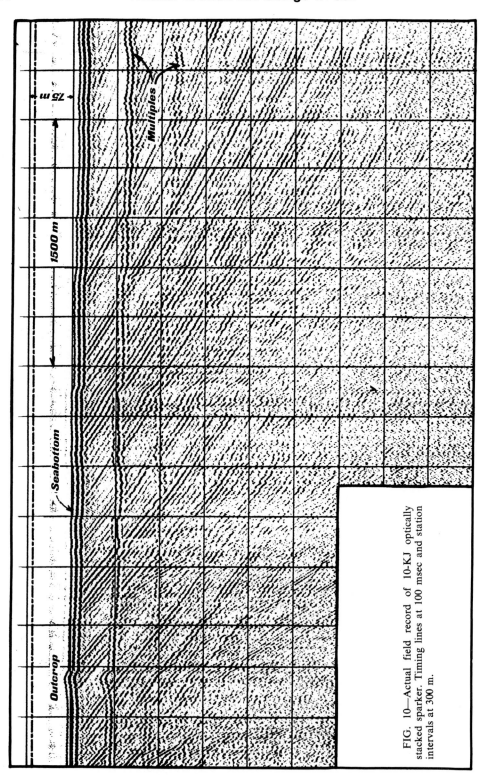

FIG. 10—Actual field record of 10-KJ optically stacked sparker. Timing lines at 100 msec and station intervals at 300 m.

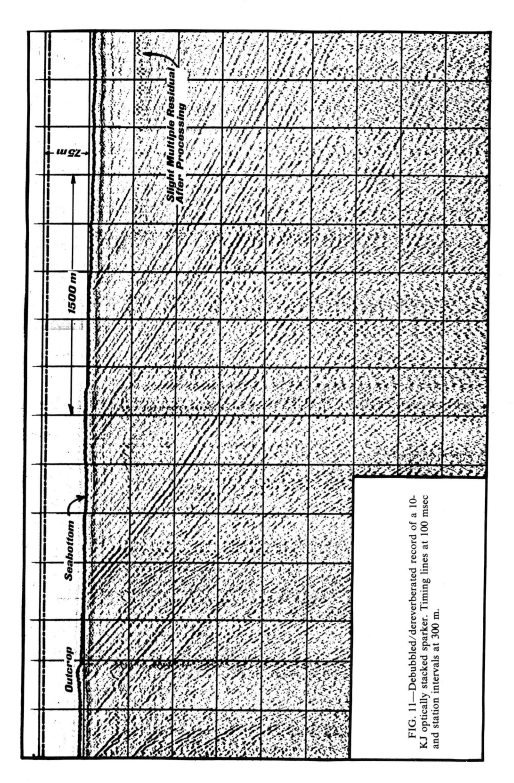

FIG. 11—Debubbled/dereverberated record of a 10-KJ optically stacked sparker. Timing lines at 100 msec and station intervals at 300 m.

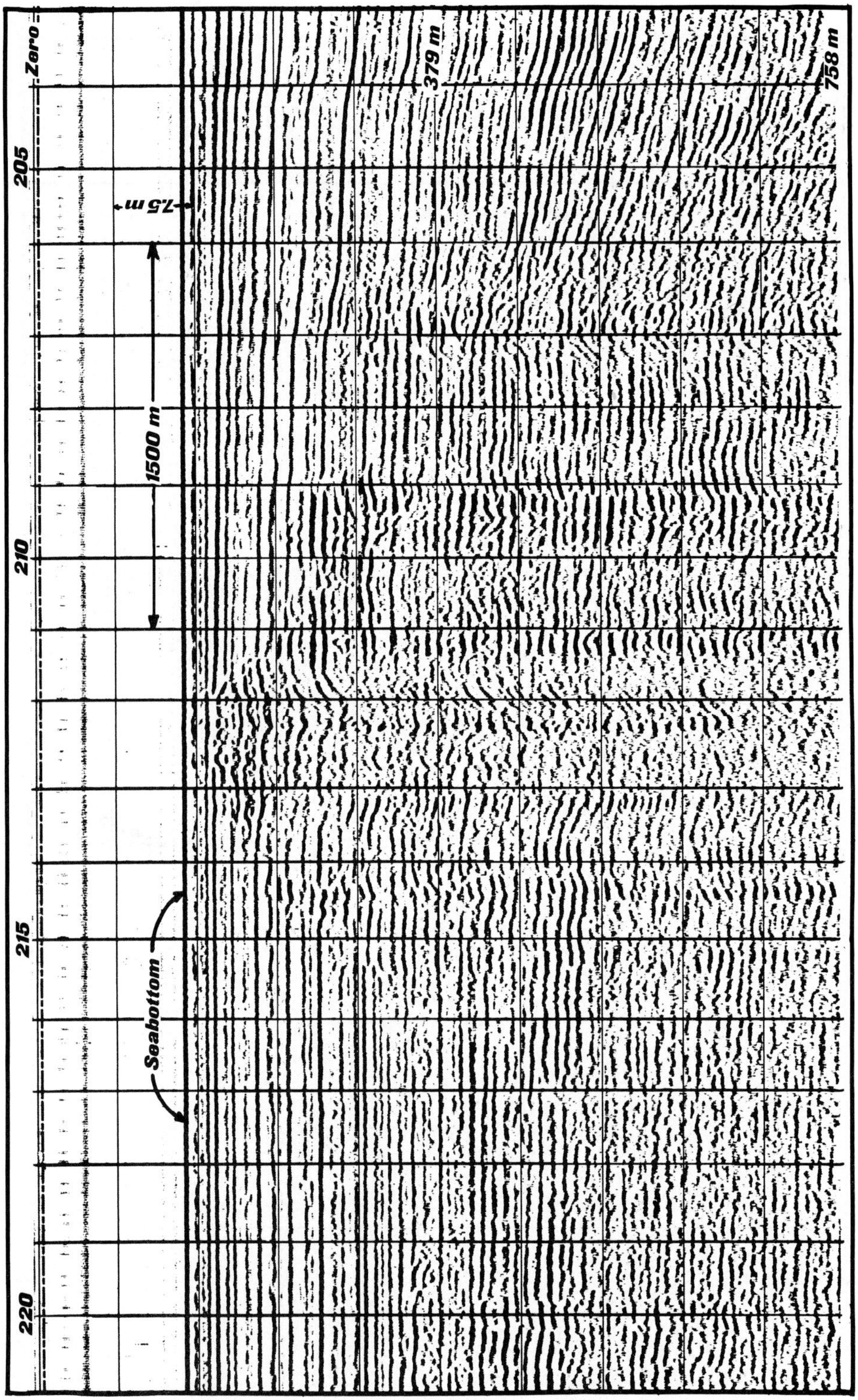

FIG. 12—Debubbled/dereverberated record of 10-KJ optically stacked sparker.

Dereverberation minimizes multiples (reverberations) of the seafloor by a subtractive process that estimates the reflection coefficient for the bottom material, computes the approximate amplitude, and subtracts it from the record. Interference due to bottom multiples is thus minimized on these dereverberated records. Because no attempt is made to remove internal reverberations, the risk of removing real data is minimized. Dereverberation works particularly well for low-order multiples in shallow water, the regime most important for high-resolution seismic processes, and where other techniques such as horizontal CDP stacking may not be effective. Concurrent with dereverberation, the direct arrival is routinely muted from the reflection data if the bottom reflection does not exactly overlap the direct arrival.

EXPLORATION ANALYSES

The high resolution geophysical systems described in this paper have both exploration and engineering applications, especially where used in conjunction with sampling programs. Some of the many exploration applications include the following:

Seafloor Geologic Mapping

The areal distribution of outcropping stratigraphic unitS having distinct acoustic characteristics, can be determined accurately using subbottom profilers and side-scan sonar as illustrated on Figures 2 and 13. Seafloor structure, including strike and dip attitudes, anticlinal and synclinal axes, and surface faults, can be determined and displayed on the seafloor outcrop map. When these features are superimposed over a bathymetric map, and the outcrop areas are documented by seafloor sampling, the result is a geologic map which is as accurate as most surface geologic maps derived from onshore exploration projects. These maps are of essential importance in most offshore exploration programs because they provide "geologic control" for deep seismic surveys,

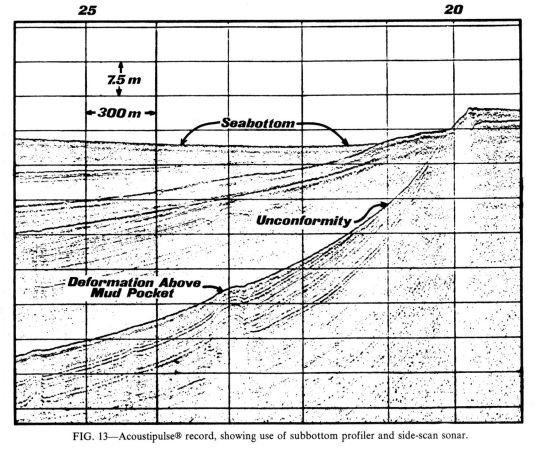

FIG. 13—Acoustipulse® record, showing use of subbottom profiler and side-scan sonar.

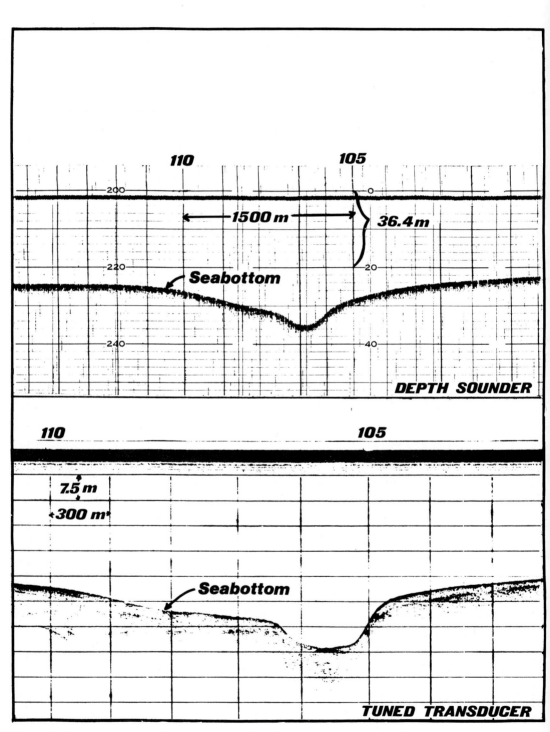

FIG. 14—Depth sounder and tuned transducer records from Lower Cook Inlet, Alaska, showing a sea bottom depression

and in some cases actually define individual drilling prospects.

Subcrop Geologic Mapping

In areas where few or no outcrops of bedrock occur (such as shown on Fig. 7), a subcrop map showing the areal distribution of stratigraphic units lying beneath a regional unconformity provides a more meaningful geologic map than the seafloor outcrop map described above. This type of map is especially useful in an area such as the Gulf of Alaska and Lower Cook Inlet, where veneers of Quaternary sediments blanket much of the continental shelf (Fig. 7). Subcropping units can be documented by shallow core holes or by lateral projection of outcrop sampling control, and the resulting map can be used in the same manner as an outcrop map.

Isopach Mapping of Shallow Stratigraphic Units

Isopach maps of stratigraphic units present at the seafloor (and to a depth of several hundred meters) are easily constructed from high-resolution acoustic data and have several important exploration applications:

1. They provide accurate stratigraphic control for evaluating potentially prospective petroleum reservoir and source beds within the basin or area of interest;

2. They provide important clues to aid in presenting the geologic history of the area and in determining whether or not the tectonic history was conducive to the entrapment of commercial hydrocarbons or influenced the depositional history;

3. They can be used in velocity programs to help convert deep seismic interpretations from time-to-depth displays by revealing the areal distribution and varying thickness of stratigraphic units having distinct interval velocities.

Shallow Structural Mapping

Shallow subsurface structural mapping using high resolution data is important in an exploration program for obvious reasons: (1) to help define the structural history and style, (2) to enable the projection of sampling control from outcrop areas into deep seismic survey results, and (3) to delineate shallow structural anomalies for more detailed investigations, etc.

The accuracy and the extent to which this mapping can be conducted using high resolution data has not been obvious. Many interpreters are frustrated by the fine details of structural features which appear on the acoustic records because they often require a more time-consuming and concise interpretation than with the same mileage

of deep seismic data. However, this fine detail is the very thing that makes high resolution acoustic data so valuable, because structural phenomena not appearing on deep seismic records (such as small faults, steeply dipping beds, subtle unconformities, etc.) may prove to be very significant in the evaluation of a prospective area.

Structure Form Contour Mapping

Also frustrating in interpreting high resolution data are the multiples which are commonly present, especially in shallow water areas. In structurally complex areas, these multiples can prevent the successful mapping of a continuous subsurface reflection, as is usually the case with deep seismic data. However, one structural mapping technique that has been used successfully in numerous complex areas offshore California and Alaska, is *structure form contouring*. In this mapping method, the apparent dip within a stratigraphic unit is measured along the line of profile, then contoured. As long as care is taken not to cross unconformities and to tie loops accurately, the resulting map is a close approximation to the structural configuration of the beds within the stratigraphic unit mapped. Even though contour values can not be assigned to the map, the amount of closure present over specific anomalies (the regional structural style) and the amount of throw on faults are accurately shown.

Locating Seeps and Shallow Hydrocarbon Accumulations

Information, useful in predicting commercial hydrocarbon production, results when high resolution data are analyzed for active seeps and subsurface accumulations. However, geochemical sampling is needed to document the reliability of these "acoustic anomalies" because of the ambiguities of direct hydrocarbon detection using geophysical devices.

Figures 14 and 15 illustrate a problem commonly arising from seep detection. Shown on the fathometer and tuned transducer records (Fig. 14) from Lower Cook Inlet, Alaska, is a sea bottom depression between stations 105 and 106. Figure 15 is an Acoustipulse® record showing the same sea bottom depression and a signal rising out of the depression 27 m into the water column. The question is whether or not the signal coming from the sea bottom depression on the Acoustipulse® is a gas seep. If so, why is it not evident on the tuned transducer and depth sounder records? If the signal on the Acoustipulse® is from a side echo, why does it appear to be coming from a depression? Side echoes are expected where there

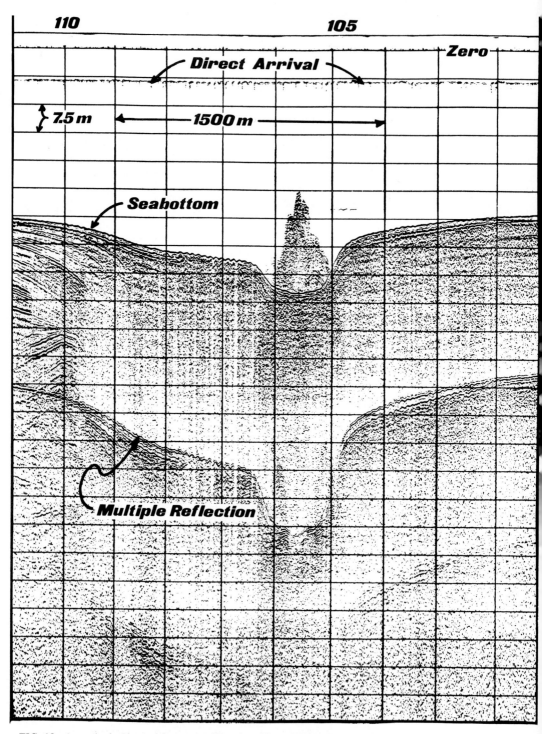

FIG. 15—Acoustipulse® record from same line of profile as in Figure 14. Note presence of signal rising from sea
bottom depression, 90 ft (27 m) into water column.

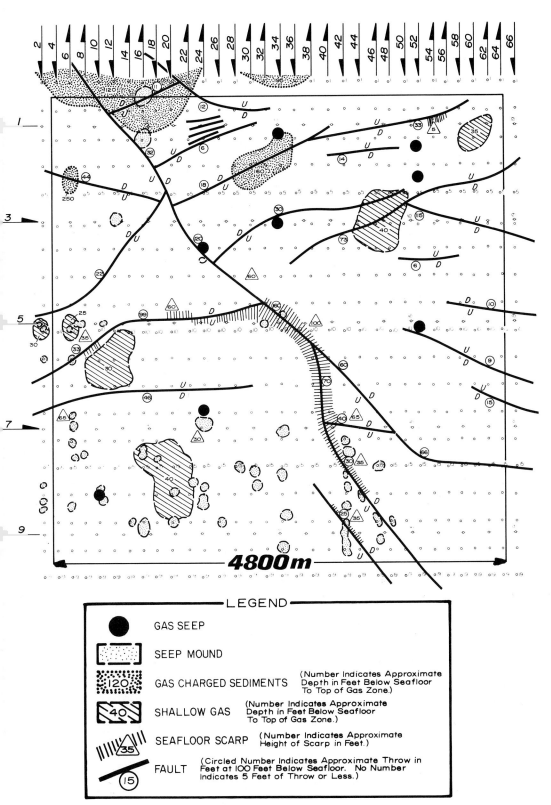

FIG. 16—Drilling and construction hazards map, based on high resolution acoustic profiling data from a drilling lease in Gulf of Mexico.

is an irregular sea bottom represented by large mounds, not depressions.

In reality, the vessel was approaching a feature so high that it was sticking out of the water, thus the vessel was forced off course to pass the obstacle. Large currents in the area had eroded a circular depression around the base of the obstacle. Where the boat turned, the apparent sea bottom depression was noted on all records. The depth sounder and tuned transducer, being directional sources, did not record the out-of-line obstacle (off to the side); whereas the Acoustipulse®, which is omnidirectional, received a signal from the obstacle. Side-scan sonar coverage of the area, which looks to both sides of the traverse line, confirmed that the signal was from an out-of-line obstacle, and not a gas seep.

Determination of Lithology and Depositional Environment

Qualitative lithologic and gross facies interpretations of stratigraphic units can often be inferred from acoustic data, even where sampling data is unavailable. These interpretations are based on the acoustic characteristics of the reflectors within the unit, and on the nature of the depositional features (such as fluvial crossbedding, glacial deposits, marine terraces, delta-front foreset bedding, etc.) that can be seen only on high resolution data, and can be invaluable in resolving the stratigraphic history of the area.

DATA INTERPRETATION FOR ENGINEERING ANALYSES

Some of the more important engineering applications of high resolution data include the following:

Accurate Bathymetric Maps

Accurate bathymetry maps constructed from high resolution data are important for drilling and production engineering for many reasons, including: (1) determination of the type of drilling rig needed at a particular site (i.e., jackup rig, semisubmersible, etc.); (2) determination of proper design criteria for marine production facilities; and (3) determination of areas where seafloor topography is unfavorable or dangerous for production platforms, pipelines, jackups, etc.

Evaluation of Potential Drilling and Construction Hazards

Drilling and construction hazards maps are a compilation of all natural and man-made hazards detected by a multisensor survey for a particular area. The hazards include active seeps, gas-charged sediments, active and inactive faults, seep mounds, steep slopes and escarpments, buried river channels, and submarine slides. Many features are dangerous only where they are encountered unexpectedly, and therein lies another value of high resolution multisensor hazards surveys. Figure 16 is a hazards map prepared for a drilling lease in the Gulf of Mexico. Several types of hazards which should be avoided completely or which require special precautions are shown.

Geotechnical Information for Foundation Design

High resolution acoustic data, combined with geotechnical sampling and in-situ testing data, can provide a complete picture of the soil conditions in an area. Using such data, a suite of maps and cross sections can be prepared which depicts foundation design parameters such as: sediment composition and thickness of distinct sediment layers, density, shear strength, compressibility, liquefaction potential, anchor holding properties, and slope stability.

A CASE HISTORY

An example of high resolution data analysis, as used to its fullest extent, is shown on Figures 17-24. The prominent bathymetric bank shown on Figures 17 and 18 was surveyed as part of a regional-reconnaissance multisensor acoustic profiling program conducted off the California coast in 1974, where deep seismic shooting resulted in fair to poor data. The program utilized a 10-KJ sparker system as the primary profiling tool and surveyed a 5- by 16-km grid. This initial survey revealed the presence of a large anticlinal feature underlying the bathymetric bank, which appeared to have possible seeps and significant areas of outcrop (Fig. 19). A reconnaissance seafloor sampling program confirmed the presence of Tertiary outcrops corresponding to distinct acoustic units. Subsequent lithologic, paleontologic, and geochemical analyses provided data which further increased the exploration interest. Preliminary maps and cross sections were then constructed to guide the second phase of the field program.

Phase 2 of the field exploration program required a detailed multisensor profile over the structural anomaly using a 4-KJ sparker as the primary tool, and detailed seafloor sampling of suspected outcrops using a gravity dart sampling system. The samples were described lithologically and analyzed for free-hydrocarbon content using a gas chromatograph onboard ship. Samples were sent to a laboratory for additional geochemical source-rock potential analysis. Cuts of the samples were examined in a Foraminifera study to

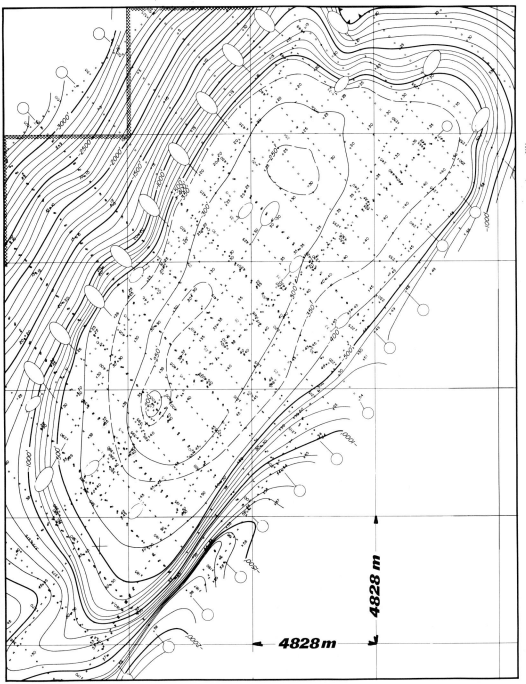

FIG. 17—Bathymetry map, showing bottom off California by multisensor acoustic profiling.

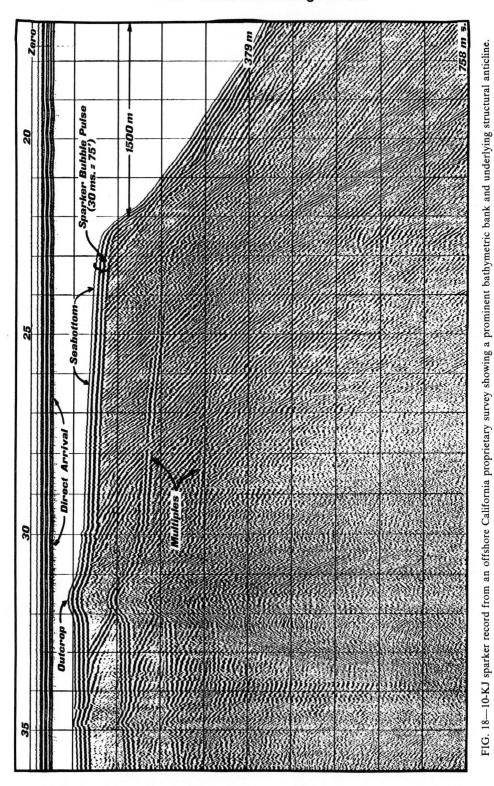

FIG. 18—10-KJ sparker record from an offshore California proprietary survey showing a prominent bathymetric bank and underlying structural anticline.

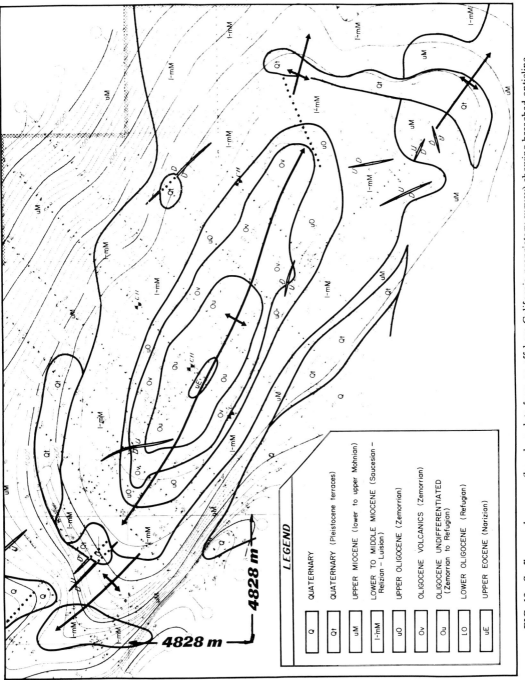

FIG. 19—Seafloor geology map (based on data from an offshore California proprietary survey) showing breached anticline.

LEGEND

Q	QUATERNARY
Qt	QUATERNARY (Pleistocene terraces)
uM	UPPER MIOCENE (lower to upper Mohnian)
l-mM	LOWER TO MIDDLE MIOCENE (Saucesian – Relizian – Luisian)
uO	UPPER OLIGOCENE (Zemorrian)
Ov	OLIGOCENE VOLCANICS (Zemorrian)
Ou	OLIGOCENE UNDIFFERENTIATED (Zemorrian to Refugian)
LO	LOWER OLIGOCENE (Refugian)
uE	UPPER EOCENE (Narizian)

4828 m

4828 m

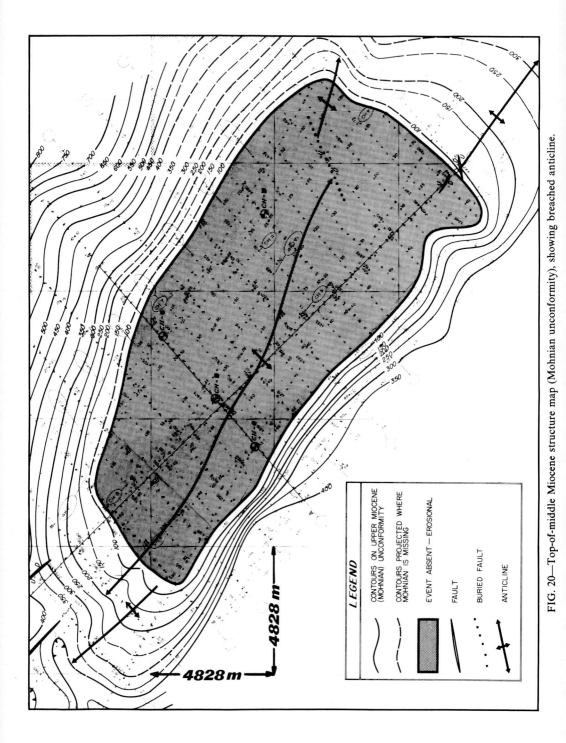

FIG. 20—Top-of-middle Miocene structure map (Mohnian unconformity), showing breached anticline.

LEGEND

CONTOURS ON UPPER MIOCENE (MOHNIAN) UNCONFORMITY

CONTOURS PROJECTED WHERE MOHNIAN IS MISSING

EVENT ABSENT — EROSIONAL

FAULT

BURIED FAULT

ANTICLINE

4828 m

4828 m

determine the geologic ages and depositional environments of the outcrops seen on the sparker records (Fig. 18). Using these data, a suite of reliable geologic maps and cross sections were constructed from the high resolution acoustic records.

Phase 3 of the program required drilling shallow core holes to provide additional paleontologic and geochemical information, and data on porosity and permeability, grain size distribution, radiometric age (of volcanics), and for evaluating the foundation conditions for production facilities. The maps and cross sections were finalized and the area's exploration potential was more fully evaluated than with deep seismic data alone.

As shown by Figure 19, the structural anomaly is a deeply eroded anticline and appears slightly asymmetrical with outcrops ranging in age from Holocene to Eocene (Pliocene sediments are notably absent in the area). Minor surface faulting occurs in the area, but appears to have little effect on the overall closure in the structure.

On the basis of sparker data, two structure maps were drafted to define the prospect—a top of middle Miocene structure map (Fig. 20), and a top of Oligocene volcanics structure map (Fig. 21). The similarity between these two maps (Figs. 20, 21) indicates that most structural growth of the anticline took place after early middle Miocene. late Miocene–Pliocene structural growth is suggested by the thinning of the upper Miocene section over the structure (Fig. 22), and by the absence of Pliocene sediments.

As demonstrated by the structure and isopach maps and the interpretive cross sections (Fig. 23), Oligocene and younger sediments can not be considered prospective because of their erosion and subsequent exposure at the seafloor. However, structural closure within Eocene sediments is apparently preserved, and these sediments must be considered structurally more prospective. The interest in the Eocene is further enhanced by data from a shallow core hole, which indicates the presence of free hydrocarbons and suitable reservoir rocks. Regional studies of the Eocene section indicate that it is probably more than 3,000 m thick in this area and that it contains shale beds, which will form the necessary seal for migrating hydrocarbons. The prospect has been defined and evaluated and is awaiting exploration drilling. At that time, the drilling and construction hazards map (Fig. 24) and the bathymetry map will be used to avoid such hazards as rugged topography and steep slopes.

APPENDIX A—DEFINITIONS

Firing Rate—(Kilojoules ÷ number of power supplies) × 0.8.

Power Supply—220 volts rectified to 4,000 volts.

Trigger Capacitor Bank—Ten 16-microfarad capacitors with parallel railgap electrodes for discharge.

Capacitor Bank—Twenty 16-microfarad capacitors.

Joule—equals Watt-second; (Capacitance × voltage²) ÷ 2.

Capacitance is expressed in microfarads, voltage in kilojoules; thus a Trigger Capacitor Bank with ten 16-microfarad capacitors (by the definition of a joule) yields 1,280 watt-seconds: $10 \times (16 \times [4]^2 \div 2) = 1,280$.

A Capacitor Bank = 2,560 joules whereas it contains twice the number of capacitors as the Trigger Capacitor Bank.

Sparker—One type of Sparker sound source is a metal cage approximately 8 ft (2.4 m) long, with a 1-ft (0.3 m) triangular shape, each side approximately 12 in. (32 cm). Nine electrodes in three groups of three are equally spaced within the cage. Stored electrical energy discharged into the electrodes shorts to the metal cage which acts as an electrical ground, and creates a super-heated bubble. This bubble grows and collapses several times until the super-heated steam is dissipated. Each time the bubble grows and collapses another pulse is generated. The fundamental frequency of a Sparker is approximately 135 Hz. The bubble pulse is approximately 30 msec in duration and is composed of three or four bubble-growth and collapse pulses. Pulse duration is approximately 75 ft (22 m).

Acoustipulse®—An Acoustipulse® transducer is a coil approximately 16 in. (40 cm) in diameter potted in epoxy which is approximately 2 in. (5 cm) thick and a 20-in. (50 cm) square. Underneath this transducer is attached a rubber diaphragm with an aluminum plate. When stored electrical energy is discharged into the coil, the eddy effect repels the aluminum plate from the epoxy mold. The diaphragm causes the plate to collapse against the mold, and the single action of the plate being repelled and returning creates a 1-msec pulse.

High-Resolution Hydrophone—A hydrophone is a 1 to 2-in. (2 to 5 cm) diameter hose 10 to 40 ft (3 to 12 m) long. Within the hose are 1 to 30 crystal elements spaced approximately 1 ft (30 cm) apart. These hydrophones can be coupled together either directly or by inserting an isolator section which is a 25 to 50-ft (6 to 15 m) oil-filled hose with no elements between the two active sections, thus any combination of channels and length can be derived.

Optically Stacked Sparker—The optically stacked 10,000 joule sparker is a basic system including a graphic recorder, three power supplies,

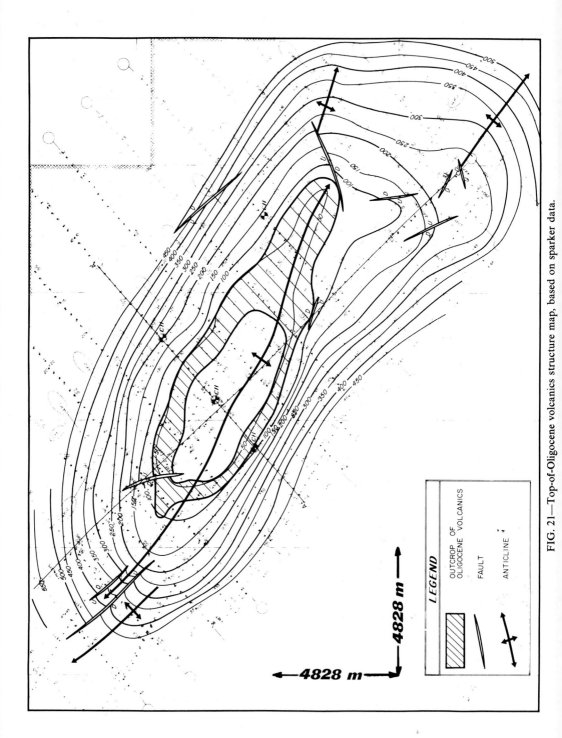

FIG. 21—Top-of-Oligocene volcanics structure map, based on sparker data.

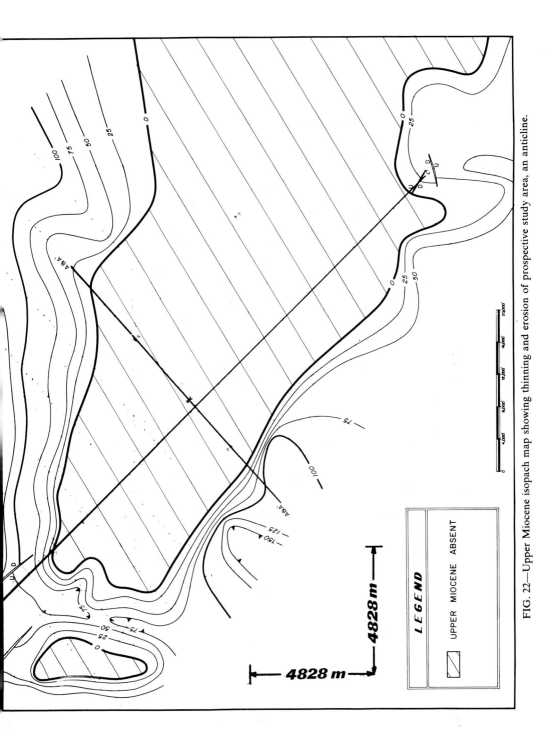

FIG. 22—Upper Miocene isopach map showing thinning and erosion of prospective study area, an anticline.

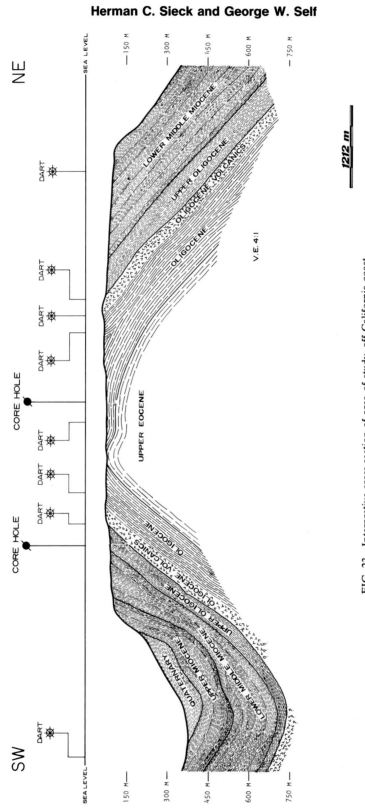

FIG. 23—Interpretive cross section of area of study, off California coast.

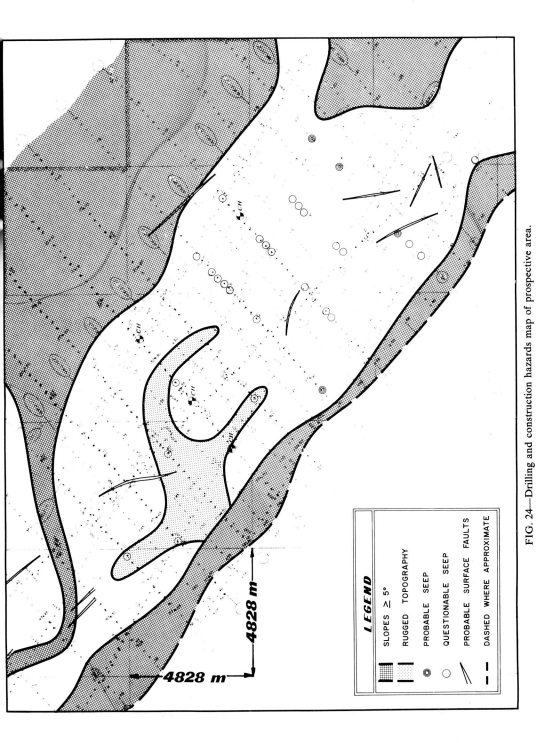

LEGEND

SLOPES ≥ 5°

RUGGED TOPOGRAPHY

PROBABLE SEEP

QUESTIONABLE SEEP

PROBABLE SURFACE FAULTS

DASHED WHERE APPROXIMATE

4828 m

4828 m

FIG. 24—Drilling and construction hazards map of prospective area.

three trigger capacitor banks, and three capacitor banks. By inputting the power supply with 220 volts, a rectified signal of 4,000 volts per power supply can be obtained to charge the capacitors. The ten 16-microfarad capacitors in the trigger bank and the twenty 16-microfarad capacitors in the capacitor bank at 4,000 volts can be charged to approximately 3,900 joules; thus the maximum output of the system can be three times 3,900—or ll,700 joules. Whereas there is variation in voltage output, meaning that the voltage discharged is not always 4,000 volts, we average the output and refer to the system as a 10,000 joule system.

Stored energy is discharged into a nine-electrode sparkarray, and in the optically stacked system, signals reflecting from the bottom and sub-bottom interfaces are received through a two-channel hydrophone system. Each channel consists of 20 elements in a 20-ft (6 m) section with the center of each section 20 ft (6 m) apart (the preset distance of 20-ft (6 m) centers, assumes that a boat travels at 6 kts, which equals 10 ft (3 m)/sec and will travel 20 (6 m) during the two-second firing cycle). Returning signals go into a shipboard electronics package consisting of a digital time delay and a stacker circuit (and as applicable, through a variable bandpass filter), and finally into a 19 in. (50 cm) graphic recorder. Prior to the signal being filtered, it can be "picked" off and a broad band signal taped on ¼ in. (0.4 cm) continuous analog tape.

The philosophy of operation in the dual-channel sparker system is that a one-second record display and a two-second firing rate is used. Thus, when the 10,000 joules is discharged, it takes 2 sec to fully charge the 10,000 joules again. Upon discharging, the reflected signal is received into trace 1, which is nearest the boat, and simultaneously, (with some normal move-out) into trace 2. The signal from trace 2 enters the shipboard electronics package, is filtered, and is recorded for 1 sec on the graphic recorder. During this time the signal from trace 1 is delayed through the digital time delay. Upon completing the recording on the paper record of trace 2, trace 1 (which was delayed) is then recorded beside trace 2. During this time the vessel moved 20 ft (6 m) and the energy stored is again discharged and trace 2 is imprinted on the graphic recorder first while trace 1 is being delayed. Trace 2 signals, which are now the common-depth point of the signal from the original trace 1, are printed side by side—thus the term "optical stacking" versus the commonly accepted practice of electrical stacking.

Where normal move-out is involved, especially in shallow water, there is an adjustment on the stacking unit that allows the move-out correction to be applied to the signal. This is accomplished by the operator looking at the trace displacement and adjusting the "focus" knob until the bottom reflection from both receivers is laid side by side. Where the shallow move-out is taken care of, and in light of the deep geometry and the small separation of hydrophone traces, the deep normal move-out is negligible.

Information is from one of the traces and is recorded for two seconds, instead of recording each trace for one second. The purpose here is that, in interpreting an area, if information exists to a depth greater than one second, it can be retrieved for interpretational use and, of course, the one-second data can be obtained if one wants to do any analog processing and playback.

The operational philosophy of optically stacked sparker systems allows more energy to be discharged in the same time as a small system, thus at the higher energy level providing the same horizontal coverage.

Debubbled Sparker—In the debubbling process, bubble pulses and surface ghosts are removed from reflection data by digital filtering, done by predictive deconvolution (Wiener filtering) using the Levinson algorithm. The resulting seismic record displays each reflector as a single, rather than multiple, line. Thus, structural features and thin beds are no longer obscured. The debubbled data also has true amplitude recovery, which means that the amplitude of reflected signals is not destroyed in the signal enhancement process. This permits clear differentiation of low-impedance horizons (phase reversals) such as horizons containing gas.

The debubbling system uses a computer to design an inverse shaping filter from a digitized source wavelet. The desired or shaped wavelet is the zero-phase equivalent of a Ricker wavelet. The digitized reflection data is then deconvolved in real time using this filter, corrected for spherical spreading and propagation path attenuation losses (by approximating the attenuation factor for the sediments), and is displayed in an analog format. Contrary to most current practices, the deconvolution filter is computed from a precise knowledge of the wavelet transmitted from the sound source, from the known depth of source and hydrophones, and from an estimate of the sea surface reflection coefficient. This wavelet is not inferred from the bottom trace—the filter is not influenced by the reflection characteristics of the bottom materials themselves. Because the sparker source is stable, the filter does not have to be recomputed for each shot. However, it is recomputed periodically or whenever variations in source/receiver geometry cause a degradation of the deconvolved data.

Dereverberated Sparker—Dereverberation minimizes multiples (reverberations) of the seafloor by a subtractive process that estimates the reflection coefficient for bottom material, computes the approximate amplitude and arrival time for the first, second, third, etc. multiple, then subtracts them from the record. Interference due to bottom multiples is thus minimized on these dereverberated records. No attempt is made to remove internal reverberations, and thus the risk of removing real data is minimized. Dereverberation works well for low-order multiples in shallow water, the regime most important for high-resolution seismic operations, and where other techniques, such as horizontal CDP stacking, may not be very effective. Concurrently with dereverberation, the direct arrival is routinely muted from the reflection data if the bottom reflection does not directly overlap the direct arrival.

section 3:
stratigraphic models
from seismic data

Stratigraphic Modeling and Interpretation—Geophysical Principles and Techniques[1]

NORMAN S. NEIDELL and ELIO POGGIAGLIOLMI[2]

Abstract A simplistic view of seismic data and their relation to stratigraphy is adopted. Each event or waveform on each data trace is assumed to relate to a sharp acoustic impedance change in the subsurface directly below the trace location. The incorporation of geologic observations and principles thus should permit interpretation of the acoustic parameter changes in regard to lithology and stratigraphy.

Several departures of the real world from the simplistic view can occur, and these seriously complicate the proposed interpretive sequence. Seismic processing methods, however, act to transform the data so that the simplistic view may be adopted. Further, the preservation of seismic amplitudes and the ability to transform seismic waveforms to more beneficial character add a new and quantitative interpretive dimension to the data.

Seismic model studies are considered as a means of establishing precise requirements for making stratigraphic correlations and describing the seismic character of specific exploration objectives. Such studies further enable the development of a quantitative approach to stratigraphic correlation, not only for the thicker lithologic units, but for thin units—under 20 m— as well. Seismic amplitudes and waveform manipulations prove to be the foundation of these analytic procedures.

Quantitative methods of stratigraphic correlation in concert with the newly developed work on interpreting seismic reflection patterns offer significant tools for stratigraphic interpretation. Not only are results of improved resolution and certainty to be anticipated, but the prospect of making stratigraphic correlations in the presence of complex structure draws closer as a routine exploration practice.

RELATING SEISMIC DATA TO STRATIGRAPHY

Stratigraphic interpretation of seismic data requires that the seismic information be expressed in geologic terms. A strictly geologic view of the earth is developed from surface observations, the guiding principles of geologic evolution, and subsurface information from boreholes. Subsurface information includes well-log measurements from a variety of physical sensors. We can correlate seismic data with the geologic view in terms of geometric description and a subset of the log measurements (namely, velocity and density). Such correlation, used in context with geological and geophysical principles, is at the heart of stratigraphic interpretation.

Typical well-log measurements include the spontaneous potential, resistivity, radioactivity parameters, sonic travel time (velocity), and density. Although the vertical detailing of such measurements is excellent, they are limited in their definition of lateral variation in subsurface parameters. Hence, a strong interpretive element which relies on fundamental geologic concepts and principles is required. Still, log measurements provide insight regarding the porosity and fluid content of the rocks and the lithology in general, and thereby enhance stratigraphic conclusions based solely on a more limited field-observation approach.

Seismic measurements of travel time and amplitudes define the subsurface geometry and give estimates of the acoustic impedances related to rock velocities and densities. Vertical detail is limited owing to the lengthy duration of the individual seismic wavelets and the occurrence of overlapping wavelets from closely spaced reflectors. Lateral definition is good although averaged over regions known as "Fresnel zones." Specialized analyses and interpretive methods can make use of indirect seismic information such as waveforms and velocity to provide insights to the nature of porosity, fluid content, and lithology in general. Here again, fundamental principles and strong interpretive elements are applied in concert.

With these backgrounds, we need a specific vehicle by which the geologic view and the seismic information can be correlated. The most elementary form of such a vehicle has been available for some time—the synthetic seismogram (see Peterson et al, 1955; Wuenschel, 1960).

The synthetic seismogram represents the viewpoint of a laterally restricted segment of the earth taken as a horizontally layered medium. A plane wave is assumed to propagate in this segment, and the seismogram is constructed from the partial reflections of the plane wave from each of the planar boundaries.

[1]Manuscript received, September 21, 1976; accepted, March 25, 1977.

[2]Consultants, GeoQuest International, Ltd., Houston, Texas 77027.

This paper represents a geophysical distillation of an approach to stratigraphic interpretation developed over a number of years by several individuals at GeoQuest International, Ltd. It is important that these contributions be recognized; most of the discussions document their work. In particular, we cite J. P. Lindsey, E. V. Dedman, M. W. Schramm, Jr., and A. K. Nath.

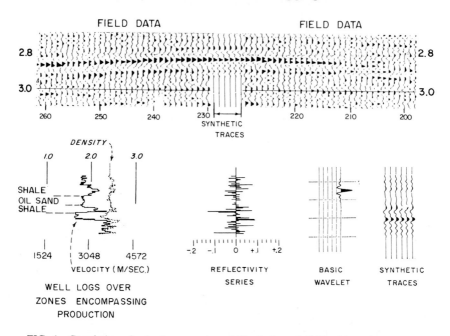

FIG. 1—Correlation of seismic expression of North Sea oil field with geologic data via seismogram synthesis. Synthetic traces inserted into field data at trace 225 which is nearest well location.

Figure 1 illustrates seismic data across an oil field in a North Sea Jurassic sandstone, and also shows a definitive synthetic-seismogram correlation. In this instance, both velocity and density logs were used to calculate reflection coefficients. The particular waveform used in developing the synthetic seismic data matches the propagating waveform of the seismic data. For this comparison, the processed field data were separated at the trace closest to the well location and six repetitions of the synthetic seismic trace were inserted. The agreement in detail is remarkably good. Two important points must be emphasized:

1. Density cannot be neglected as a physical parameter without degrading the effectiveness of our correlation; and

2. The agreement in waveform between the synthetic data and the field data is an essential ingredient of the correlation.

A synthetic-seismogram correlation as in Figure 1 requires:

1. An essentially flat subsurface having lateral continuity and homogeneity over at least a Fresnel zone (the effective subsurface area giving rise to a reflection event),

2. Lithologic boundaries that are well defined rather than transitional, and

3. A common waveform for the seismic data and the synthetic data.

We also make the unstated assumption that each event on the seismic data corresponds to a lithologic boundary, and thus rule out the possibility of noise and other events of no direct geologic significance.

Developing only the geometric viewpoint of the subsurface from a seismic profile necessitates moving beyond the limitations of the synthetic-seismogram model. More realistic models of the subsurface geology are needed for study; such models are accompanied by a variety of wave-propagation phenomena which complicate the interpretive process.

Consider the schematic overthrust-fault model shown in Figure 2 along with its seismic response as computed using wave theory. The hazards of stratigraphic interpretation in such a province are evident. Fully half of the seismic events observed do not arise by reflection but are in fact diffraction contributions. Diffraction events occur from terminating acoustic impedance boundaries or tightly curved convex surfaces and cannot be simply interpreted in the manner of reflection events. The reflection events corresponding to the hydro-carbon-water contact appear only on traces 23

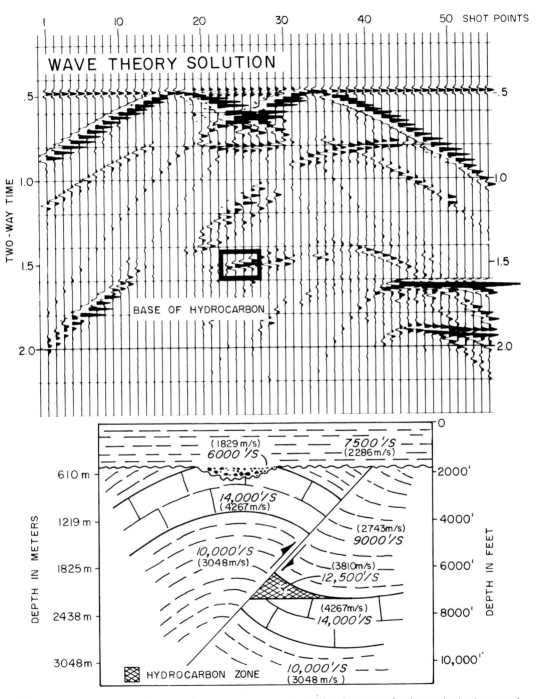

FIG. 2—Overthrust-fault model and computed seismic expression. Diffraction events dominate seismic picture, and bending of seismic rays obscures identification of reflections from hydrocarbon zone.

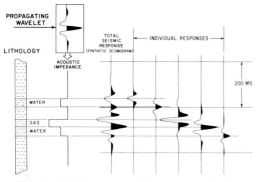

FIG. 3—Idealistic view—relation between lithology, propagating wavelet, and seismic response (after Dedman, et al, 1975).

through 26 as a slightly dipping horizon at 1.5 sec. Displacement of the contact with respect to the eroded crest of the anticline is caused by refraction or the bending of the reflection ray paths in conjunction with plotting seismic traces halfway between the sources and receiver positions (vertical plotting).

Still other difficulties exist which are not illustrated by this figure. At an elementary level, for example, the presence of noise-event alignments or events which have been reflected more than once from shallow horizons (multiple reflections) can also complicate a seismic interpretation.

Seismic data processing, apart from its more routine bookkeeping function, attempts to present the interpreter with data more closely resembling the ideal circumstances which have been assumed, as in the generation of synthetic seismograms. In this light, data processing should be regarded as an extension of the interpretive function.

Distortion of subsurface reflectors by vertical plotting and refraction effects, as well as the presence of diffraction events (as in Fig. 2), has a processing treatment in the form of migration. Migration of seismic sections portrays the subsurface with reflection events corresponding to their points of origin. In migrated sections, anticlines and synclines should appear in their true dimensions, and tight curvatures and bed terminations should be portrayed accurately without "masking" or "smearing" caused by diffractions. Similarly, the elimination of multiply reflected seismic events can be attempted both by stacking and deconvolution procedures (see Peacock and Treitel, 1969, for an account of the latter technique). Since appropriate processing may not always be accomplished, the interpreter frequently is forced to look beyond these obstacles in seeking to define stratigraphy.

The nature of the seismic waveform has been cited as an essential requirement for stratigraphic correlation and is clearly an important factor in controlling seismic resolution.

Figure 3 (see also Schramm et al, this volume) shows a portion of a lithologic log and a corresponding acoustic impedance log. Each contrast in acoustic impedance will be marked by a reflection event having a simple waveform. The polarity or sense of the reflection and its size will indicate the nature of the contrast. Individual reflection events for the model are shown along with their superposition in the resulting seismic trace.

Stratigraphic interpretation would begin with the development from each trace of an acoustic impedance log or, equivalently, a reflectivity series. These results would be correlated from trace to trace to provide the structural considerations, and also would be correlated with available geologic information. By use of geologic principles and insights appropriate to the region, lithologic estimates would be inferred, and, from these estimates and the indicated changes, geometry and depositional patterns, sequences, and history might be interpreted.

Departures from this ideal have been emphasized, and the objective may not always be directly accomplished. Nevertheless, tools such as seismic data processing can assist us in our goal, and the importance of their specific roles can be appreciated.

Good data acquisition and processing are necessary prerequisites for any seismic stratigraphic work. It has been indicated that some ability to capture, manipulate, and generally transform seismic waveforms is needed in order to facilitate the correlations which must be made. Such tools and techniques and their effects on waveform polarities and amplitudes are described in the next section. We also understand that geometric effects, diffractions, and other complicating factors and elements must be included in our interpretations. The role of modeling and, in particular, two- and three-dimensional wave-theory modeling will be discussed specifically in this regard. Finally, the problems of developing principles for making lithologic estimates, improving the vehicles for the correlations, and quantifying the resolution inherent in the seismic expressions must be addressed.

UTILIZATION OF SEISMIC AMPLITUDES AND WAVEFORMS

The rediscovery of the exploration significance of seismic-event amplitudes (documented in the symposium held by the Geophysical Society of

Houston, 1973) is not only a key step in recent progress toward direct detection of hydrocarbons, but is also recognized as one of the cornerstones in the quantitative analysis of thin stratigraphic intervals. In the recent past, difficulties in utilizing seismic amplitudes arose from mechanisms like divergence, reflection loss, and attenuation—which cause the seismogram to become rapidly weaker with increasing reflection time or depth. In the presence of this strong amplitude decay, the interpretive amplitude distinction between a strong reflection and a weak one becomes difficult to discern. Advances in seismic recording technology and new approaches for scaling data prior to display overcame this difficulty.

Attention may now be focused on seismic waveforms. Typical seismic wavelets, particularly as acquired in marine surveys, tend to be lengthy, "leggy," and generally complex in form. The specific waveform for a survey tends to remain fairly consistent from one source discharge to the next, and it propagates in the earth largely unchanged in significant character under most circumstances. This last statement must be qualified by noting that whatever changes are introduced by propagation do not affect interpretive applications when the data are presented to the interpreter at normal plotting scales, where each individual wavelet can have a maximum amplitude no greater than 2 cm.

Recently it has been possible to capture seismic waveforms. Documentation has been accomplished by direct measurements in deep water or by extraction of the waveform from the seismic response of simple lithologic complexes. Appropriate complexes include water-bottom reflections and responses from thick or thin beds having unusually high or low acoustic impedance contrasts—as, for example, Gulf Coast gas sandstones which have low contrasts with respect to the surrounding rocks.

Figure 4 shows first a typical marine wavelet captured by a direct measurement. It is next transformed to a symmetric form approximating a desired form by Wiener filtering. The Wiener filter operation is a well-known signal-processing method which has been used for many years and accomplishes transformations of signals from a given form to a desired form in the best least-squares sense (see Robinson and Treitel [1967] for a discussion of Wiener filters). Application of the Wiener filter to each of the traces from the same seismic survey having the particular propagating wavelet will result in replacement of that wavelet by the symmetric one to a good approximation. Such treatment of data will be called "wavelet signature processing" or simply "wavelet processing."

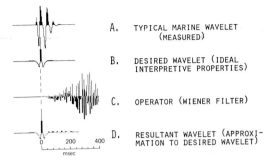

A. TYPICAL MARINE WAVELET (MEASURED)

B. DESIRED WAVELET (IDEAL INTERPRETIVE PROPERTIES)

C. OPERATOR (WIENER FILTER)

D. RESULTANT WAVELET (APPROXIMATION TO DESIRED WAVELET)

FIG 4.—Wavelet signature-processing transformation of basic marine wavelet to ideal interpretable form by Wiener filtering.

All the advantages of replacing the propagating waveform by one having zero-phase or symmetric properties are not obvious. It is clear from an interpretational viewpoint that each change in acoustic impedance will now be marked by a simple, readily understood waveform, but there are other tangible benefits of the transformation. The shortening of the waveform causes improvement in the effectiveness of standard processing methods for noise and multiple suppression. Similarly, the increased resolution of the shortened waveform permits more highly resolved seismic-velocity estimates to be made.

The resultant wavelet in Figure 4D has precisely the same range of frequency content as the original basic wavelet, and no requirement has been made for an extrapolation of information content. Berkhout (1973, 1974) and Schoenberger (1974) have noted the enhanced resolution and information potential inherent in changing waveforms without broadening the frequency band. Remaining within the well-defined frequency band allows the introduction of improved resolution without the introduction of noise. By contrast, other approaches seeking to increase resolution through broadening of the frequency band (e.g., spiking deconvolution) introduce much noise and generally produce less acceptable results.

A transformation of a Teledyne air-gun waveform is noted in Figure 5. The original waveform (Fig. 5C) has been measured by a direct waterborne arrival to a deeply submerged hydrophone. A surface receiver ghosting operator has been simulated so that this result may be compared to Figure 5A, which is the waveform as extracted from a strong water-bottom reflection in normally recorded data. Despite the effects of the hydrophone array present in Figure 5A, the comparison is most favorable, suggesting the validity of several of the ideas expressed earlier. Note that the

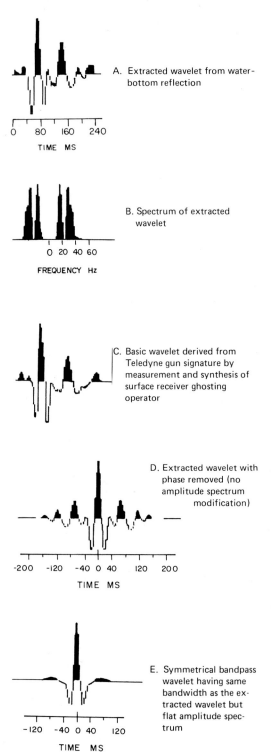

A. Extracted wavelet from water-bottom reflection

B. Spectrum of extracted wavelet

C. Basic wavelet derived from Teledyne gun signature by measurement and synthesis of surface receiver ghosting operator

D. Extracted wavelet with phase removed (no amplitude spectrum modification)

E. Symmetrical bandpass wavelet having same bandwidth as the extracted wavelet but flat amplitude spectrum

FIG. 5—Verification of wavelet consistency and role of amplitude correction in wavelet processing.

amplitude spectrum of the wavelet (Fig. 5B) shows a deep notch at about 25 Hz. Wavelet processing must fill this notch and generally flatten the amplitude spectrum to achieve an ideally interpretable waveform, as is illustrated by contrasting Figures 5D and 5E. In Figure 5D the amplitude spectrum, including the deep notch, remains unchanged, whereas it is smoothed and flattened in Figure 5E.

Examination of representative examples of wavelet-processed data reveals the improved resolution and interpretability. Compare similarly processed seismic time sections over the North Sea oil field considered in Figure 1. In one panel of Figure 6 the original waveform is present, whereas, in the second, a shorter waveform of simple symmetric character has been introduced in place of the original.

The wavelet-processed data have fewer events owing to the removal of multiple reflections. Processing techniques designed to remove multiples perform better where the shorter, more highly resolved wavelet has been introduced. Also, a more definitive interpretation which shows some indication of an oil-water contact appears to be possible.

Hence, the waveforms inherent in seismic data can be manipulated and transformed to simpler waveforms which are more amenable to interpretation. These transformations have been more readily accomplished on marine data, although R. O. Lindseth (1976) has presented results indicating analogous success with land data. In consequence, correlations between seismic data and geologic inputs can be effected with greater rigor to produce more definitive results.

CONTRIBUTIONS OF MODELING AND WAVE-THEORY MODELING

Simulation of exploration seismic data, or seismic modeling, is in fact the successor to the synthetic seismogram, as the vehicle by which proposed seismic correlations with geologic data may be verified, and also as the tool by which geologic hypotheses may be tested. Although publications describing modeling techniques and applications are few in number (see Taner et al, 1970; Shah, 1973), the technology has reached a rather sophisticated level. Neidell (1975) described some of the more advanced considerations and limitations associated with the use of seismic modeling.

Unlike the synthetic seismogram, seismic modeling accepts a description of a subsurface condition in terms of geometry and acoustic parameters, including compressional-wave velocity, density, and an attenuation factor. Two-dimensional modeling systems are designed to view the subsurface condition as having perfect out-of-the-

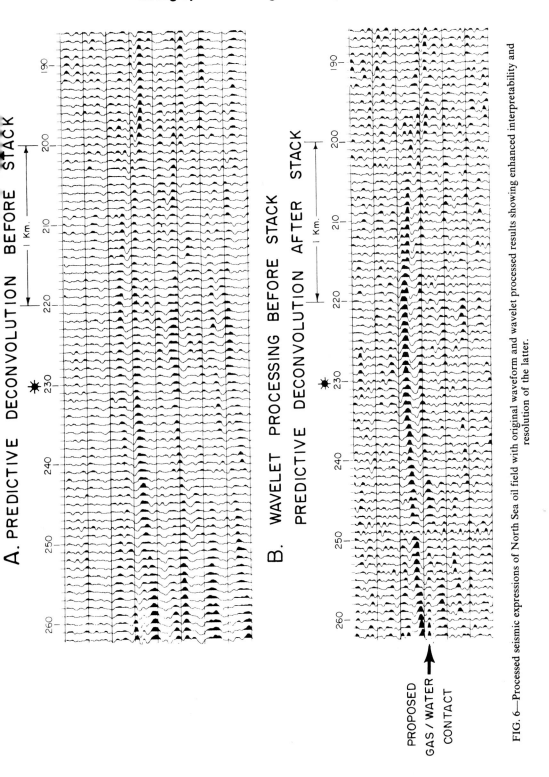

FIG. 6—Processed seismic expressions of North Sea oil field with original waveform and wavelet processed results showing enhanced interpretability and resolution of the latter.

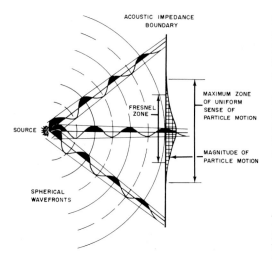

FIG. 7—Spherical wave motion encountering a flat reflector illustrating Fresnel zone region which is principally responsible for reflection event.

plane lateral continuity and homogeneity. A geometry of virtually any complexity may be treated. The seismic parameters can be permitted to vary both horizontally and vertically to represent lithologic transitions and similar subtle stratigraphic effects.

In the discussion regarding a wave-theory modeling illustration for an overthrust fault (Fig. 2), modeling was portrayed as a tool for overcoming vertical plotting and other geometric effects on the seismic section which might complicate the interpretation of stratigraphic objectives, and also for treating diffractions. The following discussion will describe modeling from a more advanced viewpoint. In particular, we can use it to develop insights to the spatial resolution inherent in seismic data and the very origin of reflections from a three-dimensional subsurface representation. Furthermore, model studies are important to the general characterization of specific exploration targets in seismic terms.

Consider the results of a two-dimensional wave-theory modeling study seeking to determine how features comparable in extent with a Fresnel zone will appear on a seismic section. The specific calculations illustrated are developed from the basic theories and concepts proposed and developed by Hilterman (1970).

Figure 7 illustrates the Fresnel zone. Spherically radiating wave motion encounters a flat acoustic impedance boundary. The maximum-sized region of this boundary, over which the particle motion is all in the same direction, is indicated. A

Fresnel zone, although directly related to the size of this region, has a somewhat smaller value since some allowance must also be made for the respective magnitudes of particle motion within this region.

If the source is at a distance R from the reflecting boundary where R represents travel distance at a velocity V over a two-way time of t seconds, and the dominant frequency content of the seismic data is f_c, then a Fresnel zone radius r_f is approximated by

$$r_f \simeq \frac{V}{4}\sqrt{\frac{t}{f_c}}$$

This mathematical relation is derived using basic geometry and an empirical weighting. For the values $V = 3{,}000$ m/sec, $t = 1$ sec, and $f_c = 25$ Hz, $r_f = 150$ m.

Seismic illumination of the subsurface may then be considered analogous to a "searchlight beam," and the subsurface area illuminated by this beam is considered to be the Fresnel zone. With such an intuitive point of view, we can consider fundamental questions such as how small in extent a subsurface feature must be as compared to a Fresnel zone in order to be detectable, and what the nature of its seismic expression must be in order to facilitate detection.

Note in Figure 8 that, as a sandstone body approaches the lateral dimension of a Fresnel zone and becomes still smaller, the seismic signature loses all reflection character and appears in the diffraction form corresponding to a reflecting point. The lessons are clear: (1) thin beds of small spatial extent have a seismic expression, but not dominantly of reflective nature; and (2) although the form of the signatures are alike for sandstone bodies smaller than the Fresnel zone, their size can be gauged by noting the differences in the strengths of the seismic events. The smaller extent bodies produce weaker events by occupying only portions of the full Fresnel zone.

Amplitude processing is concerned with the preservation of seismic strengths. With the mechanism just observed, it was seen that, *where relative amplitudes have significance, they may be related to the identification of small lithologic inhomogeneities.*

As a practical matter, the size of the Fresnel zone is initially related to the quarter wavelength of the dominant frequency component of the seismic waveform and the near-surface velocity. For the expression for r_f, the size of the Fresnel zone grows in proportion to the stacking velocity, which may be determined readily from seismic data. The case study examined by Figure 8 is a

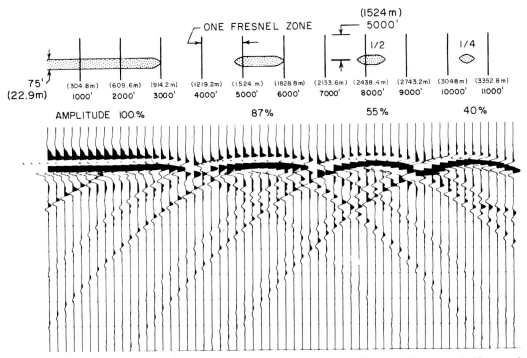

FIG. 8—Wave-theory model response for sandstone bodies of varying lateral extents illustrating significance of Fresnel zone size.

typical one for intermediate-depth Gulf Coast sandstones.

Next we can investigate the role of the Fresnel zone from the point of view of a three-dimensional wave-theory model. Figure 9 shows a reeflike box structure over which two seismic profiles are calculated. Line 1 crosses the structure through its long axis; line 4 parallels line 1 but is 152.4 m from the edge of the structure. The structure is 1,524 m below the surface.

Line 1 shows the structure clearly, along with all diffraction events at its termination. The superposition of the diffraction events from the horizon at the base of the reef almost suggests the continuity of this horizon under the structure. Observe that reflection amplitudes over the reef are comparable in magnitude with those of the reflective horizon on which it was developed.

Line 4 suggests a structure above and below (though more subtle here) the reference horizon. The reflections above the reference horizon are from the flat-topped portion of the structure (Fig. 9). The Fresnel zone which gives rise to reflection events lies partially on the reef top and partially off of it. In fact, the reduced-amplitude event seen for the reef top is proportional to that part of the Fresnel zone area which is on top of the reef.

The reference-horizon reflection amplitude seen below the reef-top reflection is also reduced since it, too, represents only a part of the Fresnel zone (although this effect is more difficult to see). Below the reference horizon the weak events are formed by re-enforcement of diffractions from the base edges of the reef acting out of the profile plane.

Three-dimensional model studies show that all seismic interpretations must be visualized in a three-dimensional environment. Such model studies are probably the most powerful means presently available for teaching the principles of seismic interpretation.

Another Fresnel zone-related model study of stratigraphic nature provides insight to the circumstances in which the synthetic seismogram may be reliably employed as a correlation tool. The particular case study is called the "sand-shale interfingering model," and the applicable illustrations are Figures 10 through 15 (see Lindsey et al, 1976).

Subsurface lithologic boundaries may range from sharp contacts to gradational ones. The lack of a well-defined boundary will understandably influence the seismic reflections. Additionally, there will be a lateral averaging over the Fresnel

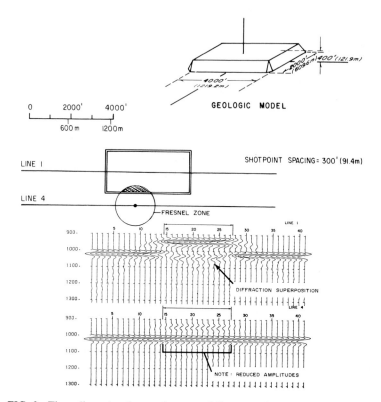

FIG. 9—Three-dimensional wave-theory modeling over "box car" reef illustrating significance of Fresnel zone in producing reflections.

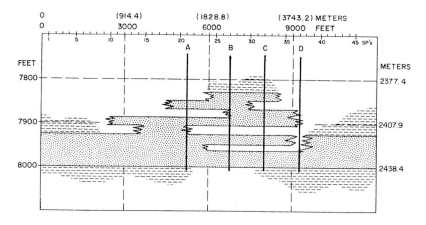

FIG. 10—Sandstone-shale interfingering model illustrating lateral lithologic transition.

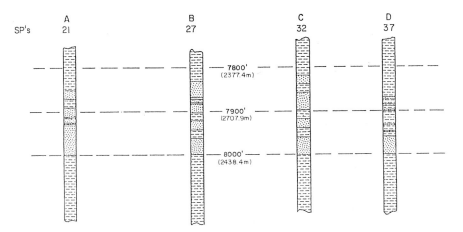

FIG. 11—Lithologic logs for four wells (Fig. 10) of sandstone-shale seismic section.

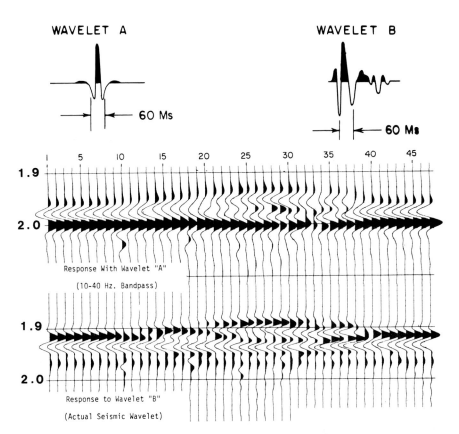

FIG. 12—Sandstone-shale interfingering-model seismic response.

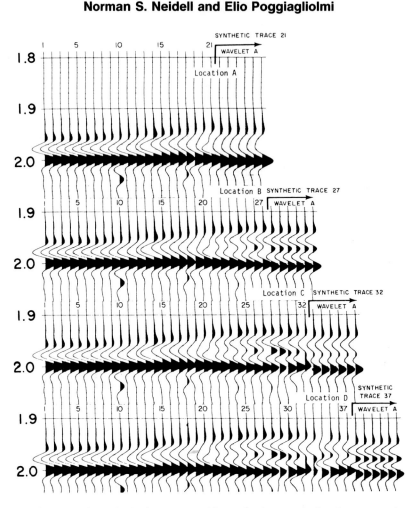

FIG. 13—Comparison of seismic response with wavelet A to synthetic seismogram using wavelet A.

zone. Where a diffuse boundary is involved, there is some weakening of the reflection response; there is also weakening in lateral continuity. Further, the seismic response may differ from a response predicted by the synthetic seismogram developed from well-log measurements even if a common seismic waveform is used. The well log measures the vertical lithologic sequence within a few inches of the borehole; seismic waves average several hundred feet laterally and thus may differ from the well log in describing the boundary character by introducing spatial variations.

The sandstone-shale interfingering model attempts to show some of the effects of a diffuse boundary. A sandstone of approximately 30.5 m thickness is embedded in a shale. The sandstone base is uniform over a large area, but the sand-

stone top is locally interfingered with the overlying shale. Four holes are shown at locations A through D (Fig. 10). The lithologic logs for these holes are shown in Figure 11.

A seismic wave-theory response to the complete sandstone unit is shown in Figure 12. The upper version uses a seismic wavelet that is symmetrical and polarized such that a positive reflection (transition from a "soft" or low-acoustic-impedance rock to a "hard" rock) will be displayed as a black or right-hand deflection. This is an ideal wavelet for visual detection of detailed stratigraphy and acoustic lithology. The lower seismic response is the same model using a wavelet actually present in some marine seismic data; the wavelet is not symmetrical or regular in any sense. Figure 13 through 15 compare both wave-

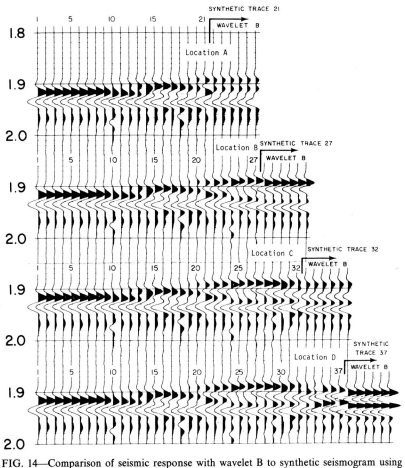

FIG. 14—Comparison of seismic response with wavelet B to synthetic seismogram using wavelet B.

lets A and B for the wave-theory seismic response to the synthetic seismogram derived from the logs using these same wavelets. The synthetic trace repeated six times is shown adjacent to the well location in each case. Contrast the good agreement at locations A and C with location D (indicated on Fig. 13 and 14). Note also the complete and obvious lack of correlation resulting from the use of differing waveforms as in Figure 15.

This study offers important guidelines for using simple synthetic seismograms. First, it is essential that the seismic waveform be common to the synthetic and the actual data. Then, we understand that the synthetic seismogram and the actual data should correlate only to the extent that the sequence of the log extrapolates laterally for at least a Fresnel zone distance (about 304.8 m for this case; Fig. 10). It is interesting to see how the application of more advanced tools does not rule

out the use of the more elementary ones, but rather instructs us in their proper and appropriate use.

The well-defined base of the sandstone section shows a consistency in amplitude and arrival time that sets it apart from the transitional upper boundary. The effects of the transition appear as a generally reduced amplitude or, more prominently, as a waveform distortion.

The search for reefs in Michigan involved elements present in a model study executed by Nath (1975). The remarkable improvement in exploration success for Silurian reefs in Michigan was the result of seismic data-processing techniques which first treat the severe statics problem that had previously degraded the data quality (see especially McClintock, 1977). Next, the preservation of seismic-amplitude information for land data constituted a significant further step; and

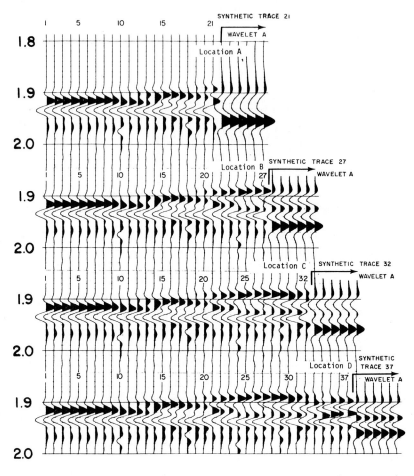

FIG. 15—Comparison of seismic response with wavelet B to synthetic seismogram using wavelet A.

the seismic characterization of the reef objective through modeling completed the interpretive sequence.

As Nath's summary concludes, the reef signature is identified as a break in continuity of certain reflections. After consideration of lithologic differences in acoustic terms, we note from the ray-trace display of Figure 16 that geometrical effects of the tight curvature disperse the reef reflection response over a broad ground surface area, causing weak responses. Hence, geometrical considerations prevail, making consideration of the acoustic differences between the anhydrite, evaporite, and carbonates of secondary importance. The reef, which is by definition a stratigraphic trap, has a seismic indication which seems stratigraphic in appearance, yet is structur-

al in origin. The indicated seismic extent of the reef has also been much exaggerated by the same ray-spreading mechanism.

Figure 17 illustrates a model primary-reflection time section developed from a North Sea location. A strong indication of a stratigraphic prospect, a pinchout, is suggested at 3.3 sec. The particular study is based on a schematic representation of a North Sea horst block, which is shown in Figure 18.

The seismic expression of the model with a simple waveform as shown in Figure 19 is readily interpreted in structural terms. The typical, longer waveform section (Fig. 17), however, shows a strong indication of a pinchout near the center of the horst block. In this case, the superposition of the lengthy waveforms and the slightly discordant

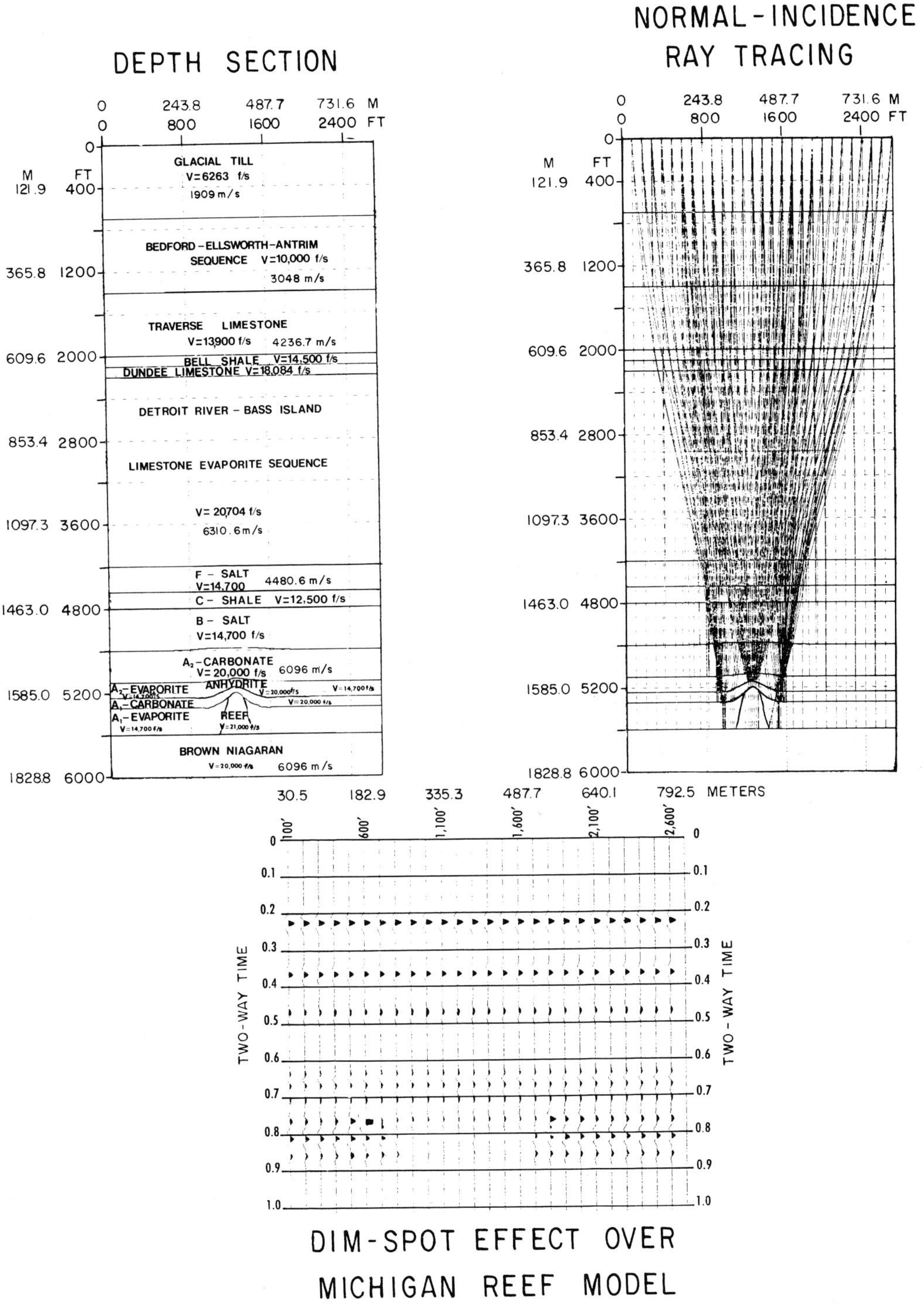

FIG. 16—Model study delineating seismic expression of pinnacle reef (after Nath, 1975).

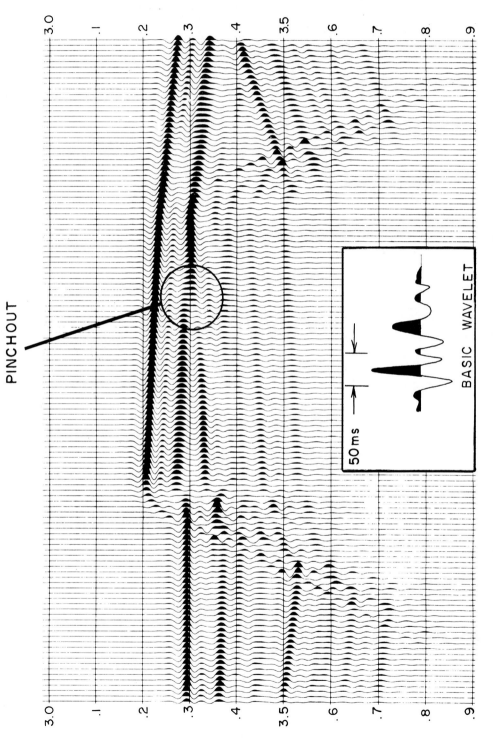

FIG. 17—North Sea horst-fault model, wave-theory solution (primaries only).

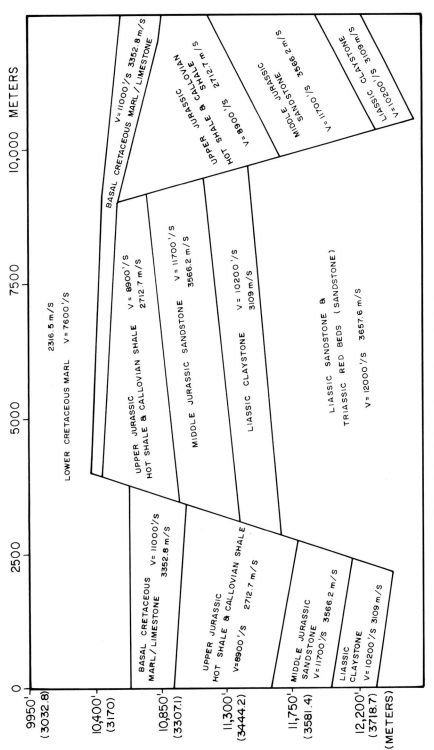

FIG. 18—North Sea model, geology and seismic parameters.

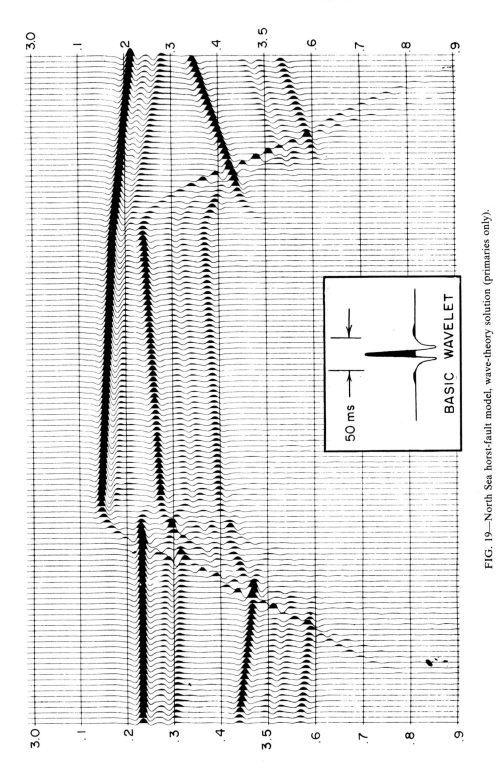

FIG. 19—North Sea horst-fault model, wave-theory solution (primaries only).

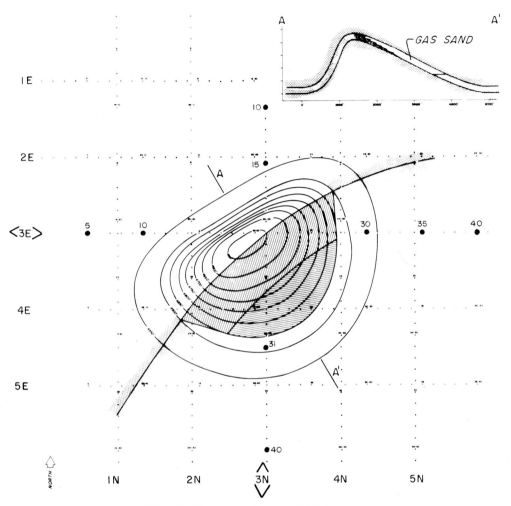

FIG. 20—Three-dimensional model-C depth map.

geometry have conspired to suggest a stratigraphic element which does not exist.

In spite of conventional processing methods, the indication of a stratigraphic prospect exists, even when yet another typical but lengthy waveform is introduced. Only the introduction of the symmetric shortened wavelet clarified the seismic expression. Clearly, then, we must be cautious of some stratigraphic-prospect indications on the grounds that the prospect itself is a product of the seismic expression and does not actually exist geologically.

The final study is a three-dimensional wave-theory model having both structural and stratigraphic elements. The structural contours and an indication of the lithology are posted on a base map (Fig. 20), which shows also the trace loca-tions for the computed seismic profiles. The structure consists of a sandstone body which has a facies change to shale short of the crest. The sandstone is partially gas-filled.

Figure 21 shows north-south and east-west seismic profiles across the structure. Since the principal axes of the feature are rotated by about 45° from these directions, the picture is far from straight forward. Still, the most pronounced features of the prospect are the amplitude anomalies to one side of the crest. The profiles of the principal axes are shown in Figure 22. They help considerably in clarifying the interpretation, but they usually are not available in exploration studies.

One of the principal lessons of this study is obvious. If we honor the structural interpreter's maxim of "drilling on the highs," the pay zone

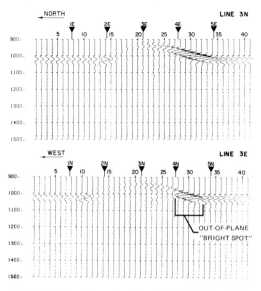

FIG. 21—Three-dimensional model-C seismic profiles.

will be missed. Instead, we must be guided by stratigraphic precepts; and these must be knowledgeably applied. The drilling of the "bright spot" on profile 3E of Figure 21 will result in a dry hole. In this case we are looking at reflections from out-of-the-plane of the profile.

Experience with seismic modeling has shown that it is unlikely for significantly differing hypotheses of the subsurface geology to lead to similar seismic expressions, particularly where we are willing to model with separated source and receiver positions, thus more closely simulating the information content of stacked seismic data. Most of the ambiguity in the model studies is concerned with subtle geologic elements rather than the gross conceptualization of the subsurface condition. These subtle elements are, of course, quite important in consideration of stratigraphic situations; hence, case studies help to suggest the full potential for stratigraphic interpretation based on available tools and techniques.

Only our ingenuity and understanding of geological and geophysical principles truly limits exploration based on the technology which has been developed.

QUALITATIVE AND QUANTITATIVE STRATIGRAPHIC CORRELATIONS

Variations in the characteristics of the stratigraphic targets which we can identify from seismic data require a refined use of those data and the available interpretation tools. The studies previously described in some sense document this

viewpoint. In such refined approaches, information may often be forthcoming which bears significantly on still other hydrocarbon-related disciplines. We can demonstrate such a circumstance by considering in another context Figures 1 and 6.

The wavelet-processing discussion used two data panels (Fig. 6) taken from a North Sea oil field. Panel A shows a typical processed section which was subjected to predictive deconvolution before stack. In panel B, wavelet-processed results have been similarly deconvolved. At the left-hand edge of the wavelet-processed panel B, there is an event having geologic discontinuity (see arrow) which strongly suggests an oil-water contact. In this environment—a Jurassic sandstone under the Kimmeridgian unconformity—an unusual sequence of fluid mechanisms is required to develop sufficient acoustic contrast between the oil- and water-filled portions so that a seismic event becomes visible.

The top of the sandstone is clearly visible as a trough, but, to the right of trace number 205, some change in lithology is taking place because we lose sight of this trough. Beyond the portion of data shown and continuing to the right, the sandstone top becomes well defined once again with yet another possible indication of an oil-water interface. If the implied lithologic change is indicative of a change in porosity, then serious consequences of a reservoir-engineering nature may be implied.

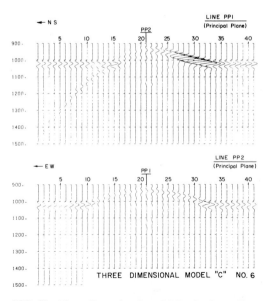

FIG. 22—Three-dimensional model-C seismic profiles.

The ability to define reservoirs with precision as indicated and even to suggest porosity changes from seismic data is well beyond the scope of usual stratigraphic interpretations. Correlation of the synthetic seismogram with seismic data for the North Sea as shown in Figure 1 corresponds also to this oil field. The Jurassic sandstone is clearly visible as a low-velocity low-density zone on the well logs and is correspondingly denoted by a trough-and-peak sequence in the seismic data.

Where thick lithologic units exist in the subsurface, quantitative approaches to stratigraphic correlations using the tools and methods described can be readily conceived. Thick units are those whose boundaries are marked by reflection events having sufficient separation in time to be clearly resolvable. In such a case, the necessary calibrations between arrival time and depth, and between arrival-time difference and thickness, are established using velocity information from nearby well-log measurements. If the lithologic unit is sufficiently well defined and thick, it may even be possible to develop the velocity information from the seismic data alone (see Taner and Koehler, 1969, for a basic discussion of seismic-velocity determination).

For normal seismic data, thick lithologic units typically encompass 20 m or more depending on depth of burial, regional velocity variation with depth, and specific characteristics of the effective seismic wavelet. This value is, in fact, very much in line with our usual view of the thickness resolution inherent in our seismic data (see Sheriff, 1976). Because many exploration situations are concerned with beds having less than 20-m thickness, the matter of their resolution in quantitative terms is far from academic. Hence, we must consider in analytical terms the thin-bed stratigraphic resolution potential inherent in seismic data.

The exposition of wavelet-processing concepts and practices gave an indication that the resolution of seismic data, at least in qualitative terms, was related to waveform (see Fig. 6). Later, a model study relating to Fresnel zones indicated that seismic amplitudes held the key to seismic resolution in the spatial dimension (see Fig. 8). We may thus anticipate that these same factors will play analogous, but analytical roles in establishing seismic resolution in the travel-time sense.

Ricker (1953), in an early work on resolution, set the pattern for most subsequent thinking. According to Ricker, the breadth of a seismic waveform entirely controlled its resolution potential. Widess (1973) carried this philosophy to a logical conclusion. Widess cited this limit of resolution as being ⅛th of the wavelength of the waveform central frequency. He noted from a synthetic-seis-

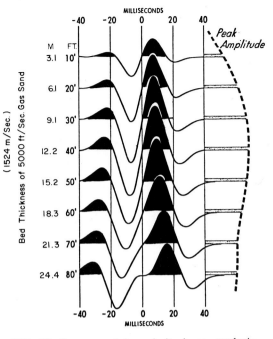

FIG. 23—Response of low-velocity layer—synthetic-seismogram study for 25-Hz Ricker wavelet.

mogram study that, as the thickness of a high-velocity bed dropped below ⅛th of the dominant wavelength, the peak-to-trough separation of the reflection signature became invariant.

Figure 23 shows computed synthetic seismograms for a single low-velocity bed for a variety of thicknesses. A 25-Hz Ricker wavelet is taken as the seismic waveform, and invariance is observed below 12.2-m thickness, in accord with results of Widess. Peak-to-trough amplitudes are now also indicated. At the thickness where peak-to-trough separation becomes invariant, "tuning" occurs and maximum reflection amplitude through event reinforcement is seen. For bed thicknesses smaller than the tuning thickness, all resolution information appears to become encoded in the event amplitudes. Widess' results exhibited the same phenomenon, which went unrecognized.

J. P. Lindsey (Geophysical Society of Houston, 1973) explicitly appreciated and utilized the principle of amplitude encoding of thin beds. Of course, in our earlier discussion of the Fresnel zones, we encountered a similar phenomenon in that the amplitudes of seismic signatures for point reflectors related to the spatial extent of small subsurface reflectors.

Hence, wavelet-processed data, taken in conjunction with data in which reflection strengths

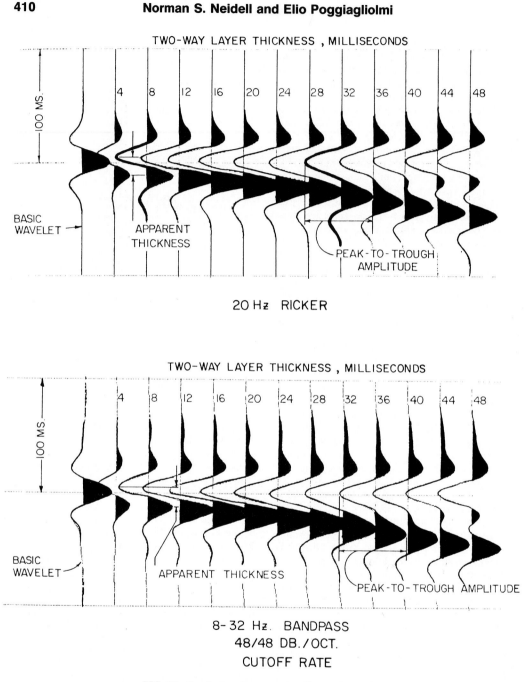

TWO-WAY LAYER THICKNESS , MILLISECONDS

100 MS.

4 8 12 16 20 24 28 32 36 40 44 48

BASIC
WAVELET

APPARENT
THICKNESS

PEAK-TO-TROUGH
AMPLITUDE

20 Hz RICKER

TWO-WAY LAYER THICKNESS , MILLISECONDS

100 MS.

4 8 12 16 20 24 28 32 36 40 44 48

BASIC
WAVELET

APPARENT THICKNESS

PEAK-TO-TROUGH AMPLITUDE

8-32 Hz. BANDPASS
48/48 DB./OCT.
CUTOFF RATE

FIG. 24—Synthetic-seismogram studies of thin beds.

are controlled so as to have significance, offer a powerful stratigraphic approach to the resolution of thin beds. The practical bounds of the method and specific mechanisms remain to be explained. Still, we must not overlook the profound economic impact on exploration for stratigraphic objec-

tives of this new viewpoint toward seismic resolution.

It is important to stress again that our new viewpoint toward resolution rests in good measure on our knowledge of the propagating waveform and the simplicity of its structure. Where we have no such reference, our insights have far less effect, as is illustrated by the work of Meissner and Meixner (1969). Although thin-bed effects were studied, intuitions received little guidance owing to use of a rather realistic, complex, lengthy seismic source pulse. Clearly, then, we shall want to apply this quantitative technology only to wavelet-processed results in which our requirements for a waveform reference are satisfied.

In Figure 24 some synthetic-seismogram studies for thin beds are repeated, this time using two differing symmetric waveforms—a 20-Hz Ricker wavelet and an 8 to 32-Hz Butterworth bandpass wavelet. The thicknesses are now presented in two-way travel-time units. For every such example a characteristic amplitude tuning thickness is noted, whereas, for lesser thicknesses, all thickness information is amplitude encoded.

From synthetic-seismogram studies of this type, we can develop sets of calibration curves as are illustrated by Figure 25. The vertical scale presents actual bed thickness in milliseconds, and the two horizontal scales, in turn, give apparent thickness measured or peak-to-trough time separation and the measured peak-to-trough amplitude in arbitrary scale units. The specific curves shown are taken directly from the synthetic-seismogram study using the 20-Hz Ricker wavelet. If the actual thickness of the bed and the apparent thickness, as determined by peak-to-trough separation, were the same value, the time-resolution calibration curve would follow the diagonal dashed line. For the thicker lithologic units or beds, this dashed line is in fact followed, showing that the resolution in travel time can be accurately accomplished.

For the particular 20-Hz Ricker wavelet under consideration, the smallest possible peak-to-trough separation which can be observed approaches an asymptote of about 17.3 msec (see Time Resolution Limit on Fig. 25). Also, at a bed thickness of 19 msec, the indicated seismic strength or amplitude tunes to a value about 40% greater than the normal amplitude level observed for bed thickness of 45 msec or more. It follows that, for seismic data processed to have a 20-Hz Ricker wavelet as the propagating wavelet, thickness estimates for beds having apparent thickness (from peak-to-trough observations) of less than 25 msec are best determined by using calibrated amplitude values.

Recall that the technology for transforming and manipulating seismic waveforms gives us the ability to impress any waveform we choose into the data so long as it does not include frequency components not present in the originally propagating waveform. Calibration curves for amplitude and travel-time measurements as shown may

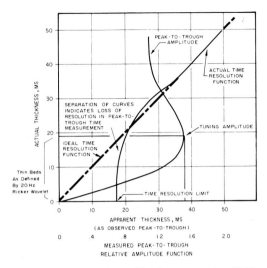

FIG. 25—Resolution calibration curves for 20-Hz Ricker wavelet developed from synthetic-seismogram study.

be developed for *any* specific symmetric wavelet used for wavelet processing. Further, an analytic means now exists for distinguishing between thick and thin beds and resolving thin beds as a function of a known propagating wavelet.

Several practical applications are apparent. First, it is clear that a thinning bed, or one which is pinching out, will have its clearest seismic expression when the tuning thickness is reached. The interpretive significance of this observation is profound for several reasons. At a most elementary level, we see that a thin gas sandstone will have the highest amplitude reflection not necessarily at its thickest portion or where there is most gas, but, rather, where the tuning thickness is attained. This is usually near the edges of the gas-filled zone. Of more consequence to this discussion is the fact that the observation of tuning amplitude levels on the processed seismic section gives us one of two calibration values needed to relate amplitudes on the seismic section to the arbitrary scale of amplitudes on our calibration-curve plot. A second calibration value is normally determined from observation of the normal amplitude level for a thick lithologic unit. The need for preservation of the significance of seismic amplitudes through processing is clearly indicated.

Next, we consider a wave-theory response from a schematic model after Lindsey et al (1976) which shows a thin shale stringer in a sandstone body of uniform thickness (Fig. 26). The 15.2 m of the gross body character leads to an invariance of waveform. Net shale content can be estimated directly by noting the peak-to-trough amplitudes.

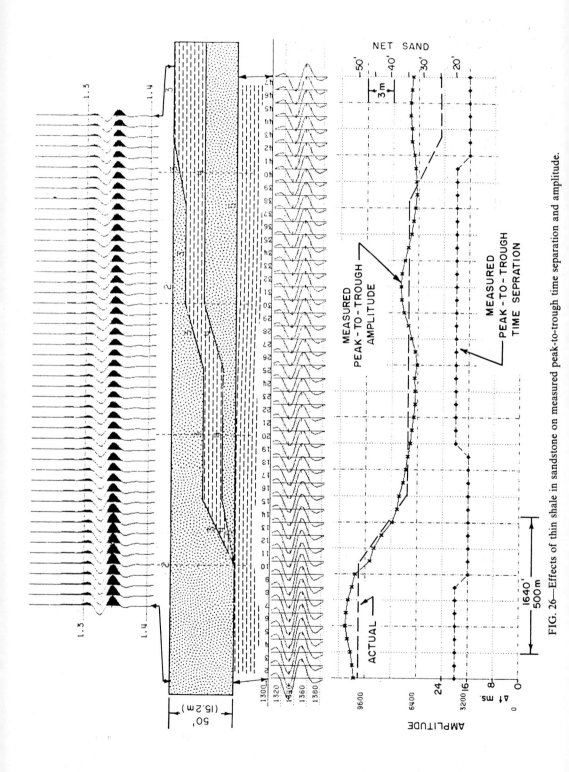

FIG. 26—Effects of thin shale in sandstone on measured peak-to-trough time separation and amplitude.

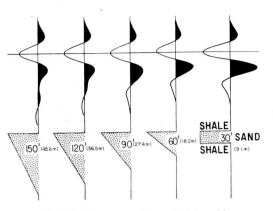

FIG. 27—Seismic response for vertical transition contacts of variable thickness.

At the bottom of the figure, an output from a computer analysis monitors the measured peak-to-trough time separation and peak-to-trough amplitude on a trace-by-trace basis. Excellent agreement is noted between the net sandstone estimates and the measured amplitudes. The peak-to-trough time separation in this case tells us nothing because we are dealing with a thin bed. Note that the most serious departures of the measured net sandstone curve from the actual curve correspond to places in the model where beds terminate. These terminations give rise to diffraction events and contaminate the reflections, causing departures from the theoretical considerations which were presented, as the latter were based on the more simplified theory of synthetic seismograms.

The model study of Figure 26 presents both a limitation on the method proposed and one of a series of cautions which must accompany its use. Although a net shale estimate may be derived, the position or distribution of that shale is not indicated in the seismic response. In an ideal case we might be able to identify 5 m in a 15.2-m unit, but we could not determine its position or distribution in the unit (i.e., five 0.5-m stringers would appear precisely as one 2.5-m stringer, etc.). We further see that diffractions which were not included in our considerations can disturb the desired quantitative correlations.

Other cautions can be derived from elements of the model. In developing Figures 23 and 24, for example, the low-velocity unit is encased between similar materials and waveform calibrations are computed accordingly. For the circumstance where the encasing lithology above and below differ significantly in acoustic properties, a correspondingly adjusted amplitude calibration would be needed. Here geologic information de-

veloped principally from nearby well logs would be necessary to develop such amplitude calibrations in anticipation of the drill.

Further, transitional contacts over depth have not been included in the model considered. We may readily appreciate the effect of such contacts by synthetic-seismogram studies as shown in Figures 27 and 28. Figure 27 shows that the principal effect of an intermediate-thickness transition zone is a decrease in amplitude, the black peak reflection being somewhat smaller than the preceding trough. A similar effect was observed for spatially derived transitions as indicated by the study of Figures 10 through 14. The ability to resolve specific transitional effects as suggested by Figure 28 is quite limited. Hence, if these effects are to be included in a quantitative analysis, a geologic basis for their incorporation must first be developed.

By way of summary, we consider next another model by Lindsey et al (1976) encompassing a transitional lithology against a structural background similar in concept to the three-dimensional model study of Figures 20 through 22. A fault edge has been introduced in Figure 25. We put into practice the computer analysis of peak-to-trough measurements and determine gross lithology, but now in the presence of a structural component. Note that amplitude level of the water sandstone has been interpreted and that all amplitude variation exhibited is of lithologic origin and not related to thinning or thickening of the particular unit.

The quantitative techniques described have been in use now for some time in a variety of practical contexts. In all their applications, however, the geophysical tools and principles described here play prominent roles. These methods are significant, but there is still more to the seismic study of stratigraphy than the delineation of

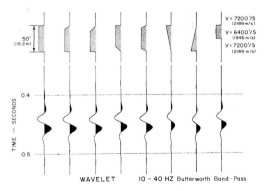

FIG. 28—Seismic response to abruptness of vertical contacts (after Meckel and Nath, this volume).

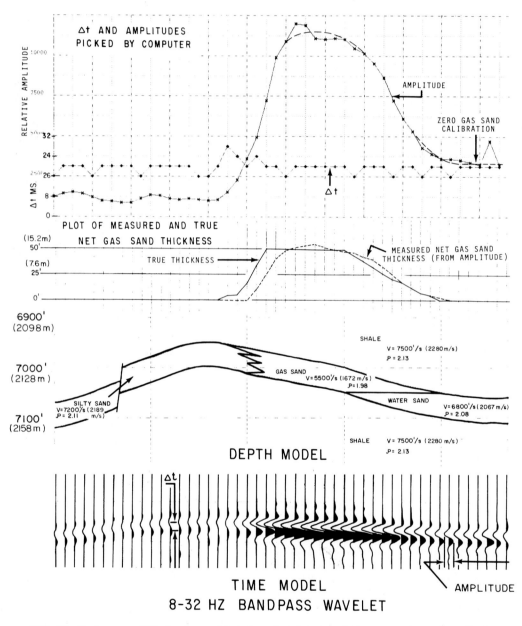

FIG. 29—Updip-permeability-barrier model study and peak-to-trough time separation and amplitude resolution analysis.

thin beds. We must cite, in particular, the pioneering work of Harms and Tackenberg (1972), Sangree and Widmier (1974), and Vail et al (1976), who set out to achieve stratigraphic objectives by identification of sedimentary processes from seismic patterns. Spatial and temporal resolution in concert with these approaches offers a comprehensive and most powerful technology for stratigraphic interpretation in a more complete sense.

SUMMARY

The relation of seismic data to stratigraphy can be understood conceptually starting from a rather naive point of view. Taking a single seismic trace, we may assume that each reflection event consists of a simple symmetric wavelet and denotes a change in lithology corresponding to a change in acoustic impedance. The polarity and size of the reflection events allow the development of an underlying reflectivity series and acoustic impedance log. These types of seismic-derived information, when interpreted from trace to trace and in light of geologic principles and available subsurface information, permit accurate correlation of seismic data with the subsurface information where these coincide, and permit extrapolations elsewhere.

The difficulties cited represent departures of real seismic data from the ideal data of the naive approach. For each such problem, some technique of seismic data processing is available as a mechanism by which the real data can be transformed or improved to have characteristics approaching the ideal.

Seismic modeling is understood to be the successor to the synthetic-seismogram calculation. Illustrations of modeling in two and three dimensions have brought forth a variety of considerations of important interpretational significance. Lessons from the model studies warn the interpreter to account constantly for the three-dimensional nature of the subsurface in both structural and stratigraphic terms.

Finally, the stratigraphic resolution potential of seismic data can be expressed in quantitative terms. The computer-based procedure which was developed uses both time measurements and amplitudes in the amplitude-processed, wavelet-processed environment to achieve high-resolution lithologic delineation.

The future for stratigraphic interpretation and finding hydrocarbons in stratigraphic environments seems particularly bright at this time in regard to the available tools and methods. Newer tools and procedures may improve our capabilities further. The practical applications of these concepts and techniques as they pertain to actual exploratory and development projects are discussed and illustrated in an accompanying paper by Schramm et al.

REFERENCES CITED

Berkhout, A. J., 1973, On the minimum-length of one-sided signals: Geophysics, v. 38, p. 657-672.
——— 1974, Related properties of minimum-phase and zero-phase time functions: Geophys. Prosp. (Netherlands), v. 22, p. 683-703.
Dedman, E. V., J. P. Lindsey, and M. W. Schramm, Jr., 1975, Stratigraphic modeling: a step beyond bright spot: World Oil, v. 180, no. 6, p. 61-65.
Harms, J. C., and P. Tackenberg, 1972, Seismic signatures of sedimentation models: Geophysics, v. 37, p. 45-58.
Hilterman, F. J., 1970, Three-dimensional seismic modeling: Geophysics, v. 35, p. 1020-1037.
Geophysical Society of Houston, 1973, A symposium: lithology and direct detection of hydrocarbons using geophysical methods: Houston.
Lindseth, R. O., 1976, Mapping stratigraphic traps with seislog (abs.): 46th Ann. Mtg. Soc. Exploration Geophys., Houston.
Lindsey, J. P., M. W. Schramm, Jr., and L. K. Nemeth, 1976, New seismic technology can guide field development: World Oil, v. 183, no. 7, p. 59-63.
McClintock, P. L., 1977, Seismic data-processing techniques for northern Michigan reefs: AAPG Stud. Geol. No. 5 (in press).
Meckel, L. D., and A. K. Nath, 1977, Geologic considerations for stratigraphic modeling and interpretation: this volume.
Meissner, R., and E. Meixner, 1969, Deformation of seismic wavelets by thin layers and layered boundaries: Geophys. Prosp. (Netherlands), v. 17, p. 1-27.
Nath, A. K., 1975, Reflection amplitude, modeling can help locate Michigan reefs: Oil and Gas Jour., v. 73, no. 11, p. 180-182.
Neidell, N. S., 1975, What are the limits in specifying seismic models?: Oil and Gas Jour., v. 73, no. 7, 144-147.
Peacock, K. L., and S. Treitel, 1969, Predictive deconvolution theory and practice: Geophysics, v. 34, p. 155-169.
Peterson, R. A., W. R. Fillipone, and F. B. Coker, 1955, The synthesis of seismograms from well log data: Geophysics, v. 20, p. 516-538.
Ricker, N., 1953, Wavelet contraction, wavelet expansion and the control of seismic resolution: Geophysics, v. 18, p. 769-792.
Robinson, E. A., and S. Treitel, 1967, Principles of digital Wiener filtering: Geophys. Prosp. (Netherlands), v. 15, p. 311-333.
Sangree, J. B., and J. M. Widmier, 1974, Interpretation of depositional facies from seismic data: Continuing Education Symposium, Geophysical Society of Houston.
Schoenberger, M., 1974, Resolution comparison of minimum-phase and zero-phase signals: Geophysics, v. 39, p. 826-833.
Schramm, M. W., Jr., E. V. Dedman, and J. P. Lindsey,

1977, Practical stratigraphic modeling and interpretation: this volume.

Shah, P. M. 1973, Ray tracing in three-dimensions: Geophysics, v. 38, p. 600-604.

Sheriff, R. E., 1976, Inferring stratigraphy from seismic data: AAPG Bull., v. 60, p. 528-542.

Taner, M. T., and F. Koehler, 1969, Velocity spectra—digital computer derivation and applications of velocity functions: Geophysics, v. 34, p. 859-881.

———— E. E. Cook, and N. S. Neidell, 1970, Limitations of the reflection seismic method; lessons from computer simulations: Geophysics, v. 35, p. 551-573.

Vail, P. R., et al, 1976, Interpretation of seismic sequences from reflection patterns: Preprint, 29th Ann. Mtg., Midwest Sect. Soc. Exploration Geophysicists, Dallas.

Widess, M. B., 1973, How thin is a thin bed?: Geophysics, v. 38, p. 1176-1180. Wuenschel, P. E., 1960, Seismogram synthesis including multiples and transmission coefficients: Geophysics, v. 25, p. 106-129.

Geologic Considerations for Stratigraphic Modeling and Interpretation[1]

L. D. MECKEL, JR.,[2] and A. K. NATH[3]

Abstract Seismic modeling techniques are attempts to mathematically and geometrically represent subsurface geology and to depict the seismic response of that geology to a propagating wavefront. In this context, modeling has become an important exploration tool (1) to test the mappability of a geologic concept, (2) to analyze the impact of expected geologic variability (porosity, thickness, etc.) on the seismic response, and (3) to evaluate the significance of event reflectivity changes, or anomalies, on uncalibrated seismic data.

Traditionally seismic data were used to identify events to map subsurface structure (faults, folds, noses) or large scale depositional geometries (pinnacle reefs, unconformities). To accomplish this, it was important to strengthen weak seismic events during processing of the recorded signals. Therefore, variations in signal strength (true amplitude) were purposely eliminated to accentuate event visibility.

Today one realizes that valuable geologic information is encoded into the shape, polarity, and true amplitude of the reflection. Where *calibrated*, it is possible to deduce important rock-fluid information from true amplitude seismic data. The information may be lithology changes indicating the reservoir boundary (and thus the trap), or fluid changes directly indicating the actual hydrocarbon accumulation. For the explorationist, this becomes another measurement tool and one with significant predictive value. The predictive value is vastly improved where subsurface response can be calibrated to rock-fluid data, preferably logs and rock samples. Modeling becomes an important vehicle to establish this calibration.

INTRODUCTION

This study examines those geologic factors which most influence seismic waveform character and therefore become important input variables in constructing seismic models which depict subsurface stratigraphic relations. These models can be used both to interpret reservoir distribution and quality on structural prospects, and to map reservoir discontinuities for stratigraphic prospects. Thus modeling is likely to play an increasingly important part in seismic interpretation. This paper is organized into three parts:

Resolution—What can we measure laterally and vertically?

Depositional factors that influence reflection character—Isolate individual geologic attributes of both clastic and carbonate rocks to evaluate their effect on seismic data.

Subsurface examples—(1) Use models to show examples of waveform changes across several trap geometries typical of many oil and gas fields; and (2) point out some pitfalls in lithologic interpretation from seismic data.

Geometric models were constructed using typical subsurface geometries across several stratigraphic trap margins. The velocity values were taken from "blocked" velocity logs[4] through the intervals studied. For clastics, the density values were calculated from Gardner's equation ($\rho = 0.23V^{0.25}$) relating density and velocity. For carbonates, representative density values were taken from logs in or near the trap being modeled. There are inaccuracies in this gross method, but nevertheless the models serve to illustrate both the principles involved and the type of subtle waveform changes that can be significant (indicative of trap boundary) where calibrated. The more sophisticated approach, which should be used in exploration, would be to use values from properly edited and digitized sonic and density logs, taken from surface to total depth.

Various computational modeling techniques have been described and used to characterize stratigraphic variability (Dedman et al, 1975; Hilterman, 1970; Nath, 1975; Neidell, 1975; Peterson et al, 1955; Trorey, 1970; Wuenschel, 1960). The reader is referred to these references for a review of modeling procedures. In this paper, the models illustrated are wave-theory solutions using one of the three wavelets shown in Figure 1. Two of these, the Butterworth bandpass 10 to 40-Hz and Ricker 25-Hz, are symmetric zero-phase wavelets similar to what might be obtained with wavelet processed data (Nath and Patch, 1977). The other model is a more complicated (undeconvolved) sleeve-exploder basic wavelet measured in an offshore province.

Each example used is shown in a geologic cross section which shows the acoustic properties of the rock-fluid system. The seismic models are displayed in time (seconds) and shot point number. Ties between the cross section and seismic model are shown where the two occur on the same page. All models are true amplitude displays.

[1]Manuscript received, November 5, 1976; accepted, March 22, 1977.

[2]Sneider & Meckel Associates, Inc., Houston, Texas 77024.

[3]GeoQuest International, Ltd., Houston, Texas 77027.

[4]Due to log availability, some logs from nearby wells through the same rock types were used.

WAVELETS USED

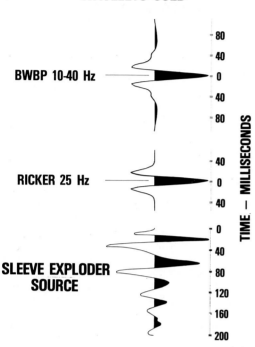

BWBP 10-40 Hz

RICKER 25 Hz

SLEEVE EXPLODER
SOURCE

TIME — MILLISECONDS

FIG. 1—Example wavelets used in this study.

RESOLUTION—WHAT CAN WE MEASURE?

A recurrent question of major concern to the explorationist is: What is the resolving ability of modern seismic reflection data? It is significant to note that the science is improving so steadily that today's answer will probably be obsolete tomorrow. Nevertheless, the question has two parts: (1)

How thick (or thin) can a reservoir be and still be identified seismically? And, (2) how wide (or narrow) does a unit have to be in order to be visible seismically?

In both cases the question of resolution arises when the dimensions of the geologic feature become small with respect to the wavelengths used. These questions have been addressed by Widess (1973), Nath and Meckel (1976), and Neidell and Poggiagliolmi (this volume). The following summarizes their observations.

Thickness (or Thinness)

The frequency band width of a seismic wavelet propagating through the earth controls resolution of the thickness of a unit. Separate reflections from the top and bottom of a unit can be identified down to a thickness of about ⅛ the wavelength of the central frequency of the wavelet (Widess, 1973). This thickness has been termed the *critical resolution thickness*. If the thickness is less than this critical thickness, the peak-to-trough time separation of the reflections (normally used to measure bed thickness) becomes invariant. For typical wavelets and representative subsurface velocities, the critical thickness turns out to be in the range of 50 to 80 ft (15 to 24 m). Many, perhaps most, reservoirs being explored are less than 60 ft (18.2 m) thick. It therefore becomes important to infer thickness information from the seismic data for units thinner than the critical thickness.

Figure 2 shows the reflection geometries associated with a thinning (thickness shown in milliseconds), low-acoustic impedance unit surrounded by a uniform lithology of higher acoustic impedance. The bed thickness is measured in terms of two-way time, and the top and bottom interfaces

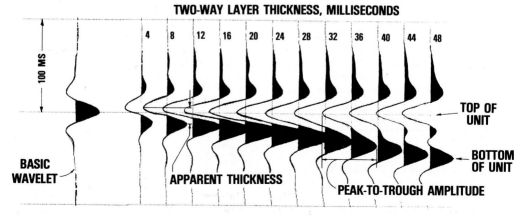

TWO-WAY LAYER THICKNESS, MILLISECONDS

FIG. 2—Seismic response of unit which thins laterally from right to left.

are varied in steps of 4 msec. The seismic response has been generated by convolving the two-termed reflectivity series with an 8 to 32-Hz Butterworth bandpass wavelet. Figure 3 is an accompanying quantitative plot showing the variation of measured peak-to-trough thickness (in msec) and relative amplitude as a function of the actual thickness for the example in Figure 2. For acoustically thick units the time-separation between the responses from the top and bottom interfaces of the unit are measurable, and this time measurement is valid to a critical bed thickness.

For acoustically thin units (less than ⅛ of the dominant wavelength) peak-to-trough timing is constant and information on bedding thickness is expressed in the amplitude of the event. There is first a noticeable amplitude increase with decrease in bed thickness. The maximum amplitude occurs at the tuning thickness which corresponds to approximately where the peak-to-trough time separation becomes invariant. Below this point, amplitude decreases nonlinearly with a decrease in unit thickness. Thus for thin units, thickness information is encoded in the amplitude of the event.

The above analysis assumes that the reservoir is encased in the same lithology, such that the top and bottom reflection coefficients are the same. However, there are many geologic situations where the unit overlying the reservoir does not have the same acoustic impedance as the unit underlying the reservoir; thus the top and bottom reflection coefficients of the reservoir will differ. Figure 4 shows a quantitative plot of three additional cases where the bottom coefficient differs from the upper one. A change in reflection coefficient shifts the relative amplitude curve to lower values, and the difference in shape between them is only a matter of relative scaling factors. Tuning thickness remains the same for each example.

In summary, peak-to-trough time separation can be used to estimate bed thickness for units thicker than the tuning thickness. Below tuning thickness, amplitude response can be used to measure the thickness of the unit where the reflection coefficients have been calibrated. The examples shown are noise-free. The actual signal-to-noise ratio of the seismic data will determine and limit interpretive ability for low-amplitude responses.

Lateral Resolution

Reflections are not responses from specific points of subsurface acoustic boundaries but are a summation of responses from an area surrounding the geometrically predicted reflection "point." This area is called the first Fresnel zone. The size

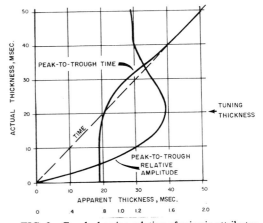

FIG. 3—Graph showing relation of seismic attributes to changing bed thickness for 8 to 32-Hz bandpass wavelet.

of a Fresnel zone depends on several parameters: distance from source, effective velocity, and the frequency composition of a wavelet. For example, a Fresnel zone is about 1,000 ft (304 m) wide at a depth of 5,000 ft (1,524 m) for typical seismic wavelengths in the Gulf of Mexico.

Lateral resolution addresses the question of how wide a unit has to be for it to be seismically visible and identifiable. Figure 5 shows the expected two-dimensional seismic response from several different-sized sandstone bodies whose widths are measured in terms of a Fresnel zone. Where the reflecting body approaches the width of one Fresnel zone, the identifiable seismic reflection character begins to change, and at still smaller dimensions takes on a point-diffraction type response. However, the amplitudes of the seismic responses decrease with a reduction in the lateral width of the bed. This criterion becomes an interpretive tool in making bed-width estimations. Fresnel zone effect is a three-dimensional phenomenon and should be treated as such. Hilterman (1976) has extended this analysis to three-dimensional model studies.

Qualification

True-amplitude seismic data which is wavelet processed becomes a significant tool for determining the thickness and estimated width of exploration targets. Where appropriately calibrated, resolution of thickness and width can be extended to smaller dimensions than traditionally envisioned. The practical boundaries of this potential resolution are yet to be evaluated, and depend on signal-to-noise ratio, frequency content of the seismic wavelet, and quality of seismic data

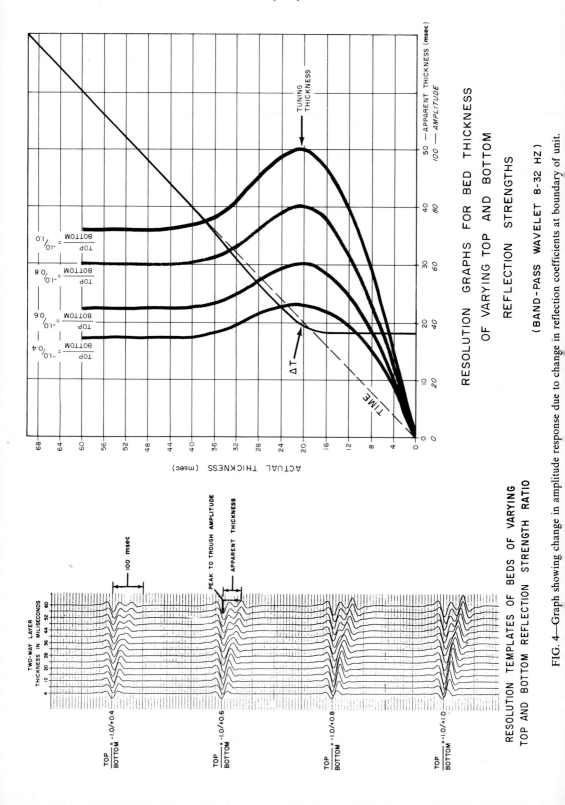

FIG. 4—Graph showing change in amplitude response due to change in reflection coefficients at boundary of unit.

FRESNEL ZONE MODEL AND WAVE THEORY RESPONSE

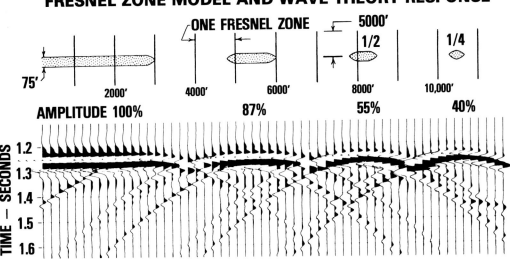

FIG. 5—Lateral resolution model showing sandstone body of varying width.

processing. However, the exploration impact of this emerging resolution ability is very large. Estimating the volume of a trap is very dependent on obtaining an estimate of the thickness and lateral extent of the reservoir unit.

DEPOSITIONAL FACTORS THAT INFLUENCE SEISMIC CHARACTER

Suppose we were to lower a geologist into a borehole to describe those factors in the subsurface which are most important to the reflectivity of rocks. What would he describe? Keep in mind that geologists have developed extensive vocabularies to describe rocks for a variety of purposes. Much of that vocabulary may be of little or no value in describing the acoustic properties of rocks.

It is important to identify those rock characteristics most significant in establishing rock reflectivity, to construct the most realistic models for comparison with actual seismic data for interpretative purposes. Rock information is grouped into three categories for consideration in this paper:

1. Properties inherent within the reservoir unit (such as lithology, porosity, etc.);

2. Properties associated with the reservoir-seal contact (e.g., acoustic impedance contrasts, types of contacts); and

3. Properties related to the total stratigraphic context of the unit (e.g., proximity and spacing of adjacent acoustic interfaces).

Properties Within the Reservoir Unit

Geologic attributes of the reservoir considered in this section include: lithology, porosity, type of pore fluid, effective stress, and amount and distribution of nonreservoir interbeds within the reservoir. Acoustic information for many of these geologic attributes is available in a number of published papers. This section attempts to bring the geologically significant information together in one place and also introduce some new concepts.

Lithology—Porosity—Lithology is the composition (type) of rock framework. Porosity is a measure of the amount of holes in a rock. Both of these parameters significantly effect the density and velocity of a rock. For most rock types there is a systematic relation between density and velocity. Figure 6 (after Gardner et al, 1974) shows an empirical plot of density-velocity relations for many common rock types which span a range of basins, depths, and geologic ages. Lines of equal acoustic impedance have been added to Gardner's basic illustration in order to establish the effect of lithology and porosity on the acoustic impedance (ρV) of rocks. Rocks that fall along one of these lines have the same acoustic impedance and will not generate a reflection if in contact, even though their densities and velocities may differ markedly.

For each rock type shown, the systematic increase of density and velocity relates to a de-

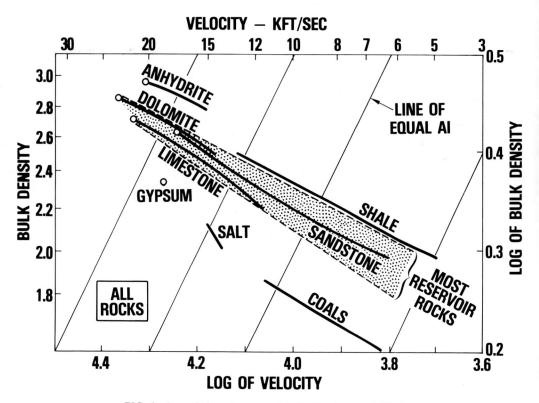

FIG. 6—Acoustic impedance graph (after Gardner et al, 1974).

crease in porosity (Fig. 7). Lithology and porosity can be empirically related to velocity by the time-average equation. The equation is most applicable for reasonably consolidated rocks containing brine.

Figure 6, an acoustic-impedance graph, shows that the resulting rock property curves trend normal to lines of equal acoustic impedance of a rock. There are several significant exploration implications of this graph:

1. An accurate description of rock lithology and estimate of porosity help define density and velocity, the two critical parameters for establishing the acoustic impedance of a rock.

2. Rocks of quite different densities and velocities can have the same acoustic impedance. Thus, significant changes in lithology may not be acoustically visible. It is also true that any one of several geologic models can give the same seismic response.

3. Small changes in porosity significantly change a rock's acoustic impedance.

4. The empirical relations shown (Gardner et al, 1974, eq. 8, p. 779) permit the estimation of reflection coefficients from velocity information alone.

$$R_c = \tfrac{1}{2}\left(1.25 \ln \frac{V_1}{V_2}\right)$$

The effect of density on the reflection coefficient is taken into account by multiplying the coefficient (due to the velocity contrast) by 1.25.

Type of Pore Fluid—Changes in the type of pore fluid can appreciably alter the acoustic impedance of a reservoir. The way pore fluids influence compressional seismic velocity has been mathematically described by Gassman (1951). Effects of hydrocarbon replacement of brine as the pore fluid have been well-documented by Domenico (1974) and Gardner et al (1974). The presence of hydrocarbons of lower density and velocity than water, decreases the acoustic impedance of the reservoir rock. The two most important implications of this, as shown in Figure 8, are:

1. The way in which a change in pore fluid content is manifested seismically depends on the relative acoustic impedances of the reservoir and the lithologies encasing the reservoir; and

2. Where hydrocarbons replace brine as the pore fluid, the reflection event (depending on the relative acoustic impedance contrast with the surrounding lithologies) may show an increase in

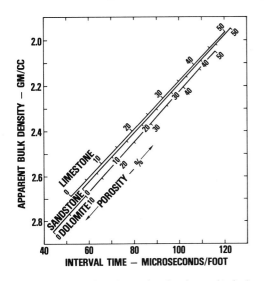

FIG. 7—Interrelation of porosity, density, and velocity for various reservoir lithologies.

trough amplitude (or bright spot) if the surrounding lithologies have a higher acoustic impedance; a decrease in peak amplitude (or dim spot) if the encasing lithologies have a lower acoustic impedance; or a polarity reversal if the surrounding lithologies have a slightly lower impedance than the encased brine reservoir.

Effective Stress—Overburden pressure, produced by the total weight of the overlying rock-fluid column, is approximately 1 psi (0.07 kg/sq cm) per ft. The fluid pressure gradient resulting

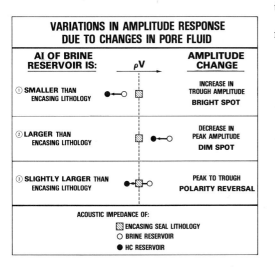

FIG. 8—Variations in amplitude response due to changes in pore fluid of reservoir.

from the weight of the overlying fluid in pore space averages 0.465 psi (0.032 kg/sq cm) per ft. The effective pressure on the rock framework is the difference between the overburden and fluid pressures, and is approximately ½ psi (0.03 kg/sq cm) per ft.

Compressional velocities of porous rocks have been shown to increase with an increase in effective pressure (Gardner et al, 1974; Todd and Simmons, 1972; Wyllie et al, 1960). This is not surprising since the velocity of fluids in the pore system increases with increased pressure. But a similar velocity increase, with an increase in effective pressure, is observed for igneous rocks (granite) with no porosity, and thus no internal fluid system (Gardner et al, 1974, p. 773). This velocity increase has been related to the closing of microcracks in the rock, a property which may be inherent in all rocks which have rebounded from their maximum burial depth.

Figure 9 shows the velocity increase due to pressure is more rapid initially (at low pressures) for a wide range of rock types. This rapid initial increase in velocity has been related to the initial closing of microcracks in the rocks to produce a more "continuous" rock framework. At greater pressures this factor becomes less significant and there is only a small increase in velocity.

For most deeply buried rocks, the velocity change due to effective pressure is low relative to changes produced by other properties, such as porosity. However, where obtaining velocity data from rock samples (core or outcrop), the most accurate values will be obtained if measurements are made at estimated in-situ pressures. Some acoustic impedance anomalies may relate directly to abrupt changes in effective stress.

Nonreservoir Interbeds—Reservoir units commonly contain interbeds of nonreservoir litholo-

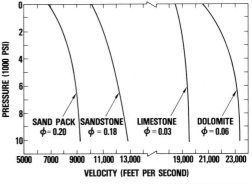

FIG. 9—Compressional velocity as function of effective stress for several rock types.

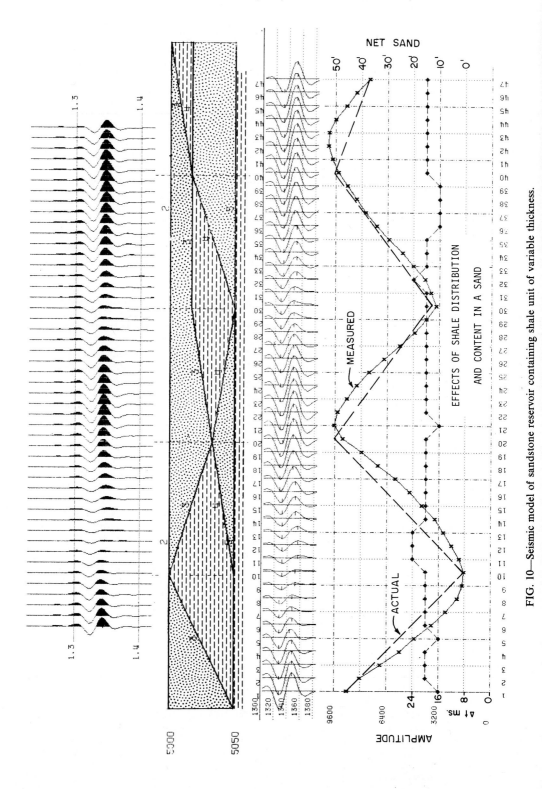

FIG. 10—Seismic model of sandstone reservoir containing shale unit of variable thickness.

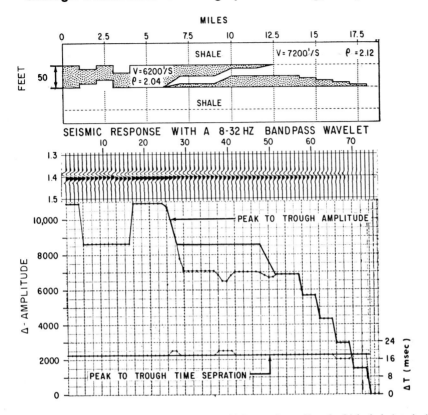

FIG. 11—Seismic model of sandstone reservoir which contains uniformly thick shale interbed.

gy—for example shale interbeds within a sandstone reservoir or tight limestone interbeds within a porous dolomite unit. The nonreservoir interbeds can vary in both total thickness and distribution (position and number of interbeds) within the reservoir. The following material examines the effect of these "contaminants" on the reflection character of an acoustically thin reservoir unit. The models used show shale interbeds within a 50-ft (15.2 m) reservoir sandstone.

Figure 10 shows a single shale interbed which varies in thickness within a 50-ft (15.2 m) sandstone body. The timing measure is invariant as the thickness of the shale interbed changes. However, the amplitude varies proportionally to the thickness of shale. Where calibrated, Lindsey et al (1976) have shown that amplitude is a good quantitative measure of net sandstone thickness.

Figure 11 shows a single shale interbed which varies in position within a 50-ft (15.2 m) sandstone. The shale interbed produces an amplitude decrease relative to that for a sandstone unit with no shale. The timing measure is insensitive to the presence or position of the shale interbed. The peak-to-trough amplitude is responsive to the presence of the shale interbed, but reveals nothing about its position within the reservoir unit. Thus the seismic process measures only net acoustic impedance difference, or in other words, net sandstone thickness in acoustically thin reservoirs.

Figure 12 shows a constant amount of shale within a 50-ft (15.2 m) sandstone, but the shale occurs in a variable number of discrete interbeds. The timing measure remains invariant and the amplitude again measures the net acoustic impedance contrast (or net sandstone in this case).

Where the thickness of the reservoir unit is less than the tuning thickness of the wavelet, amplitude is a good measure of net reservoir thickness; the amplitude is insensitive to the position and number of nonreservoir interbeds. Where calibrated, amplitude is useful in assessing the gross lithologic characteristic of the unit. This is helpful for the geologist, who can commonly estimate net reservoir thickness based on knowledge of the

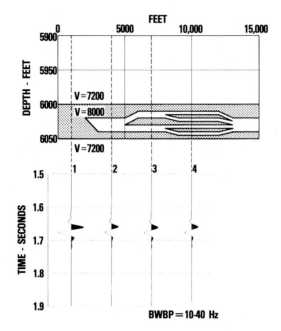

FIG. 12—Seismic model of sandstone reservoir containing number of shale interbeds.

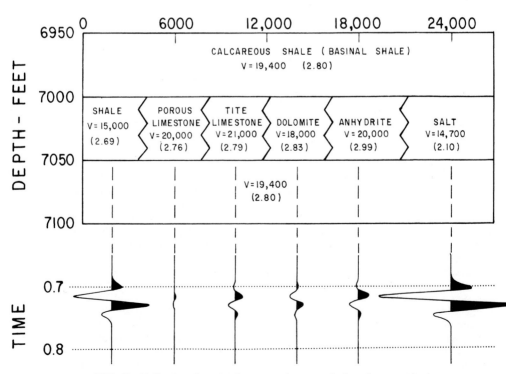

FIG. 13—Reflection character due to varying acoustic impedance contrast.

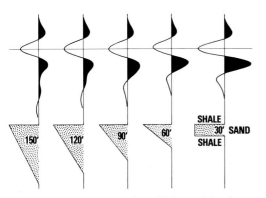

SHALE
30' SAND
SHALE

150' 120' 90' 60'

FIG. 14—Seismic response from thick transitional contacts (wavelet 10 to 40-Hz BWBP).

genesis of the unit. But it is difficult to estimate the exact number, spacing, and position of nonreservoir interbeds within the gross reservoir package.

Properties Associated with Reservoir-Seal Contact

The boundaries of the reservoir are important contributors to the waveform character of the unit. Geologic aspects of these boundaries are the acoustic impedance contrast, abruptness of vertical contacts, and nature of lateral contacts.

Acoustic Impedance Contrast—Geologic properties (such as lithology, porosity) discussed previously, significantly affect the absolute impedance values for rocks. Well logs measure this absolute value of density or velocity. However, seismic processes measure the differences in acoustic impedance (ρV) of rocks.

Referring to Figure 6, the acoustic impedance graph, we see that many lithologies can have the same acoustic impedance (thus no reflectivity contrast) in spite of very different density and velocity values. This is illustrated in Figure 13 where several typical carbonate lithologies (shale, porous limestone, tight limestone, porous dolomite, anhydrite, salt) are encased in the same calcareous shale. The reflections vary for each lithology shown. Those associated with the two porous rock types, which would be of prime exploration interest, provide the weakest reflections. The best reflections are associated with contacts between two nonreservoir rock types.

Abruptness of Vertical Contacts—Vertical contacts between two different rock types can be sharp or, as is more common in nature, transitional to some degree. These transitional zones can vary in thickness (from several feet to more than 100 ft or 31 m) and in position (upper or lower contact). How do transitional boundaries affect the character of the seismic reflection?

Figure 14 illustrates the effect of transitional contacts in acoustically thick beds. For the example shown, as the transitional lower contact becomes thicker, there is an associated decrease in the amplitude and loss of higher frequency. The model can be visually flipped to evaluate the effect of a thick transitional upper contact.

For acoustically thin beds (Fig. 15), there is an amplitude reduction as the transitional contact thickens. The amplitude information again provides information on the "net" reservoir; it does not differentiate the position (upper or lower contact) of the transitional boundary. Amplitude decreases as the thickness of the transitional zone increases, indicating a loss of net reservoir.

Nature of Lateral Contacts—Lateral contacts, from reservoir to seal, can be: (1) Abrupt (as for a channel deposit or reef); (2) transitional—intertonguing of two different lithologies resulting in the alteration of rock properties as in the facies change shown on the left side of the model in Figure 16; or (3) transitional—a gradual change from one lithology to another as shown on the right side of the model in Figure 16.

Abrupt boundaries usually terminate an event and can create diffraction patterns.

If the thickness of the unit is less than the tuning thickness (as in Fig. 16) the seismic character does not distinguish the type of transitional contact. Again, the amplitude measures gross lithology of the interval and not the detailed lithologic arrangement. Intertonguing or gradational contacts both show the same effect—a gradual decrease of amplitude from a good reservoir, laterally, to good seal.

Properties Related to Total Stratigraphic Context

The reflection character for a single unit (reservoir) is also dependent on the proximity and spacing of acoustic interfaces above and below the unit under consideration. The seismic signal is the interference pattern from all reflections, approximately one wavelength above and below the unit under study. For reservoir units isolated in a thick homogenous section, the resulting waveform involves the interference of only two wavelets, one from the top and one from the base of the unit. Figure 2 shows the result of this interference pattern which allows interpretation of unit thickness.

To the extent that there are nearby acoustic interfaces above and below the reservoir unit (within a distance of about one wavelength), the wavelets from each of these interfaces contribute to the response for the unit under consideration. If the surrounding acoustic interfaces remain uniform in spacing and value in a lateral direction, then a change in lithology at the studied horizon may be

SEISMIC   RESPONSE   TO   ABRUPTNESS
OF   VERTICAL   CONTACTS   IN   THIN   BEDS

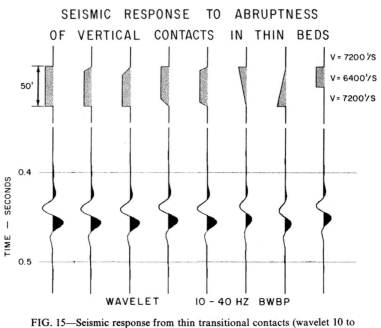

FIG. 15—Seismic response from thin transitional contacts (wavelet 10 to
40-Hz BWBP).

detected. But if the surrounding acoustic interface change (Fig. 17), these changes will manifest themselves at the event of interest and it will be difficult distinguishing changes that *could* occur in the unit of study from changes that *do* occur in surrounding units (Fig. 18).

Figure 19 (top) shows a cross section of the regressive coastal sequence in the Upper Creta-

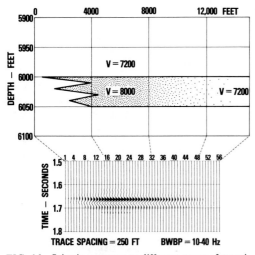

FIG. 16—Seismic response to different types of transitional lateral contacts.

ceous Belly River Formation of the Alberta basin. The section shows three offlapping barrier sandstones, each about 25 to 35 ft (8.6 to 10.6 m) thick. Basinward, the bars are transitional to marine shales and silts (white area). Landward, these bars change to nonmarine and brackish shales and silts (diagonal-lined area) which contain a few thin sandstone lenses. Though each reservoir is similar in thickness, porosity, and type of contacts, a seismic model shows each to have a very different seismic fingerprint. Each barrier sandstone is denoted by a peak event (Fig. 19), but the three reservoir events differ in both amplitude and waveform as a result of interference from adjacent acoustic interfaces. This example shows that the total stratigraphic context—proximity and geometry of adjacent acoustic interfaces—becomes a very important consideration in analyzing the reflection character of a reservoir unit.

SEISMIC MODELS OF SUBSURFACE TRAPS

The following clastic and carbonate examples, based on subsurface rock geometries and acoustic properties, show the expected seismic response of several recurrent trap styles of many oil and gas fields. Specific geometries and rock properties for similar targets can vary considerably depending on depth of burial, age, position in basin, and exact thickness, so these responses do not uniquely represent the seismic signature for the trap type

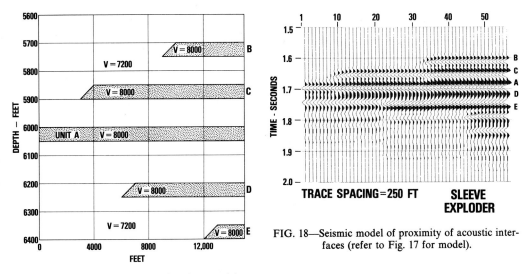

FIG. 17—Proximity of acoustic interface model.

FIG. 18—Seismic model of proximity of acoustic interfaces (refer to Fig. 17 for model).

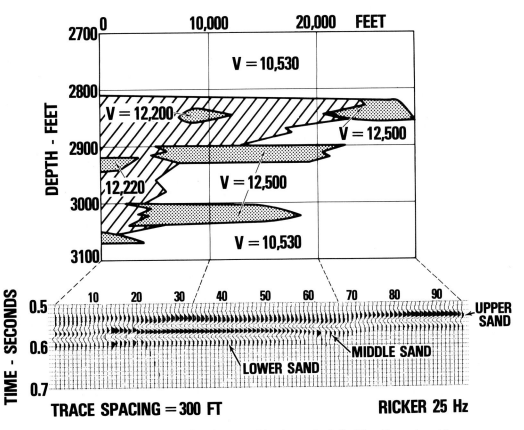

FIG. 19—Seismic model for offlapping coastal barrier sands, Belly River Formation, Alberta.

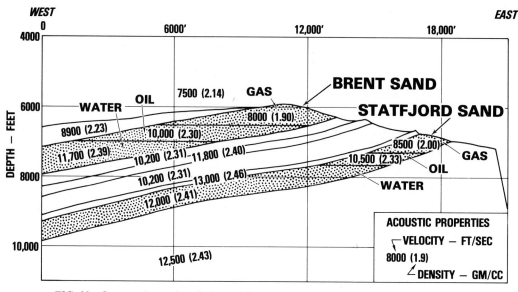

FIG. 20—Cross section and geologic model of Brent field, North Sea (after Seaten, 1976).

modeled. The following examples are used to show: (1) the stratigraphic significance that can be related to subtle waveform changes, and (2) the importance of calibrating with subsurface rock-log data.

Fluid Level Determination in Unconformity Trap

Example—Jurassic sandstones, Brent field, North Sea.

Setting—The Brent field is in the extreme northern part of the United Kingdom sector of the North Sea, close to the median boundary with Norway. The field is located in the northern part of the Viking graben, a major north to south-trending structure separating Scotland and Norway.

Geology—The field is a combination structural-stratigraphic truncation trap which consists of a westerly tilted and partially eroded fault block containing two Jurassic reservoirs, overlain unconformably by sealing shales of Cretaceous age (Fig. 20). The two reservoirs, the Brent and Statfjord sandstones, have porosities which range from 7 to 37%, and permeability from 5.5 to 8 darcys.

The Middle Jurassic Brent sandstone, first discovered in 1971, is about 845 ft (265.6 m) thick, with a 700-ft (213 m) total hydrocarbon column. The Lower Jurassic Statfjord sandstone, discovered in 1973, has a total hydrocarbon column of 611 ft (182.2 m) and an average porosity of 23%. The expected reserves of the two sandstones is indicated to be over 1.3 billion bbl of oil and 3.4

Tcf of gas. The geometries and acoustic properties of these rocks are shown in Figure 20, a dip cross-section.

Seismic Response—Figure 21 is a wave theory solution of the model using a 10 to 40-Hz zero-phase Butterworth bandpass wavelet. The gas-oil and oil-water contacts within the Brent sandstone can be easily seen as modeled. However this is not true on real seismic data due to the transitional character of the fluid contacts. There are two marked changes in the reflection character of the event marking the top of the Brent sandstone. A strong peak first weakens, then goes to a trough (polarity reversal); these changes correlate with the fluid levels in the reservoir. The updip change in reflection character represents the lateral change in the reflection coefficient at the top of the Brent sandstone due to the updip change in pore fluid: water, to oil, to gas.

Identifying fluid levels in the Statfjord sandstone is a more difficult problem. The high acoustic impedance of the overlying Bressy limestone has a dominating influence and "masks" the top of the sandstone. As one follows the trough underneath the strong positive peak from the Bressy unit, the trough amplitude increases updip at two places. The top of the gas sandstone (beneath the Lower Cretaceous Marl) produces a sequence of reflections which can easily be mistaken for diffractions. Real diffractions from the unconformity surface further complicate the interpreter's task. The gas-oil contact is identifiable as a positive event dipping eastward. This correlates with

WEST   EAST

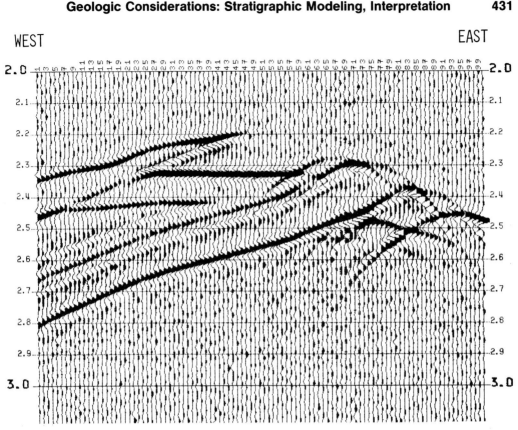

FIG. 21—Seismic model of the Brent field. Time section with 10 to 40-Hz zero-phase Butterworth bandpass wavelet (noise added).

the most updip increase in trough amplitude. The oil-water contact is masked by diffractions; however, indications of this fluid contact (using the Brent sandstone as an example) can be seen in the zone where the negative amplitude starts gaining strength below the Bressy-Statfjord boundary. The subtle seismic expression of this accumulation may explain why it was the last to be discovered in the Brent field.

Deltaic Stratigraphic Traps

Example—Muddy Formation, Powder River basin, Wyoming.

Setting—The Powder River basin is a Laramide basin between the Bighorn Mountains and the Black Hills uplift in northeastern Wyoming. The northeastern flank of the basin, a gentle homocline dipping 1° westward toward the axis of the basin, has been an area of intense stratigraphic exploration since the mid 1960s.

Geology—The Muddy Formation is a deltaic system which prograded westward into the Creta-

ceous seaway during Early Cretaceous time. The Muddy reservoirs are of two types: (1) a NE-SW trending set of bars, and (2) a group of westward trending distributary channels (Gaither and Meckel, 1972). Both reservoir types have belt or shoestring geometries which are conducive to the formation of stratigraphic traps on the homoclinal flank of the basin (Fig. 22). Most of the reservoir units are less than 50 ft (15.2 m) thick and would be considered acoustically thin. Bell Creek and Recluse fields are typical examples of these types of stratigraphic traps. Bell Creek field is a 30-ft (9 m) bar sandstone in the lower Muddy. The reservoir is underlain by marine shales and overlain by nonmarine silts, shales, and coal of the upper Muddy. Laterally, the main reservoir sandstone pinches out updip into lagoonal shale and silt (Fig. 23).

Recluse field produces from a distributary channel sandstone in the lower Muddy. Reservoir sandstone occurs only in the lower part of the channel and abuts against the channel wall (Fig.

MAJOR RESERVOIR FACIES

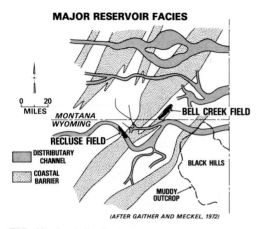

(AFTER GAITHER AND MECKEL, 1972)

FIG. 22—Sand distribution within the Muddy Formation, northern Powder River basin, Wyoming.

24). The channel is overlain by the same upper Muddy lithology that overlies Bell Creek.

Seismic Response—An acoustic model, using a 25-Hz Ricker wavelet, for these two types of stratigraphic traps, is shown in the upper part of Figure 25. The lower part of this figure shows the seismic response of these two trap types as well as a quantitative display of the lateral changes: (1) in peak-to-trough time separation, and (2) amplitude across these two traps. For each trap there is an amplitude increase associated with the reservoir sandstone in the lower Muddy relative to the adjacent nonreservoir Muddy zone. However, amplitude appears to be more sensitive to total Muddy thickness rather than presence or absence of sandstone. Because of compaction, the Muddy is generally thicker where reservoirs are present.

This is further emphasized by a variation in the channel model. In many channel systems a shale plug, or fill, may occur against the channel wall so that the good reservoir does not extend all the way across the channel. This is shown in the cross section in the upper part of Figure 26. The example is lateral and downdip from the previous model so the velocity values are higher. Figure 27

BELL CREEK FIELD - POWDER RIVER BASIN

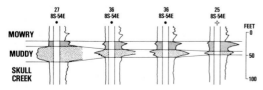

FIG. 23—Bell Creek field cross section, Montana—barrier sandstone in Muddy Formation.

RECLUSE FIELD - POWDER RIVER BASIN

FIG. 24—Recluse field cross section, Wyoming—channel sandstone in Muddy Formation.

shows the response for the same channel if the reservoir sandstone extended all the way across as it did in the Recluse field model. Again, the seismic response is most sensitive to the total Muddy thickness, not the presence of a reservoir sandstone. The small changes in amplitude and timing indicate the geomorphic edge of the channel and not the edge of the reservoir, and thus the trap boundary.

Exploration for Very Thin Reservoir Systems

Example—Cardium Formation, Alberta basin, Canada.

Setting—The Alberta basin is a large asymmetric Laramide structural basin. The very gently dipping east flank of the basin is a major stratigraphic trap province. Most of the Upper Cretaceous sands were transported eastward across the basin from an orogenic western source and pinch out eastward (now updip) into marine shale.

Geology—Discovery of stratigraphic traps in the Upper Cretaceous Cardium Formation of the Alberta basin is an exploration success story. The Cardium consists of many discrete marine sand-conglomerate reservoir units which range from several to 70 ft (1 to 21 m) thick. One of the largest oil fields in Canada, the Pembina field, produces from one of these thicker and geographically more extensive marine bars. Subsequent drilling has found many other stratigraphic accumulations in smaller bars.

Many of the thicker units have been drilled, and current exploration is directed toward thin marine conglomerates between the thicker "pods" of sandstone. These thin (up to 10 ft or 3 m thick) conglomerates are commonly fringed by "tight" silt which is laterally transitional into marine shale (see cross section, top Fig. 28).

Seismic Response—The seismic response using a Ricker 25-Hz wavelet is shown in the lower part of Figure 28. The conglomerate reservoir produces a pronounced peak event. There is a decrease in peak-to-trough amplitude as the reservoir becomes transitional into silt; however, there

STRATIGRAPHIC TRAPS IN THE POWDER RIVER BASIN OF MONTANA AND WYOMING

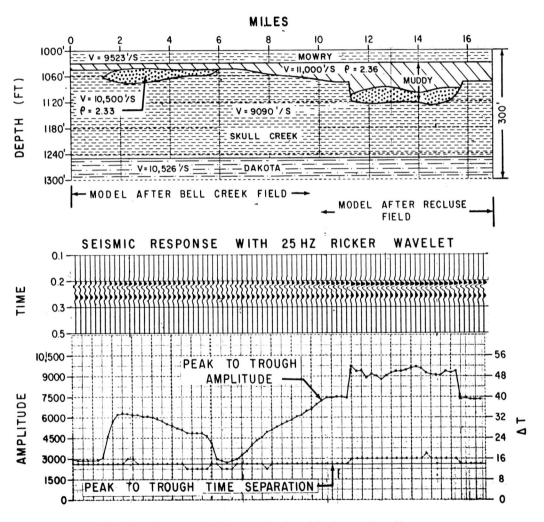

FIG. 25—Seismic model of the Bell Creek and Recluse stratigraphic traps.

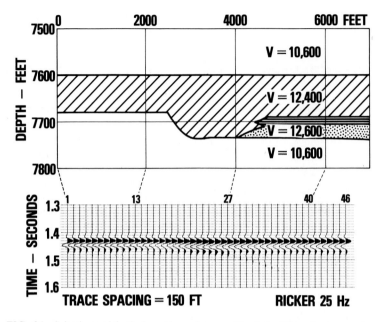

FIG. 26—Seismic model of channel sandstone with shale fill against one edge,
Muddy Formation, Wyoming.

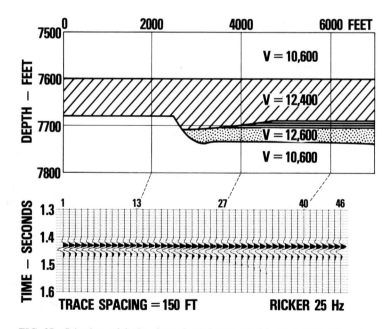

FIG. 27—Seismic model of a channel sandstone, Muddy Formation, Wyoming.

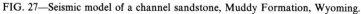

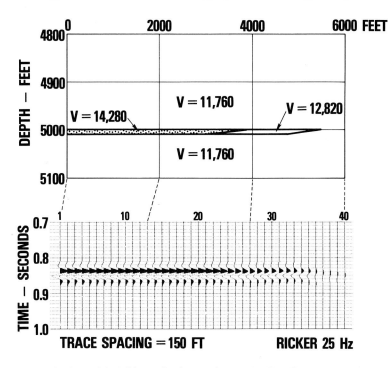

FIG. 28—Seismic model of thin marine bar sandstone, Cardium Formation, Alberta, Canada.

is no difference in the peak-to-trough timing from conglomerate bar to "tight" silt. The trap boundary between the porous bar and "tight" silt is difficult to locate seismically.

The updip (to the right) dimming of the Cardium peak event could also be interpreted as: (1) a thinning of the porous conglomerate. Such thinning would produce a decrease of amplitude as discussed in the previous section, or, (2) a gas effect in conglomerate reservoir rock. The gas would lower the acoustic impedance of the unit.

For either of these interpretations, a well at trace 27 might find a thin, gas pay zone. For the geology modeled, such a well would encounter only "tight" silt and the accumulation would still be downdip in the porous conglomerate. Thus subtle waveform changes can be stratigraphically significant, but can lead to several multiple working hypotheses. The Cardium example shows that knowledge of the expected geology has to be used to assign risk to the several probable interpretations.

Carbonate Shelf Margin and Platform Traps

Example—Permian Abo and San Andres Formations, Delaware basin, New Mexico.

Setting—The north flank of the Delaware basin in southeastern New Mexico is rimmed by a series of porous reef margins separating marine sediments on the south from platform sediments on the north. Dolomitized tidal-flat reservoirs occur updip from the Permian shelf breaks. Both reservoir systems are major exploration objectives.

Geology—Some of these Permian shelf margins, such as the Abo reef at Empire field, are very narrow targets, being only ½ to 2 mi (0.8 to 3.2 km) wide. The Abo reef lens, up to 600 ft (182 m) thick, pinches out landward into cyclical shelf carbonates (tight) and shale, and basinward into marine shale (Fig. 29). The reef is overlain by basinal marine shales (Bone Spring formation) and "tight" shelf rocks. The lithology contacts at the shelf margin are abrupt.

Seismic Response—The upper contact of the reservoir unit produces a strong peak event (Fig. 30) which depicts the geometry of the top of the reef; however, the two margins, basinward and landward, are not obvious. The landward edge is marked by a reflection character change (sudden increase in amplitude) that becomes significant where calibrated; it appears to represent the focusing of energy due to the curvature at the reef's

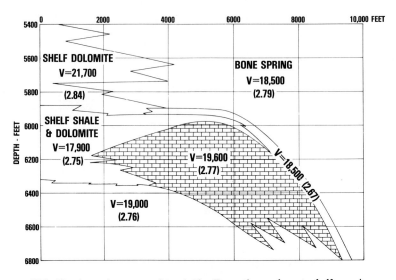

FIG. 29—Acoustic cross section of Abo Formation carbonate shelf margin, Delaware basin, New Mexico.

back edge. The reef is such a distinctive geologic zone it may surprise a geologist not to readily see its landward and seaward margins. However, in consideration of its acoustic impedance contrasts with lateral rock types, the lack of well-defined lateral margins is not surprising (refer to acoustic values in Fig. 29).

Geology—The San Andres tidal-flat sequence is positioned in an overall regressive setting and comprises three dolomitized tidal-flat units which slightly offlap each other (Fig. 31). In contrast to the Abo shelf margin reservoir, the units are thinner (30 to 50 ft or 9 to 15.2 m thick) and are gradually transitional into the lateral seal. Basinward, the reservoir units are transitional to porous shallow marine dolomite, and landward, the

units are transitional with nonporous anhydritic dolomite off the coastal plain. The optimum exploration belt, 10 to 20 mi (16 to 32 km) wide, consists of intertonguing porous and "tight" dolomite. The largest accumulation in this trend is the Slaughter–Levelland field. The trap is created by the updip interfingering of porous tidal-flat dolomites with tight continental facies. The acoustic properties of the reservoir-seal system are summarized in the cross section in the upper part of Figure 31. The uppermost unit at the top of the cross section is a low velocity marine shale which represents a regional transgression at the top of the San Andres.

Seismic Response—A major event (positive polarity) marks the transgressive top of the San An-

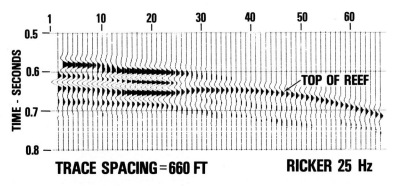

FIG. 30—Seismic model of Abo Formation carbonate shelf margin.

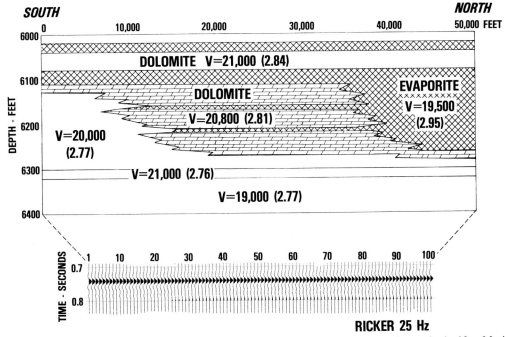

FIG. 31—Seismic model of dolomitized tidal-flat reservoirs, San Andres Formation, Delaware basin, New Mexico.

dres, the marine shale–evaporitic dolomite contact. In spite of the many interfaces between the reservoir and the seal, the very low acoustic impedance contrast between these lithologies produces no major reflection character in the interval of study. The updip (to the right in the model)

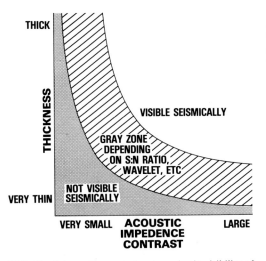

FIG. 32—Schematic graph showing seismic visibility of a unit.

pinchout of the major pay zones are not identifiable seismically. Reference to the acoustic impedance graph (Fig. 6) shows that a "tight" anhydritic dolomite can have the same acoustic impedance as a porous dolomite, even though the densities and velocities of the two rocks differ.

SUMMARY

These examples have shown that the two critical factors for seismically mapping a unit, are (1) thickness of unit, and (2) the acoustic impedance contrasts between the reservoir and adjacent (below, above, lateral) seals.

Figure 32 schematically shows the role of these two variables in determining the visibility of a unit. If units are thick enough or have a large enough acoustic impedance contrast at their margins, they can be easily identified and mapped. However, as the units thin, or the acoustic impedance contrast reduces, there are areas where the unit is not seismically visible. There is a gray zone separating these two end points; this gray zone is commonly dependent on the quality of the seismic data, for instance the exact determination of the wavelet, the signal-to-noise ratio, and the quality of amplitude preservation.

Seismic modeling has become an important tool in making stratigraphic interpretations from seismic data. Modeling serves to:

1. Test the mappability of a geologic concept. How does the expected trap manifest itself seismically?

2. Examine the impact of geologic variability of the target on the seismic response. Most geologic reservoirs can have a wide and economically acceptable range in parameters such as thickness and porosity. What variations in seismic response are equally valid indications of the target?

3. Evaluate the significance of amplitude anomalies which occur on uncalibrated lines. It may be possible here to only assign probabilities to a number of valid geologic interpretations whose model responses match the anomaly.

Modeling is useful in both mature exploration areas and in frontier provinces. Subtle changes in amplitude and waveform can be directly correlated to variations in geologic properties such as lithology, thickness, fluid content, and spacing of adjacent acoustic interfaces. Knowledge of the propagating wavelet allows one to calibrate seismic response and make interpretations of thin units—one's interpretative ability is only as good as the initial calibration from surface-subsurface rock and log data, and the quality of the recorded data.

SELECTED REFERENCES

Bowen, J. M., 1975, The Brent oil field, in A. W. Woodland, ed., Petroleum and the continental shelf of north west Europe: New York, Wiley and Sons, p. 353-361.

Dedman, E. V., J. P. Lindsey, and M. W. Schramm, 1975, Stratigraphic modeling: a step beyond bright spot: World Oil (May), v. 180, no. 6, p. 61-65.

Domenico, S. N., 1974, Effect of water saturation on seismic reflectivity of sand reservoirs encased in shale: Geophysics, v. 39, no. 6, p. 759-769.

Gaither, A., and L. D. Meckel, Jr., 1972, The Muddy Formation of northern Powder River basin—a stratigraphic paradox (abs.): AAPG Bull., v. 56, p. 618-619.

Gardner, G. H. F., L. W. Gardner, and A. R. Gregory, 1974, Formation velocity and density—the diagnostic basics for stratigraphic traps: Geophysics, v. 39, no. 6, p. 770-780.

Gassman, F., 1951, Elastic waves through a packing of spheres: Geophysics, v. 15, p. 673-685.

Geophysical Society of Houston, 1973, A symposium: Lithology and direct detection of hydrocarbons using geophysical methods: Houston.

Harms, J. C., and P. Tackenberg, 1972, Seismic signatures of sedimentation models: Geophysics, v. 37, no. 1, p. 45-58.

Hilterman, F. J., 1970, Three-dimensional seismic modeling: Geophysics, v. 35, no. 6, p. 1020-1037.

—— 1976, Interpretive lessons from three-dimensional modeling: Preprint, SEG Convention, Houston.

King, R. E., ed., 1972, Stratigraphic oil and gas fields—classification, exploration methods, and case his

tories: Soc. Exploration Geophysicists Spec. Pub. 10; AAPG Mem. 16, 687 p.

Lindsey, J. P., M. W. Schramm, Jr., and L. K. Nemeth, 1976, New seismic technology can guide field development: World Oil (June), v. 182, no. 7, p. 59-63.

Meissner, R., and E. Meixner, 1969, Deformation of seismic wavelets by thin layers and layered boundaries: Geophys. Prosp. [Netherlands], v. 17, no. 1, p. 1-27.

Nath, A. K., 1975, Reflection amplitude, modeling can help locate Michigan reefs: Oil and Gas Journal (March 17), v. 73, no. 11, p. 180-182.

—— and L. D. Meckel, Jr., 1976, Seismic modeling for structural and stratigraphic interpretation: Preprint, SEG Convention, Houston.

—— and J. R. Patch, 1977, Full utilization of seismic resolution—the key to deterministic stratigraphic interpretation from marine data: Offshore Tech. Conf. Proc., Paper No. 2784, May 1977.

Neidell, N. S., 1975, What are the limits in specifying seismic models?: Oil and Gas Jour. (Feb. 17), v. 73, no. 7, p. 144-147.

—— and E. Poggiagliolmi, 1977, Stratigraphic modeling and interpretation—geophysical principles and techniques: this volume.

Pegrum, R. M., G. Rees, and D. Naylor, 1975, Geology of the north west European continental shelf, in The North Sea: London, Graham Trotman Dudley, Ltd., v. 2.

Peterson, R. A., W. R. Fillipone, and F. B. Coker, 1955, The synthesis of seismograms from well log data: Geophysics, v. 20, no. 3, p. 516-538.

Ricker, N., 1953, Wavelet contraction, wavelet expansion and the control of seismic resolution: Geophysics, v. 18, no. 4, p. 769-792.

Sangree, J. B., and J. M. Widmier, 1974, Interpretation of depositional facies from seismic data: Continuing Education Symposium, Geophys. Soc. of Houston.

Seaton, E., 1976, Brent field nears first oil delivery: Oil and Gas Jour. (Jan. 26), v. 74, no. 4, p. 125-129.

Schoenberger, M., 1974, Resolution comparison of minimum-phase and zero-phase signals: Geophysics, v. 39, no. 6, p. 826-833.

Sheriff, R. E., 1976, Inferring stratigraphy from seismic data: AAPG Bull., v. 60, no. 4, p. 528-542.

Todd, T., and G. Simmons, 1972, Effect of pore pressure on the velocity of compressional waves in low-porosity rocks: Jour. Geophys. Research, v. 77, no. 20, p. 3731-3743.

Trorey, A. W., 1970, A simple theory for seismic diffractions: Geophysics, v. 35, no. 5, p. 762-784.

Vail, P. R., et al, 1976, Interpretation of seismic sequences from reflection patterns: Preprint, 29th Annual Midwestern SEG, Dallas.

Widess, M. B., 1973, How thin is a thin bed?: Geophysics, v. 38, no. 6, p. 1176-1180.

Wuenschel, P. C., 1960, Seismogram synthesis including multiples and transmission coefficients: Geophysics, v. 25, no. 1, p. 106-129.

Wyllie, M. R. J., G. H. F. Gardner, and A. R. Gregory, 1960, Principles underlying the interpretation of acoustic velocity logs: Preprint (paper 1639-G), Soc. Petroleum Engineers Formation Evaluation Symposium, Houston, Nov. 21-22, 1960.

Seismic Stratigraphic Model of Depositional Platform Margin, Eastern Anadarko Basin, Oklahoma[1]

WILLIAM E. GALLOWAY,[2] MARSHALL S. YANCEY,[3] and ARTHUR P. WHIPPLE[3]

Abstract Three-dimensional stratigraphic analysis of cratonic-basin margins has demonstrated complex genetic interrelations between shelf, shelf-edge, and basinal facies. Application of seismic stratigraphic modeling has proved useful in analyzing the geometry of platform-margin deposits of the Pennsylvanian Hoxbar Group (Missourian) in the eastern Anadarko basin in Oklahoma.

Seismic modeling requires four principal steps: (1) tabulation of petrophysical parameters of the lithologies included in the model; (2) construction of a series of model stratigraphic sequences along a line of section; (3) generation of synthetic seismograms for each model sequence; and (4) comparison of the synthetic traces with corresponding field traces. Results of such a model study, combined with subsurface geologic data, suggest an interpretation of Hoxbar platform evolution incorporating two outbuilding or progradational depositional episodes separated by an upbuilding depositional episode.

INTRODUCTION

The combination of seismic data and physical stratigraphic interpretation is becoming an increasingly important aspect of subsurface exploration. Significant examples of the application of seismic data to regional facies analysis are presented by Sangree and Widmier (1977), Mitchum et al (1977), and Stuart and Caughey (1977). The objective of this paper is to demonstrate the potential for detailed facies delineation through the use of iterative seismic models.

Utility of the seismic tool for stratigraphic interpretation in a mature exploration province depends directly on the explorationist's ability to relate specific waveforms to specific lithologic units. Seismic modeling is one workable method of matching waveforms and their lateral changes to corresponding stratigraphic facies and facies boundaries. Because simple seismic models can be constructed by use of a conventional synthetic seismogram program, modeling capability is now available to a large segment of the exploration industry.

Both one- and two-dimensional modeling techniques can be applied, depending on the geologic problem. One-dimensional model programs compute a synthetic seismogram of a vertically stacked series of velocity (or velocity and density) slabs. Such programs are considered one-dimensional because they assume a straight, vertical ray path at all reflective interfaces (Fig. 1A). All interfaces are assumed to be horizontal. Two-dimensional programs, which can be based on several different theoretical approaches, can reproduce the effects of bed dip and curvature (Fig. 1B). Use of a two-dimensional program is necessary when geometric characteristics of the geologic section are in question. Both one- and two-dimensional modeling programs produce only a simplified approximation of the earth response, but model results commonly compare favorably with field data.

Applications of modeling are illustrated by reviewing the use of seismic data in reconstruction and interpretation of a Pennsylvanian shelf edge in the Anadarko basin, Oklahoma. The specific problems involved in the reconstruction include: (1) correlation of shelf and basinal facies across a shelf edge; and (2) spatial delineation of the framework sandstone and limestone facies, leading to an interpreted depositional history.

GEOLOGIC FRAMEWORK AND DATA BASE

The study area lies in Canadian County, Oklahoma, on the eastern shelf of the Anadarko basin, which was a major locus of deposition during Late Pennsylvanian time (Fig. 2). Several clastic sequences prograded westward across central Oklahoma at that time, filling the basin with repetitive sequences of mudstone, sandstone, and limestone.

A major facies change occurs within the upper part of the Hoxbar Group (Missourian) along a north-south line that can be traced at least 25 mi (40 km; Fig. 2). The western platform sequence of interbedded shelf limestone, shale, and lenticular, discontinuous sandstone changes abruptly west-

This paper is reprinted from the September 1977 *BULLETIN*.

[1]Read before the Association, April 8, 1975. Manuscript received, August 23, 1976; accepted, November 10, 1976. This paper also will appear in AAPG *Memoir* 26. Published with permission of the Acting Director, Bureau of Economic Geology, The University of Texas at Austin.

[2]Texas Bureau of Economic Geology, Austin, Texas 78712.

[3]Continental Oil Company, Ponca City, Oklahoma 74601. This paper was prepared while the senior writer was a member of the Exploration Research Division, Continental Oil Company. The writers thank Continental for support and for permission to publish this paper. Figures were drafted by the cartographic section, Bureau of Economic Geology, under the supervision of James W. Macon.

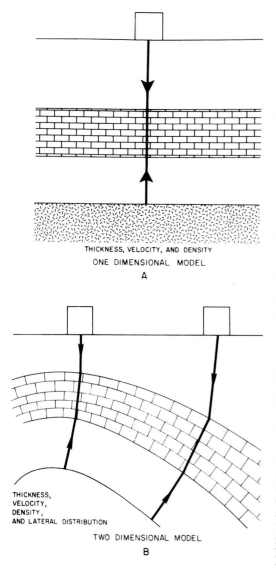

THICKNESS, VELOCITY, AND DENSITY

ONE DIMENSIONAL MODEL

A

THICKNESS,
VELOCITY,
DENSITY,
AND LATERAL DISTRIBUTION

TWO DIMENSIONAL MODEL

B

FIG. 1—Characteristics of one-dimensional, **A**, and two-dimensional seismic models, **B**. In **A**, seismic wave penetrates geologic section and returns to source/receiver along same straight-line ray path. Model uses thickness, internal velocity, and density of various layers to compute reflection character. In **B**, ray paths are refracted across dipping interfaces. Thus, effects of bed configuration are reproduced by modeling, and cross-sectional views may be obtained.

ward into an expanded sequence dominated by sandstone and mudstone (Fig. 3). Although well control is moderately dense, the abruptness of the facies change precludes definite correlation of stratigraphic units across the facies change. Consequently, subsurface nomenclature differs in

shelf and basinal areas. Named units commonly used in subsurface correlation include the Belle City, Haskell, and Tonkawa limestones on the shelf, and the Medrano, lower Wade, and upper Wade sandstones in the basin (Fig. 3).

In addition to the conventional electric-log net, a grid of 12 intersecting common-depth-point (CDP) seismic lines overlies the line of facies change (Fig. 2). Sonic logs are available for a few of the wells; two sonic logs typical of shelf and basinal stratigraphic sequences were selected for synthetic seismogram computation. Though the sonic logs selected are not of wells located directly on the log cross section or any of the seismic lines, the resultant synthetic seismograms were easily correlated with the field records. The Tonkawa limestone is a particularly good marker that extends across the map area. Thus, specific seismic events can be related to specific stratigraphic units (Fig. 3), and the geometry of the seismic reflectors can be used immediately to aid in interpretation of the facies change. As expected, the facies change produced a distinctive seismic configuration, as shown on field-record B (Fig. 3). The wedge-shaped pattern shown on this and all other lines crossing the facies change suggests a progradational shelf edge, or depositional platform margin, and supports the preliminary interpretation. Other east-west lines were used for refining the map of the shelf edge.

SEISMIC MODELING

Once stratigraphic units are correlated with specific events on the seismic field traces, the grid of seismic data becomes a significant additional source of stratigraphic data that, within the limits of data quality and band width, can be used to solve problems of correlation and facies delineation. Seismic modeling, which in its simplest form is the intentional manipulation of available sonic logs and derived synthetic seismograms, provides the tool needed to utilize this expanded data base.

One-Dimensional Modeling

The initial step in seismic stratigraphic analysis of the shelf edge consisted of the construction of several one-dimensional models of selected field traces on line B to determine relations between lithologic units and their seismic response. The interval modeled extends from the Tonkawa limestone at the top through a continuous calcareous shale unit at the base of the west-east cross section (Fig. 3). Thus all units potentially affected by

[4] Addition of density data produced little significant change in the synthetic. Velocity variation dominated the seismic response; thus density was not used as a variable in subsequent work.

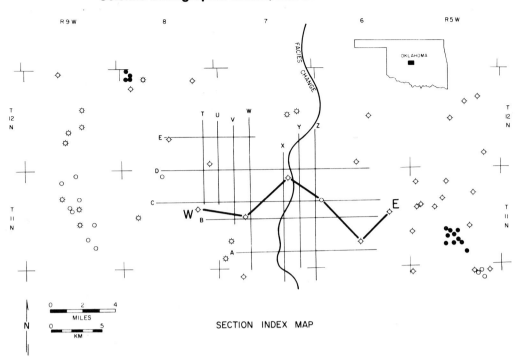

FIG. 2—Index map showing location and trend of upper Hoxbar facies change, and location of well and seismic data used in study.

the facies change are included between upper and lower reflectors.

The modeling process included three basic steps.

1. Four seismic traces in line B representative of different vertical stratigraphic sequences were selected: (MO-1) basin, (MO-2) shelf-basin transition, (MO-3) shelf edge, and (MO-4) shelf. Locations of these traces are shown in Figure 3. Wells C and E were projected into the seismic line to provide rock-unit thickness data for models 3 and 4 (MO-3, MO-4), respectively. Unit thicknesses were projected updip slightly from well A for model 1. No well data were directly applicable to the transition-zone model 2 (MO-2); bed thickness and sequences were based on the initial geologic interpretation of the geometry of the shelf-to-basin transition. In summary, thickness and vertical sequence of beds at the location of the four selected traces were tabulated.

2. Velocity-versus-lithologic data were tabulated using all available sonic logs in the study area. Velocity (V_I) is dependent on rock type (matrix) and porosity; however, in the section studied, velocity data are quite consistent for each lithologic unit. Shale velocity ranged from 12,000 to 13,000 ft/sec (3,600 to 3,960 m/sec), sandstone velocity averaged 14,000 to 15,000 ft/sec (4,270 to 4,570

m/sec), and limestone velocity ranged from 17,000 to 19,000 ft/sec (5,180 to 5,790 m/sec). The wide range of velocities provides optimum conditions for bed definition and makes the seismic response relatively insensitive to errors in velocity selection. (Tests indicated that variations in velocity of several hundred feet per second had no discernible effect on the modeled seismic response.) The velocity data were combined with bed-thickness data to produce four velocity-slab profiles, which provided the input to the synthetic seismogram program. In effect, four simple sonic logs were constructed. Data on the lithologic-sequences and velocity slabs used to produce the models are shown in Figure 4 (the vertical scale has been converted to two-way travel time by the modeling program so that the schematic sections and velocity slabs can be compared directly to the resultant seismic traces).

3. After processing and filtering to a band width comparable to field data (8 to 45 Hz), the preselected field trace was compared with the model trace, and successive iterations, with appropriate changes in slab thickness or velocity, were run where necessary to improve the match. Events on the model trace are displaced vertically from the units they represent because reflections are generated at interfaces between beds.

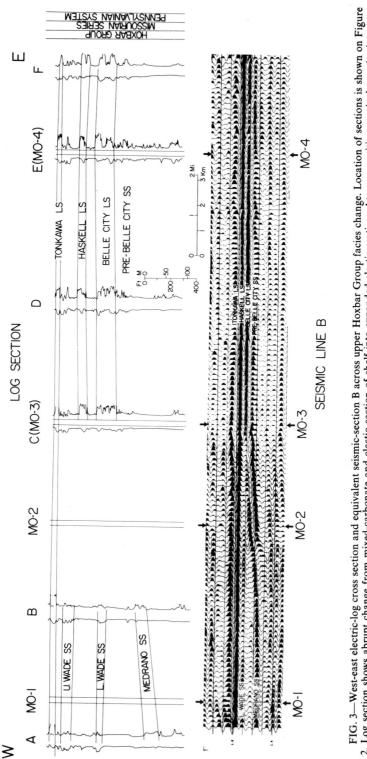

FIG. 3—West-east electric-log cross section and equivalent seismic-section B across upper Hoxbar Group facies change. Location of sections is shown on Figure 2. Log section shows abrupt change from mixed-carbonate and clastic section of shelf into expanded clastic section. At same position, equivalent seismic events display equally abrupt lateral changes. Locations of modeled traces (MO-1, etc.) are indicated by arrows on seismic section and by columnar section on geologic cross section.

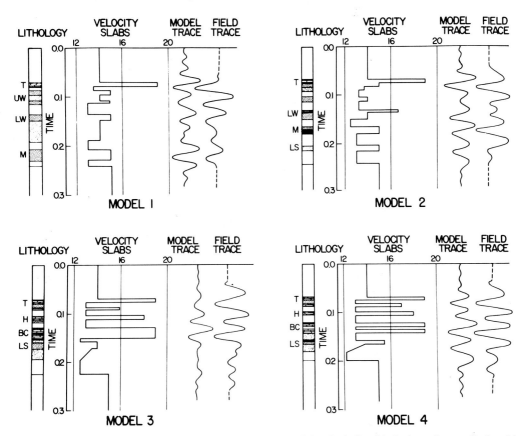

FIG. 4—One-dimensional models of selected seismic traces. All data including lithologic section are displayed in time domain for visual comparison. Velocity-slab plot graphs interval velocity (in ft/sec) used for each lithologic unit. Models are considered adequate when all major events are reproduced with correct peak-to-peak spacing. *T*, Tonkawa; *UW*, upper Wade; *LW*, lower Wade; *M*, Medrano; *LS*, lower sandstone; *H*, Haskell; *BC*, Belle City.

Review of the final models (Fig. 4) reveals several pertinent features.

Model 1—Peaks are produced by the lower Wade and Medrano sandstones, which are high-velocity slabs in the middle or lower velocity mudstone. The upper Wade also is underlain by a thick transition zone of sandy, calcareous mudstone of intermediate velocity. A strong "pseudo-peak" is produced by the upper Wade sandstone in combination with side lobes of the Tonkawa and lower Wade events. This peak is even more pronounced on the field data, suggesting further augmentation by interbed multiple energy not reproduced by the synthetic modeling program.

Model 4—At the other extreme, the shelf model also shows good agreement between model and field traces. Strong peaks are generated by the Belle City and Haskell limestones (Fig. 4). The thin limestone and lenticular sandstones just below the Tonkawa are lost between the two more

prominent reflectors. The thin, calcareous cap on the lower sandstone amplifies its response and partially negates the smoothing effect of the basal transition zone into underlying, low-velocity shale.

Model 3—Model 3 is similar to model 4 except for the absence of a shale break within the Belle City and the absence of a calcareous cap above the lower sandstone. The effect of the lower transition zone is pronounced, and a sharp trough is absent below the lower sandstone event.

Model 2—The rapid change from trace to trace makes selection of a single trace for modeling somewhat arbitrary. However, several iterations showed that the observed seismic events can be generated by four intermediate-velocity layers (about 14,000 to 15,000 ft/sec or 4,270 to 4,570 m/sec) at the stratigraphic levels shown. Tracing these events laterally indicated that the first peak directly below the Tonkawa peak grades basinward into the "pseudopeak" at about upper Wade level, the second peak becomes the lower Wade event basinward, the third peak drops in the section and becomes the Medrano event of model 1, and the lowest peak rises slightly shelfward to become the lower sandstone event of models 3 and 4. Thus, velocity data and lateral correlation suggest that the layers of the transitional model are composed of mixed sandstone-limestone beds and mudstone.

Integration of models, logs, and geometry of the seismic events at this point could reasonably answer many questions about the shelf-to-slope transition. Construction of a two-dimensional model, however, will add additional detail and test the limits of seismic stratigraphic interpretation.

Two-Dimensional Modeling

Two-dimensional seismic modeling necessitates construction of a geologic cross section in the plane of the seismic section with adequate detail to reproduce the lateral and vertical distribution of seismic events. Although this sounds like a formidable task, experience with one-dimensional models provides the needed "feel" for the seismic response of various types of changes expected within the framework of the study area. For example, thickness of limestone units has little effect on waveform because of the extreme velocity contrasts with surrounding facies. However, some sensitivity to thickness of sandstone units might be expected because of their intermediate velocities and because their associated waveforms show significant variability on the field data (compare lower Wade and Belle City events in Fig. 3).

To test seismic response to changing sandstone-bed thickness, another seismic-stratigraphic technique was employed. A sonic log from the basin sequence was digitized and a series of models constructed by substituting different thicknesses of intermediate-velocity material at the levels of the lower Wade and Medrano sandstones. Results are shown in Figure 5. The synthetic seismograms show that, in the range from 0 to 100 ft (0 to 30 m), greater thickness of lower Wade sandstone results in increasing peak amplitude of the appropriate event. Within this geologically reasonable range of thicknesses, the seismic data should reflect details of the sandstone-unit geometry. In contrast, as little as 50 ft (15 m) of Medrano sandstone produces a large peak that changes little as successive increments of high-velocity material are added. This result is supported by the continuity and uniform amplitude of the Medrano event on the field section.

Using the seismic section as a guide, and combining data from one-dimensional models and logs, it was possible to construct the geologic cross section shown at the top of Figure 6. As with all interpretive cross sections, this one is somewhat schematic; but, because the section must show the bed geometry that is digitized and put into the computer, certain geologic conventions, such as sawtooth interfingering of rock units, are not used. Most fine detail (such as interbedding of units that are inches or a few feet thick) cannot be resolved by seismic waves of the frequencies retrieved in the field (8 to 45 Hz in this example); where thick transition zones occur, separate velocity slabs were drawn. The thick wedge of 14,000-ft/sec (4,570 m/sec) material below the lower Wade sandstone is a good example of a transition zone. Interestingly, a heterogeneous sequence of sand, silt, and mud also may have a higher velocity than the overlying porous sandstone; this is seen in the lower sandstone interval, which has a velocity of 13,800 ft/sec (4,206 m/sec) and is underlain by a 15,000 ft/sec (4,572 m/sec) transitional unit.

Output from computations for the model section is shown in the lower two panels of Figure 6. The impulse-response plot displays the distribution, polarity, and strength of reflecting interfaces in the vertical time domain. Filtering at the appropriate band width produces a model section that should closely resemble the field data if the input cross section is accurate.

Comparison of this final model with the field section illustrates the detail that can be reproduced by modern modeling technology combined with appropriate geologic input (Fig. 6).

1. A strong, continuous Tonkawa peak extends across the section and provides a reference datum for other underlying events.

2. A strong, continuous peak is generated on the shelf by the Haskell limestone, but continues uninterrupted across the shelf edge and into the basin. West of the shelf edge, the peak is generated by a combination of sidelobe and multiple energy and is not related to a particular velocity interface. The model section accurately reproduces this "pseudoevent."

3. A strong peak is generated on the shelf by the Belle City Limestone.

4. A peak of variable amplitude is produced by the lower Wade sandstone, which laps up slightly against the Belle City and dies out. The peak is underlain by a broad quiet zone.

5. The Medrano sandstone produces a strong, continuous peak, which laps up into the slope and grades across a low-amplitude transition zone into the Belle City event.

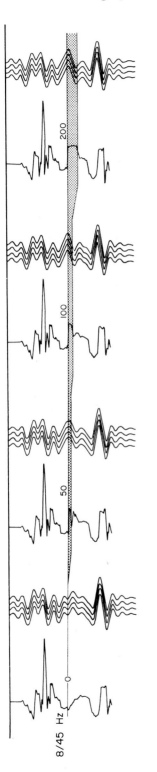

VARIABLE LOWER WADE

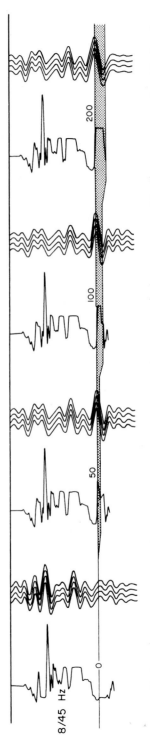

VARIABLE MEDRANO

FIG. 5—Effect of increasing sandstone thickness on seismic response, lower Wade and Medrano sandstones. Model seismograms are produced by inserting increments of material with appropriate velocity at stratigraphic position of sandstone unit on digitized sonic log.

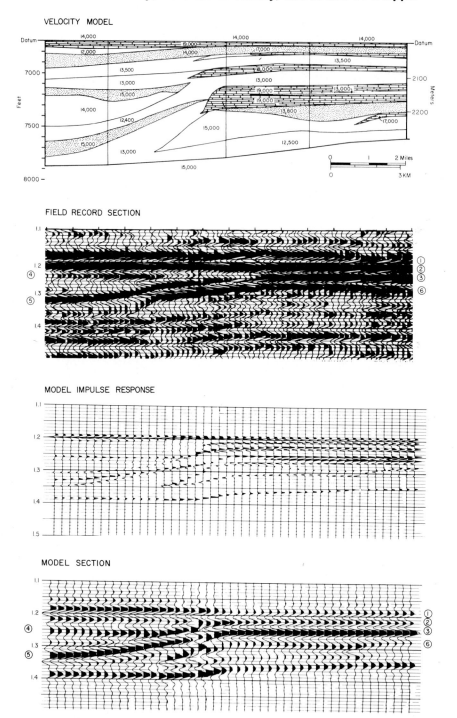

FIG. 6—Geologic-model cross section, equivalent part of field-record section, impulse-response plot of geologic section in vertical time domain, and computer-generated synthetic seismogram of geologic model. All principal events on field section are reproduced by synthetic-model section. Numbers beside seismic events correspond to usage in text.

6. A peak which dips down and dies out at the shelf edge, and which is underlain by a quiet zone, is produced by the lower, progradational sandstone.

Results of the model support the validity of the interpretive section; correlation problems are solved, and the spatial distribution of lithofacies across the shelf edge is depicted.

GEOLOGIC INTERPRETATION

The ultimate objective of modeling is the integration of seismic data with geologic information to test, refine, and document the interpreter's concepts about the three-dimensional distribution of lithofacies. Knowledge of lithofacies geometry, composition, and position in the basin of deposition is, in turn, the foundation of genetic stratigraphic analysis.

Correlation shows that the lower, progradational sandstone-shale sequence, Belle City Limestone, Medrano sandstone, and lower Wade sandstone form the shelf margin. Three-dimensional relations of these units suggest that three principal depositional episodes (terminology of Frazier, 1974) were responsible for development of the margin (Fig. 7). The lower sandstone and lower Wade episodes (I and III) are progradational or outbuilding episodes; the Medrano–Belle City episode (II) is primarily an upbuilding episode on both the shelf and basin. Isolith maps of facies within each of these episodes (Fig. 8) isolate genetically equivalent units and are the basis for the following interpretations.

Episode I

A major, fluvial-dominated delta system prograded westward into the Anadarko basin pro-ducing a progradational wedge as much as 500 ft (152 m) thick and consisting of a prodelta-mud platform capped by dip-oriented, anastomosing sandstone units (Fig. 8A). Framework sands were deposited in distributary-channel, mouth-bar, and delta-margin environments. Progradation of the delta system produced considerable differential relief between the shelf and the basin, thus forming the foundation of the shelf edge.

Episode II

Deltaic deposition ceased in the study area, perhaps as a result of upstream diversion of the fluvial system, and compactional subsidence permitted transgressive reworking of the deltaic deposits. As conditions stabilized, the shallow-water platform became the site of carbonate generation and accumulation, particularly at the shelf edge. Strike-oriented isopach trends (Fig. 8B) suggest biohermal growth. During the same episode, additional terrigenous sediment was introduced into the basin, primarily south of the study area. However, the rate of influx was slow and, instead of building a progradational mud platform, finer sediments were dispersed into the basin; sands, which also contain significant amounts of carbonate detritus, were transported downslope to form an extensive submarine apron at the base of the slope. Geometry (Fig. 8C), stratigraphic position, and compositional differences all support interpretation of the Medrano as a submarine-fan complex. Thin sandstones that trend across the shelf within the Belle City (as shown on Fig. 8C) may have provided some sand to the shelf edge, but the Medrano depocenter was south of the study area.

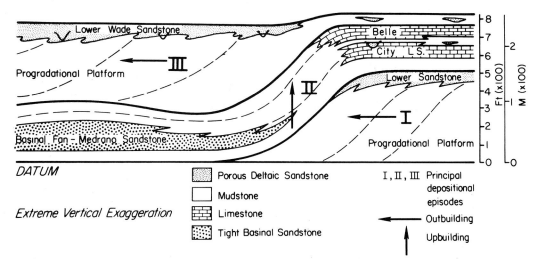

FIG. 7—Three depositional episodes producing upper Hoxbar shelf margin. No horizontal scale.

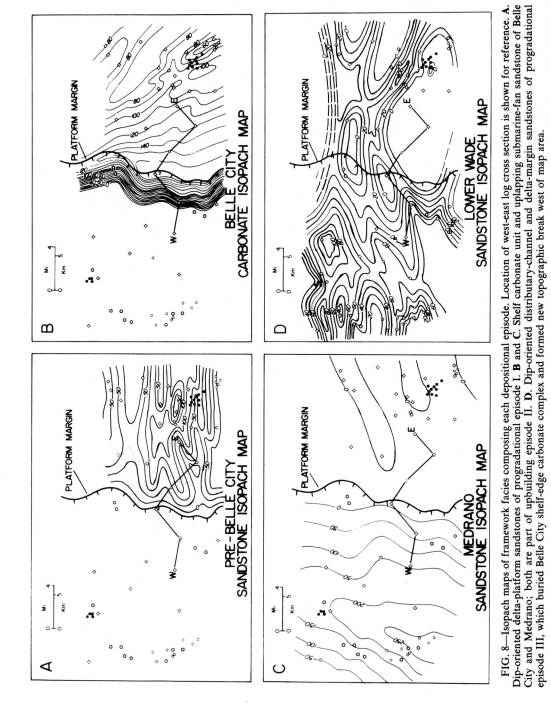

FIG. 8.—Isopach maps of framework facies composing each depositional episode. Location of west-east log cross section is shown for reference. A. Dip-oriented delta-platform sandstones of progradational episode I. B and C. Shelf carbonate unit and uplapping submarine-fan sandstone of Belle City and Medrano; both are part of upbuilding episode II. D. Dip-oriented distributary-channel and delta-margin sandstones of progradational episode III, which buried Belle City shelf-edge carbonate complex and formed new topographic break west of map area.

Episode III

Influx of terrigenous sediment resulted in a renewed period of delta growth and progradation beyond the old platform edge (Fig. 8D). Water remained shallow across the platform, resulting in thin delta-margin sequences and consequent incision by major distributary channels into underlying limestone and shale formed during episode II. Beyond the shelf, the delta system was more intensely reworked by waves and progradation was slower—a common evolutionary trend in cratonic delta systems (Galloway, 1975). The abundance of carbonate debris in the deltaic sandstones and underlying prodelta muds further indicates that progradation was slow and interrupted locally at least twice by temporary destructional phases. Progradation during episode III extended several miles beyond the mapped area.

CONCLUSIONS

The examples given herein illustrate some of the possible techniques and applications of seismic modeling in genetic stratigraphic analysis. In the Wade-Medrano shelf edge, seismic response is complex. Reflection events may accurately portray lithologic continuity or mislead with their continuity; they may be sensitive or insensitive to changes in bed thickness; and they may be the product of a single lithologic interface or a cumulative result of side lobes, multiples, and complex interface distribution. Only when these various possibilities are sorted out does it become possible to use seismic data to test, document, and amplify detailed stratigraphic interpretation. Accurate three-dimensional description of lithofacies thus becomes a powerful tool for reconstruction of the depositional history of a basin.

REFERENCES CITED

Frazier, D. E., 1974, Depositional episodes: their relationship to the Quaternary stratigraphic framework in the northwestern portion of the Gulf basin: Texas Bur. Econ. Geology Circ. 74-1, 28 p.

Galloway, W. E., 1975, Process framework for describing the morphologic and stratigraphic evolution of deltaic depositional systems, in M. L. Broussard, ed., Deltas, models for exploration: Houston Geol. Soc., p. 87-98.

Mitchum, R. M., Jr., et al, 1977, Stratigraphic interpretation of seismic reflection patterns in depositional sequences: AAPG Mem. 26.

Sangree, J. B., and J. M. Widmier, 1977, Interpretation of clastic depositional facies from seismic data: AAPG Mem. 26.

Stuart, C. J., and C. A. Caughey, 1977, Seismic facies and sedimentology of terrigenous Pleistocene deposits in northwest and central Gulf of Mexico: AAPG Mem. 26.

A Case History of Geoseismic Modeling of Basal Morrow-Springer Sandstones, Watonga–Chickasha Trend: Geary, Oklahoma—T13N, R10W[1]

WILLIAM A. CLEMENT[2]

Abstract The project area is along the northeastern flank of the Anadarko basin, near Geary, Oklahoma. In the area, basal Pennsylvanian Morrow–Springer sandstones subcrop beneath the pre-Atokan, Wichitan unconformity and form the prolific gas condensate-producing Watonga-Chickasha trend.

Productive Morrow–Springer distributary channel sandstones form discontinuous lenticular reservoirs. Dense well control is normally required to map sandstone geometry and distribution. Integration of high resolution seismic data, with the application of geoseismic modeling techniques which use geologically derived input parameters, can better delineate the shape and extent of these reservoirs between points of well control. Vibroseis®seismic data, filtered conventional dynamite data, synthetic seismograms from sonic logs, one-dimensional geoseismic models, and geologic/lithologic data from well cuttings, cores, and logs, can be used to construct meaningful sandstone distribution maps. Geoseismic models constructed by use of a remote computer terminal demonstrate that high resolution 12 to 67-Hz seismic data can distinguish sandstones to a minimum thickness of 20 ft (6 m) when stratigraphic conditions are such that high interval velocity (15,000 to 15,500 ft/sec or 4,572 to 4,724 m/sec) sandstones are encased in lower interval velocity (10,000 to 11,500 ft/sec or 3,048 to 3,505 m/sec) shales. Reflection coefficients between sandstones and shales range from 0.1 to 0.3. Results of the modeling indicate that sandstones exceeding 20 ft (6 m) in thickness are arranged in a NW-SE trend, approximately 2,500 ft (762 m) wide, across the project area in T13N, R10W.

Subsequent development drilling has yielded both positive and negative results. Two wells encountered the predicted basal sandstone; one well encountered thin interbedded tight sandstones and shales. Lateral facies change from thick porous sandstone to interbedded high interval velocity facies (17,000 ± ft/sec or 5,182± m/sec); tight, highly indurated sandstones and shales can produce similar seismic signatures. Modeling of high resolution seismic data can accurately predict the presence of high velocity zones within the basal Morrow–Springer section, but further criteria must be applied to distinguish productive sandstones from thin, high-velocity sandstones interbedded with shales.

INTRODUCTION

The development of methods for delineating productive reservoir sandstones beyond the well bore has long been an objective for exploration geologists (Forgotson, 1969; Mannhard and Busch, 1974; Pate, 1959; Potter, 1967), and grown in importance as traps become more elusive and the search for new oil and gas becomes more difficult. This paper describes the procedures used and the results obtained using a combined seismic-subsurface approach to exploration for lenticular stratigraphic traps in the Anadarko basin of Oklahoma. This approach represents a preliminary effort to integrate subsurface geologic information with newly developed geophysical modeling techniques.

Objectives

The immediate objective in the study area was to use *geoseismic* modeling techniques to define more accurately, the thickness and distribution of productive basal Springer sandstone to aid in development drilling. Previous work in the area indicated that geologic relations meet the requirements necessary for the application of seismic modeling techniques. These relations are, in particular: a strong velocity contrast between economically important basal Springer sandstones and thicker encasing mud rocks; the presence of bracketing, regionally developed limestone marker units which can be easily correlated and mapped from subsurface well logs and seismic data; and an area of good seismic record quality and abundant well control which can be tied into the seismic data.

Setting

The study area (Fig. 1) is south and east of Geary, Oklahoma, in T13N, R10W. The Geary area lies along the northeastern flank of the Anadarko basin, where overpressured (Breeze, 1971; Davis, 1974) basal Pennsylvanian sandstones subcrop beneath the pre-Atokan, Wichitan unconformity (Fay et al, 1962) and form the prolific gas-condensate producing Watonga–Chickasha trend. Consequently, most operators in the area group all producing zones in the basal Pennsylvanian interval under the informal term "Morrow–Springer."

[1]Published with permission of Continental Oil Company. Manuscript received, November 1, 1976; accepted, January 20, 1977.

[2]Project Geologist, Exploratory Projects Section, Continental Oil Company, Oklahoma City, Oklahoma, 73112.

The writer is indebted to Continental Oil Company for permission to publish this paper. Appreciation is also expressed to personnel of Continental Oil for their help in introducing the author to seismic stratigraphic modeling techniques.

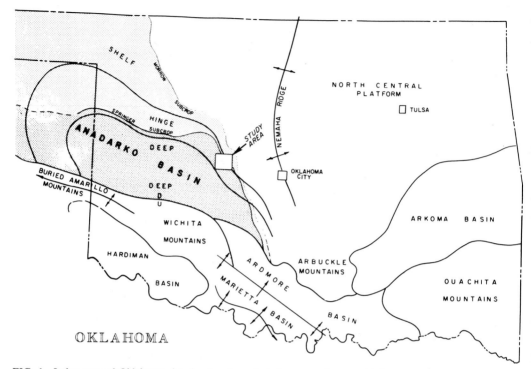

FIG. 1—Index map of Oklahoma showing location of study area, subcrops of Morrow and Springer units, and major structural and tectonic features of Oklahoma.

BASAL SPRINGER SANDSTONE

Although several sandstone zones are productive in the Geary area, the most prolific zone occurs at the base of the Springer interval, referred to as the basal Springer sandstone in Figure 2. Some doubt exists as to whether the Springer series shown on the correlation section is Pennsylvanian or Mississippian in age (Davis, 1974; Jacobsen, 1959; Swanson and West, 1968; Takken, 1967). Unpublished data indicate that floral assemblages identified in a core of the Springer interval north of the study area were those characteristic of Chesterian assemblages in other parts of the basin. For this reason the Springer and Mississippian boundaries are questionable.

Well log data from T13N, R10W indicate that the basal Springer sandstone thickens in the northwest part of the township (Fig. 3) where it joins the "Old Woman Channel" as described by Davis (1974). Elsewhere in T13N, R10W, the basal Springer sandstone rarely exceeds 45 ft (14 m) in thickness but develops porosity of economic importance where the thickness of the sandstone exceeds 10 to 15 ft (3 to 4.6 m). As shown on stratigraphic cross section A-A' and B-B' in Fig-

ures 4A and 4B, the basal Springer sandstone is distributed irregularly throughout the township, as are several younger sandstones in the Morrow–Springer interval.

Overpressuring in Morrow–Springer

Because of the overpressured nature of the Morrow–Springer section in the Geary area, as elsewhere along the Watonga–Chickasha trend (Breeze, 1971; Davis, 1971; 1974), successful well completions have been made in thinner sandstone zones than could be made in many normally pressured environments. As an example, the Conoco 1 Leck (well 7 in Fig. 4A and 4B) was completed in the basal Springer sandstone for a calculated open flow of 20.6 MMCFGPD (0.583 million cu m) plus 126 bbl (793 kl) of condensate per 6 hours from 28 ft (8.5 m) of sandstone with a 17-ft (5 m) section of 10 to 12% porosity. The well has produced almost 5 Bcf (141.5 million cu m) of gas and 182,000 bbl (1.1 million kl) of condensate. It should ultimately produce 10 to 11 Bcf (283 to 311 million cu m) of gas and greater than 250,000 bbl (1.6 million kl) of condensate. The well encountered the sandstone at 11,200 ft (3,414 m) as it drilled to a total depth of 14,146 ft (4,312 m) in the Ordovician Simpson. Shut-in pressure was 8,448 pounds (3,832 kg) per sq in. at 11,200 ft (3,414 m) and the calculated fluid pressure gradient is 0.754 pounds (.34 kg) per sq in. per ft, highly overpressured or geopressured according to Breeze (1971; normal fluid pressure gradient is defined as 0.465 pounds or 0.21 kg per sq in. per ft of depth).

The Leck well was drilled in a section formerly condemned by a dry hole in the basal Springer, the Sinclair 1 Steele, in the southeast quarter.

Predictability

Efforts to offset the Conoco 1 Leck, using conventional subsurface mapping techniques proved disappointing. There was already the dry hole, (well 6, Fig. 4A, 4B) one-half mi (0.8 km) east of the Conoco 1 Leck in the southeast quarter of Section 21, in which no basal Springer sandstone was encountered. Development spacing for Morrow–Springer gas is on 640-acre (259 ha.) spacing, consequently, the first direct offset to the Leck well was drilled one-half mi (0.8 km) south in the C/NW of Section 28 (well C, Fig. 4B). This well was completed for a calculated open flow of 1.2 MMCFGPD (0.034 million cu m) from 12 ft (3.7 m) of basal Springer sandstone with a 5-ft (1.5 m) section of 8 to 10% porosity, near the minimum for an economic completion.

A second production offset was drilled in Section 29 (well 8, Fig. 4A, 4B); the well did not encounter basal Springer sandstone but was productive in other zones. A third offset in Section 20 (well D, Fig. 4B) encountered 7 ft (2.1 m) of tight basal Springer sandstone.

It was decided that because of the poor predictability in the basal Springer sandstone using conventional sandstone mapping techniques, newly developed seismic modeling methods should be used to aid in the delineation of the sandstone.

INPUT PARAMETERS FOR SEISMIC MODELING

Before proceeding with a seismic modeling program in the Geary area, several geologic and seismic input parameters were studied.

Lithologic Character and Reservoir Geometry

The basal Springer pay section is composed of a fine to very fine-grained quartz sandstone ranging in thickness from less than 10 ft (3 m) to more than 80 ft (24 m) in the NW quarter of the township. More commonly, thickness ranges from 10 to 45 ft (3 to 14 m) with sandstones less than 10 ft (3 m) being locally calcareous and highly indurated whereas those thicker than 10 ft (3 m) develop

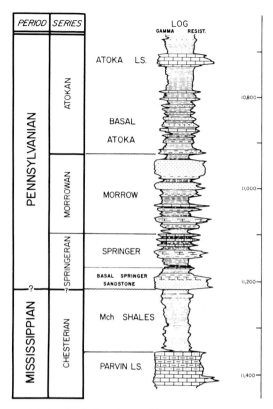

FIG. 2—Geologic column with gamma-ray resistivity log and generalized lithologic section, showing Pennsylvanian and Mississippian units in the Geary area.

zones of porosity ranging from 8 to 16%. The thicker sandstones (45 ft, or 14 m, and greater) commonly have thick zones of porosity in the 12 to 14% range. This may reflect a higher energy environment at the time of deposition than for sandstones less than 40 ft (12 m) thick which commonly have porosity zones distributed in thin intervals throughout the sandstone rather than as one thick zone.

The basal Springer sandstone is overlain by 25 to 40 ft (7.6 to 12 m) of laminated dark carbonaceous mudstone. The sandstone is underlain by 120 to 200 ft (37 to 61 m) of black mudstone with thin beds of graded, fine to very fine-grained quartz sandstones and siltstones that are interpreted as distal prodelta and shelf edge deposits.

Basal contacts of this sandstone with underlying mudstones are sharp, yet upper contacts appear to be gradational. In a core sample, this zone is a sequence of low angle, cross-bedded, fine sandstone grading upward into interbedded mud and ripple-bedded silty, very fine sandstone over-

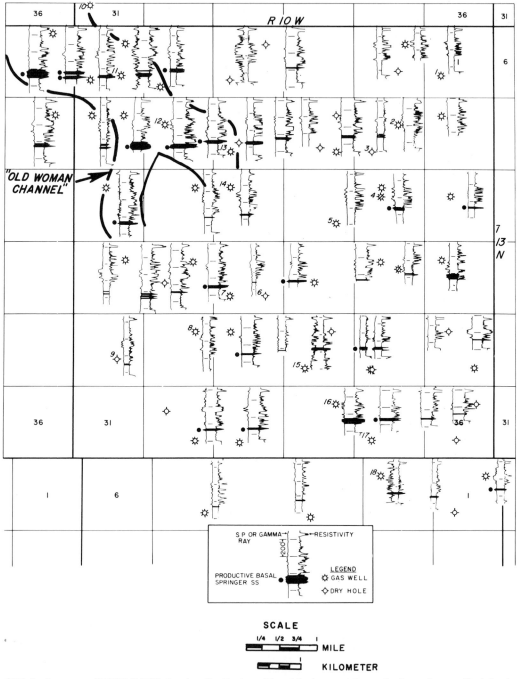

FIG. 3—Log map of T13N, R10W showing distribution of basal Springer sandstone in the study area. Each log is bounded at top by Atoka or Novi limestone, and at base by Parvin limestone.

lain by black carbonaceous mudstones. Unlike the basal sandstone, those occurring higher in the Morrow section display both gradational and sharp lower boundaries.

As interpreted from the available well control, lateral extent of the reservoir sandstones varies from a few hundred to a few thousand feet.

Structural dips range from 2 to 2.5° (180 to 230 ft per mi, or 55 to 70 m per 1.6 km) southwest on the underlying Parvin limestone (Fig. 5A) where a paleodepositional strike runs in a NW–SE direction as determined from the Atoka to Parvin isopach (Fig. 5B).

Pressure and Pore Fluids

As determined from initial shut-in pressures and bottom-hole pressure from scout data (Breeze, 1971), fluid pressure gradients range from 0.642 to 0.754 pounds (0.29 to 0.34 kg) per sq in. per ft. Corrected bottom-hole temperatures range from 200°F to 230°F (93°C to 110°C). Reservoir sandstones contain water, gas, and condensate.

At the prevailing bottom-hole temperatures and pressures, gas is present in a dense but gaseous state with an estimated hydrocarbon density of 0.25 g/cc and with an average fluid density of 0.45 g/cc.

Acoustic Stratigraphy

Well control in the area is abundant, averaging one well per section. Several wells have been logged with sonic tools, most with density tools, and all with spontaneous potential or gamma ray and resistivity tools.

Sandstone densities from logs range between 2.4 and 2.65 g/cc with accompanying interval velocities calculated from sonic logs of, from 14,000 ft to 17,000 ft (4,267 to 5,182 m) per sec (Fig. 6), depending on shale content, amount of porosity, and type of cementation. Highest velocity sandstones are very fine subangular, glassy, and highly indurated. Productive sandstones with porosities ranging from 8 to 12%, fall within the 15,000 to 15,500 ft (4,572 to 4,724 m) per sec velocity range. Shale densities range from 2.40 to 2.55 g/cc, and their accompanying interval velocities vary from 9,800 ft (2,987 m) per sec to 11,500 ft (3,505 m) per sec. The very low shale velocities are due to undercompaction and overpressuring (Breeze, 1971). Shales within the Springer section generally are the slower, having velocities of 9,800 to 10, 500 ft (2,987 to 3,200 m) per sec whereas Morrow shales range up to 11,500 ft (3,505 m) per sec.

The bracketing Atoka and Parvin limestones have interval velocities ranging from 18,000 ft

(5,486 m) per sec where shaley or porous, to 21,500 ft (6,553 m) per sec where densely crystalline. Limestone densities range from 2.65 to 2.75 g/cc. Gradational velocities between the three main lithologic groups result from the transitional lithologies encountered within the Morrow–Springer section.

The density and interval velocity data derived from the well logs were used to reconstruct the stratigraphy using parameters which could be resolved by seismic methods. The resulting generalized acoustic stratigraphic model for the Geary area comprises an interval of 350 to 800 ft (107 to 244 m; see Atoka to Parvin interval isopach map, Figure 5B) bracketed top and bottom by two very high velocity limestones with an intervening clastic sequence composed of alternating high velocity sandstone lenses of variable thickness, low velocity siltstones, and very low velocity shales. The basal Springer section is characterized as having up to 45 ft (14 m) of high velocity sandstone encased between thicker overlying and underlying intervals of low to very low velocity shales.

Reflection Coefficients

To be resolved by the reflection seismograph, the layers within the acoustic model must reflect sufficient energy to the surface to be recorded as discernable seismic events. The relative amplitude of the reflected energy can be measured by the reflection coefficient. As defined by Dobrin (1960, p. 25), the reflection coefficient (R) is the square root of the ratio of reflected energy in a longitudinal wave (Er) to incident energy (Ei) at an interface, when the angle of incidence is equal to zero degrees. The amount of energy reflected at this interface is dependent on the contrast in acoustic impedance, defined as the product of the density ρ and velocity (V) of each layer, across the interface. Therefore:

$$\text{Reflection coefficient, } R = \sqrt{\frac{Er}{Ei}} = \frac{(V_2\rho_2 - V_1\rho_1)}{(V_2\rho_2 + V_1\rho_1)}$$

As shown in Figure 6, density variations between limestones, sandstone, and shales in the Geary area are much less than accompanying velocity contrasts, indicating that velocity is the primary determinant of the reflection coefficient. Reflection coefficients were computed from several wells using both sonic and density logs, yielding values between sandstone and shales of 0.1 to 0.3, which are considered very good data for the Mid-Continent area. A minor reduction in the reflection coefficient was noted where density values were not used.

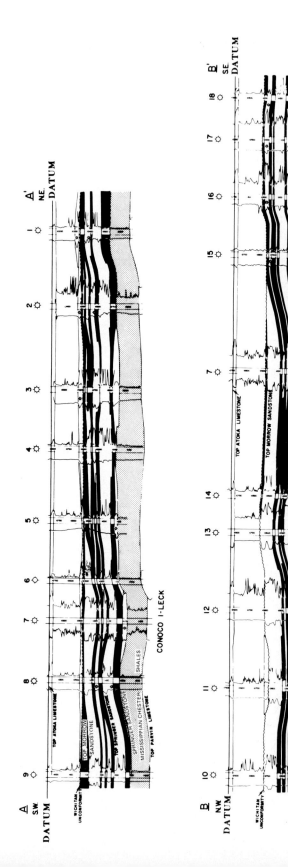

FIG. 4A—Stratigraphic cross sections A-A' and B-B' across T13N, R10W. Conoco 1 Leck, C/SW Section 21, is well 7 here.

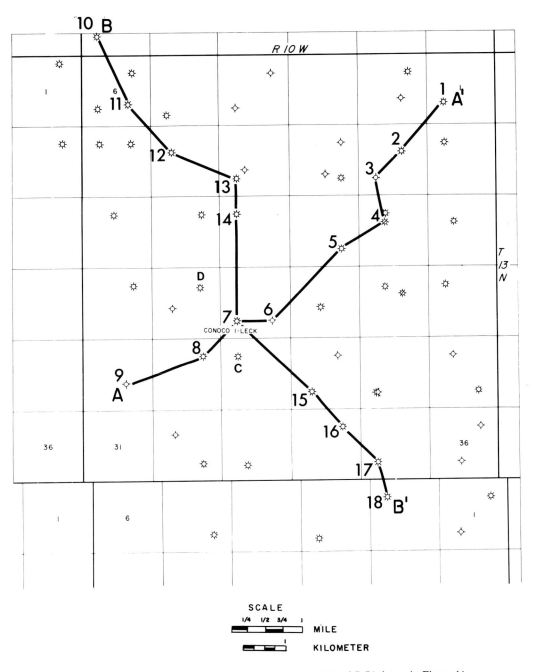

FIG. 4B—Index map for stratigraphic cross sections A-A′ and B-B′ shown in Figure 4A.

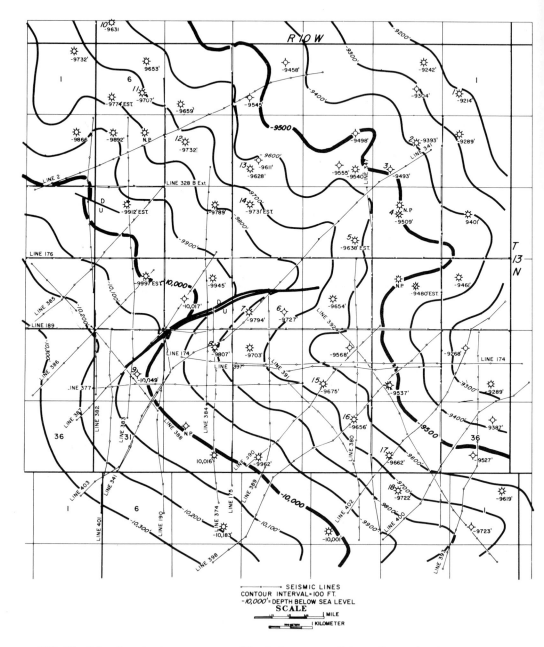

FIG. 5A—Subsurface structure map at top of Parvin limestone, utilizing well control and seismic coverage.

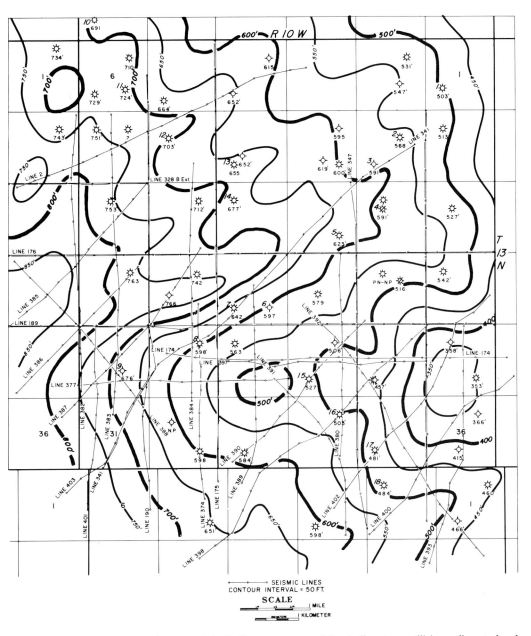

FIG. 5B—Interval isopach map from top of Atoka limestone to top of Parvin limestone, utilizing well control and seismic coverage.

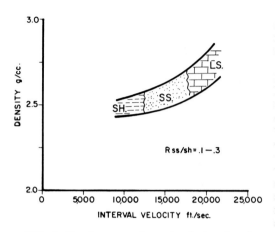

FIG. 6—Density versus interval velocity plot, for wells in Geary area, showing approximate lithologic boundaries.

GEOSEISMIC MODELING

The term *geoseismic modeling*, as applied to seismic modeling in the Geary area, evolved as a result of the close interdependence of geologic and seismic input parameters during the study.

The techniques used in this study were based on a program which allowed the construction of a predicted geologic model using an interval velocity—density versus depth distribution—from which reflection coefficients were computed. These values were convolved with an appropriate filter pulse to generate a synthetic seismogram (Fig. 7). A more complete mathematical discussion on the construction and interpretation of synthetic seismograms is given by Sengbush et al (1961).

The synthetic seismograms were then compared visually with recorded field data. The synthetic seismogram or trace in Figure 7 gives a fair representation of the velocity-density model. A better representation was achieved, however, using SAILE® (Seismic Acoustic Impedance Log Exploration) processing, developed by Continental Oil Company's Exploration Research Division (Fig. 8). The mathematical derivation (integration of seismic trace) of this filtering process will not be discussed in this paper, but by using SAILE® processing, it is possible to generate an approximation of the acoustic log from a conventional seismic trace. The advantage of SAILE® over conventional processing is that visual inspection of the data allows deflections to the right (peaks) to be interpreted as high velocity layers, while those to the left (troughs) may be interpreted as low velocity beds, as is the case with conventional

sonic logs. Consequently, all the geoseismic models and field records were processed using SAILE®.

By applying the appropriate filter pulse and SAILE® processing to properly constructed geologic models, it was anticipated that synthetic seismograms, such as those shown in Figure 9, could be generated. As a test of this hypothesis, velocity and density logs from the Conoco 1 Leck (Fig. 10A), and the Sinclair 1 Steele, were digitized, integrated into the time domain, and convolved with various SAILE® filter pulses. The resulting synthetic seismograms in Figure 10B substantiated what had been anticipated. The Conoco 1 Leck had 28 ft (8.5 m) of basal Springer sandstone and the Sinclair 1 Steele had 7 ft (2.1 m) of tight sandstone. The resulting synthetic traces show that above 55 Hz, the basal zone, where sufficiently thick, becomes identifiable as a pronounced amplitude event within a shale section which is otherwise characterized by a broad

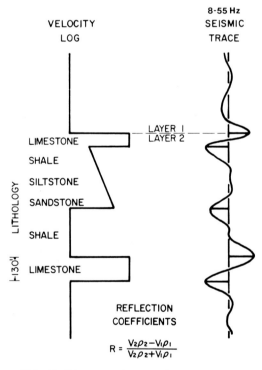

FIG. 7—Diagrammatic example, showing typical seismic trace generated by 8 to 55-Hz band-pass pulse in Geary area. Velocity log at the left is generalized, showing Atoka limestone as upper boundary, and Parvin limestone as lower boundary. As with conventional sonic logs, velocity increases to the right. The seismic trace is superimposed over the reflection coefficient log.

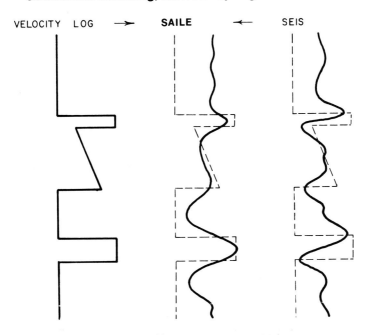

FIG. 8—Illustration showing (diagrammatically) derivation of SAILE® process. This process gives better visual approximation of velocity or sonic log.

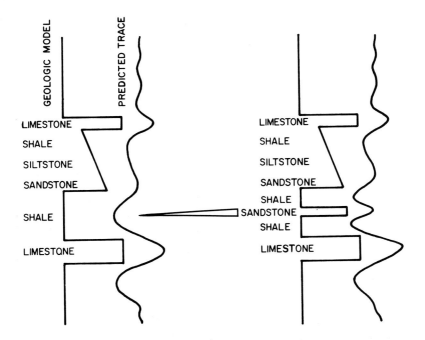

FIG. 9—Illustration showing velocity models with and without basal Springer sandstone and predicted SAILE® traces.

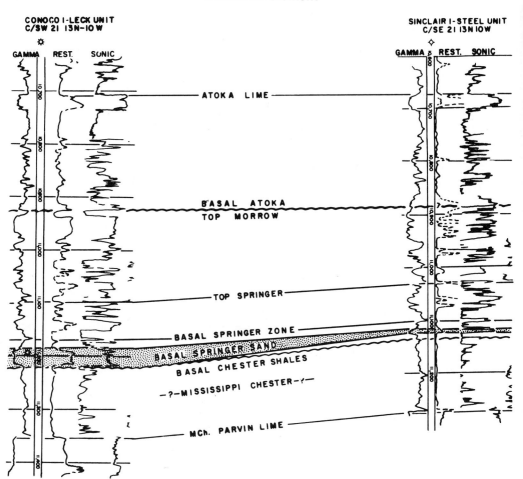

FIG. 10A—Stratigraphic cross section between Conoco 1 Leck and Sinclair 1 Steele.

trough. Models run with and without multiples showed no change in the waveform.

A series of models was constructed using increasing thicknesses of basal sandstone. These demonstrated an increasing reflection amplitude as the "tuning" thickness was approached. The phenomenon of bed tuning or greatest composite reflection amplitude, as demonstrated by Sengbush et al (1961), occurs when the bed thickness, in two-way time, is equal to one-half the basic period of the seismic pulse. The thickness at which a bed "tunes" also represents the threshold thickness at which, ideally, the upper and lower interfaces of the bed could be defined by a peak and trough relation, or in the case of SAILE®, at the inflection points. Bed tuning, more commonly stated, occurs where the thickness of the bed, in

two-way time, is equal to one-half the wavelength of the seismic pulse. True thickness would be measured in one-way time; therefore, thickness should equal one-quarter the wavelength.

Using the standard equation for wavelength (λ) written: $\lambda = V/f$, where V equals the velocity of the medium and f equals the frequency of the pulse. It can be demonstrated that with a given velocity of 15,000 ft (4,572 m) per sec for the basal sandstone, and recoverable high end frequencies approaching 60 cycles/sec, the tuning thickness would be ¼ λ or 62.5 ft (19.1 m). Available well control indicated that basal sandstone thickness did not exceed 45 to 50 ft (13.7 to 15.2 m). Consequently, measurement of bed thickness in the basal Springer by a peak-to-trough method, or in the case of SAILE®, inflection-to-inflection

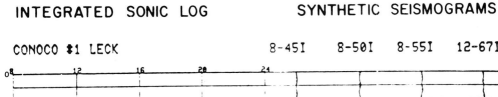

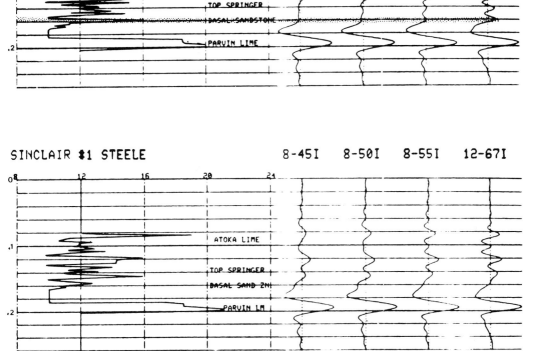

FIG. 10B—Integrated sonic logs and accompanying SAILE® synthetic seismograms generated using various filter pulses from sonic data from Conoco 1 Leck and Sinclair 1 Steele.

method, was considered unreliable and, most likely, incorrect. By precisely matching the synthetic seismograms generated from predicted geologic models to the actual field traces, it was considered possible to achieve semi-quantitative thickness values that could be integrated with the existing well control and interpreted.

This "closest fit" method was based on the assumptions that: interval velocities of the sandstone varied within the limits specified, sandstone thickness ranged from 0 to 50 ft (15.2 m) (maximum), and no changes occurred within the lithologic sequence in the lower Springer other than

those previously accounted for. A method such as this also has to assume that the results obtained from an ideal situation, represented by a computer modeling program, will not precisely reproduce the results obtained in a real life situation recorded in the field until all variables that affect seismic data can be taken into account.

Seismic Coverage

Because of the encouraging results obtained in the preliminary model study, a seismic grid was set up. Lines were laid out to include as many wells as possible which had been logged with both

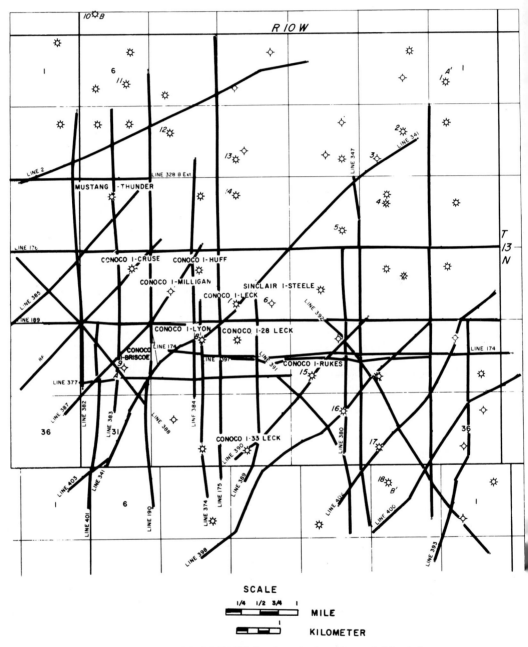

FIG. 11—Index map of T13N, R10W showing seismic grid recorded in study area.

sonic and density tools (Fig. 11). The data thus obtained included several 12-fold dynamite seismic lines recorded broadband and several 10-fold Vibroseis® lines recorded 10 to 70 Hz. All modeling interpretations were made from 8 to 55, 10 to 57, and 12 to 67 filtered, SAILE® processed versions of these lines. Some pre-existing Vibroseis® data recorded 7 to 41 Hz and 8 to 45 Hz, were used for structural mapping and qualitative stratigraphic information. In all, 20 lines up to 5 mi (8 km) long were recorded for a total of approximately 61 mi (98.15 km). Grid spacing varied from one-half to one-quarter of a mi (0.8 to 0.4 km). After the data was recorded and processed, synthetic seismograms from available well logs were compared with the field data and units were correlated and modeled.

Line 0-383

As shown in Figure 12A and 12B, on Line 0-383, the Parvin limestone is easily recognized as a strong amplitude event (the reflection coefficient is 0.32). The Atoka limestone, being much thinner, is less easily recognized. Line 0-383 crossed two wells— the Conoco 1 Briscoe, a dry hole in Section 30, and the Mustang 1 Thunder, a basal Springer producer in Section 18. The Conoco 1 Briscoe was logged with both sonic and density tools. The Mustang 1 Thunder had only a density log; therefore, interval velocities for sandstone, shales, and siltstones were derived from correlation with the Briscoe well. An obvious amplitude event is present from shot points 20 to 28 from 0.035 to 0.040 sec above the Parvin. This event is correlative with the occurrence of the basal Springer sandstone on the digitized well log from the Thunder well. The Thunder well had 22 ft (6.7 m) of gross basal Springer sandstone with 10 ft (3 m) of section with 9 to 10% porosity. No corresponding amplitude event is evident in the Briscoe well which had 7 ft (2.1 m) of tight basal Springer sandstone.

Modeling Procedure

Two sets of models were generated at shot point 20, the location of the Mustang 1 Thunder.

In the first case, it was hypothesized that there was no nearby well control, so that the stratigraphy would have to be reconstructed using only the SAILE® seismic trace and the known stratigraphic variables for the area; namely, the position on the field record of the Atoka and Parvin limestones, their thicknesses, and their velocities. The velocities of the various intervening sandstones, shales, and siltstones were also considered known values obtained from area sonic logs. The

remaining unknown was the vertical distribution and arrangement of these units within the section.

According to the SAILE® program (as shown in Fig. 8), deflections to the right are caused by high velocity layers and deflections to the left are caused by low velocity zones. Using these assumptions, an overlay was made on the SAILE® trace and the upper inflection point on each peak was marked, starting with the "Atoka lime" and continuing downward through the Parvin reflection. The overlay showed a section composed of peak-trough couplets, or theoretically, sandstone/siltstone-shale couplets, although the thin sandy limestones in the upper part of the section produce a peak, depending on their composite thickness and bed spacing. A high velocity layer composed of sandstone, sandy limestone or sandstone, and siltstone was then drawn in, followed by a layer of shale. If the peak appeared symmetrical, it was considered to represent a uniform slab of sandstone, whereas, if the peak displayed a gentle downward slope it was considered to represent a sandstone underlain by lower velocity siltstone. The shapes of the units represented the ramp, step, and square velocity functions as defined by Sengbush et al (1961).

At this stage the velocity models were still in the time domain and could not be input to the computer until they were converted to dimensions which could be described in terms of thickness in feet and interval velocity. Consequently, the two-way time interval occupied by the peak-trough couplet on the record was measured (it should be noted that the field records were expanded to scales of 10 and 20 in. or 25.4 and 50.8 cm per sec). Each sandstone or sandstone/siltstone layer was assigned an average interval velocity (the starting and ending velocities in a ramp function were averaged) and its two-way time thickness measured.

The interval velocity was then multiplied by the measured two-way time thickness to obtain a two-way thickness in feet, half of which was considered true thickness. A shale velocity was then assigned to the remaining time portion of the couplet and an accompanying thickness in feet was calculated. When values were calculated for all the "sand-shale" couplets in the trace and the thickness and velocity for the Atoka and Parvin "limes" added, the thickness-velocity model was input to the computer in terms of layer thickness in feet accompanied by an interval velocity. From these values, reflection coefficients were computed and then convolved with an appropriate SAILE® filter pulse to generate a synthetic trace. The synthetic trace was compared with the record

CONOCO "I BRISCOE   MUSTANG "I THUNDER

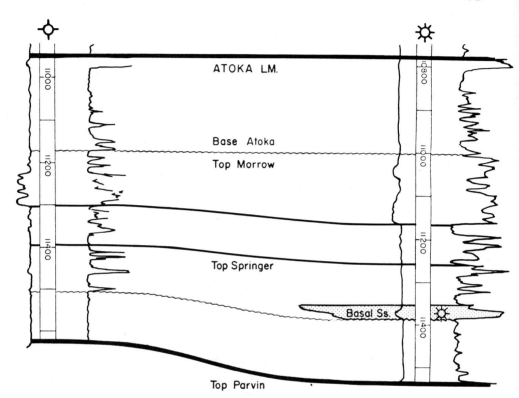

FIG. 12A—Stratigraphic cross section between Mustang 1 Thunder and Conoco 1 Briscoe.

trace and, if the fit was not acceptable the procedure was repeated by changing the bed thickness and/or interval velocity. For practical purposes, an acceptable fit was considered a reconstruction of the stratigraphy.

This procedure is obviously tedious and time consuming, and at the time the study began, remote interactive computer terminals were not available. However, late in the study a computer display terminal became operational. Use of the display terminal enables the explorationist to generate a greater and more accurate number of synthetic traces in a shorter time.

Because of the rapid turn around time from input, to screen display, to hard copy duplication, several possible geologic alternatives can be tested and compared with the field data. Needless to say, many more shot points could be modeled in much greater detail with much less time and calculation involved. Therefore, more and better fits were achieved which resulted in a more detailed stratigraphic interpretation.

The second model constructed at the Mustang 1 Thunder location was much less difficult than the first because this procedure involved assigning interpolated interval velocities to sandstone, shale, siltstone, and limestone units on the density log. The resulting synthetic traces compared favorably with the corresponding SAILE® trace at shot point 20 (Fig. 13).

Using the stratigraphic parameters derived from the models at shot point 20, several traces south (to the left in Figure 12B) of the Mustang 1 Thunder were modeled. Results indicate that the basal Springer sandstone thickens from 30 ft (9.1 m) near shot point 20 to nearly 50 ft (15.2 m) at shot point 24, then thins gradually to less than 10 ft (3 m) beyond shot point 31.

Line 0-386

The procedures used in modeling Line 0-383 were repeated on Line 0-386, a 12-fold dynamite line running NE-SW across Section 19. The line was shot over a proposed location in the NE

CONOCO
NO.1 BRISCOE

MUSTANG
NO. 1 THUNDER

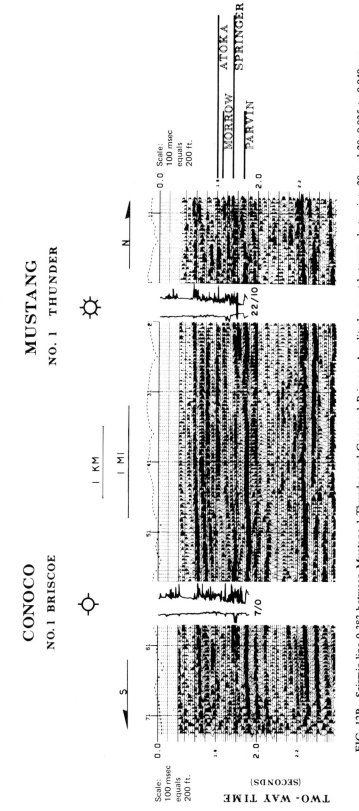

FIG. 12B—Seismic line 0-383 between Mustang 1 Thunder and Conoco 1 Briscoe. Amplitude event between shot points 20 and 28, 0.035 to 0.040 sec above Parvin, corresponds to basal Springer sandstone (in black) on digitized E-log in Thunder well.

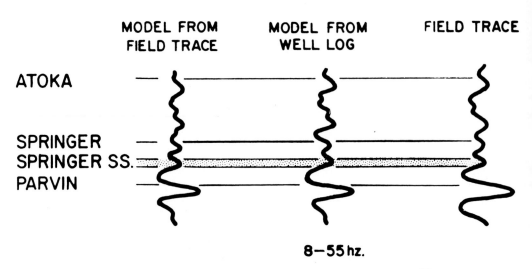

FIG. 13—Synthetic seismograms generated for Mustang 1 Thunder from models constructed, using field trace, and models constructed using data from density and electric logs at well.

quarter of Section 19. An interpretation of sub-surface well data indicated that the basal sandstone, productive in the Mustang 1 Thunder, passed through the proposed location. The well, the Conoco 1 Cruse, was spudded before full interpretation of Line 0-386 was made.

As shown in Figure 14, a strong amplitude event occurs 0.030 to 0.035 sec above the Parvin limestone. Several shot points were modeled using velocity data from the Conoco 1 Leck in Section 21 and the Conoco 1 Briscoe in Section 30. The two modeled shot points in Figure 14, shot points 9.5 and 11.5, indicated that the basal Springer sandstone, in excess of 30 ft (9.1 m) to the NE, thinned to less than 10 ft (3.0 m) at shot point 13, the Cruse location.

When the well penetrated the basal Springer section, it encountered 6 ft (1.8 m) of tight sandstone. The model trace generated at shot point 13 matched both the synthetic seismogram from the sonic log recorded in the Cruse well, and the field trace at shot point 13 (Fig. 15).

MAPPING

Geoseismic modeling used on Lines 0-383 and 0-386 demonstrated that the technique could define the lateral limits of the basal Springer sandstone and was qualitative in defining the thickness of the sandstone at any one shot point. As modeling progressed and more lines were examined and modeled, it was found that the threshold gross thickness of the basal sandstone which could be detected with confidence was between 15 and 25 ft (4.6 and 7.6 m). Below that value the sensitivity of the seismic tool was considered

questionable. Several more lines displayed strong amplitude responses in the basal Springer zone and were modeled. With increased experience using the technique, it was found that modeling proceeded more quickly where a synthetic seismogram from the nearest well was used as a preliminary stratigraphic model. Altering interval thicknesses and modifying interval velocities within the preliminary model could usually produce good matches between model and field trace quickly.

Following the interpretation and modeling of the recorded data, an amplitude-anomaly distribution map was constructed (Fig. 16A). Areas within the outline were modeled using thickness values greater than 20 to 25 ft (6.1 to 7.6 m) of basal Springer sandstone. As stated previously, the modeling technique can not distinguish porous sandstone from tight sandstone due to the thicknesses and velocities involved. As a result, the amplitude anomaly map was used as a base for mapping porosity trends from which drill sites were selected.

The gross sandstone isopach in Figure 16B was constructed using the well control to date, and therefore post-dates the amplitude anomaly map. It can be observed that the original amplitude distribution pattern in Figure 16A is fairly consistent with the 20 ft (6.1 m) gross sandstone contour in Figure 16A. The sandstone map reveals a network of interconnected sandstone trends oriented in both dip and strike. The thicker sandstone in the NW quarter of the township is interpreted as the southern end of the "Old Woman Channel" mentioned earlier. A second distributary channel

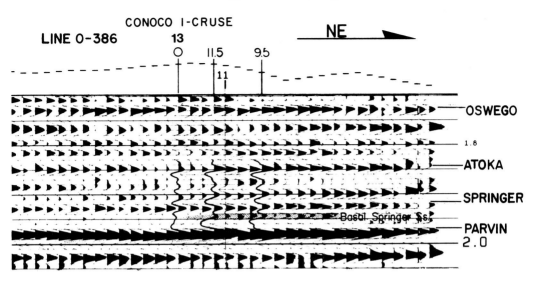

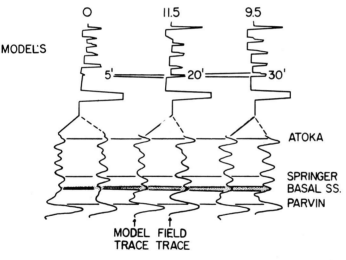

FIG. 14—Part of line 0-386 running through Conoco 1 Cruse location showing location (0) and modeled shot points 9.5 and 11.5. Velocity versus depth models constructed at each location are shown below seismic section along with respective model trace and field trace.

flowed southwestward across T13N, R10W apparently intermingling to some extent with sandstones being deposited by the "Old Woman Channel."

The presence of glauconite in many of the samples, especially those in the SW quarter, indicates that marine influence was strong and probably dominated the depositional environment at different times causing several sandstone bodies to be redistributed along the strike. This condition occurred during temporary advances and stillstands of the shoreline when the amount of mate-

rial being delivered by the distributaries could not equal or exceed the rate of subsidence. Areas in the south-central and southeastern part of the map with thin sandstones or no sandstones reflect a slight topographic high that resulted from deeper structure which existed at the time of deposition.

DEVELOPMENT DRILLING

To date, three wells have been drilled at locations recommended on the basis of sandstone distribution maps constructed using the geoseismic

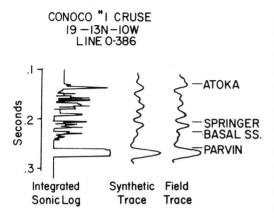

CONOCO #1 CRUSE
19−13N−10W
LINE 0-386

Integrated
Sonic Log

Synthetic Field
Trace   Trace

—ATOKA

—SPRINGER
—BASAL SS.
—PARVIN

FIG. 15—Integrated sonic log from Cruse well with accompanying synthetic seismogram generated from sonic data. Field trace at shotpoint 13, Cruse location, shown at right.

modeling technique. Two were successful completions, one a dry hole.

Development-well spacing for gas wells in the Morrow–Springer is one well per 640 acre (259 ha.), with each well being located no closer than 1,320 ft (402 m) from the section line. However, it is unlikely that any one reservoir in the Morrow–Springer in the Geary area covers an entire section or that any one well drains a full 640 acres (259 ha.). New wells have been drilled to the Morrow–Springer and have established production in sections already containing Morrow–Springer dry holes. The locations chosen on the basis of modeling were selected to be as near as possible to shot points modeled with thick basal sandstone, and still remain within the legal spacing requirements in the sections to be developed. The first two, one in Section 33 and the other in Section 27, were located as close as possible to modeled amplitude events on Line 0-390, although neither well was optimally located. The modeling indicated that the basal sandstone thickened from less than 20 ft (6.1 m) at shot point 1, to 30 ft (9.1 m) between shot points 15 and 21, and thinned again to less than 20 ft (6.1 m) at shot point 25 (Fig. 16A).

The first well drilled, the Conoco 1-33 Leck was located at approximately shot point 4 on Line 0-390 in the SW quarter of Section 33. The location was chosen on the basis of evidence from modeling indicating the presence of basal sandstone and proximity to basal Springer production in a well in the SE quarter of Section 32. Modeling indicated 20 ft (6.1 m) or less of basal Spring-

er sandstone at the location. The well encountered 14 ft (4.3 m) with 5 ft (1.5 m) of section having 12 to 15% porosity (Fig. 16B). The average interval velocity through the basal sandstone as calculated from the sonic log is 16,900 ft (5,151 m) per sec, higher than expected, with an average density of 2.58 g/cc from the density log. Samples from the well cuttings indicate that the sandstone is well cemented, subrounded to subangular, quartzitic, clean, and slightly glauconitic. Surrounding shales have average interval velocities of 10,200 ft (3,109 m) per sec and average densities of 2.47 g/cc resulting in a reflection coefficient of 0.268. The well was completed in the basal Springer for an initial production of 1.1 MMcf (0.031 million cu m) and 9 bbl (56.6 kl) of condensate per day.

The second well, the Conoco 1-27 Rukes was drilled near shot point 26 on Line 0-390 in the SW quarter of Section 27 in an attempt to locate as near as possible to the thick basal sandstone, indicated by modeling in the NE quarter of Section 33. The amplitude event present at this location was not as strong as that observed to the south and could only be duplicated using values less than 20 ft (6.1 m) of gross sandstone. The Rukes well penetrated 16 ft (4.9 m) of basal Springer sandstone with 4 ft (1.2 m) of section having up to 9% porosity. Well cuttings show the sand grains to be subangular to subrounded, glassy, well cemented with traces of glauconite, and very similar to that in the Leck 1-33. No sonic log was run in this well, but because of the similar lithologic description and similar density of 2.57 g/cc, it is assumed that the average interval velocity through the sandstone was close to 17,000 ft (5,182 m) per sec. The basal sandstone was not tested in the Rukes well. Instead a completion was made in a porous lower Morrow sandstone for a calculated open flow of 6.1 MMCFGPD (0.173 million cu m).

Neither of the wells drilled to this point was in an optimum location to test a well-defined, strong-amplitude event which had been modeled as a thick basal Springer sandstone. The opportunity arose when a well in the NE quarter of Section 20, the Conoco 1 Huff was plugged due to declining production. Plugging the well permitted a redrill of the section.

Section 20 is crossed in a NE to SW direction by Line 0-387 which displays a strong, well-developed amplitude event in the basal Springer zone between shot points 34 and 41, a distance of some 2,300 ft (701 m; Fig. 17). The amplitude event occurs between 0.035 and 0.040 sec above the Parvin reflection, and was modeled at several shot points. The Conoco 1 Huff was used as a preliminary stratigraphic model because it was the closest well control to the anomaly. Because no sonic log had been run in the well, velocities were interpolated using the sonic log from the Conoco 1 Leck in Section 21. In the basal Springer zone, the Huff had encountered 7 ft (2.1 m) of tight, very fine-grained, hard sandstone. Modeling of the line indicated that southwest of the Huff well the basal sandstone thickened from 7 ft (2.1 m) to approximately 50 ft (15.2 m) at shot point 38, then thinned again to less than 10 ft (3 m) at shot point 33.

A drillsite was located at shot point 38 in the SW quarter of Section 20. The well, the Conoco 1 Milligan, was drilled to a total depth of 11,660 ft (3,554 m) in the Parvin limestone. Instead of encountering the predicted 50 ft (15.2 m) of basal Springer sandstone, the Milligan encountered 18 ft (5.5 m) of very fine-grained, tightly cemented, slightly glauconitic basal sandstone. Alone, this sandstone thickness would have been predicted by the modeling, but the stratigraphic circumstances were complicated by the development of two thin (6 and 7 ft, or 1.8 and 2.1 m), high velocity (16,100 ft/sec or 4,907 m/sec; and 16,800 ft/sec or 5,121 m/sec), calcareous sandstones within the section immediately below the basal sandstone. Reflections from these units reinforced reflections from the thin basal sandstone and created a strong combined amplitude response. Constructive interference or constructive addition (Sengbush et al, 1961) within evenly spaced thin beds can produce the same results as achieved by bed "tuning" in thicker units. As shown in Figure 18, a 56 ft (17.1 m) zone of evenly spaced, thin, high velocity layers can generate a seismic signature very similar to a 50 ft (15.2 m) zone of 15,500 ft/sec (4,724 m/sec) sandstone.

The recoverable high-end frequencies in the field data recorded in the study area reach a maximum (not greater than 70 Hz) which would be insufficient to define a stratigraphic situation as occurs in the Milligan well. The frequencies necessary to distinguish sandstones as thin as 15 ft (4.6 m) would be in the order of 422 Hz, which must be considered beyond present seismic capabilities considering the depth and surface conditions involved.

The results obtained from the Milligan example clearly illustrate the consequences one faces when applying stratigraphic modeling techniques to a complex depositional environment in which several stratigraphic alternatives may exist and be recorded as very similar seismic events on band-limited data. If circumstances such as these occur within an area to be modeled, then, without additional distinguishing geologic or geophysical criteria, some question must remain as to the validity of the interpretation. Detailed facies identification and mapping, or additional diagenetic criteria, may be helpful in isolating productive from nonproductive areas before modeling. Also, broader band width seismic data could possibly be recovered to provide better thin-bed resolution to help distinguish prospective from nonprospective areas.

It follows, alternatively, that the greater the number of stratigraphic possibilities which can be diagnosed seismically, using modeling techniques, the higher the degree of confidence that can be placed on a geoseismic interpretation.

SUMMARY AND CONCLUSIONS

The following statements and conclusions have been listed to briefly summarize the results obtained with the geoseismic modeling program in the Geary area:

1. The Geary area encompasses T13N, R10W and is along the NE flank of the Anadarko basin within the updip limits of the stratigraphically complex, overpressured Lower Pennsylvanian Morrow–Springer sequence. The section, composed of alternating fine-grained sandstones, siltstones, and shales, forms a regional gas-condensate producing stratigraphic trap.

2. Sandstones within this sequence display great diversity in orientation and origin, leading to the conclusion that deposition took place under deltaic to transitional shallow-marine conditions.

3. In the Geary area, the basal Springer sandstone is a prolific producer and is the prime exploration target in T13N, R10W. Conventional subsurface mapping techniques, however, were unable to successfully predict the occurrence of the basal sandstone in wells drilled, offsetting production.

4. Although the Morrow–Springer section is stratigraphically complex, significant velocity contrasts between low velocity shales (9,800 to 11,500 ft/sec, or 2,987 to 3,505 m/sec) and high velocity (14,000 to 16,500 ft/sec, or 4,267 to 5,029 m/sec) sandstones create reflection coefficients between 0.1 and 0.3. This makes the section, especially the basal sandstone, amenable to geoseismic modeling.

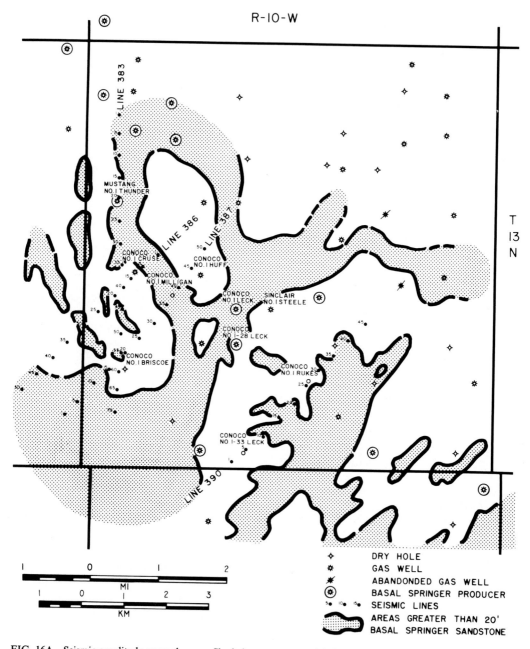

FIG. 16A—Seismic amplitude anomaly map. Shaded areas were modeled using thickness values—20 ft (6.1 m) for basal Springer sandstone.

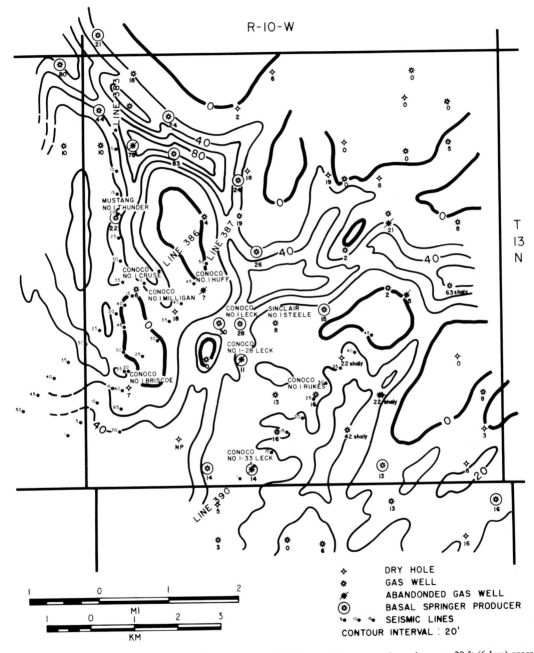

FIG. 16B—Present basal Springer gross sandstone isopach. Note good correspondence between 20 ft (6.1 m) gross sandstone contour and original 20 ft (6.1 m) amplitude anomaly outline.

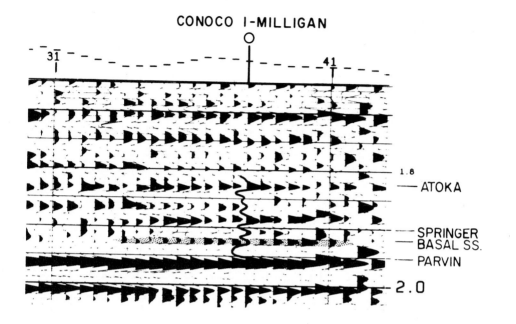

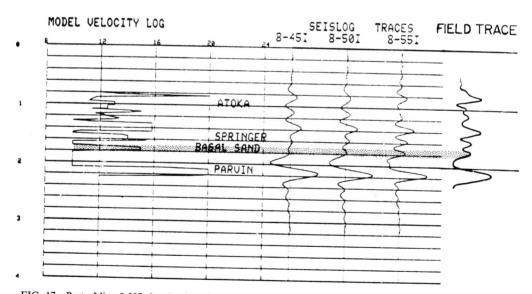

FIG. 17—Part of line 0-387 showing location of Conoco 1 Milligan and amplitude anomaly, model velocity log constructed at Milligan location using 50 ft (15.2 m) of basal sandstone and resulting synthetic seismograms run at various band widths.

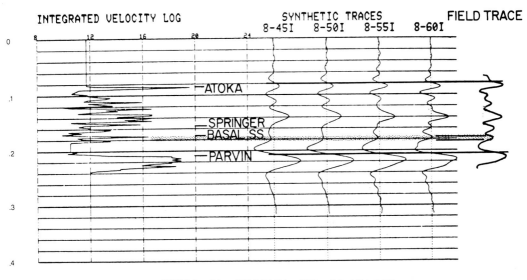

FIG. 18—Integrated sonic log from Conoco 1 Milligan and resulting SAILE® synthetic seismograms generated from it.

5. Geoseismic modeling techniques, together with SAILE® processed field data and geologic/lithologic input parameters utilizing well data and sample data, were combined in an effort to correctly predict the distribution and thickness of the basal Springer sandstone in the study area to aid exploitation drilling.

6. Velocity-density models were constructed using known geologic parameters to generate synthetic seismograms and demonstrate results which should be expected on the field data. These results were substantiated by synthetic seismograms derived from sonic and density logs within the study area.

7. Modeling indicated that when present in thicknesses of more than 20 ft (6.1 m) the basal Springer sandstone could be identified on 8 to 55-Hz SAILE® processed field data by a strong amplitude event within the low velocity shale section above the Parvin limestone.

8. A strong amplitude event is present on several dynamite source and Vibroseis® lines which were recorded, broadband, on a grid over the study area. The amplitude events are coincident with wells encountering significant thicknesses of basal Springer sandstone. Synthetic seismograms generated from sonic and density logs from these wells duplicate field data and seismograms generated from velocity models interpreted from the field record.

9. Geoseismic modeling techniques were applied to obtain semi-quantitative thickness values for the basal Springer sandstone at shot points on lines that display the amplitude anomaly. These values were combined with thickness values from the available well data and used to construct a gross sandstone isopach of the basal sandstone in the study area.

10. The Conoco 1 Cruse was drilled at a location on Line 0-386 where no amplitude event is present. Models constructed from the line indicated that the basal sandstone thinned to less than 10 ft (3 m) at the Cruse location. The well encountered 6 ft (1.8 m) of tight sandstone, as predicted by the models.

11. Three wells were drilled on the basis of geoseismic modeling. Two encountered thin basal sandstones as predicted; the third, the Conoco 1 Milligan, was drilled on a well-developed amplitude anomaly modeled as representing a maximum of 50 ft (15.2 m) of basal Springer sandstone. The Milligan did not penetrate

the predicted thickness of basal Springer sandstone but instead encountered a stratigraphic configuration which was not expected, but nevertheless produced the same seismic signature as the model.

12. The stratigraphic configuration that occurs in the Conoco 1 Milligan cannot be distinguished from that which was modeled using conventional methods. Alternative stratigraphic solutions to seismic events will have to be considered in modeling any complex depositional environment.

REFERENCES CITED

Adams, W. L., 1964, Diagenetic aspects of lower Morrowan, Pennsylvanian, sandstones, northwestern Oklahoma: AAPG Bull., vol. 48, no. 9, p. 1568-1580.

Benton, J. W., 1972, Subsurface stratigraphic analysis, Morrow (Pennsylvanian), north-central Texas County, Oklahoma: Shale Shaker, vols. 21-23, 1970-1973, p. 1-29.

Breeze, A. F., 1971, Abnormal-subnormal pressure relationships in the Morrow sands of northwestern Oklahoma: Shale Shaker, vols. 21-23, 1970-1973, p. 45-66.

Busch, D. A., 1974, Stratigraphic traps in sandstones—exploration techniques: AAPG, Memoir 21, 174 p.

Davis, H. G., 1971, the Morrow–Springer trend, Anadarko basin, target for the 70s: Shale Shaker, vols. 21-23, 1970-1973, p. 83-93.

——— 1974, High pressure Morrow-Springer gas trend, Blaine and Canadian counties, Oklahoma: Shale Shaker, vol. 24, no. 6, p. 104-118.

Dobrin, M. B., 1960, Introduction to geophysical prospecting: New York, McGraw-Hill, 446 p.

Dwight's Natural Gas Well Production Histories, 1976, Gas graph report 7, northwestern Oklahoma: v. 1, Dallas, Dwight's.

Fay, R. O., W. E. Ham, J. T. Bado, and L. Jordan, 1962, Geology and mineral resources, Blaine County, Oklahoma: Oklahoma Geol. Survey Bull. 89, 258 p.

Forgotson, J. M., Jr., A. T. Statler, and M. David, 1966, Influence of regional tectonics and local structure on deposition of Morrow Formation in western Anadarko basin: AAPG Bull., v. 50, no. 3, p. 518-532.

——— 1969, Factors controlling occurrence of Morrow sandstones and their relation to production in the Anadarko basin: Shale Shaker, vols. 17-20, 1967-1970, p. 135-150.

Jacobsen, L., 1959, Sedimentation of some Springer sandstone (Mississippian–Pennsylvanian) reservoirs, southern Oklahoma: AAPG Bull., v. 43, no. 11, p. 2575-2591.

Khaiwka, M. H., 1973, Geometry and depositional environment of Morrow reservoir sandstones, northwestern Oklahoma: Shale Shaker, vols. 21-23, 1970-1973, p. 170-193.

Mannhard, G. W., and D. A. Busch, 1974, Stratigraphic trap accumulation in southwestern Kansas and northwestern Oklahoma: AAPG Bull., v. 58, no. 3, p. 447-463.

Pate, J. D., 1959, Stratigraphic traps along north shelf of Anadarko basin, Oklahoma: AAPG Bull., v. 43, no. 1, p. 39-59.

Petroleum Information Corp., 1976, Vance Rowe Oklahoma oil reports, northern Oklahoma and panhandle-northwestern: Oklahoma City, May, 1976.

Potter, P. E., 1967, Sand bodies and sedimentary environments—a review: AAPG Bull., v. 51, no. 3, p. 337-365.

Sengbush, R. L., P. L. Lawrence, and F. J. McDonal, 1961, Interpretation of synthetic seismograms: Geophysics, v. 26, no. 2, p. 138-157.

Swanson, D. C., and R. R. West, 1968, Anomalous Morrowan–Chesterian correlations in western Anadarko basin (Abstract): AAPG Bull., v. 52, no. 3, p. 551.

Takken, S., 1967, Subsurface geology of the north Gotebo area, Oklahoma: Shale Shaker vols. 15-17, 1964-1967, p. 330-338.

Practical Stratigraphic Modeling and Interpretation[1]

M. W. SCHRAMM, JR., E. V. DEDMAN, and J. P. LINDSEY[2]

Abstract Advances in the past 5 years in geologic modeling and seismic processing now make it possible to perform stratigraphic interpretation to a degree not previously possible. These techniques provide the geologist and geophysicist with an ability to correlate well data with seismic data to a higher degree of reliability, particularly in stratigraphic situations that were previously difficult to interpret. The ability to detect lateral facies changes, vertical lithology transitions, reefs, channel sands, barrier bars, pinchouts, and gas/liquid contacts is greatly enhanced.

These techniques have been used successfully and practically to aid exploration in diverse geologic provinces; and these methods have been used not only in an exploration context, but also in exploitation situations. In exploitation, methods are continually being evolved which incorporate the use of seismic, geologic, and engineering data for more refined estimates of oil and gas in place, as well as recoverable reserves.

INTRODUCTION

Recent technologic advances in both geology and geophysics as applied to the search for stratigraphic traps have led to a relatively new and very practical interpretive modeling technique which enables the geologist and geophysicist (1) to resolve stratigraphic problems interactively, (2) to predict lithologies and types of traps, (3) to predict, in some instances, the occurrence of hydrocarbons, and (4) under special circumstances, to quantify hydrocarbon accumulations.

A model is a concept from which one can deduce effects which can be compared to observations, then used in understanding the significance of the observations. A model may be physical, mathematical, or, as it is generally used in geology, conceptual. Models may range from simple examples (such as a "delta model" or a "basin model" which can be studied and projected along with one's thoughts while exploring for a delta or evaluating a basin) to extremely complex reservoir, economic, or geoseismic models which are based on mathematics and require the use of sophisticated computer programs.

Geoseismic modeling is essentially a computational procedure which simulates the seismic response that would be generated from an assumed geologic situation or subsurface model. This subsurface model is normally defined in terms of interface geometry, interval properties (such as velocity, density, and attenuation), and other acoustically significant parameters. Further, the subsurface definition should incorporate all pertinent geological and geophysical information and

assumptions. Specifically, stratigraphic modeling can be defined as being related to the interpretive problems in which the identification of lithologic changes is the primary objective; that is, the hydrocarbon trapping mechanism is primarily stratigraphic.

Stratigraphic modeling probably should be dated from the early 1950s, when the synthetic seismogram was first introduced. With the advent of the sonic logging tool, a key acoustic property of rocks could be measured in some detail, and the relations between the rock sequence and the corresponding seismic data could be studied. The computation of the synthetic seismic response was a one-dimensional stratigraphic model. At first, only primary reflections were considered; later, interbed multiples were included, and then the effects of attenuation. Today, it would seem that the sophistication of the synthetic seismogram calculation leaves little more to consider. So, what gives us hope that stratigraphic modeling can be expanded to become a more useful exploration tool than the synthetic seismogram?

Two new observations are the basis for reviewing stratigraphic modeling: (1) the observed correlation between gas content in offshore Gulf of Mexico subsurface sandstones and seismic-event amplitudes, and (2) the ability to measure and use a basic propagating seismic wavelet to interpretive advantage. The first observation is more generally known as the "bright spot" phenomenon. From its introduction to the exploration community several years ago, the so-called bright-spot technology progressed from merely recognizing "bright spots" on seismic records to mapping their areal extent and calculating the thicknesses and vertical distribution of the sandstones that they represent. More correctly, so-called bright spots are amplitude anomalies which may or may not be significant in terms of occurrence of hydrocarbons.

The greatest initial use of stratigraphic modeling was in connection with Gulf Coast sandstone reservoir problems. The reason it works well and is useful there, is illustrated by the log in Figure 1, which represents a sandstone body nearly filled

[1]Manuscript received, September 1, 1976; revised and accepted, December 10, 1976.

[2]Consultants, GeoQuest International, Ltd., Houston, Texas 77027.

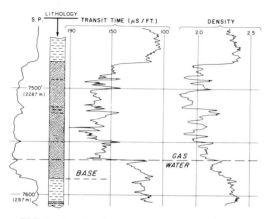

FIG. 1—Log showing acoustic parameters for a gas sandstone.

with gas. The velocity-log response clearly marks the gas-water contact. The contact of the shale and the gas-bearing sandstone is also clearly seen, as both velocity and density change. The boundaries produce a 22% reflection coefficient, and cause a strong seismic response. The important aspect of this log is that the acoustic anomaly it demonstrates is related to thickness of the productive zone.

VERTICAL LITHOLOGY AND SEISMIC RESPONSE

The fundamental concepts of stratigraphic modeling are rooted in the basic relation between the lithology and its seismic response. In simplest form, a seismic wavelet moves as an acoustic wave through layers of dissimilar rock, and each interface produces a reflected replica of the incident wave. If the acoustic impedance decreases at the interface, the reflected basic wavelet is negative to the incident wavelet; whereas an increase in impedance produces the positive reflected wavelet. The magnitudes of these reflections are a function of the acoustic impedance contrast. Large changes produce large reflections.

Figure 2 illustrates seismic responses to a subsurface lithologic sequence. In Figure 2A a marine air-gun signature is used as the propagating waveform, and in Figure 2B a zero-phase bandpass wavelet is used. Traces 1 through 5 represent the individual responses of each acoustic interface to the respective wavelets, and trace 6 in each of the two figures, 2A and 2B, represents the total response of all the acoustic interfaces.

If stratigraphic interpretations are to be made of seismic data, Figures 2A and 2B illustrate in a simplified form what must be known—the basic propagating wavelet. The correct description of the propagating wavelet for a specified suite of seismic data can be used to generate models of vertical lithologic sequences. Comparing these models with actual seismic data will then permit stratigraphic interpretation of the modeled data.

A critical problem exists, however, in stratigraphic modeling, and it becomes readily apparent when a direct comparison is made between the total-response traces of Figures 2A and 2B. The marine-airgun wavelet produces a very complex response not easily synthesized. When an attempt is made, using the total-response trace of Figure 2A, to identify major acoustic interfaces with corresponding positive or negative peaks on the modeled trace, a time lag occurs, causing the expected response characteristics to be poorly defined. The zero-phase band-pass wavelet, in contrast, shows no time lag in comparison to the acoustic interface. Its response characteristics are more easily recognized.

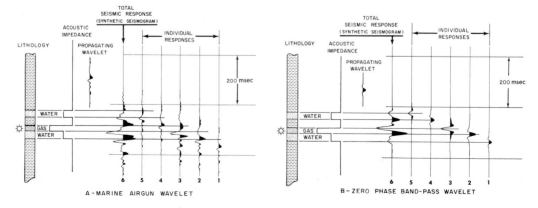

FIG. 2—Relation between lithology, fluid content, and seismic response.

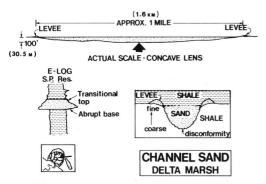

FIG. 3—Geologic model for a channel sandstone.

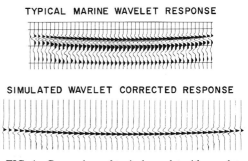

FIG. 4—Comparison of typical wavelet with wavelet-corrected seismic response to geologic model of channel sandstone in Figure 3.

WAVELET CORRECTION

A marine basic wavelet is a complex time series of perhaps 0.25 to 0.30 sec in length. It is composed of an energy-source signature, ghosts, and cable and instrumentation responses. This basic wavelet can be extracted from seismic data or synthesized from an accurate knowledge of the various response characteristics mentioned above[3] Once the basic wavelet shape is known, a mathematical operator can be designed to correct it to a shorter (0.06 to 0.07 sec) zero-phase wavelet with the same frequency bandwidth as the original wavelet. This filtering process accomplishes the data transformation from a complex to a simple propagating wavelet (shown in Fig. 2), thus increasing the resolution and making stratigraphic modeling more definitive.

SCHEMATIC STRATIGRAPHIC MODELS

The following schematic model examples illustrate the tutorial and problem-solving use of stratigraphic modeling. They were conceived as representing a few of the more common stratigraphic deposits which exhibit somewhat different geometries. They were computed by using a modified wave-theory modeling approach which combines ray tracing with the integration of the total response from a subsurface interface at each receiving point.

Channel Sandstones

Figure 3 shows the geologic conditions related to formation of a channel sandstone. As indicated in the small inset map, such sands are characteristic of deltaic marsh areas related to the distribu-

taries of major rivers. The cross-sectional view of the sandstone is shown both exaggerated in vertical scale (box inset) and in actual scale. The E-log response indicates a sandstone top that is transitional from the overlying shale. This transition is the result of clay contamination or shale interbedding in the upper portion of the sandstone body. The base of the sandstone is relatively clean and provides an abrupt contact with the underlying shale.

Because of the transitional properties, the reflection from the sandstone top is weaker than that from the bottom. This is incorporated in the wave-theory solution shown in Figure 4. The response, using a "typical marine wavelet," is observed to be somewhat complex, although relatively uniform laterally. A time lag is observed because the principal return is from the base of the channel. It is difficult to make interpretive judgments about the sandstone thickness for this response, since the wavelet is so complex.

If the original basic wavelet had been corrected to a zero-phase band-pass equivalent, the channel-sandstone response would appear as indicated

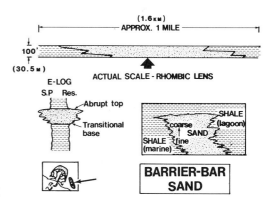

FIG. 5—Geologic model for a barrier-bar sandstone.

[3]A discussion of methods of wavelet extraction or estimation is beyond the scope of this paper. For a more detailed discussion of principles, refer to a companion paper in this *Memoir* by N. S. Neidell and E. Poggiagliolmi.

TYPICAL MARINE WAVELET RESPONSE

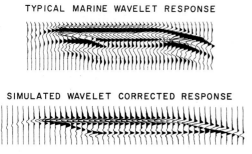

SIMULATED WAVELET CORRECTED RESPONSE

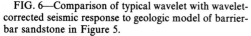

FIG. 6—Comparison of typical wavelet with wavelet-corrected seismic response to geologic model of barrier-bar sandstone in Figure 5.

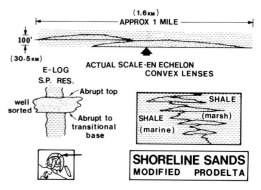

FIG. 7—Geologic model for shoreline sandstones.

in the alternate response of Figure 4. Precise knowledge of the waveform and sufficiently noise-free conditions permit the top of the sandstone, although acoustically weaker than the bottom by a factor or two, to be read and the sandstone thickness to be estimated accurately. Also, the body may be reviewed as a channel sandstone, as depicted on the geologic model.

Barrier-Bar Sandstone

Figure 5 is a geologic diagram of a barrier-bar sandstone found outside the deltaic environment, as indicated in the map inset. Such sandstone deposits separate lagoonal shale deposits from marine shale deposits, as indicated in the depth cross sections. The E-log shows an abrupt sandstone top where grains tend to be coarse, and a transitional base where sand grains are fine and interbedding with shale occurs.

This model response is shown in Figure 6 for both the typical marine wavelet and its band-pass equivalent symmetrical wavelet. The clarity of the band-pass wavelet response again needs wavelet-correction processing in order to qualify the data for stratigraphic interpretation. Note the difference in reflection strength for the top and bottom of the sandstone because of the transitional base.

Shoreline Sandstones

The geologic model shown in Figure 7 depicts a series of shoreline sandstones which were regressively deposited en echelon on the seaward rim of a delta complex. The vertical grain-size distribution is similar to that of the barrier sandstone bodies, and the sheetlike bodies may transitionally assume the convex lens shape of offshore bars. The contact of the sandstone bodies with overlying shale is distinctively abrupt, and the basal contact may be abrupt to transitional.

Figure 8 shows the comparative seismic responses using a typical marine wavelet and a cor-

rected wavelet. In this example, the offset in the upper response may lead to an incorrect interpretation of a fault, particularly if, as in a real case, the section is obscured by noise. The wavelet-corrected response, however, clearly shows the existence of two distinct sandstone bodies, as depicted on the geologic model. The sandstones are lenslike, discrete, en echelon bodies which pinch out in two directions. From an interpretive standpoint, a fault can be readily discounted. Significantly, the correct assessment of the degree of communication between the sandstones that are indicated on the seismic section as amplitude anomalies has important impact on exploration and exploitation economics.

Facies Change and Sandstone Content

Figure 9 shows a model designed to illustrate the general case of facies changes. Here, a porous sandstone unit varies laterally in shale content. The relative amount of sandstone, as well as its vertical distribution, has been modeled for lateral changes.

TYPICAL MARINE WAVELET RESPONSE

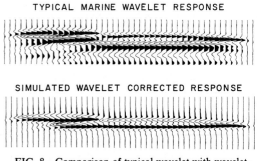

SIMULATED WAVELET CORRECTED RESPONSE

FIG. 8—Comparison of typical wavelet with wavelet-corrected seismic response to geologic model of shoreline sandstones in Figure 7.

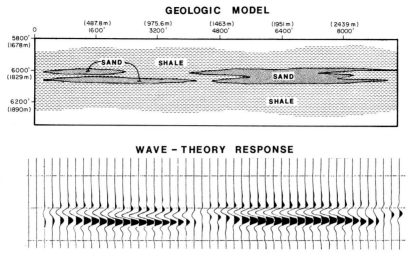

FIG 9.—Geologic model and wave-theory response for porous sandstone lenses in shale.

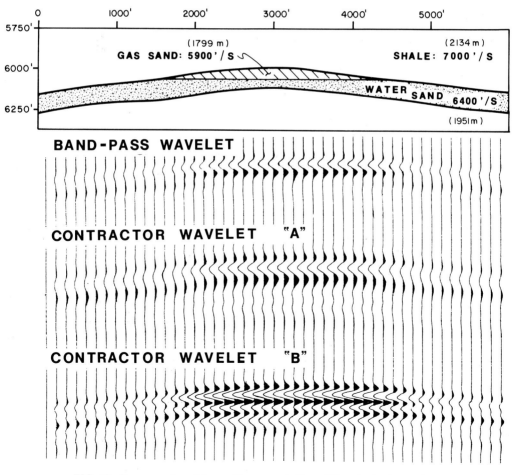

FIG. 10—Response of sandstone with gas cap to three different seismic wavelets.

Amplitude is the principal key to interpreting sand content. The weak seismic zone corresponds precisely to the 100% shale location—a permeability barrier in this case—and the strongest responses reflect the greatest sandstone content. To some extent, the vertical distribution of sandstone may be inferred from the seismic response. The major white deflection in the left portion of seismic response has marked time relief corresponding to the vertical position of the sandstone top. The farthest right responses, where sandstone is concentrated at the top and bottom of the vertical unit, also have waveform characters which differ from those seen elsewhere in the model.

Use of the model response would provide a good measure of sandstone thickness. It is not generally possible to determine the number of closely spaced, thin sandstones as in this model, since the seismic bandwidth does not permit such definition.

Water Sandstone with a Gas Cap

Figure 10 illustrates a model derived from an actual case history. This is a sandstone of rela-

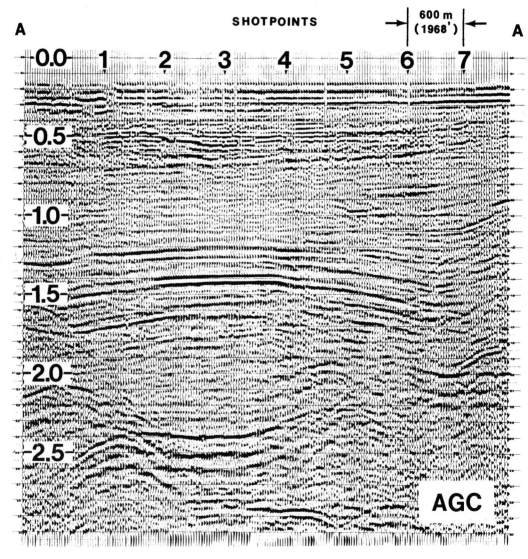

FIG. 11—Standard AGC seismic section showing indications of an amplitude anomaly (bright spot) between shotpoints 2 and 4 at about 1.4 sec.

tively uniform thickness and mild structural closure. The upper 60 ft (18.3 m) of this 120-ft (36.6 m) sandstone is gas saturated. The seismic data exhibited a bright spot about equal to the amplitude relief shown in the figure. This model was originally computed with theoretically derived rock velocities and densities for the sandsone and shale values from well-log calibration data. An amplitude relief of only 1.25:1 was obtained from these values, and the model was altered to have the values shown in Figure 10. These values were derived by assigning a 4% reflection coefficient to

the shale–water-sandstone interface and adjusting the gas-sandstone velocity to provide the amplitude relief shown by seismic data. Densities were included in the reflection calculation and generally follow the equation of Gardner et al (1974).

First, reference is made to the model responses shown for two different, but documented, marine wavelets. These might be thought of as coming from two different contractors. It is not our point to choose between these two wavelets. It is apparent that the seismic sections look quite different, and, if they were members of a grid of data on the

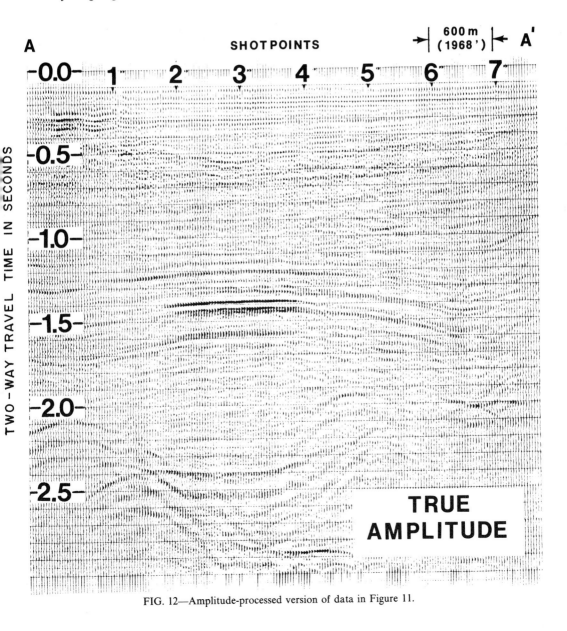

FIG. 12—Amplitude-processed version of data in Figure 11.

same prospect, it would be difficult to "time-tie" the data at their intersections.

The second point is one already made: it is generally unreliable to attempt only to pick the top and bottom of the gas sandstone for thickness estimates without using relative-amplitude information. Without this information, however, it is still possible to detect the probable presence of the sandstone and to outline it on a map. Marine wavelet "A" and "B" have a bandwidth approximately equal to the 8 to 32-Hz, band-pass wavelet response shown in Figure 10. Thus, each of the other two could have been converted to this latter model response with appropriate processing. No problems of tying the data between contractors would then exist.

Although the gas entrapment in this particular schematic model is structurally controlled, the effects of the gas-bearing part of the formation with regard to amplitude variation are fundamental to practical stratigraphic interpretation.

PRACTICAL APPLICATION

Block X, Offshore Texas

Let us study a simple case history of stratigraphic modeling and the interpretive lessons it

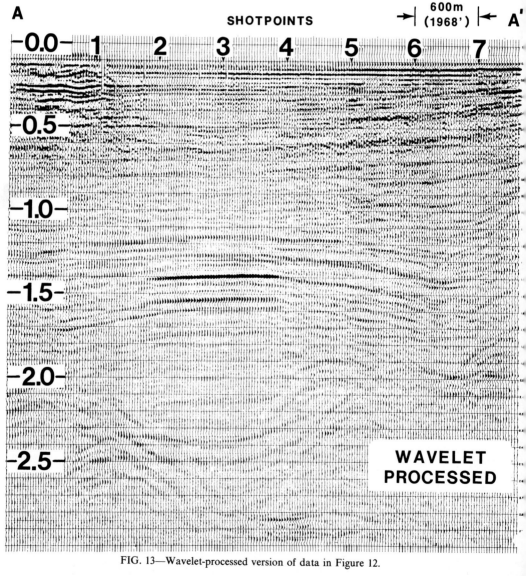

FIG. 13—Wavelet-processed version of data in Figure 12.

A   A'

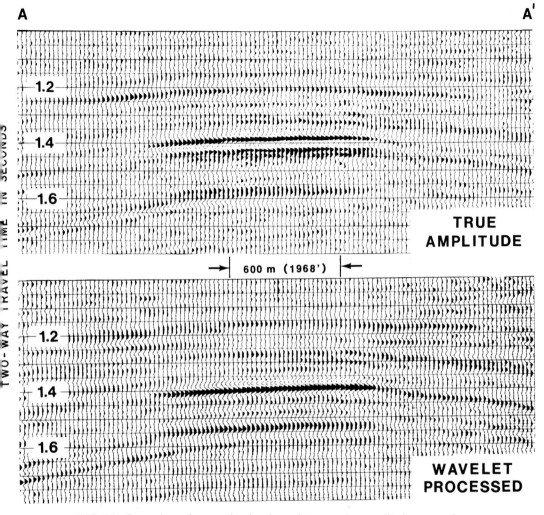

FIG. 14—Comparison of conventional and wavelet processing, amplitude preserved.

provides. Figure 11 is a standard seismic section[4] which gives the typical indications of an amplitude anomaly between shotpoints 2 and 4 at about 1.4 sec. The anomaly lies offshore Texas in the Gulf of Mexico, but the specific location cannot be divulged for proprietary reasons. Structural closure is evident, and we would expect to have a seismic section, processed to maintain true amplitudes, for stratigraphic analysis.

Figure 12 shows the amplitude version. The suspected zone is indeed bright, and it now becomes important to determine the thickness or thicknesses of gas-productive zones involved. Presumably the areal extent of the anomaly can be

mapped from the data grid available, of which this section is but one line. The problem is that the wavelet complex constituting the amplitude anomaly is as much a function of the basic wavelet as it is of the lithology. Consequently, it is necessary to produce a section processed to correct the wavelet shape to something more desirable; that is, a symmetrical band-pass wavelet having the same bandwidth as the original wavelet.

Figure 13 is the result of this processing on the true-amplitude section. The complex is simpler than in the true-amplitude version; this simplification, if we could study the data in detail, is a characteristic of all event complexes, bright or not. This is the principal badge of reliability for any wavelet-correction processing. It is a signifi-

[4]Sections in Figures 11-15 were provided by the courtesy of Teledyne Exploration, Inc.

cant test of having the correct basic wavelet and, consequently, the correct operator design. It constitutes a necessary but not sufficient condition of valid processing.

Figure 14 provides an amplified comparison of the anomalous amplitude for conventional and wavelet-processed data. The change in overall character of all the data elements is quite visible. Not only is the amplitude-anomaly complex obviously simplified, but so is a less-than-"bright" zone below at 1.55 sec.

Some may think, with some validity, that the major change that has taken place is merely a reversal in plot polarity. This particular amplitude anomaly does benefit in visual simplicity by reversing the polarity, but more than this has taken place. Although it is arbitrary to assign white deflections a negative reflection coefficient and black deflections a positive coefficient, thus giving most offshore Gulf Coast gas sandstones the general appearance of Figure 12, this generally does not produce a simplification of wavelet character. Situations occur, however, where merely reversing the plot polarity provides help in either appearance or stratigraphic interpretation.

Figure 15 is the same comparison as in Figure 14 except that both sections have automatic gain control. The purpose of this process is to make more visible the depth of wavelet character at locations other than the amplitude anomaly event. Using the assigned rule for reflection-coefficient polarity, low-acoustic-impedance layers that are

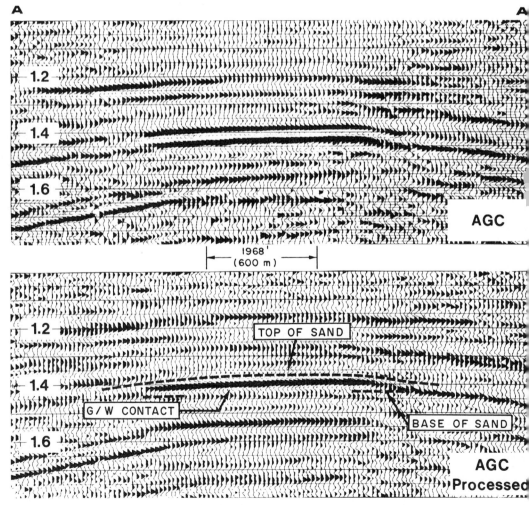

FIG. 15—Comparison of conventional and wavelet processing, AGC.

relatively thin may be interpreted. The response character to identify is an asymmetric sequence of two major lobes, the first deflecting to the left (white) and the second deflecting to the right (black). These zones turn out to be the porous sandstone of Pliocene-Pleistocene and late Miocene age, for which porosity in excess of 25% and sandstone quality suitable for commercial accumulations are inferred.

Sandstone thickness is determined from the timing between these major identifying wavelet lobes. Response characteristics for such sandstones are documented by modeling studies such as the sandstone-shale facies model of Figure 9. In this particular problem, two-way time thickness was converted to feet by using calibration curves for sandstone velocities derived from well logs measured in similar rock types. Reasonably good estimates are obtained if only shale velocities are known for the depths in question. Amplitudes were then measured for reflection-coefficient magnitude, and these in turn predicted sandstone velocities relative to those of shale.

Figure 15 strongly suggests that, with true amplitude and wavelet processing, coupled with stratigraphic modeling and the lessons it teaches, porous sandstones might be reliably identified on the AGC (automatic gain control) wavelet-processed section. This indeed is possible and provides a new tool in the mapping of sandstone/shale ratios in Tertiary clastic sections.

The AGC wavelet-processed section in Figure 15 shows a gas-bearing sandstone at the structurally high position of a sandstone unit extending down both flanks of the structure. The black lobe marking the base of the gas sandstone appears depressed in time in the water zone as compared to the gas-bearing portion. Close inspection would reveal that the white lobe marking the top of the sandstone is not time-disturbed at the edge of the gas-bearing section, thus showing the time thickness of the water sandstone to be greater than the time thickness of the gas sandstone. This is not a reasonable result for a uniformly thick sandstone fully loaded with gas if we assume gas-sandstone interval velocity to be lower than water-sandstone interval velocity. If Figure 1 is at all representative, this seems to be the case. These observations lead us to the conclusion that the sandstone is only gas saturated in the upper portion and a gas water contact is present at the black lobe of the gas-sandstone response. The alternative is to assume that the gas-saturated portion abruptly thins at the water contact downdip, which is unlikely. Figure 16 is a portion of an electric log at the position of the amplitude anomaly, which confirms interpretive predictions such

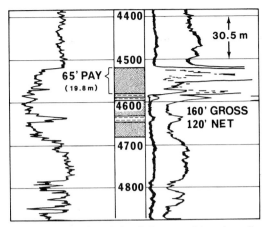

FIG. 16—Portion of electric log at position of amplitude anomaly in Figures 11 to 15, showing 65 ft (19.8m) of net gas saturation.

as gas-sandstone velocities, thicknesses, and portions saturated. The zone of interest has 120 net ft (36.6 m) of sandstone with the top 65 ft (19.8 m) gas-saturated. The two shale stringers in the sandstone are not visible on the seismic record, because they lie below the resolution afforded by the data bandwidth.

Viewing the entire field from a stratigraphic standpoint using wavelet-processed conventional (AGC) data as shown in Figure 17, it is possible to distinguish most porous sandstones whether or not they contain gas. Superimposed on this section is a lithologic-log overlay with the log depths converted to time through use of a suitable time-depth chart. The sandstones on the log do, in fact, correspond to the position and relative thickness of sandstones interpreted from the seismic section. In making such a comparison, a bulk time adjustment allows for differences in seismic and log measurement data as well as an "instrumentation lag" for the seismic record. Even local deviations for the time-depth curve are valid relaxations as long as they do not require local interval velocities to get out of reasonable bounds.

The results of this comparison are quite satisfying. Although thicknesses are not a perfect match in some places, and the number of individual sandstones do not always correspond, it is possible to detect major sandstone units and a good measure of gross total sandstone in the column before drilling. The use of this type of data, together with sound geologic judgment, permits the construction of maps showing the sandstone-shale ratio and net sandstone thickness of selected stratigraphic sequences in unexplored areas with little or no well control.

It is important to note the following points related to judging the direct match between properly processed seismic data and a corresponding log:

1. Seismic responses cannot show all beds seen on every log. The log is inspecting a column with a radius of a few feet, whereas the seismic record is responding to a Fresnel zone dimension— i.e., 300 to 1,200 ft (91.4 to 365.8 m).

2. Precise calibrations and lithology-to-response correlations are nearly impossible to accomplish using conventional seismic data.

3. The sandstones of commercial interest exhibited by seismic data are those with sufficient porosity, thickness, lateral dimensions, and continuity to constitute adequate hydrocarbon storage volume.

This case history is completed with the generation of two maps: the areal extent of the amplitude anomaly and the isolith map for the entire sandstone. Figure 18 shows the areal map of the amplitude anomaly, which in this case represents the gas-bearing portion of the sandstone. The northwest-southeast line (A-A′) crossing the

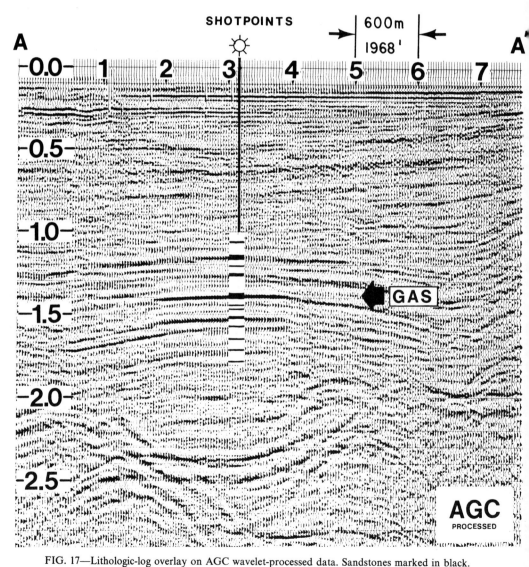

FIG. 17—Lithologic-log overlay on AGC wavelet-processed data. Sandstones marked in black.

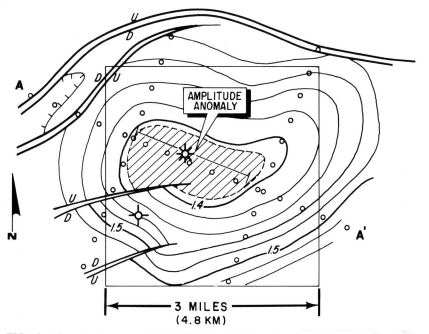

FIG. 18—Time structure map of Block X, Gulf of Mexico, showing position of well, amplitude anomaly, and extent of gas-productive zone. Line A-A'is same for section in Figure 11 to 17.

block was the seismic line used in Figures 11–15 and 17, and the gas-well location is also shown on this line. Neither intersecting line shows the amplitude anomaly. Plotting its extent on the structural map for that same horizon shows the likely areal shape. The anomaly is made to follow the contour level at the indicated gas/water contact. This produces a picture commensurate with the failure of the anomaly to appear on the intersecting lines and thus correlates well with structure. It should be pointed out again that this particular example represents a simple case of structural entrapment, though the procedures used to determine the position of sandstones and the presence of gas are those used in stratigraphic modeling and interpretation. Similar but more advanced techniques are used to decipher and delineate combined structural-stratigraphic and stratigraphic traps, and to quantify their hydrocarbon content. As a matter of fact, de-emphasis of and digression from thinking in purely structural terms, particularly in oil and gas exploitation and reserves engineering, should be encouraged.

Figure 19 shows the isolith map for the entire sandstone bed in which gas is locally trapped. Thicknesses were read from the AGC wavelet-

processed data away from the amplitude anomaly and posted on the base map for contouring. The areal extent of the anomaly (gas) was placed on this map for comparison with the thickness distribution of the sandstone. This was experimentally accomplished without prior knowledge of, or cali-

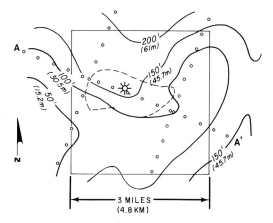

FIG. 19—Net-sandstone isolith map, Block X, Gulf of Mexico, showing position of gas well and outline of amplitude anomaly.

bration with, the thickness of the sandstone unit in the well. The well datum was added to the map after the map was prepared; it is readily observed that the approximate map thickness of the sandstone at the well site corresponds to the 160 gross ft (48.8 m) of sandstone measured on the log (Fig. 16). Consequently, it must be assumed that, in this case, what is being measured and mapped over the area from seismic data is gross, not net, sandstone thickness.

Notably, the gas-bearing portion of the sandstone does not relate to gross-thickness patterns. It can be interpreted that the axis of depositional thinning from less than 50 ft (15.2 m) west of the block to slightly less than 100 ft (30.5 m) on the east side of the block represents an early depositional structure that existed prior to development of the present structural crest where the gas was finally emplaced.

Effects on Seismic Response of Shale Distribution and Content in a Sandstone

No modeling was done in the development of the foregoing case history. This neither implies that modeling is not required nor that it is not of valid use. Actually, many of the assertions made about seismic data characteristics were derived

from modeling experiences previous to this episode. Also, we have not addressed the problem of precise prediction of reservoir volume and the variance in this calculation, an exercise which would require careful modeling.

In order to address the problem of stratigraphic interpretation of properly processed seismic data preparatory to quantitative (and thus economic) consideration, first consider and model the effects on seismic responses of shale distribution and content in a sandstone. The following two models are hypothetical, if not unrealistic, and are shown diagrammatically for illustrative purposes only.

An acoustically thin sandstone (less than one-quarter wavelength) gives a seismic response that varies in amplitude but not in shape or character with thickness. Consequently, the thickness of a thin sandstone can be measured from the amplitude of its seismic response if a suitable calibration can be established. The two models shown here illustrate how the seismic response is dependent upon a contaminant in the sandstone, such as shale. The vertical distribution of the shale and relative thickness are varied to observe the consequences on seismic response. It must be emphasized that in practical applications insights to lithologic distribution in thin units must be derived from geologic concepts and principles applicable

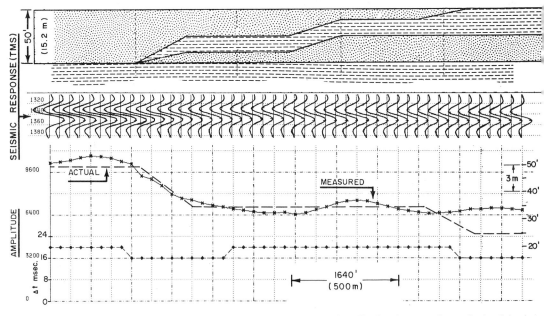

FIG. 20—Effects of a shale stringer in a thin sandstone on seismic reflection from sandstone body. Seismic-wavelet shape does not vary with shale thickness or distribution in the sandstone. Amplitude, however, is directly related to net amount of sandstone.

to the circumstances. They are not directly provided by the seismic data.

The first model (Fig. 20) has a 17-ft (5.2 m) shale stringer embedded in a 50-ft (15.2 m) sandstone and is changed in its vertical placement. The seismic response shows only an amplitude variation with a uniform waveshape. Trough-to-peak amplitude and time difference are measured and plotted using the computer, as shown at the bottom of the figure. The amplitude variation is seen to correlate very well with net sandstone thickness measured from the model. The trough-to-peak times remain constant if the sample rate is assumed to be 0.004 sec.

The second model (Fig. 21) was designed to vary both the thickness and the vertical position of the embedded shale. Once again the amplitude measure correlates with net sandstone thickness, whereas the timing measure is uniform and provides no information about thickness.

The conclusion drawn from these models is that, where a rock layer is thin, the presence of host-rock contaminants reduces the seismic response amplitude in proportion to the amount of contaminant. The amplitude measure then becomes proportional to the net rock thickness and, when properly calibrated, provides a degree of net-thickness resolution that cannot be obtained from a timing measurement. This has implications for using seismic reflection amplitude to measure the volume of reservoir rocks that are too thin to resolve by time-difference measurements.

Updip Permeability Barrier

Figure 22 shows a 50-ft-thick (15.2 m) gas-filled sandstone which loses porosity and permeability in the updip direction over a mild-relief structure. Closure is less than 200 ft (60.8 m) in about 3 mi (4.8 km), and the gas-bearing zone is not at the crest of the structure. Seismic response to this primarily stratigraphic situation shows a typical amplitude anomaly.

Seismic amplitude and time-difference plots were made with the computer as in Figures 20 and 21. Because the sandstone is thin, time difference has no change across the structure; but amplitude is related to the gas zone. The manner in which amplitude reduces downdip (as shown in the time model in the lower part of Fig. 22) indicates a gas/water contact, even though an expected "flat spot" is not visible because the sandstone is too thin. Updip amplitude behavior is more characteristic of loss of porosity and/or permeability. This behavior might also indicate the pinchout of the sandstone on the flank if the extent of the sandstone was not known.

Correspondence between net thickness of the gas sandstone and seismic amplitude is shown in the top panel of the figure. Migration of reflections would make the amplitude measurements coincident horizontally with the actual thickness

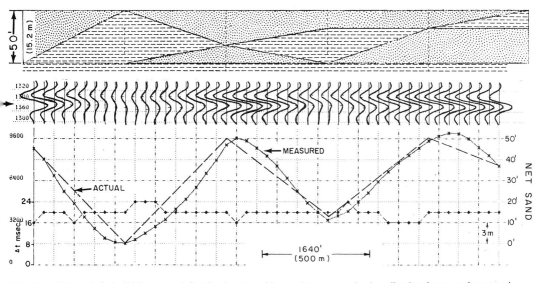

FIG. 21—Effect of shale thickness and distribution in a thin sandstone on seismic reflection from sandstone unit. Seismic-wave shape is invariant, but seismic applitude is directly correlated with net sandstone thickness.

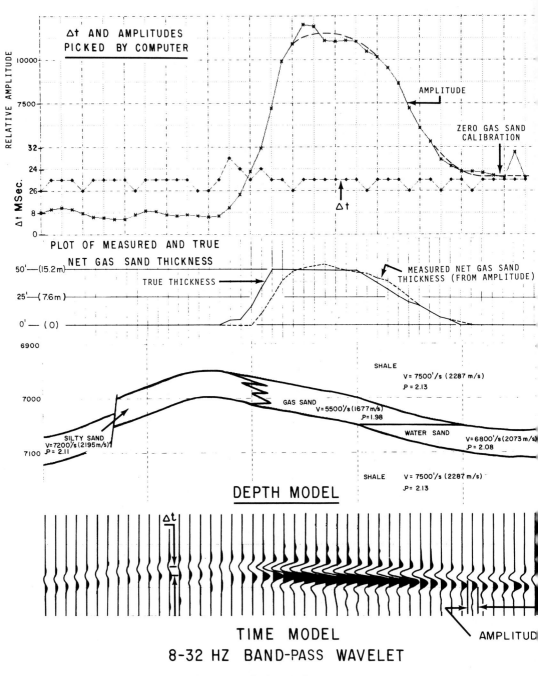

FIG. 22—Updip Permeability Barrier: seismic amplitude as indicator of net hydrocarbon thickness, loss of porosity, and gas/liquid contact.

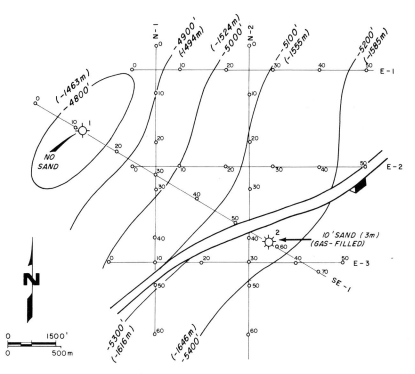

FIG. 23—Structure map, top of silty shale section.

measurements. If this seismic profile represents one of several which define the gas-bearing portion of the sandstone, it is apparent that the calibrated measure of amplitude for each profile would make it possible to contour the net thickness of the gas-sandstone pocket on a base map. Volumetric measurement of the reservoir rock then is made.

Geologists have been aware of this type of entrapment, or of similar situations where a sandstone pinches out or loses porosity and permeability updip. The problem lies in the ability to demonstrate a stratigraphic change, and to delineate the position at which the change takes place. Stratigraphic modeling affords an interpretation dimension whereby such suspected entrapment situations can be proved or disproved, thus avoiding excessive drilling or the condemnation of good prospects by "testing" the structure anomaly with a dry hole.

QUANTIFICATION OF A STRATIGRAPHIC TRAP

General

The following case study is an actual example of the practical use of stratigraphic modeling and interpretation as it applies to exploration for and

quantification of a gas-filled lens of sandstone. All of the concepts and techniques previously discussed were employed in the evaluation of this sandstone lens, lying in the offshore Louisiana portion of the Gulf of Mexico. This general location and the fact that we are dealing with Pleistocene sandstones are the only identification permitted at client request.

The prospect area includes two wells, the first of which was drilled on a structural closure (Fig. 23) and the second of which was drilled in a downdip position on the south side of a regional down-to-the-coast fault. Well No. 1 encountered a 100-ft (30.5 m) silty shale section at about -4, 780 ft ($-1,457$ m). Well No. 2 encountered a silty zone at about $-5,350$ ft ($-1,631$ m) and a 10-ft (3 m) gas sandstone at about $-5,400$ ft ($-1,646$ m). Available to the original interpreters of the prospect were standard AGC sections for the north-south lines N-1 and N-2, and for the east-west lines E-1, E-2, and E-3, plus some additional old coverage in the area of well No. 1 which is irrelevant to this problem.

In an attempt to define the position, extent, and ultimately the thickness and potential of a possible large gas-bearing sandstone lens, the

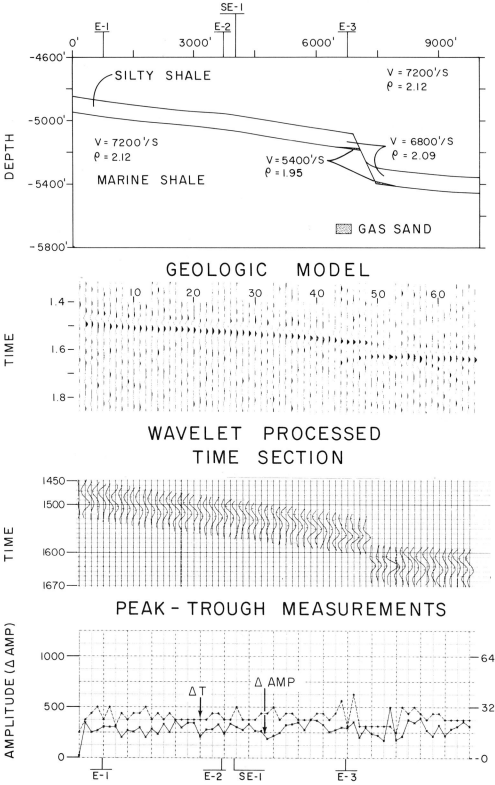

GEOLOGIC MODEL

SE-1

E-1   E-2   E-3

SILTY SHALE

V = 7200'/S
ρ = 2.12

V = 7200'/S
ρ = 2.12

V = 6800'/S
ρ = 2.09

V = 5400'/S
ρ = 1.95

MARINE SHALE

GAS SAND

WAVELET PROCESSED TIME SECTION

PEAK - TROUGH MEASUREMENTS

ΔT

Δ AMP

FIG. 24—Quantitative stratigraphic analysis of line N-1.

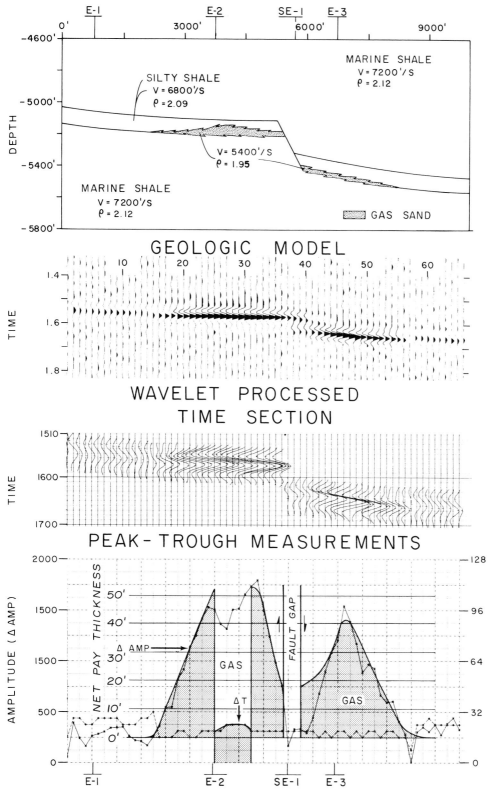

FIG. 25—Quantitative stratigraphic analysis of line N-2.

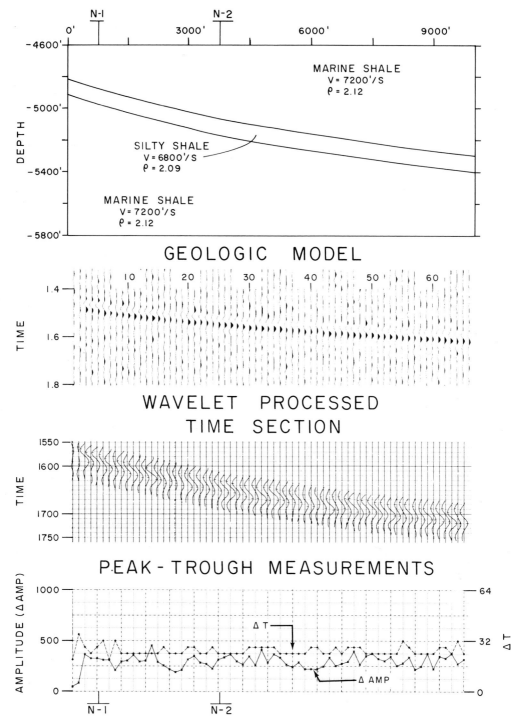

FIG. 26—Quantitative stratigraphic analysis of line E-l.

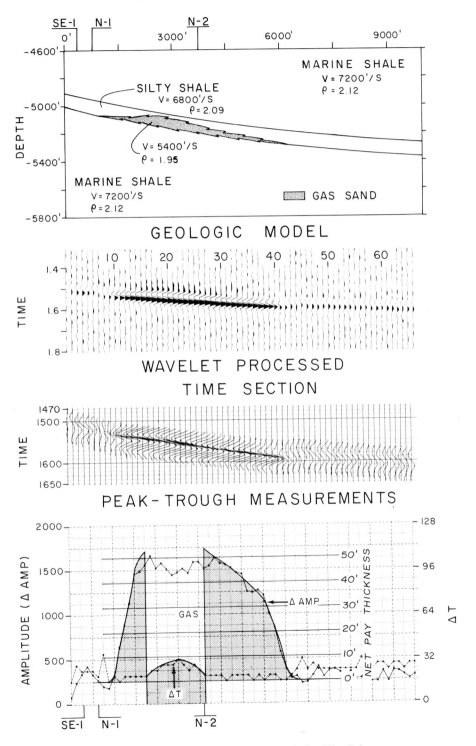

FIG. 27—Quantitative stratigraphic analysis of line E-2.

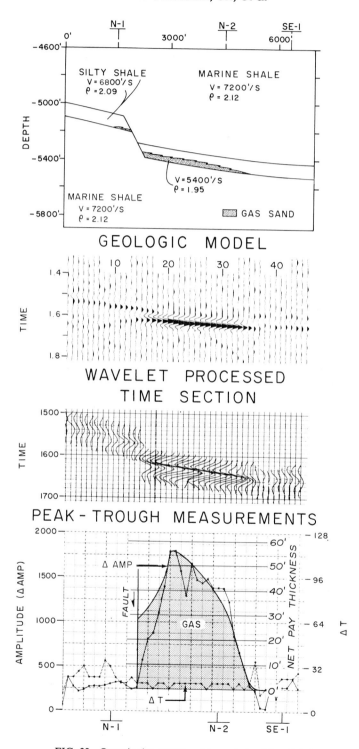

FIG. 28—Quantitative stratigraphic analysis of line E-3.

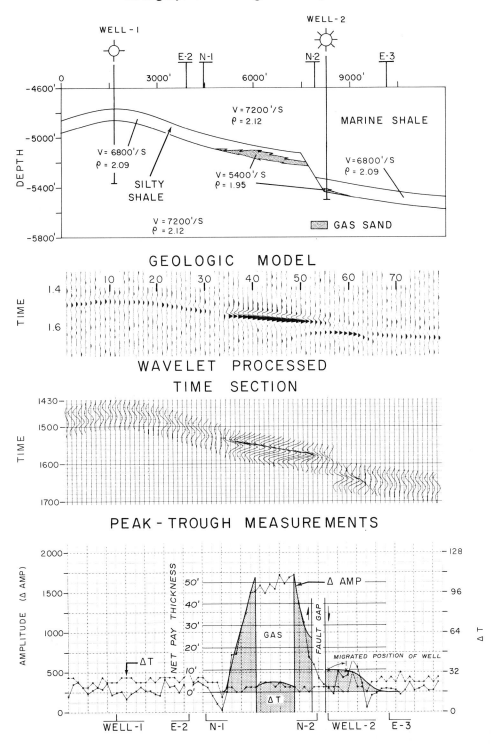

FIG. 29—Quantitative stratigraphic analysis of line SE-1.

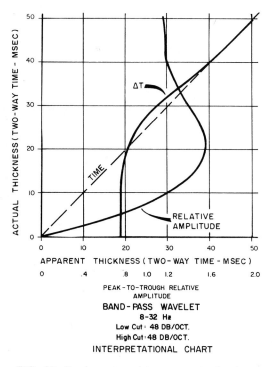

BAND-PASS WAVELET
8-32 Hz
Low Cut : 48 DB/OCT.
High Cut : 48 DB/OCT.
INTERPRETATIONAL CHART

FIG. 30—Band-pass wavelet response to low-impedance thin bed.

data for the N and E lines were obtained, a wavelet was extracted, and the lines were reprocessed for wavelet correction. Concurrently, shooting of line SE-1 was performed for the purpose of correlating the two wells and calibrating seismic with geologic data at well No. 2. Prominent amplitude anomalies are immediately apparent on lines N-2, E-2, and E-3, and they obviously represent one coherent anomalous stratigraphic unit which is cut by a fault. The largest or most extensive segment of the anomaly occurs on the undrilled, upthrown side of the fault.

Although the existence of an anomaly is obvious from these data, it was not obvious from the original seismic data. Relative-amplitude sections would have shown the anomaly, but would not have provided the best wavelet and, hence, information necessary to determine with assurance the existence of gas or paramenters necessary for quantification.

Procedural Interpretation

All of the information necessary for interpretation has been condensed for each seismic line and displayed in Figures 24 through 29. Figure 31 is the contoured net-producing-zone isopach superposed on a structural base which resulted from the interpretation of these data.

Figures 24 through 29 each contain (1) a depth cross-section display of the object horizon, (2) a portion of the seismic cross section which is in true amplitude and has been wavelet-processed, (3) an expanded display of the seismic event which has been extracted from the complete traces for which peak-to-trough time differentials, ΔT, and relative amplitudes have been automatically measured, and (4) a display of the measured time and amplitudes from (3).

Six seismic lines (Fig. 23) were used to make this interpretation, two north-south, three east-west, and one southeast-northwest line. Interpretations have been made of each line and annotated on the amplitude ΔT plots. Two lines, N-1 and E-1, showed no abnormal amplitudes and were assumed to indicate the absence of a gas sandstone. The remaining lines did show high amplitudes and were used to measure gas-sandstone thickness. For demonstration purposes, line SE-1 (Fig. 29) will be used and the logic of interpretation explained.

Well 2 was drilled and showed the presence of 10 ft (3 m) of net gas sandstone. The migrated position of the well was accounted for, as noted, to calibrate the amplitude data properly. Fault locations for both the upthrown and downthrown horizon positions were made and the amplitude curve and ΔT plots were hand-smoothed. A calibration plot was made for the propagating wavelet used in the wavelet processing (Fig. 30) to determine the tuning point and amplitude decay as a function of bed thickness (milliseconds). Two separate curves are presented in Figure 30; the vertical axis is actual two-way time thickness in milliseconds for low-acoustic-impedance sandstone beds in a higher acoustic-impedance shale matrix. Two scales are plotted on the horizontal axis. One scale is in milliseconds of two-way time of the peak-to-trough time separation as measured on an 8 to 32-Hz, zero-phase, band-pass wavelet convolved with varying thicknesses of low-acoustic-impedance sandstone embedded in a shale matrix. If the wavelet had infinite bandwidth, time separations of the peak-to-trough response wavelet would fall on a 45° line representing time. Since the bandwidth of this particular wavelet is finite, the curve labeled "ΔT" describes the peak-to-trough response to varying bed thickness. The measured ΔT becomes invariant for bed thicknesses below about 0.018 sec. The other scale, the relative-amplitude response of this wavelet, is plotted on the horizontal axis against actual thickness on the vertical axis.

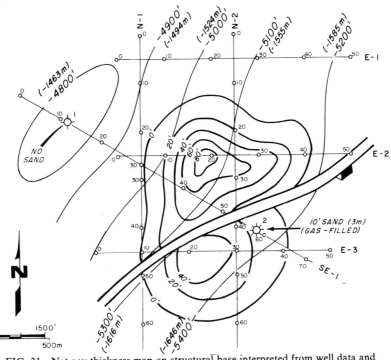

FIG. 31—Net-pay thickness map on structural base interpreted from well data and wavelet-processed seismic data.

Above 0.040 sec in actual bed thickness, the amplitude is fairly constant and increases to a maximum at about 0.018 sec, or the two-way time below which the apparent time thickness becomes invariant. This curve demonstrates the approximate linear decay of peak-to-trough amplitudes as the actual bed thickness approaches zero.

An examination of the data on SE-1 after smoothing indicates the existence of two tuning points, one at shotpoint 40 and another near shotpoint 50. The ΔT plot shows an increase between these two points which indicates that the objective net pay is resolved in time. Notice that this ΔT is less than the ΔT measurements of the silty shale bed on either side of the high-amplitude data. This is due to the fact that a reflection is being generated from the top of the gas sandstone. A phase inversion also indicates this abrupt change from a weak negative reflection at the top of the gas sandstone. Measurements of ΔT are used in the interval between shotpoints 40 and 50 as shown on Figure 29. The gas-sandstone velocity, as extracted from the sonic log of Well No. 2, of 5,400 ft (1,646 m) per second was multiplied by the tuning thickness (0.020 sec) to compute a thickness of 54 ft (16.5 m). These two values of 10 ft (3 m) and 54 ft (16.5 m) were used to construct a linear scale for calibrating the amplitude values in feet. The thickness values using ΔT's are simply computed by multiplying one-half the ΔT by the gas-sandstone velocities. The base line for the zero, net gas-sandstone thickness is chosen where the amplitude drops to that of the silty shale, which occurs at an approximate relative amplitude of 250.

The interpretation is completed by tying loops on amplitude values, where possible, and/or using the same amplitude scale in feet, as presented in Figure 29. The final interpreted map of the gas sandstone is shown in Figure 31. The thicker part of gas sandstone exists to the southwest of Well No. 2, where it is interpreted to exceed 40 ft (12.2 m) in thickness on the downthrown block, and on the north side of the fault, where it is interpreted to attain a thickness of 80 ft (24.4 m). If all other economic parameters and conditions were favorable, well locations could readily be made to test both segments of the lens, and the decision could be made as to where to set the production platform.

Amplitude measurements, as shown by this example, give the geologist and geophysicist an additional parameter for use in stratigraphic mapping in both exploratory and development projects. However, interpretation and experience are still very important ingredients in the use of these new data.

CONCLUSIONS

On the basis of experience to date, the following conclusions can be drawn about the practical use of stratigraphic modeling:

1. Modeling of stratigraphic situations of interest is an invaluable tool for helping the geophysicist and geologist understand more fully the phenomena he observes, particularly in relation to anomalous amplitude events.

2. It is necessary to measure and allow for the exact shape of the basic seismic wavelet, either through its use in generating interpretation templates or, preferably, through wavelet-correction processing.

3. Wavelet-processed data open the door to stratigraphic interpretation of porous, commerical sandstones from the seismic record, in great detail and with good reliability. This is equivalent to a new sandstone/shale ratio technique of much greater precision and local reliability than any previous method.

4. Stratigraphic modeling is specifically useful in some instances for making careful estimates of pore volume and focusing on the key parameters upon which these estimates depend. This assumes, of course, that pore volume can be related to relative changes in acoustic impedance.

5. The method is applicable to geologic provinces other than the Gulf Coast and has the general capability of providing lithologic information ahead of drilling.

The models and actual case studies in this paper are intended to illustrate some of the key principles that now can be applied to stratigraphic interpretation. Modeling and knowledge of wavelet shape will play an increasingly significant role in defining calibration techniques, particularly in relating well data to seismic data. Acoustically meaningful parameters and their translation into reservoir-descriptive parameters are complex processes which will require careful integration of geophysical, geological, and engineering disciplines, and analytical techniques which facilitate such integration.

The reader can extrapolate these concepts to see how they might be used to guide field development once a field has been found. The key, of course, is calibration. The more well data and engineering data that exist, the more we are able to refine the comparison between seismic response and reservoir definition.

REFERENCES CITED

Dedman, E. V., J. P. Lindsey, and M. W. Schramm, Jr., 1975, Stratigraphic modeling: New Trends in Seismic Interpretation, Denver Geophys. Soc. Cont. Ed. Seminar, Golden, Colorado, April 17-18.

Domenico, S. N., 1974, Effect of water saturation on seismic reflectivity of sand reservoirs encased in shale: Geophysics, v. 39, p. 759-769.

——— 1975, Rock characteristics and pore fluid content: New Trends in Seismic Interpretation, Denver Geophys. Soc. Cont. Ed. Seminar, Golden, Colorado, April 17-18.

Gardner, G. H. F., L. W. Gardner, and A. R. Gregory, 1974, Formation velocity and density; the diagnostic basis for stratigraphic traps: Geophysics, v. 39, p. 770-780.

Harms, J. C. and P. Tackenburg, 1972, Seismic signatures of sedimentation models: Geophysics, v. 37, p. 45-58.

Lindseth, R. O., 1975, Lithology determination (sands, shales, reefs, etc.): New Trends in Seismic Interpretation, Denver Geophys. Soc. Cont. Ed. Seminar, Golden, Colorado, April 17-18.

Lindsey, J. P., E. V. Dedman, and M. W. Schramm, Jr., 1975, Stratigraphic modeling: a step beyond bright spot: World Oil, v. 180, p. 61-64.

——— M. W. Schramm, Jr. and L. K. Nemeth, 1976, New seismic technology can guide field development: World Oil, v. 182, p. 59-63.

Sengbush, R. L., P. L. Lawrence, and F. J. McDonal, 1961, Interpretation of synthetic seismograms: Geophysics, v. 26, p. 45-58.

Tegland, E. R., 1970, Sand-shale ratio determination from seismic interval velocity: 23d Annual Midwestern Reg. Mtg., SEG and AAPG, Dallas, March 8.

Index

Explanation of Indexing

A reference is indexed according to its important, or "key" words. Authors and titles are also represented here; where more than one author has contributed to a paper, each person is cited, alphabetically, according to his last name.

Three columns are to the left of the keyword entries. The first column, a letter entry, represents the AAPG book series from which the reference originated. In this case, M stands for Memoir Series. Every five years, AAPG merges all its indexes together, and the letter M will differentiate this reference from those of the AAPG Studies in Geology Series (S) or from the AAPG Bulletin (B).

The following number is the series number. In this case, 26 represents a reference from Memoir 26.

The last column entry is the page number in this volume where the reference will be found.

A small dagger symbol (†) is used to highlight a manuscript title entry.

Note: This index is set up for single-line entry. Where entries exceed one line of type, the line is terminated. (This is especially evident with manuscript titles, which tend to be long and descriptive.) The reader sometimes must be able to realize keywords, although commonly taken out of context.

Index